09

The Theo

The Theory of Island Biogeography Revisited

EDITED BY

Jonathan B. Losos
and Robert E. Ricklefs

PRINCETON UNIVERSITY PRESS
PRINCETON AND OXFORD

Published by Princeton University Press, 41 William Street, Princeton, New Jersey 08540
In the United Kingdom: Princeton University Press, 6 Oxford Street, Woodstock, Oxfordshire OX20 1TW

Library of Congress Cataloging-in-Publication Data

The theory of island biogeography revisited / edited by Jonathan B. Losos and Robert E. Ricklefs.
p. cm.
Includes index.
Derived from a meeting held at Harvard in October 2007 to celebrate the fortieth anniversary of the publication of The theory of island biogeography, by Robert H. MacArthur and Edward O. Wilson.
ISBN 978-0-691-13652-3 (hardcover : alk. paper)—ISBN 978-0-691-13653-0 (pbk. : alk. paper) 1. Biogeography—Congresses. 2. Island ecology—Congresses. I. Losos, Jonathan B. II. Ricklefs, Robert E. III. MacArthur, Robert H. Theory of island biogeography.
QH85.T44 2010
578.75'2—dc22

2009010056

British Library Cataloging-in-Publication Data is available

This book has been composed in Sabon

Printed on acid-free paper. ∞

press.princeton.edu

Printed in the United States of America

10 9 8 7 6 5 4 3 2

Contents

Foreword

Robert M. May

INSOFAR AS ANY ONE EVENT can be said to mark the coming of age of ecological science as a discipline with a theoretical/conceptual base, it is the publication in 1967 of MacArthur and Wilson's *Theory of Island Biogeography,* the inaugural "Monograph in Population Biology" in the Princeton University Press series.

It is easy to forget how young a science ecology is. We did not start a systematic naming and codification of the plants and animals we share the world with until a century after Newton and the founding of the world's major scientific academies (the canonical date for Linnaeus's *De Rerum Naturae* is 1758; for the founding of the Royal Society, 1660). The very word ecology is not much more than a century old, and in 2009 neither of the two oldest ecological societies has yet attained its century (the British Ecological Society was established in 1913, the Ecological Society of America in 1915).

One way of accounting for the development of any particular area of the natural sciences comes from the classic sequence of Brahe, Kepler, Newton: systematic observation and description; tentative patterns that give coherence to the observed facts; fundamental ideas or laws that explain the patterns. This characterization of the quest for real understanding as a journey from asking "what" questions to asking "why" questions is a deliberate oversimplification, but I think it is nevertheless useful.

The early years of ecological science are largely Brahe, verging into Kepler. Up to the 1960s the textbooks clearly reflect this. There are, of course, exceptions. These reach as far back as the late 1700s, when Gilbert White first looked beyond the "cabinets of curiosities" of his time to ask questions such as why the swift population of Selborne was so very steady at eight breeding pairs per year. The work of Lotka and Volterra in the 1920s—itself partly anticipated by earlier work in the 1880s—raises significant theoretical issues about competitive and predator-prey relations. This being acknowledged, the fact remains that up into the 1960s the leading ecology texts, such as Andrewartha and Birch's *The Distribution and Abundance of Animals,* were at best like earlier descriptive chemistry texts in which the empirically derived Periodic Table gave coherence, but before the underlying quantum mechanical basis of atomic structure had illuminated the Periodic Table itself.

In marked contrast, today's ecology texts present a richer view of the world. Of course there is a factual foundation of natural history observations along with careful idea-testing experiments in field and laboratory. Many of these field and laboratory experiments themselves play off against theoretical ideas and "why is it so" questions. While some of the theory is verbal (as, let us not forget, Darwin's influential theory was!), much of it is—when necessary—explicitly mathematical, and sometimes sophisticatedly mathematical. After all, mathematics is ultimately no more, although no less, than a way of thinking clearly.

This volume derives from a meeting held at Harvard to celebrate the fortieth anniversary of the publication of *Theory of Island Biogeography*. Happily, Ed Wilson was with us to enjoy it. Sadly, Robert McArthur was not, having died very young only five years after its publication; had he lived, I believe we would be further down the road than we are.

One notable feature of this lively meeting was the size of the audience, reflecting the huge growth in the national and global community of ecological researchers. When, around fifty years ago, ecologists gathered to celebrate Evelyn Hutchinson's *Festschrift*, the ecological community numbered less than a tenth that of today. Hutchinson's impact was summed up by a picture, showing a tree whose trunk was Hutchinson, branches his graduate students, leaves his postdocs, and circumambient butterflies and other insects associates; the total assembly was small, yet it represented a fair fraction of the world's ecological theorists. The number present at the symposium associated with the present volume, although small relative to the current global population of ecologists, was roughly ten times that around Hutchinson's tree.

Given the environmental problems that currently loom over the planet, this large and rapid growth in what might be called the ecological task force is greatly and unreservedly to be welcomed. Almost forty years ago, in the Preface to *Stability and Complexity in Model Ecosystems*, I wrote that "I have been struck by the attitude of constructive interest in others' work which seems to prevail among ecologists. The competition and predation which characterise many other disciplines seem relatively absent, possibly because the field has not yet reached (or exceeded) its natural carrying capacity"; this has the implicit corollary that physics was, perhaps, a bit less civil (a theme elaborated much more recently, and in a constructive and interesting way, by Lee Smolin in *The Trouble with Physics*). Be this as it may, my belief—reinforced by the contents of the present book—is that ecological science has achieved much over the past forty years, with the remarkable growth in the research community reflecting both advances in understanding on many fronts (most of which pose further questions and open further avenues for research) and increasing recognition of the pressing problems which need to be addressed. I also, perhaps Polyannaishly, believe the ecological community has largely succeeded in preserving its collegial character despite such increases in numbers.

As good ecologists and/or evolutionary biologists, we all recognize that dispersal strategies are one of the key issues in life history choices. Effective application of ecological knowledge to environmental problems requires not only teachers and researchers in schools and universities, but also professional ecologists in NGOs, in consultancies, in local, state, and federal government offices, and elsewhere. Too often, Ph.D. supervisors unintentionally suggest career paths confined to universities. This is understandable but unfortunate: we need ecological expertise more widely disseminated and applied.

The *Theory of Island Biogeography* has recently and justly been reprinted as the first volume in Princeton University Press's (PUP) series of "Landmarks in Biology." And it is a true landmark among landmarks. The PUP series of "Monographs in Population Biology," which it led off, under Robert MacArthur's editorial direction, has continued strongly. Just before leaving PUP, the Commissioning Editor in Life Science, Sam Elworthy, made an informal list of the thirty most cited monographs in ecology and evolution. This is headed, as you would expect, by Darwin's *Origin of Species*, and books by Fisher, Mayr, and so on. But twelve of the thirty are in the series MacArthur and Wilson started and set the standard for.[1] Citations can, of course, be misleading. For example, Darwin's *Origin*—although deservedly top—actually owes more of its citations to the history of science Darwin industry than to science as such.

The fact remains that the MacArthur and Wilson book marks a true turning point in the advance of ecological science, and thence in our understanding of how the natural world works. The extraordinary scope and diversity of the contributions in the present book testify to this. This is no ordinary collection of symposium papers. Although multiauthored, I would call it a metalevel monograph, illustrating the many doors that MacArthur and Wilson opened for us.

[1]S. Elworthy, *Bulletin of the British Ecological Society* 38(2):55–57 (2007). I suspect that an exhaustive search of Google Scholar might turn up some "top thirty" titles missed by Elworthy, but I think his list is basically sound. I cannot resist adding that my wife, Judith May, who was earlier at PUP and later at Oxford University Press, commissioned no fewer than fifteen of Elworthy's thirty books (some in various series at Princeton and at Oxford, others as stand-alone texts).

Preface

Jonathan B. Losos and Robert E. Ricklefs

ROBERT MACARTHUR AND EDWARD WILSON'S 1967 book, *The Theory of Island Biogeography*, is the dominant symbol of a transition that took place four decades ago from descriptive to analytical approaches in ecology and biogeography. Change was in the air during the dynamic decade of the 1960s, and, both together and independently, MacArthur and Wilson made seminal contributions to ecology and evolution. Had they not written *The Theory of Island Biogeography*, MacArthur and Wilson would still be recognized as two of the most influential figures of this period.

Every contemporary student is taught MacArthur and Wilson's graph with the crossed colonization and extinction curves, along with the astonishing implication that island biotas assume a dynamic steady state in which species continually disappear from islands only to be replaced at an equal rate by new colonists. Few of these students realize that *The Theory of Island Biogeography* also was a compelling call for a comprehensive refashioning of biogeographical thinking. Inescapably, biogeography theory fully integrates much of ecology, population biology, evolution, and paleontology, with important implications for conservation of species. Islands and archipelagoes are, in many respects, microcosms of the rest of the world.

The symposium held at Harvard University during the fortieth anniversary year of *The Theory of Island Biogeography* gave both of us an excuse to read this wonderful book (yet again!) and to reflect, as many of the authors in this volume have done, on its legacy. Two aspects of the book stood out for us. First, so much of what we take for granted about the modern disciplines of ecology, evolution, and conservation biology can be traced directly back to one or several of the seven chapters. For example, the relationship between species number and area, the subject of chapters 2 and 3, and certainly one of the dominant empirical patterns in all of biology, has been fully assimilated into theory relating loss of species to habitat destruction, underlies much of spatial ecology, and is a foundational observation for neutral theory. Topics discussed in "The Strategy of Colonization" (chapter 4) are fundamental to present-day areas as diverse as life-history evolution and population viability analysis. Chapter 5, "Invasibility and the Variable Niche," presented a general theory of community assembly and introduced the concept of ecological saturation. "Stepping Stones and Biotic Interchange" (chapter 6) has metamorphosed into metapopulation biology and landscape ecology. "Evolutionary Changes Following Colonization" (chapter 7) presaged much contemporary research on the success of invasive species.

Second, in contrast, some of the areas emphasized by MacArthur and Wilson remain relatively unexplored or their promise unfulfilled. As the authors pointed out in their first chapter, "the fundamental processes, namely dispersal, invasion, competition, adaptation, and extinction, are among the most difficult in biology to study and understand." This remains true today. By their nature, the processes underlying biogeographic distributions and evolution within the geographic context occur on vast scales of time and space, at least relative to individual human experience. By way of contrast, most tests of equilibrium theory have depended on observations on small islands close to sources of colonists over relatively short periods. The evolutionary dimension is largely missing; the study of haphazard events over long distances has only recently gained ascendancy—partly as vicariant explanations for biogeographic patterns have lost their luster—and the promise of understanding the emergence of biotas de novo in remote archipelagoes has yet to be realized. In particular, Ed Wilson's call for the "biogeography of the species" to take a central place in understanding pattern and process in the natural world is just beginning to receive the attention it deserves.

As this volume is published in 2009, the bicentennial of the birth of Charles Darwin, we are reminded of the crucial influence of islands on this most observant and thoughtful of biogeographers. We also are reminded that much of the momentum of Darwin's original insights concerning the origin, distribution, and evolution of species had been lost by the middle of the last century. MacArthur and Wilson's *The Theory of Island Biogeography* was arguably one of the pivotal points in restoring Darwinian traditions of careful observation and reflection to ecology and evolutionary biology, and conveying the excitement of its study. It was the nature of the time, to be sure, but *The Theory of Island Biogeography* made the single most persuasive case for integrating population and evolutionary thinking into biogeographic analysis and interpretation.

This book, and the symposium upon which it was based, sprang from a casual lunch-time realization early in 2007 that the year marked the fortieth anniversary of the publication of MacArthur and Wilson's opus. Encouraged by Harvard's Center for the Environment and Museum of Comparative Zoology, we invited sixteen scholars to participate, including a mixture of older biologists, some of whom began their careers in the buoyant waters pouring forth from *The Theory of Island Biogeography* and the exciting change it represented, and younger investigators who continue to feel the influence of that work. To our amazement, fifteen accepted our offer. All but one symposium participant have contributed chapters, and one additional contributor has been added.

The participants and approximately three hundred symposium attendees endured an unseasonably warm October, 2007, weekend in the un-air-conditioned Geological Lecture Hall at Harvard. They were enchanted by Ed Wilson, who joined us to celebrate the occasion. In his talk, he recounted the origins of the partnership between himself and Robert MacArthur, who died in 1972, and regaled us with stories about the early days of experimental biogeography. We were also pleased that Lord May of Oxford (formerly just Bob to many of us) was available to address the symposium and write a perceptive foreword to this book. We were also gratified that so many of the packed audience were graduate students and postdocs, some of whom came from great distances and, hopefully, left inspired.

The sixteen contributions in this book are loosely grouped into three sections: the history of island biogeography theory, ecology, and evolution. In the first section, Wilson recounts the early days from personal experience, Lomolino, Brown, and Sax review the development of biogeography theory more generally and outline areas of future synthesis, Schoener examines the famous equilibrium model and some of its early tests, while Whittaker, Triantis, and Ladle expand the theory by incorporating the life stages of islands themselves.

Islands, of course, are ecological systems, and many ecological systems have island attributes. These themes are explored with respect to trophic cascades on islands of different size (Terborgh), food web ecology (Holt), metapopulation dynamics (Hanski), conservation in a fragmented world (Laurance), equilibrium theory and assembly rules for island biotas (Simberloff and Collins), and the neutral theory of metacommunity diversity (Hubbell).

Finally, since Darwin's time, islands have provided laboratories for the study of evolution, including changes following colonization (Clegg), species formation (Grant and Grant), the special circumstances of remote archipelagoes (Gillespie and Baldwin), Lesser Antillean birds as a case study (Ricklefs), the role of speciation in building diversity on large islands (Losos and Parent), and the parallels between island biogeography theory and population genetics theory (Vellend and Orrock).

In reading these articles and reviewing the literature on island biogeography, we were struck by two observations. First, the legacy of *The Theory of Island Biogeography* is alive and thriving. When we first envisioned this book, we expected most contributions to be retrospective, reviewing the ideas laid forth in that book and assessing how they had fared. By contrast, a glance at this book will indicate that many of the chapters are looking primarily forward, rather than backward. Some of the most exciting areas in ecology and evolutionary biology—metapopulation theory, the neutral theory of biological diversity, trophic cascade theory, the synthesis of ecological and phylogenetic evolutionary approaches, to name a few—were inspired by or are being integrated with island biogeography.

Much of this work was at most only hinted at by MacArthur and Wilson, yet exciting developments today have a clear intellectual thread leading back to that work, as many articles attest.

Conversely, the field seems to have passed by some of the issues that were at the heart of debate concerning island biogeographic theory in the 1970s and 1980s. As Schoener's article indicates, even though the equilibrium theory was central to the excitement and controversy surrounding the book, its status is currently uncertain. In part, this reflects studies that suggest that the domain of circumstances to which the theory applies is more limited than originally suggested. More generally, though, it simply reflects the fact that few researchers today are measuring rates of colonization, extinction, and species turnover. The crossing-lines diagram may be the most enduring icon of MacArthur and Wilson's book, but work devoted to quantifying such curves and assessing their significance no longer appears to be a high priority.

Similarly, the field of conservation biology was founded when island biogeographic thinking was applied to questions of nature reserve design. The ensuing bitter debate over SLOSS (single large or several small protected areas) played itself out through journal pages and led to the design of many experiments, the most large scale being the "Biological Dynamics of Forest Fragments" project still ongoing in Amazonian Brazil. But, as Laurance's chapter indicates, the field here, too, has moved on, not because the debate has been settled definitively, but because researchers recognize that other issues are more directly relevant in shaping conservation policy.

Books such as this—and the symposia on which they are sometimes based—represent the combined efforts of many people behind the scenes. The symposium held at Harvard University was underwritten by the Harvard University Center for the Environment and the Museum of Comparative Zoology. We thank the directors of these institutions—Dan Schrag and James Hanken—for their support, and Jim Clem, Jenny MacGregor, and Lisa Matthews of HUCE for their tireless efforts to organize and pull off the event. In turn, the quality of this volume was immeasurably improved by the review process. All manuscripts were reviewed by at least two colleagues; in most cases, one was a book contributor and the other an outside reviewer. In addition to the efforts of the contributors, we thank A. Badyaev, J. Chase, B. Emerson, R. Ewers, J. Foufopoulos, N. Gotelli, L. Harmon, L. Heaney, I. Lovette, M. McPeek, T. Price, and D. Spiller. This book could not have been produced without the help of Princeton University Press. Many thanks to J. Chan, K. Cioffi, A. Kalett, R. Kirk, and J. Slater.

Contributors

Bruce G. Baldwin, Department of Integrative Biology,
University of California, Berkeley
James H. Brown, Department of Biology,
University of New Mexico
Sonya Clegg, Division of Biology,
Imperial College London
Michael D. Collins, Department of Biology,
Hampden-Sydney College
Rosemary G. Gillespie, Department of Environmental Science,
University of California, Berkeley
B. Rosemary Grant, Department of Ecology and Evolutionary Biology,
Princeton University
Peter Grant, Department of Ecology and Evolutionary Biology,
Princeton University
Ilkka Hanski, Department of Biological and Environmental Sciences,
University of Helsinki
Robert D. Holt, Department of Zoology,
University of Florida
Stephen P. Hubbell, Department of Ecology and Evolutionary Biology,
University of California, Los Angeles
Richard J. Ladle, Biodiversity Research Group,
Oxford University Centre for the Environment
William F. Laurance,
Smithsonian Tropical Research Institute
Mark V. Lomolino, Department of Environmental and Forest Biology,
SUNY College of Environmental Science and Forestry
Jonathan B. Losos, Museum of Comparative Zoology and Department of Organismic and Evolutionary Biology,
Harvard University
Robert M. May, Department of Zoology,
University of Oxford
John L. Orrock, Department of Biology,
Washington University

Christine E. Parent, Section of Integrative Biology,
University of Texas at Austin
Robert E. Ricklefs, Department of Biology,
University of Missouri, St. Louis
Dov F. Sax, Department of Ecology and Evolutionary Biology,
Brown University
Thomas W. Schoener, Section of Ecology and Evolution,
University of California, Davis
Daniel Simberloff, Department of Ecology and Evolutionary Biology,
University of Tennessee
John Terborgh, Center for Tropical Conservation, Nicholas School of the Environment and Earth Sciences,
Duke University
Kostas A. Triantis, Biodiversity Research Group,
Oxford University Centre for the Environment
Mark Vellend, Departments of Botany and Zoology, and Biodiversity Research Centre,
University of British Columbia
Robert J. Whittaker, Biodiversity Research Group,
Oxford University Centre for the Environment
Edward O. Wilson, Museum of Comparative Zoology,
Harvard University

The Theory of Island Biogeography Revisited

Island Biogeography in the 1960s

THEORY AND EXPERIMENT

Edward O. Wilson

Intellectual Origins

When I was still a graduate student, in the early 1950s, an idea was circulating that I found inspirational. It originated with William Diller Matthew, a vertebrate paleontologist at the American Museum of Natural History. In 1915 he had suggested that over long periods of Cenozoic time, the most successful of new mammalia genera and families have been arising from a central headquarters of macroevolution. Matthew concluded that the north temperate zone was that geographic cradle. The new clades were by and large intrinsically dominant over those originating in the southern continents. Radiating into diverse adaptive types, they spread outward into the peripheral land masses respectively of Africa, tropical Asia, Australia, and tropical America. As they expanded, they tended to displace early prominent genera and families that were ecologically similar, first from the north temperate evolutionary headquarters and then the southern land masses. The ruggedness of the species originated from a challenging climate, Matthew thought.

For example, rhinocerotids, once dominant elements of the north temperate regions, have fallen back before groups such as deer and other cervids, while early dominant carnivores have retreated before the currently dominant canids and felids. What people living in the north temperate zone think of as "typical" mammals are just the dominants presiding at macroevolutionary headquarters at the present time.

In 1948 and later, in 1957, Philip J. Darlington, then Curator of Entomology at Harvard's Museum of Comparative Zoology, pressed on with Matthew's idea. But he altered it fundamentally, at least for the nonmammalian land vertebrates. In a study of the cold-blooded land and freshwater vertebrates—reptiles, amphibians, and fish—Darlington identified the headquarters as the Old World tropics.

By the 1980s, with much richer fossil data in hand than available to Matthew and Darlington, researchers had shifted placement of the Cenozoic headquarters to the "World Continent," a biogeographically historical construct comprising Africa, Eurasia, and North America, and in particular the vast tropical regions within them. Evidence supporting this view came from the phenomenon of the Great American Interchange, the mingling of the independent adaptive radiations of North and South America made possible by the emergence of the Panamanian land bridge about three million years ago. The pattern of the exchange supported the view that competitive displacement among land vertebrates has been a reality. It also suggested that the evolutionary products of the World Continent, represented by North America during the Interchange, were generally superior to those of South America—as revealed by replacement at the levels of genus and family (Simpson 1980, Marshall 1988).

The Taxon Cycle

In 1954–55 the Matthew-Darlington epic view of global territorial biogeography was in the back of my mind, although not to any pressing degree, when I undertook field work on the ant fauna of part of the Melanesian archipelagic chain, from New Guinea to Vanuatu, Fiji, and New Caledonia. I had been elected for a three-year term as a Junior Fellow of Harvard's Society of Fellows, which gave me complete support and freedom to go anywhere to study anything I chose. (I wish this kind of opportunity were available to all new postdoctoral scholars—the world would benefit enormously.) My main goal was to collect and classify the ants of this still poorly known part of the world ant fauna (figure 1.1). Within three years after returning, during which I began an assistant professorship at Harvard, I had managed to publish or put in press monographs on a large minority of the species, many of which were previously undescribed.

While in the field I took as many notes on the natural history of the species as I could. Back home, combining systematics and ecology, I looked for patterns that might shed light on the origins of that classic archipelagic fauna. One day, in a eureka moment consuming only a few minutes, I saw a relation between the spread of species between islands and archipelagoes, on the one hand, to within-island speciation and shifts in habitat preference during evolution, on the other. This was in 1958. I believe I was the first to see such a connection; at least I was not guided by any other work I knew at the time.

These connections were summarized in what I later called the taxon cycle (figure 1.2). The taxon cycle comprises the following steps, at least as displayed by the Melanesian ant fauna. Species enter the Melanesian chain of archipelagoes primarily through New Guinea out of tropical Asia and, less so, out of Australia. Those judged to be in an early stage of expansion possess a continuous distribution and a relatively small amount

Figure 1.1. E. O. Wilson with guard crossing the lower Mongi River, Papua New Guinea, April 1955.

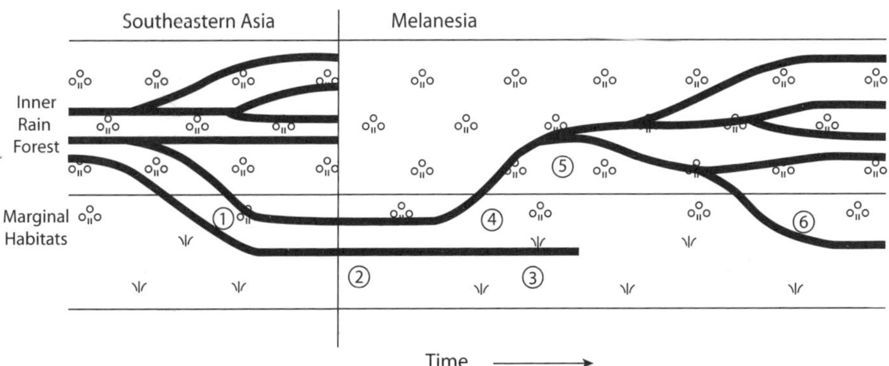

Figure 1.2. The taxon cycle in the Melanesian ant fauna (Wilson 1965, modified from Wilson 1959).

of geographic variation. They turned out to be mostly specialized on marginal habitats, those inhabited by relatively small numbers of species. In Melanesia, the marginal habitats include littoral environments of the coastal shore, river-edge forests, and savannas. Such are places that are happenstance staging areas for between-island dispersal. Local populations on individual islands are not adapted by natural selection for overseas dispersal. Rather, they are preadapted for overseas dispersal by virtue of the greater probability of an overseas launch followed by survival in the habitats of the islands they reach, which are similar to the marginal habitats from which they departed.

When such a preadapted species colonizes a more distant or smaller island, it encounters smaller ant faunas. The species then often experiences what I have called "ecological release." This means that its populations, in addition to holding the beachhead (so to speak), are able to spread inland and occupy habitats less well filled by potential competitors than in the more species-rich islands from which they came. By moving into central habitats, including lowland and mid-mountain rainforests of the interior, the colonies adapt to new conditions. In time they diverge sufficiently to be called a different race or species. During speciation and adaptive radiation, the colonist clades sometimes also generate new, endemic species adapted to the marginal habitats, and the taxon cycle is set to begin again.

By the time I had finished this first round of research on Melanesia I was a nesiophile, if I may be allowed to coin a term. Nesiophilia, the inordinate fondness and hungering for islands, may be a genetic condition. But, whether hereditary or not, I believe it is shared by many, if not all, who gave lectures at the 2007 island biogeography symposium held at Harvard. Even today, over fifty years following my early visits to Cuba and the South Pacific, I continue sporadic field research on the ants of the West Indies, as much just to visit islands as to conduct scientific research.

The Species Equilibrium

In 1959 I met Robert H. MacArthur, a powerful and charismatic intellect and a naturalist of the first rank. Robert, as he preferred to be called, died of cancer in 1972 at the very premature age of 42, when he was at the height of his productivity. All who know his work will agree it was a huge loss for both ecology and evolutionary biology (see figure 1.3). We became friends, and one of our common concerns was the growing decrepitude of our specialties (as we saw it), in dismaying contrast to the newly triumphant emergence of molecular biology. Ecology and evolutionary biology seemed like the aforementioned rhinos and archaic carnivores, surrendering university chairs and grants to the new wave of biologists coming out of the physical sciences. It was clear in the 1960s that their achievements were to be the hallmark of twentieth-century biology.

Figure 1.3. Robert H. MacArthur (left), with Richard Levins during visit with E. O. Wilson, Dry Tortugas, Florida, 1968.

Being both ambitious and purpose-driven, we soon narrowed our conversations down to the following question: How could our seemingly old-fashioned subjects achieve new intellectual rigor and originality compared to molecular biology? What can we learn from molecular biology on how to advance our own science? We agreed that the basic problem was that ecology and evolutionary biology were still mostly unrooted. They needed foundations from which explanations can be developed bottom-up. Theory has to work from lower to higher levels of biological organization. Either alone will not do. Population biology was the discipline we thought could serve as base to reinvigorate the theory of ecology and evolutionary biology. (Such was the line of reasoning by which I later produced the first syntheses of sociobiology, in *The Insect Societies*, in 1971, and *Sociobiology: The New Synthesis*, in 1975.)

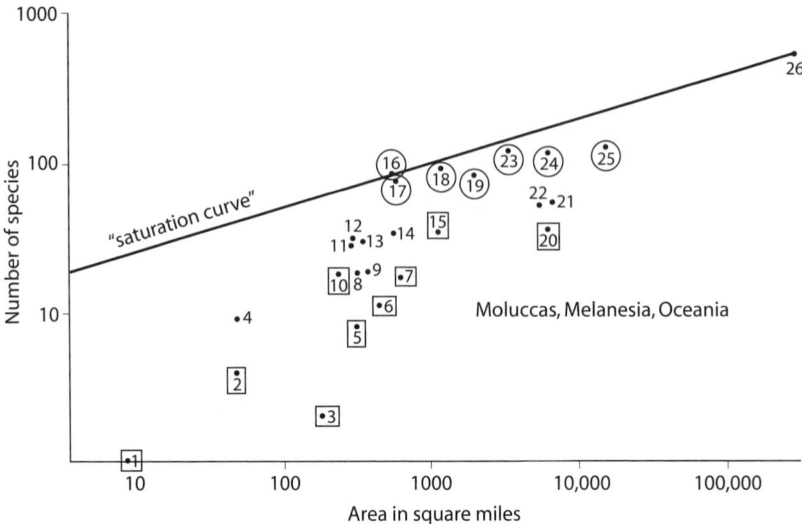

Figure 1.4. Area-species curves, birds, showing areas and distance effects (MacArthur and Wilson 1967).

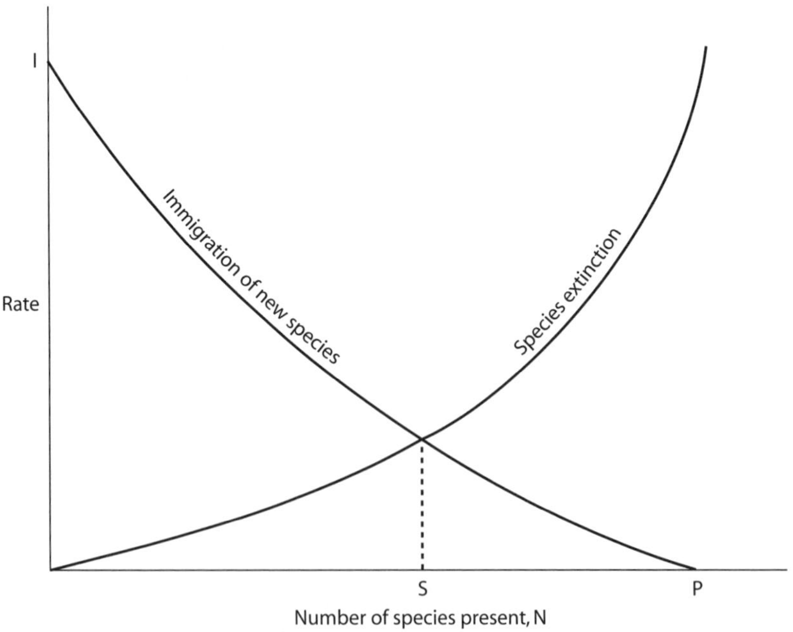

Figure 1.5. Crossed immigration and extinction curves, with the changing intersections (equilibria) predicting the area and distance effects (MacArthur and Wilson 1963).

During our first meeting in early 1960, I urged the prospect of island biogeography on MacArthur. Islands are the logical laboratories of biogeography and evolution, I said. There are thousands of them, for example the Ten Thousand Islands of Florida Bay. There are vast arrays of at least partly isolated faunas and floras living on them. Each is an experiment awaiting the analyses of evolution and ecology.

I showed MacArthur a set of area-species curves I had collected, including one for the ants of Melanesia. With echoes of Matthew, Darlington, and the taxon cycle in my head, I conjured up images of competition, geographic displacement, and equilibrium—in those days we spoke of equilibrated faunas as being "saturated" (equilibrial) or unsaturated (below equilibrium) (figure 1.4). In short time, MacArthur came back with the crossed curves of immigration and extinction rates of species on an island as functions of numbers of species already on the island. Where they crossed was our equilibrium (figure 1.5)!

We were both very pleased with this abstract representation. It seemed the logical portal to the real and complex world of islands and archipelagoes. It invited ideas from population biology, including the demography of growth and decline, the response of populations to density-dependent or -independent factors, and the way species fit together in configurations that allowed more or fewer to coexist. We published the main outlines of what we had found in 1963. Then we began a series of more extensive discussions, mostly by correspondence, about how to tie the processes of immigration and extinction to the data and derivable principles of population ecology and genetic evolution. The result of the back-and-forth was *The Theory of Island Biogeography* in 1967. It was published as the first book of the still flourishing Princeton University Press monograph series on population biology and evolutionary theory.

Experimental Island Biogeography

That was all well and good for the goals we had set, but it was all book work, and talk. Waves of nesiophilia still washed over me. I yearned to keep up what I enjoyed in Melanesia, by physically exploring faunas, especially ant faunas, from island to island. But I couldn't go back to Melanesia due to the long visits required. I was now married with a teaching job at Harvard. So I conceived the idea of a natural laboratory of island biology, close to home, where experiments in biogeography and ecology could be performed and then monitored during frequent but relatively brief periods. I had an advantage in choosing that option: I studied insects. Insects and other arthropods are relatively very small and live in large populations that inhabit very small places. Therefore the islands could be relatively small, and the generation times of the inhabitants could be expected to be conveniently short.

Beguiled by this dream, I pored over maps of islands, particularly very small islands forming micro-archipelagoes, that lie all around the Atlantic and Gulf coasts of the United States. Soon I hit upon the Florida Keys as the logical place to go. That choice was made easier by the fact that much of my childhood had been spent on or close to the coasts of South Alabama and the panhandle of Florida. It would be like going home.

The best approach to experimental island biogeography, I thought, would be to start with many islets that are ecologically similar but vary in area and distance, then turn them into miniature Krakatoas. That is, find a way to eliminate the faunas and then follow the process of recolonization. If the islands were small enough, they would have resident breeding populations of insects and other arthropods, but constitute no more than a small part of the home ranges of birds and mammals. And if the islands were numerous enough, or at least if their natural environments were sufficiently transient, the experiment would have no significant effect on the island system as a whole. In other words, it should not scandalize my fellow conservationists.

The site I first picked was the Dry Tortugas, at the very tip of the Florida Keys. In the summer of 1965, with a small group of graduate students, I visited all of the smallest of these islands and identified the meager array of plants and arthropods on them. The idea was to continue the process until a hurricane wiped the islands clean, then observe their subsequent recolonization by plants and arthropods. I knew that we might have to wait for several years for such a storm to pass over. Providentially, in the 1965 season not one but two hurricanes swept the Dry Tortugas. When we returned in 1966, we found the smallest islands bare of the terrestrial life we had observed just months earlier. Our study could then begin.

However, by this time I had grown dissatisfied with the prospects for these particular miniature Krakatoas. There were too few such islands, the faunas and floras seemed too small, hurricanes were too few and unpredictable, and there was no way to run controls.

So I next turned to the red mangrove islets of Florida Bay. They had none of the shortcomings of the Dry Tortugas. But they did have one large disadvantage: hurricanes would not be able to strip away all the arthropods from the dense mangrove foliage. That had to be done as part of the experimental procedure. At this point Daniel S. Simberloff, who had begun his doctoral studies under my direction, joined me in the enterprise. The year was 1965.

Figure 1.6. Mangrove islet covered by rubberized nylon tent for fumigation (1968).

Dan and I quickly became colleagues more than student and teacher (after all, we were trying something completely new). We chose the islands that seemed most favorably located and visited them to be sure of their suitability. Next we set out to meet two daunting goals: first, locate a professional exterminator who would undertake the admittedly bizarre job of eliminating all the arthropods without harming the vegetation; and second, line up the help of the few systematists able to identify, to the species level, the beetles, bark lice, moths, spiders, mites, and other arthropods of the Florida Keys.

After a lengthy search in the Miami area, we turned up one professional exterminator, Steve Tendrich, who was intrigued by the eccentricity of the project and willing to take the job. After Dan and I had surveyed the arthropods on one of the islands ("E1"), Tendrich sprayed it with a short-lived insecticide. Our follow-up survey revealed that all of the arthropods on the surface had been eliminated, but a few still survived in the beetle burrows of the branches and stems. Tendrich then turned to fumigation with methyl bromide, a gas that dissipates rapidly after application. He experimented with cockroach egg cases and red mangrove saplings to determine the dosage strong enough to kill resistant arthropods but not so strong it would harm the mangrove (figure 1.6). We then proceeded to census four more islands, "defaunate" them, and begin the

Figure 1.7. E. O. Wilson, in red mangrove tree with osprey nest, Florida Keys, 1968.

Figure 1.8. Daniel Simberloff, near E7, October 10, 1966.

monitoring process (figures 1.7 and 1.8). After a successful start, Dan began the grueling process of monthly centimeter-by-centimeter inspection of each island, while I managed the process of consulting the taxonomic experts who could identify the arthropod species (Simberloff and Wilson 1969).

Within two years, the numbers of species on all the islands had returned to their preextermination levels. The most distant island (E1), which began with a low number as expected, returned to its same low level. Thus the existence of species equilibria was demonstrated. To an amazing degree, however, the composition of the species differed from island to island, and on the same island before and after defaunation (Simberloff and Wilson 1971). Also, the rapidity of the recolonization and the extensive and frequent turnover of most species, were consistent with the basic MacArthur-Wilson equilibrium model applied to small islands. Finally, the protocols for individual species and groups of species revealed important details of the natural history of colonization. For example, spiders arrived early, in many cases almost certainly by ballooning with silken threads, but suffered rapid turnover. In contrast, mites generally arrived later and persisted with less turnover.

Epilogue

I am very pleased that the research I have recalled here has not become entirely obsolete, yet it has been greatly exceeded during the ensuing four decades in ways I could not have imagined. What we found and said in the 1960s appears to be generally true, and that is the best for which any scientist can ever hope.

Literature Cited

Darlington, P. J. 1948a. The geographical distribution of cold-blooded vertebrates. *Quarterly Review of Biology* 23:1–26.

———. 1948b. The geographical distribution of cold-blooded vertebrates (concluded). *Quarterly Review of Biology* 23:105–23.

———. 1957. *Zoogeography: The Geographic Distribution of Animals.* New York: Wiley.

Marshall, L. G. 1988. Land mammals and the Great American Interchange. *American Scientist* 76:380–88.

Matthew, W. D. 1915. Climate and evolution. *Annals of the New York Academy of Science* 24:171–318.

MacArthur, R. H., and E. O. Wilson. 1963. An equilibrium theory of insular zoogeography. *Evolution* 17:373–83.

———. 1967. *The Theory of Island Biogeography.* Princeton, NJ: Princeton University Press.

Simberloff, D. S., and E. O. Wilson. 1969. Experimental zoogeography of islands: defaunation and monitoring techniques. *Ecology* 50:267–78.

———. 1971. Experimental zoogeography of islands: a two-year record of colonization. *Ecology* 51:934–37.

Simpson, G. G. 1980. *Splendid Isolation: The Curious History of South American Mammals.* New Haven, CT: Yale University Press.

Wilson, E. O. 1959. Adaptive shift and dispersal in a tropical ant fauna. *Evolution* 13:122–44.

———. 1965. The challenge from related species. In *The Genetics of Colonizing Species,* ed. H. G. Baker and G. L. Stebbins, 7–27. New York: Academic Press.

Island Biogeography Theory

RETICULATIONS AND REINTEGRATION OF "A BIOGEOGRAPHY OF THE SPECIES"

Mark V. Lomolino, James H. Brown, and Dov F. Sax

The history of biogeography, like that of all natural sciences, is one whose exact origins are incredibly difficult if not impossible to pinpoint, and its conceptual threads split and again intertwine in a captivating, dynamic tapestry chronicling the geographic, ecological and evolutionary history of the world's biota. While fascinating accounts in their own right, studies of the historical development of scientific theories (e.g., "discoveries" of the theory of natural selection by Charles Darwin and Alfred Russel Wallace, of continental drift by Alfred Lothar Wegener, or of the structure of DNA by James Watson and Francis Crick), also provide valuable lessons for developing some truly transformative advances in the future. Here we review the historical development of island biogeography theory, with special emphasis on MacArthur and Wilson's equilibrium theory, to demonstrate how the science of biogeography develops, not just as a regular accumulation of facts and succession of paradigms, but through a reticulating phylogeny of insights and ideas often marked by alternating episodes of diversification and reintegration.

In the following section we present a brief history of island theory, in general, and summarize foundational insights that were available to scientists by the middle decades of the twentieth century in their attempts to explain patterns in geographic variation among insular biotas. Because MacArthur and Wilson's seminal contributions are the focus of all chapters in this volume, we see little need to describe their theory in detail here, beyond noting that their intent was to develop a theory with a much broader domain than is generally appreciated. Thus, in the third section of this chapter we describe the ontogeny and contraction in the conceptual domain of MacArthur and Wilson's theory, from the wealth of ecological and evolutionary phenomena comprising their general theory and monograph to an increasingly more narrow focus on the equilibrium model of species richness that came to preoccupy much of the field during the

1970s and 1980s. In the final sections of this chapter we observe that, like other disciplines in contemporary biogeography, evolution, and ecology, island theory may again be entering an exciting and perhaps transformative period of advance through consilience and reintegration. Toward this end, we conclude with a case study on biogeography, ecology, and evolution of insular mammals to illustrate an approach toward integration of island biogeography, which may ultimately lead to a more comprehensive and insightful understanding of the ecological and evolutionary development of insular biotas.

Insights Foundational to MacArthur and Wilson's Theory

Below we summarize seven advancements or approaches developed by the early decades of the twentieth century that were integral to the final articulation of MacArthur and Wilson's equilibrium theory.

1. *Encyclopedia of patterns.* Island research has a distinguished history of providing insights that have either fundamentally transformed existing fields of science, or spawned new ones. Indeed, that environmentally similar but geographically isolated regions are comprised of distinct biotas (Buffon's law) was a discovery fundamental to the realization that life was dynamic—species evolved in isolation (Buffon 1761; for summaries on the historical development of biogeography, see also Briggs 1995, Lomolino et al. 2004, Lomolino et al. 2006:13–38). Following Buffon's articulation of biogeography's first law, others (e.g., Candolle 1820) would provide cogent arguments on the geographic and temporal dynamics of biotas, and how their distributions and evolution were strongly influenced by interactions among the species. Thus, the early naturalists of the Age of European Explorations—visionaries whom today we recognize as the founders of the fields of biogeography, evolution and ecology—set out to describe the diversity and the geographic and temporal variation of life across an expanding spectrum of domains from the local and short-term scales to global and geological (evolutionary) ones.

Certainly the most distinctive types of newly discovered biotas, and of unrivaled importance to development of theories in biogeography, evolution, and ecology, were those inhabiting isolated islands. The seminal works of Darwin and Wallace are legendary in this respect, but these nineteenth-century naturalists were far from the first to appreciate the heuristic value of studying insular biotas (see summaries in Berry 1984, Wagner and Funk 1995, Grant 1998, Whittaker and Fernandez-Palacios 2007). During the eighteenth century, Carolus Linnaeus's explanation for the origin, diversity, and distribution of life on earth was premised on the existence of an insular Paradise of creation and, later, an isolated mountain range where the world's biota persisted during the biblical deluge and then dispersed to occupy their current ranges (Linnaeus 1781). Given the difficulty of accommodating this single center of origin/persistence theory with Buffon's discovery of the distinctiveness of regional biotas, Karl Ludwig Willdenow proposed that, rather than just one, there were many centers of origin, each situated in montane regions across the globe, where regional biotas were created or persisted during catastrophic periods (Willdenow 1792).

Perhaps most foundational to the origins of island biogeography theory were the accounts of Johann Reinhold Forster's (1778) circumnavigational voyage with Captain James Cook on the *H.M.S. Resolution* (1772–75). Not only did he find compelling evidence to support the generality of Buffon's law for plants as well as mammals and birds, and for other regional biotas beside those of the tropics, Forster also described patterns that continue to be at the core of research on the geographic, evolutionary, and ecological development of isolated biotas. He described the general tendency for isolated biotas to be less diverse than those on the mainland, and for the diversity of plants to increase with island area, availability of resources, variety of habitats, and heat energy from the sun. Thus, two fundamental patterns which island theory attempts to explain—the species-isolation and species-area relationships—along with basic explanations for those patterns (precursors of area per se and habitat diversity hypotheses, and species-energy theory; Hutchinson [1959], Preston [1960], Williams, [1964], MacArthur and Wilson [1967], Brown [1981], Wright [1983], Currie [1991], Ricklefs and Lovette [1999], Hawkins et al. [2003], Kalmar and Currie [2006]) were well established early in the historical development of these disciplines.

Charles Darwin, Alfred Russel Wallace, Joseph Dalton Hooker and many other naturalists of the late eighteenth and early nineteenth centuries would continue to add to the already voluminous accounts and explanations for the diversity and geography of island life. As we now well know, their efforts to explain this immense and ever-expanding encyclopedia of patterns would shake the very foundations of established doctrine and eventually lead to identification of the fundamental, dynamic processes influencing the diversity and geography of nature.

2. *Dynamics of nature (global to regional scales).* The Age of European Exploration and, indeed, the first globalization of the natural sciences, provided scientists with far more than just a fascinating and continually expanding catalogue of the marvels of nature. As engrossed as they may have been with describing empirical patterns, these early global

explorers and naturalists must have also felt compelled to explain them. Thus, Buffon's (1761) explanation for the distinctiveness of biotas included long distance dispersal and adaptive evolution of populations as their ranges shifted in response to changes in Earth's regional climates and environmental conditions. Again, Forster's (1778) explanation for gradients in diversity of plants among islands and across the continents was based on his understanding of the abilities of these species to respond to geographic variation in resources, habitat diversity, and solar energy. Thus, comparisons of the diversity and composition of biotas across regions and along geographic clines would eventually become irrefutable evidence that the natural world—its climate, geology, and species—was mutable, challenging those early naturalists to develop dynamic, causal explanations. Their theories of the historical development of regional biotas would focus on factors influencing the fundamental processes of biogeography—*extinction*, *immigration,* and *evolution*. That is, biotas responded to the regional- to global-scale dynamics of land and sea by suffering extinctions, by dispersing to other areas, or by evolving and adapting in place.

3. *Ecological interactions and emergence of ecology*. While the early global naturalists—the first "biogeographers"—continued to explore broad-scale and long-term patterns in biological diversity, others focused on the dynamics of biotas at more local spatial and shorter temporal scales. With each new revelation, it became increasingly more clear that patterns in distribution and abundance of species at these scales were strongly influenced, not just by the three fundamental biogeographic processes, but by interactions among species themselves. Thus, just as evolutionary theory diverged from that of biogeography during the early decades of the twentieth century, the field of ecology would diverge from other studies of the geography of life to become a distinctive and respected science in its own right. In fact, MacArthur and Wilson would include *ecological interactions* (in particular, "competition") as one of the fundamental, albeit challenging processes to study.

> Biogeography is a subject hitherto little touched by quantitative theory. The main reason is that the fundamental processes, namely dispersal, invasion, competition, adaptation and extinction, are among the most difficult in biology to study and to understand. (MacArthur and Wilson 1967, p. 4)

4. *Advances in theoretical and mathematical ecology*. Challenges in understanding dynamic systems led scientists to become increasingly more sophisticated and adept in their abilities to translate ideas and assumptions into graphic and mathematical models that would thus make them testable within an objective, logical framework. Theoretical and mathematical scientists from a broad diversity of disciplines realized that the

system properties they studied, whether they were geological formations, climatic conditions, chemical concentrations, gene frequencies, population abundance, or species distributions, resulted from interactions among opposing processes (e.g., orogeny and erosion; precipitation and evaporation; oxidation and reduction; or mutations, drift, birth, and death; e.g., Hardy [1908], Weinberg [1908], Lotka [1925], Pearl [1925], Volterra [1926, 1931], Fisher [1930], Gause [1934]). Often, the mathematical solutions to such problems would be simplified by assuming dynamic steady states, or equilibrial conditions, which could also be visualized in associated graphical models as the intersection of a system of curves describing opposing processes. The emerging discipline of mathematical ecology, lead by such distinguished scientists as G. Evelyn Hutchinson and his students (including Robert H. MacArthur), were quick to apply the tools developed by colleagues modeling the dynamics of other systems to their own studies of dynamics in the distributions and diversity of life.

5. *Earlier syntheses and integrations.* As we observed above, throughout the history of biogeography, and likely that of all other disciplines of science, its early explorers not just reported, but almost simultaneously and perhaps irresistibly attempted to synthesize the accumulated facts and ideas to provide a comprehensive description of how nature works. Monographs and treatises of Wallace (1857, 1869, 1876), Darwin (1859, 1860), and Hooker (1853, 1867) are familiar, if not legendary, attempts at such syntheses and integrations of patterns and developing theory in biogeography. Less well known and seldom read, but arguably as impressive if not influential, were the earlier works of Buffon (1761), Forster (1778), Humboldt (1805), Candolle (1820), and Agassiz (1840), and later those of Sclater (1858, 1897), Raunkiaer (1904, 1934), Dammerman (1922, 1948), Elton (1927, 1958), Docters van Leeuwen (1936), Simpson (1940, 1943, 1956, 1980), Mayr (1942), Lack (1947), and Darlington (1957).

Brown and Lomolino (1989) described the early and independent development by Eugene Gordon Munroe of an equilibrium theory of island biogeography—one with predictions of species richness based on island characteristics and opposing processes of immigration, extinction, and evolution (excerpted pages of Munroe's dissertation are available at www.biogeography.org/resources.htm). Unfortunately, he was unsuccessful in publishing his theory (outside of his 1948 dissertation, there is an abstract published in the 1953 *Proceedings of the Seventh Pacific Science Congress,* and a paper published in *The Canadian Naturalist* [Munroe 1963, pp. 304–305], which included a brief summary of his equilibrium theory), so there is no evidence that this work directly contributed to

MacArthur and Wilson's development of their theory. This episode of multiple discoveries in the history of science (*sensu* Merton [1961]) does, however, demonstrate the reticulating nature of island theory and that nearly all the requisites for an equilibrium theory of island biogeography were available over a decade before MacArthur and Wilson's seminal collaboration.

Nearly simultaneously with the completion of Munroe's dissertation, Karel Willem Dammerman published his comprehensive classic comparing the faunal dynamics of Krakatau to those of two continental islands (Durian and Berhala) and two oceanic islands (Christmas and Cocos-Keeling). While, as Thornton (1992) noted, Dammerman actually used the term "equilibrium," his extensive and meticulous account of the fauna of these islands was almost purely descriptive, lacking any attempt at a conceptual synthesis of underlying, causal processes. Rather, his goal was to develop a detailed and comprehensive description of the faunas inhabiting these islands and to explain why certain species but not others were successful at colonizing these environments (Dammerman 1948, p. vii). He did attribute variation in number of species among islands, again not the focus of his monograph, to proximate factors including island isolation, island size, tropical versus arctic climates, elevation, topographic relief, and development and variety of the vegetative communities (described by Docters van Leeuwen 1936), but his concept of "equilibrium" is mentioned only in brief and only in a phenomenological sense. That is, he used this term to characterize the apparently asymptotic slowing of species accumulation on certain islands, but said nothing about a possible balance among opposing processes. Thus, his concept of equilibrium was more similar to that envisioned by John Willis (1922, p. 229) and later by David Lack (1947, 1976), with islands accumulating species until all ecological space was filled (perhaps also presaging Wilson's [1959, 1961] concept of ecological "saturation" of islands).

Interestingly, early publications and insights from studies of the faunal dynamics of Krakatau had no obvious impact on Munroe's development of his equilibrium theory (Munroe 1948 and 1953; personal communication to MVL, 2007), which may be somewhat understandable given that Dammerman's book was not yet published, and that Munroe's field research focused on the biota of a different and distant part of the globe (i.e., the Caribbean archipelagoes versus those of Indonesia). In contrast, reports from Docters van Leeuwen (1936), Dammerman (1948), and others studying colonization following the 1883 eruption of Krakatau provided key empirical insights for future syntheses on the subject, including those first developed by E. O. Wilson and, eventually, in his transformative collaborations with Robert MacArthur as well (see MacArthur and Wilson 1967, pp. 43–51).

Roughly one decade after Munroe developed his theory, the field would witness another confluence of ideas attempting to synthesize the encyclopedic accumulation of island patterns and existing theory. In this case, however, the synthesis was a genuine precursor to MacArthur and Wilson's future theory—one presented in E. O. Wilson's papers on the ecological and evolutionary development of ant communities across Melanesia, wherein Wilson described his theory of the *taxon cycle* (1959, 1961; see Ricklefs, this volume). While few would argue that these papers were not influential, we believe their impact on the field, in general, and on the theory MacArthur and Wilson were about to develop, in particular, may still be largely underappreciated. Indeed, careful study of Wilson's taxon cycle papers reveals that they presented the first clear articulation of what would become the stated goal of MacArthur and Wilson's collaboration: "to examine the possibility of a theory of biogeography at the species level" (MacArthur and Wilson 1967, p. 5). Thus, Wilson's 1959 paper identified the concept of *a biogeography of the species* as being central to his theory of the ecological and evolutionary development of insular biotas.

> There is a need for a "biogeography of the species" [quotes his], oriented with respect to the broad background of biogeographic theory but drawn at the species level and correlated with studies on ecology, speciation, and genetics. (Wilson 1959, p. 122)

It may well be that his theory of taxon cycles, and in particular the concept of *a biogeography of the species*, may again become foundational to emerging and more integrative theories of island biogeography (see our discussion in the final section of this chapter). Indeed, although the heuristic promise of the research agenda outlined in the above quotes was unappreciated by many biogeographers caught up in the "normal science" (*sensu* Kuhn 1994) of the 1970s and 1980s, a selection of insightful research programs continued to study the ecological and evolutionary development of insular communities as interrelated phenomena (e.g., Ricklefs and Cox 1972, 1978, Diamond 1975, 1977, Erwin 1981, Roughgarden and Pacala 1989).

6. *Dynamics of nature at finer scales (from global and regional down to archipelago and island)*. Wilson, like Munroe before him, was strongly influenced by the theories of William Diller Matthew, George Gaylord Simpson, and Phillip J. Darlington (incidentally, Darlington provided advice to both Munroe and later Wilson during their early development as scientists). Matthew (1915), Simpson (1940, 1943, 1944) and Darlington

(1938, 1943, 1957) each cogently asserted that the earth, its land and sea, its climate and its species were dynamic; with biotas expanding from their centers of origin, dispersing across new regions and then adapting, evolving and, in most cases, suffering eventual extinction depending on the vagaries of regional to global environments (views overlapping to some degree, but also in some ways contradicting those central to Willis's [1915, 1922] age and area theory). Wilson was able to telescope Darwin and Wallace's center of origin-dispersal-adaptation (CODA) perspective from global and geological scales down to more local spatial and short-term temporal scales. That is, his theory described the dynamic development of biotas on particular archipelagoes and islands in evolutionary and ecological time. Wilson recounted his scientific epiphany in his autobiography (1994, pp. 214–15).

> It dawned on me that the whole cycle of evolution, from expansion and invasion to evolution into endemic status and finally into either retreat or renewed expansion, was a microcosm of the worldwide cycle envisioned by Matthew and Darlington. To find the same biogeographic pattern in miniature was a surprise then. . . . It came within a few minutes one January morning in 1959 as I sat in my first-floor office . . . sorting my newly sketched maps into different possible sequences—early evolution to late evolution. . . . Discovery of the cycle of advance and retreat was followed immediately by recognition of another ecological cycle. . . . I knew I had a candidate for a new principle of biogeography.

Thus, Wilson's independent synthesis produced a "new principle" —*a biogeography of the species*, which was a process- and species-based theory that explained the dynamic distributions of species and the geographic variation in biodiversity among islands. Patterns in insular community structure among regions, archipelagoes, and islands were functions of the dynamics of processes operating across global and geological scales down to local and ecological ones. These processes included immigration and range expansion, evolutionary divergence and diversification, extinction, and ecological interactions; the latter affecting each of these more fundamental processes.

7. *Advancing science through collaborative synthesis*. Despite all its prescience and promise, the impact of Wilson's independent synthesis developed in his taxon cycle papers was soon to be overshadowed by his future collaboration with Robert Helmer MacArthur. As noted earlier, Wilson's theory of taxon cycles and his concept of *a biogeography of the species* arguably constituted an integral and precursory stage in the development of their equilibrium theory. Perhaps the most fundamental

reason for the success of their collaboration is just that—it was a genuine collaboration, which melded and expanded the complementary strengths and visions of each beyond what they were capable of in their independent, albeit distinguished, research programs.

Exemplary cases of transforming science through collaborative syntheses included Watson and Crick's legendary deciphering of the structure of DNA, achieved some ten years prior to MacArthur and Wilson's first paper (see Watson 1968). The synergistic benefits of this and other, earlier collaborations in the natural sciences were not lost on Wilson and MacArthur, as evidenced, for example, by Wilson's earlier collaboration with William Brown on the phenomenon of character release (one that would later be integrated into Wilson's theory on taxon cycles; see Brown and Wilson [1956]), and those of MacArthur with his mentor, G. E. Hutchinson, and their students and colleagues (e.g., Hutchinson and MacArthur 1959, MacArthur and Levins 1964, 1967, MacArthur and Connell 1966). As Robert J. Whittaker (personal communication, 2008) observes, it seems ironic but perhaps fitting that the collaboration which contributed to the dominance of molecular biology in the 1950s and 1960s—for some time marginalizing whole-organism biology and community ecology—would be answered by the collaboration between MacArthur and Wilson, which reenergized ecology and biogeography by providing , as Whittaker puts it, a "radically updated framework for this branch of science" (see Wilson 1994, chap. 12,"The Molecular Wars").

Rather than being satisfied with their first collaboration—the relatively focused, albeit intriguing, joint paper they published in 1963—MacArthur and Wilson were determined to develop a full-scale, integrative synthesis of island theory. At first rather humbly stated at the end of their 1963 paper, their goal was "to deal with the general equilibrium criteria, which might be applied to other faunas, together with some of the biological implications of the equilibrium condition." But, fully realizing the revolutionary potential of their first collaboration, they had agreed by December of 1964 to once again join forces, this time to "write a full-scale book on island biogeography, with [the] aim of creating new models and extending [their] mode of reasoning into as many domains of ecology as [they] could manage" (Wilson 1994, p. 255).

In summary, the cumulative knowledge of the geography and diversity of nature and, more importantly, the deepening understanding of and ability to model the dynamics of the natural world and the underlying, scale-dependent causal processes, rendered the development of an equilibrium theory of island biogeography not only possible, but likely, if not inevitable. This appears to be a relatively common phenomenon, with the classic and best-known example in the biological sciences being the convergent and nearly simultaneous "discovery" or rediscovery of the theory of natural selection by Alfred Russell Wallace and Charles Darwin, providing some invaluable lessons on how transformative advances in the natural sciences are achieved (see also Merton's [1961] review of episodes of multiple, independent discoveries in science).

As with other disciplines, biogeography advanced not just as a regular accumulation of facts and succession of alternative and increasingly more accurate concepts, but through syntheses and re-integrations in a reticulating phylogeny of sometimes convergent if not equivalent theories. Munroe's independent development of an equilibrium theory, Lack's (1947) concept of the filling of ecological space, and Wilson's concept of "saturation" of insular biotas (as part of his taxon cycle theory), are illustrations of this phenomenon (in this case, incarnations of similar if not equivalent concepts of island biogeography). Yet these revolutionary advances in biogeography, along with its descendant disciplines of ecology and evolution, were ultimately achieved by addition of the final component in the above list of foundational elements—a genuine collaborative synthesis between two of the field's established visionaries.

Success and Subsequent Evolution of MacArthur and Wilson's Theory

Despite some interesting and sometimes heated debate over the merits of the equilibrium model of species richness during the four decades since its initial articulation, there should be little question that MacArthur and Wilson's theory has had a revolutionary influence on biogeography and related disciplines, and they certainly achieved one of their primary goals: "creating new models and extending [their] mode of reasoning into as many domains of ecology [and other disciplines] as [they] could manage" (Wilson 1994, p. 255).

Our purpose in this section is not to chronicle the hundreds if not thousands of studies that were stimulated by their theory: indeed, much of our own earlier research was developed to evaluate the tenets of their theory or to modify it to create other means of analyzing and understanding the ecological and evolutionary assembly of isolated biotas (Brown 1971, 1978, Brown and Kodric-Brown 1977, Lomolino 1986, 1990, 1994, 1996, 2000, Sax et al. 2002). Rather than focus here on how the theory influenced other research programs in these areas (which we believe is well covered in other chapters of this book), our purpose in the following paragraphs is to describe how the theory MacArthur and Wilson presented in their 1967 monograph was substantially transformed, at least in its predominant development and applications during the normal science (*sensu* Kuhn 1996) of the next two decades.

As we described earlier, the intended domain of MacArthur and Wilson's theory was quite broad: again, in the introduction to their book, they made their ultimate goal quite clear.

> The purpose of this book is to examine the possibility of a theory of biogeography at the species level. We believe that such a development can take place by looking at species distributions and relating them to population concepts, both known and still to be invented.(MacArthur and Wilson 1967, pp. 5–6)

In their conclusion (MacArthur and Wilson 1967, p. 183), they returned to this very general theme of a process- and species-based reintegration by calling for the field of biogeography to

> be reformulated in terms of the first principles of population ecology and genetics . . . to deemphasize for the moment traditional problems concerning the distribution of higher taxa and the role of geological change . . . and to turn instead to detailed studies of selected species. A "biogeography of the species" [quotes theirs] requires both theory and experiments that must be in large part novel.

Despite these goals of developing a very general, species- and process-based theory—one covering not just patterns in richness, but including a host of other ecological and evolutionary phenomena (including *r*/*k* selection, niche dynamics, geometry and strategies of colonization, and evolution), the research agenda during the 1970s and 1980s seemed so captivated with the equilibrium model of species richness that it often lost sight of the broader agenda of *a biogeography of the species*. During this period, ecological biogeographers became intrigued with the abilities to model species as though they were "atoms in a gas law context" (personal communication, R. Ricklefs 2008): the very general theory could be recast in a more narrow sense—as a model of how richness of equivalent, noninteracting, and nonevolving species varies with island area and isolation ("mere curve-fitting," *sensu* Haila [1986]; "a numbers game" *sensu* Whittaker [1998], Whittaker and Fernandez-Palacios [2007]). As we noted earlier, the heuristic promise of Wilson's theory of taxon cycles and a biogeography of the species was not lost on everyone, as a group of distinguished ecologist and biogeographers continued to pursue and develop these concepts throughout this period. Eventually, their insights would be integrated into a set of now emerging theories that promise to provide some genuinely transformative advances in island theory (see other chapters in this volume, and the final sections of this chapter).

As Stuart Pickett and his colleagues explain in their important book *Ecological Understanding: The Nature of Theory and the Theory of Nature*, theories are far from static, but typically if not invariably undergo an ontogeny of their own (Pickett et al. 2007; see also Kuhn 1996). Most theories are first described in a premature form, well before the requisite knowledge and conceptual tools necessary to fully appreciate and develop their potential import. Wegener's (1912a, 1912b, 1915) theory of continental drift—first proposed some five decades before the scientific community fully embraced it—is one of the most striking cases of delayed acceptance of a truly prescient and potentially transformative theory in natural science. Early articulations of equilibrium concepts by Munroe, and of Wilson's theory of taxon cycles and his concept of species saturation and a biogeography of the species, represent similar episodes of unappreciated prescience in biogeography. By the time MacArthur and Wilson collaborated to develop their theory, however, the empirical and conceptual foundations of island biogeography, and in particular the abilities of scientists to visualize and model dynamic processes, had progressed to the point that a genuinely paradigmatic advance could be achieved and widely appreciated.

The ontogeny of MacArthur and Wilson's equilibrium theory weaves a tapestry whose fabric and modified forms are just beginning to become clear after four decades of maturation and retrospection. One perhaps key factor, which was actually lacking from its subsequent development, was the continued involvement of its creators. Tragically, MacArthur died of renal cancer just five years after he and Wilson published their monograph. Wilson conducted some fascinating experiments in island biogeography in the late 1960s, again a collaboration (this time with his distinguished student—Daniel Simberloff (see Simberloff, this volume), but Wilson's interests and energies soon turned to other demanding and highly successful endeavors, including evolutionary biology, sociobiology, and conservation of biological diversity. The subsequent period of over three decades of the theory's maturation, then, were left to a rapidly growing community of biogeographers and ecologists, including critics as well as champions.

While it may appear that the theory's subsequent development can be characterized by an expansion of the domain of its *applications* (e.g., application of the equilibrium model of species richness to a broad diversity of isolated ecosystems, including lakes, mountaintops, and other patches of terrestrial ecosystems, as clearly anticipated by MacArthur and Wilson [1967, pp. 3–4]; see Pickett et al. 2007, p. 104), we believe that just the opposite has occurred at least in terms of the theory's *conceptual* domain. According to Yrjö Haila, during the 1970s and 1980s the theory suffered a "reification" (*sensu* Levins and Lewontin 1980) with an increasingly more narrow focus on species richness correlations and on the explanatory performance of the iconic, equilibrium model, with an apparent waning of appreciation for the broader value of "the theory as a research programme that directs attention to the dynamic nature of island communities in general, and to mechanisms that determine the colonization process in specific situations" (Haila 1986, p. 379; see also Sismondo 2000). A review of MacArthur and Wilson's monograph, including the various excerpts included above which described their stated goals, makes it clear that the equilibrium model of species richness was just one component (albeit one of the most central, compelling, and easiest to visualize and remember) of their attempt to develop a truly comprehensive theory of island biogeography ("a biogeography of the species," again, first articulated by Wilson in his original, taxon cycle paper of 1959).

Contraction in the conceptual domain of MacArthur and Wilson's theory (at least as practiced by many biogeographers through the 1970s and 1980s) was symptomatic of concurrent specialization and splintering across the very broad domain of biogeography itself, including widening divisions between, as well as within, ecological and historical biogeography. We are, however, encouraged by the more recent groundswell of biogeographers now calling for a reexpansion in the domain of island theory and a reintegration of the field (e.g., Brown and Lomolino 2000, Brooks 2004, Brown 2004, Lieberman 2004, Lomolino and Heaney 2004, Riddle and Hafner 2004, Ebach and Tangney 2007, Stuessy, 2007; see also chapters in this volume, especially those by Grant and Grant, by Whittaker et al., by Losos and Parent, and by Ricklefs). We agree that this can best be accomplished by developing more integrative theories of island biogeography—those that encompass the full breadth of patterns in geographic variation among insular biotas, and are based on the premise that those patterns result from predictable variation in the fundamental biogeographic processes among islands and species, and across scales of space, time, and biological complexity.

Toward Consilience and Integrative Theories of Island Biogeography

Here we outline the fundamental components of one approach for developing theories that may advance the field through consilience and integration in order to achieve a new biogeography of the species, i.e., a process- and species-based explanation for the very broad diversity of interrelated patterns and underlying processes affecting insular biotas. First, we describe the conceptual domain of an integrative theory of island biogeography, and then list the tenets that are fundamental to this approach and, in combination, requisite to a genuinely transformative advance in the field. We then conclude with a case study illustrating how two apparently disparate phenomena (patterns of insular distributions and those of microevolution on islands) can be more fully understood within the context of the same, integrative theory.

Conceptual Domain and General Statement of the Theory

Integration not only provides a means of expanding the variety of phenomena studied, but also provides us with a means of better understanding the causal nature of intriguing and interdependent phenomena, given that each is influenced by processes that operate across interdependent domains of space, time, and biological complexity. For example, interactions among species not only influences their abundance and distributions at local scales, but can strongly influence fundamental biogeographic processes, thus modifying patterns in distributions, diversity, and distinctiveness at regional to global scales as well.

The conceptual domain of an integrative theory of island biogeography should include a broad diversity of patterns in geographic variation in the characteristics of insular individuals, populations, and communities. One fundamental premise of this theory is that these patterns result from the regular and predictable variation among islands and among species in characteristics that influence the fundamental biogeographic processes—immigration, extinction, and evolution. That is, the fundamental capacities of species (to immigrate to islands, and survive and evolve there) should vary in a nonrandom manner among species (e.g., when those species are ordered by body size or energetic requirements), while rates of immigration, extinction, and evolution of those species should vary in a nonrandom manner among islands (e.g., when islands are ordered by area, isolation, primary productivity, or carrying capacity). Therefore, the successful integration, or reintegration, of island theory will depend on our abilities to evaluate the generality and validity of its fundamental tenets (described in the next section), to further develop its integration with theory in other domains of science, and to assess its potential applications for conserving the evolutionary and geographic context of isolated biotas (see Haila 1986, p. 385).

Among the most valuable approaches for discovering and understanding patterns emergent across multiple scales of space, time, and biological complexity are those developed by macroecologists (see Brown 1995, Gaston and Blackburn 2000). Thus, macroecology may well provide a useful conceptual and analytical framework for reintegration across the broad domain of island biogeography theory (*sensu latissimo*; i.e., all patterns in geographic variation among insular biotas). Below, we list and briefly describe seven tenets and conceptual elements that seem requisite to integrative theories of island biogeography. Taken separately, none of the assertions described in the following list is revolutionary, but in combination they comprise a conceptual framework that has much promise for achieving the species- and process-based theory at the core of Wilson's biogeography of the species.

Fundamental Tenets of an Integrative Theory of Island Biogeography

SCALE DEPENDENCE

1. The relative importance of each of the fundamental biogeographic processes (immigration, extinction, and evolution) and of ecological interactions varies in a predictable manner across spatial and temporal scales and among species. For example, the relative importance of evolution in terms of its influence on patterns of diversity and distinctiveness among insular biotas likely increases as we consider broader spatial and temporal scales (e.g., archipelagoes spanning greater degrees of isolation and those including larger islands (figure 2.1); see also Lomolino 1999, 2000, Heaney 2000, Losos and Schluter 2000, Whittaker 2004, Whittaker et al. 2008).

NATURE OF INFLUENCE

2. Island biogeographic patterns result from both independent and interactive influences of immigration, evolution, and extinction, which should be functions of the system (island and archipelago) and species traits affecting those processes (see tenets 3 and 4, respectively). Distributions of particular species among islands, in turn, should be functions of their immigration capacities *relative to* their abilities to maintain populations on those islands: i.e., populations of a focal species are most likely to occur on those islands where conditions (e.g., isolation and area) are such that the probability of immigration by that species is high relative to its likelihood of extirpation following colonization of that island (figure 2.2). A species can inhabit even the most isolated islands of an archipelago if those islands are relatively large (such that extirpation probabilities for its populations are compensatorily low). Similarly, evolutionary divergence is also dependent on the combined effects of these processes—being most prevalent on those islands that are both isolated and large, such that gene flow is relatively low and persistence times and within-island barriers (e.g., major rivers and mountain chains) provide the requisite conditions for divergence among and within large islands (e.g., see Wagner and Funk 1995, Heaney 2000, Losos and Schluter, 2000).

SYSTEM AND SPECIES TRAITS OF PRIMACY

3. *System traits of primacy.* Most important among the geographic or system variables influencing the fundamental biogeographic processes and feedback mechanisms (listed in tenet 7, below) are

- area, isolation, topographic relief, age and disturbance history of the islands, and

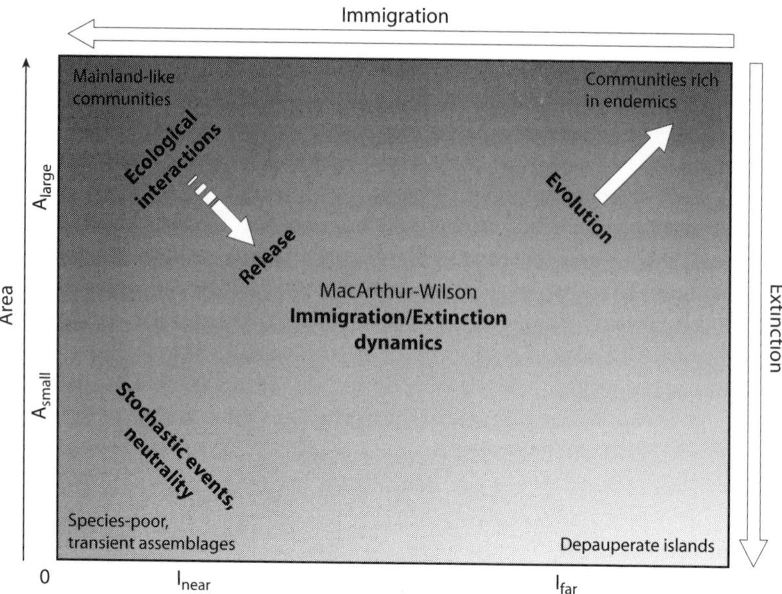

Figure 2.1. Scale dependence of the biogeographic and ecological processes (*Immigration, Extinction, Evolution; Ecological Interactions,* and *Ecological Release*) influencing community structure of insular biotas: here placed within the geographic context of two principal characteristics of island ecosystems (Area and Isolation). Because immigrations and extinctions of nearly all species in the focal biota are so frequent at relatively fine scales (i.e., on islands $< I_{near}$ and A_{small}), community structure on these islands tends to be driven by stochastic events, which produce apparently random assemblages of species, with richness and species composition varying independent of island isolation and area (i.e., the near-island effect, and the small-island effect [see MacArthur and Wilson 1967, pp. 30–32; Lomolino and Weiser 2001], respectively). On somewhat more isolated and larger islands, the structure and dynamics of insular communities should approach those envisioned by MacArthur and Wilson's equilibrium model, although differences in immigration abilities and resource requirements among the species may result in non-random assemblages of communities on these islands (e.g., producing community nestedness across gradients of isolation and area [see Darlington 1957, p. 485, figure 57, Wilson 1959, p. 128, figure 2, Patterson and Atmar 1986, Lomolino 1996]). On islands that are very isolated and very large with respect to immigration abilities and resource requirements of most species in the focal biota (i.e., on islands $> I_{far}$ and A_{large}), evolution becomes an important force influencing the diversity and distinctiveness of their communities (see Losos and Schluter 2000). Finally, the relative importance of ecological interactions and ecological release varies with diversity of insular communities (shown here as a gradient of decreasing shading from species-rich to depauperate islands; note that speciation within isolated archipelagoes comprised of relatively large islands [top, right-hand corner of the figure] can promote relatively high diversity as well as endemicity). Note also that the effects of geological dynamics of the islands (Whittaker et al. 2008) are not included in this version of the model.

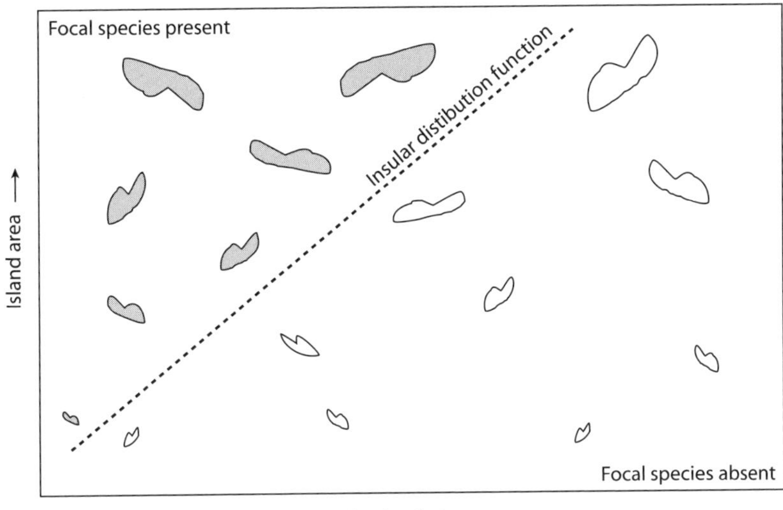

Figure 2.2. The insular distribution function (dashed line) can serve as a fundamental level in an integrative and hierarchical approach to island biogeography theory, providing a means of placing a diversity of patterns of variation among insular biotas within a geographic context (here, as described by island area and isolation). The insular distribution function is essentially a constraint line (*sensu* Brown 1995), whose slope and intercept should vary in a predictable manner with characteristics of the archipelagoes (tenet 3) and focal species (tenet 4; see explanation in the text; see also Lomolino 1986, 1999, 2000, Hanski 1986, 1992, and this volume).

- latitudinal position, and nature of the immigration filters (characteristics of the intervening seascapes) of the archipelagoes.

These correlates of biogeographic variation among islands have been discussed throughout the history of the field, from the early studies of Forster (1778), through those of Darwin and Wallace, to current research in all aspects of island theory (see Lomolino et al. 2006, chapters 13 and 14, Whittaker and Fernandez-Palacios 2007).

4. *Species traits of primacy*. Most important among the species traits influencing the fundamental processes and capacities of species (i.e., their immigration abilities, and their abilities to survive, evolve, and dominate other species on islands) are those that most strongly influence resource requirements and how those resources are utilized for dispersal, survival, and ecological interactions, and are transformed into offspring. In animals, most important among these traits are body size, *bauplän* (i.e., the body plan common to particular groups of organisms, including such features as the degree of symmetry, specialization among body segments,

or number of limbs) and trophic strategy (e.g., foliage gleaning insectivore, grazing herbivore, or cursorial, top carnivore). For plants, traits of primacy likely include size of gametophyte or sporophyte, growth form (e.g., epiphytic, herbaceous, shrub, or tree), propagule dispersal mechanisms, and principal energetic and metabolic pathways (e.g., parasitic, nitrogen fixing, C3, C4, and CAM).

COVARIATION OF FUNDAMENTAL PROCESSES

5. *Among systems.* Along with exhibiting predictable patterns of variation along geographic gradients (e.g., along those of increasing area, isolation, or latitude), the fundamental processes also exhibit significant covariation among islands and archipelagoes. For example, larger islands may experience more immigrations (the target area effect; Gilpin and Diamond 1976, Hanski and Peltonen 1988, Lomolino 1990), fewer extinctions (Macarthur and Wilson 1963, 1967), and a greater degree of evolutionary divergence (e.g., see Lomolino et al. 2006, figure 14.19a, after Mayr and Diamond 2001); archipelagoes located in higher latitudes may experience fewer immigrations (except when those waters freeze over; Lomolino 1988, 1993), lower persistence times (due to lower ambient temperatures, productivity, and carrying capacities), and lower rates of evolutionary divergence (due to the decelerating effects of cooler temperatures on life history processes) (Rohde 1992, Cardillo 1999, Allen et al. 2002, Brown et al. 2004, Wright et al. 2006).

6. *Among species.* Given that natural selection operates on *combinations* of interdependent traits which comprise entire organisms, then the fundamental capacities of insular biotas (abilities to colonize, survive, dominate other species in ecological interactions, and evolve on islands) should exhibit significant covariation among species. For example, along a gradient of increasing body size of vertebrates, vagilities (for active immigration), resource requirements (and therefore their susceptibility to extirpation), and abilities to dominate other species in ecological interactions should increase, while rates of evolutionary divergence should decline (Lomolino 1989, 1985, 1993, McNab 2002, Millien 2006, Millien and Damuth 2004, Millien et al. 2006). In invertebrates, while resource requirements, ecological dominance, and evolutionary rates may exhibit similar trends, pagility (capacity for passive immigration) of at least some species groups (e.g., land snails; Vagvolygi 1975) may actually decline with increasing body size.

FEEDBACK

7. The generality of biogeographic patterns and the interdependence among underlying, fundamental processes are affected, and possibly enhanced by three important feedback mechanisms.

a. *Ecological interactions* among species, which can influence each of the fundamental capacities of other species (i.e., their abilities to immigrate to, and survive and evolve on, islands). Included here are well-demonstrated effects of competition (Brown and Wilson 1956, Grant 1968, 1971, 1996, 1998, Crowell 1962, Grant and Grant 2007 and their chapter in this volume, Losos and Queiroz 1997), predation (e.g., Lomolino 1984, Schoener et al. 2001, Schoener et al. 2002), parasitism (Apanius et al. 2000, Fallon et al. 2003), mutualism, commensalism, and succession driven by prior colonists (Thornton 1996, Whittaker et al. 1989) on immigration, establishment and extinction of insular plants and animals
b. *Microevolution*, which can substantially alter life histories and fundamental capacities of species. Perhaps most striking among these insular phenomena are the innumerable and intriguing cases of evolutionary divergence associated with reduced dispersal abilities of insular forms, including the development of flightlessness in thousands of species of insular invertebrates and birds (McNab 1994a,b, 2002, Steadman 2006) and reduced capacities for flight and enhanced terrestrial nature in many other species (e.g., the short-tailed bats of New Zealand—family Mystacinidae), and reduced dispersal mechanisms, and increased woodiness and arboreal growth forms in otherwise herbaceous plants (Carlquist 1974, Givnish 1998).
c. *Macroevolution (speciation)*, which can strongly influence patterns in diversity and distinctiveness among insular communities. This is another scale-dependent process (tenet 1; figure 2.1) and, because it influences fundamental properties of insular communities (i.e., the number and types of species), it can have cascading effects by influencing each of the other fundamental biogeographic processes (immigration and extinction) and the above feedback mechanisms (ecological interactions and microevolution) as well (see Emerson and Kolm 2005). Where important (i.e., on very large and very isolated islands), macroevolution can play a predominant role in determining the structure of insular biotas, creating hotspots of diversity and distinctiveness rivaling and in some cases exceeding those of the richest mainland communities (e.g., mammals of the Philippines [Heaney 2004, Heaney and Regalado 1998]; ferns, drosophilids, snails and honeycreepers of Hawaii [Wagner and Funk 1995]; asters and *Anolis* lizards of the Caribbean [Losos and Schluter 2000, Losos and Thorpe 2004, Francisco-Ortegal et al. 2008]; cichlids of Africa's Rift Valley Lakes [Meyer 1993]).

An Illustration of the Integrative Approach in Island Theory

Transformative advances in science are often achieved by novel approaches for visualizing fundamental, underlying processes and their variation across scales (in this case, those of biogeographic, evolutionary, and biological complexity). Following MacArthur and Wilson's (1967) exemplary graphical models, the developments in the field of macroecology also provide some compelling demonstrations of the utility of these transitional-scale models, or "macroscopes" (Brown 1995, Gaston and Blackburn 2000). Here, we utilize such graphical models to demonstrate how two sets of what have traditionally been viewed as intriguing but unrelated phenomena—ecological assembly (distributional patterns) and evolution of insular body size—can be better understood within the context of a more integrative approach to island biogeography theory.

As we pointed out earlier, the graphical model of insular species distributions illustrated in figure 2.2 can serve as a geographic template for integration among the scale-dependent processes influencing the ecological and evolutionary development of insular biotas (tenets 1 and 2). Lomolino (1999, 2000) presented an earlier version of this approach to island biogeography theory, which was hierarchical but also species-based because it was premised on the assumption that many patterns in assembly of insular communities derive from predictable variation among their focal species. Again, we are assuming that insular distributions of each focal species are functions of the combined effects of immigration and extinction (tenet 2). Therefore, islands whose coordinates (isolation and area) fall above the dashed constraint line (the insular distribution function) of figure 2.2 are more likely to be inhabited by the focal species. Elsewhere, we have shown how variation and covariation among important system and species traits (tenets 3–6) and ecological interactions among insular populations (tenet 7a) can be integrated into this hierarchical approach to explain ecological assembly and geographic variation among insular biotas (including intra- and interarchipelago patterns in species richness and species composition; see Lomolino 2000, figures 3–5, 9–11; see also Simberloff and Collins, this volume). Here, we demonstrate how evolutionary divergence among insular populations (tenet 7b) can be added to the theory to explain some intriguing insular patterns—in this case, the truly remarkable phenomenon of body size evolution on islands.

The "island rule" describes a graded trend away from norms of body size observed in species-rich, continental environments, such that on islands small species exhibit gigantism, whereas large species exhibit dwarfism (figures 2.3a and 2.3b). We describe this as a "graded" trend because the tendency toward gigantism or dwarfism declines as we move from species of extreme to those of more modal size.

The generality of the "rule" is, of course, not universal but still surprising given that it is now reported not just for terrestrial, nonvolant mammals (as in its original articulations by Foster [1964] and modifications

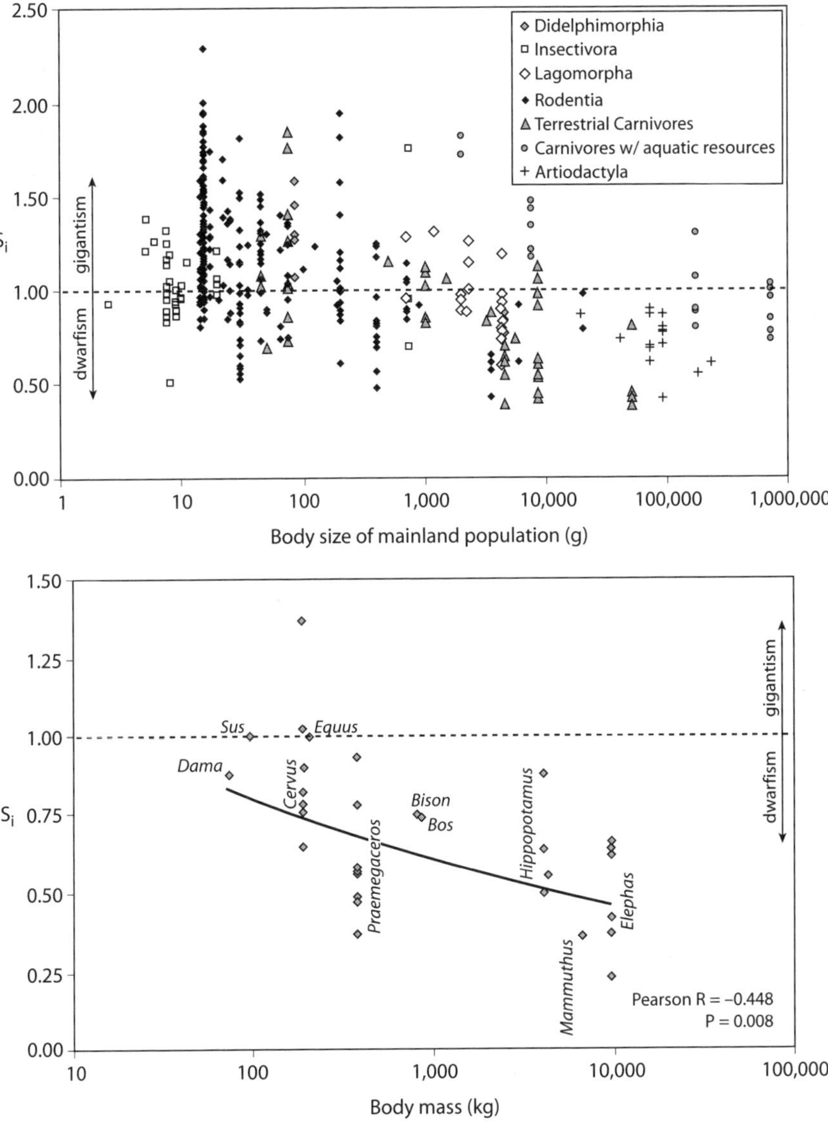

Figure 2.3. (Top) Body size trends for insular mammals. S_i=relative size of insular forms expressed as a proportion of body mass of their mainland relative (see Lomolino 1985, 2005). (Bottom) Antiquity of the island rule: body size trends for ungulates and "elephants" (orders Artiodactyla, Perisodactyla, and Proboscidea) of Mediterranean islands during the Pliocene and Pleistocene (reanalysis of data from Raia and Meiri [2006], Raia [personal communication 2008]; body mass estimates from Palombo [personal communication 2008]). S_i=insular body size as proportion of body size of the mainland population (as linear dimensions of metatarsus, metacarpal, humerus, or tibia).

by Van Valen [1973], Heaney [1978], and Lomolino [1985]), but also for a broad diversity of vertebrates and invertebrates (see reviews by Lomolino 2005, and Lomolino et al. 2006; see also Meiri et al. 2004, 2007, 2008a,b, Meiri 2007, Price and Phillimore 2007). Other reports of patterns consistent with the island rule include those for groups as varied as recent, deep-sea gastropods (McClain et al. 2006), Pliocene-Pleistocene ungulates (figure 2.3b), sauropod dinosaurs (Jianu and Weishampel 1999, Sander et al. 2006), and Pleistocene hominins (Brown et al. 2004, Morwood et al. 2004, Morwood 2005).

On the other hand, some species groups appear anomalous or at least equivocal with respect to the patterns predicted by the island rule, and all show substantial variation about the general trendlines of figure 2.3 (i.e., beyond that accounted for simply by ancestral body size). This residual variation is at least partly a function of the fact that this relatively simple model does not take into account variation in key traits of the islands (tenet 3) or focal species (tenet 4), nor does it consider the possible effects of covariation (tenets 5 and 6) and feedback (tenet 7) among biogeographic processes. Yet, as we asserted above, at least some of these shortcomings can be addressed by using the model of scale dependence (figure 2.1) and the insular distribution function (figure 2.2) to place these evolutionary patterns in an ecological and geographic context. Our goals in this section are, therefore, threefold:

1. to provide an explanation for the island rule which is based on the tenets of the general theory, described above,
2. to place this explanation within the context of the geographic template provided by insular distribution functions, and
3. to explain some apparently anomalous trends in insular body size, including the tendency for carnivorous mammals to exhibit equivocal patterns (figure 2.4) and for rodents to exhibit dwarfism on some very disparate islands—i.e., on nearshore and on oceanic islands (in the latter case, with dwarfed elephants), but not on those of intermediate isolation.

As the Indonesian paleobiologist, Dirk Albert Hooijer observed in a paper published the same year as MacArthur and Wilson's classic monograph, "wherever we find elephants we also have giant rodents. . . . we have no means of knowing how many generations were involved, it is, however, likely that evolutionary velocity has been higher under these conditions than is usual" (Hooijer 1967, p. 143).

Consistent with the tenets of an integrative theory of island biogeography, the explanation for the island rule featured here centers on the scale dependence of fundamental, causal processes (tenets 1 and 2)—in this case, how they vary between insular and mainland environments, among

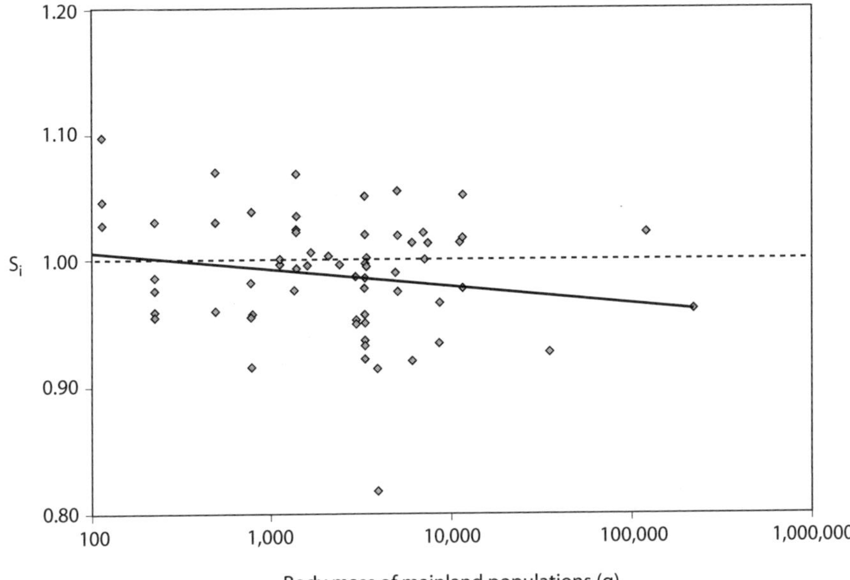

Figure 2.4. Body size of insular carnivores (Mammalia, Carnivora). It exhibits substantial variation about the trend, although the relationship appears to be statistically significant (*P* [one-tailed test that the slope is not$<$0.0] is$<$0.05) and in the direction consistent with the island rule (after Lomolino's [2005, pp. 1684–85, figure 2] reanalysis of Meiri et al's [2004] data; see also Price and Phillimore [2007]). S_i is body mass of insular populations expressed as a proportion of that of their mainland relatives.

islands within the same archipelago, and among species. Body size influences all physiological process and life history characteristics of animals (Calder 1984, McNab 2002), in turn producing some very regular patterns of variation and covariation among the fundamental capacities of organisms (tenet 6); i.e., in their abilities to immigrate to islands, and survive, evolve, and dominate other species in ecological interactions there. The result is that there may be an optimal size (associated with an optimal combination of fundamental traits and capacities) for organisms with a given *bauplän* and trophic strategy (represented by the shaded triangle in figure 2.5). This optimum, however, should vary with characteristics of the insular environments that influence fundamental capacities of the species (i.e., with *isolation, latitude,* and *area* of the islands, affecting immigration, survival, and evolution; tenets 2, 3, and 4), and with *diversity* and *species composition* of particular insular communities,

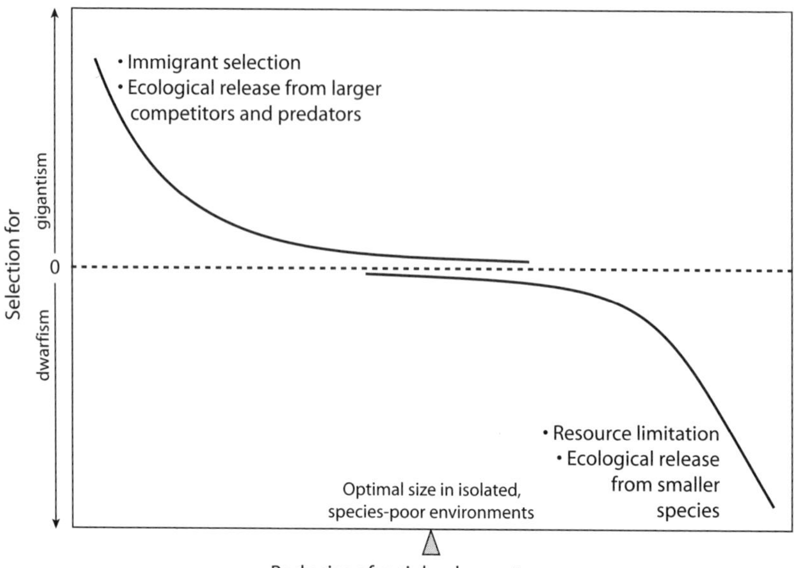

Figure 2.5. An explanation for body size evolution of insular vertebrates (i.e., the island rule). It is based on how selective pressures and fundamental capacities of species (to immigrate to, and survive, evolve, and dominate other species in ecological interactions on islands) varies with body size of their ancestral forms (see explanation in the text along with that in Lomolino [1985, 2005], Lomolino et al. [2006], and references therein).

affecting ecological interactions (tenet 7a), which in turn drive niche and character dynamics (tenet 7b; Brown and Wilson 1956). Thus, in species-rich mainland communities, pressures from a diversity of predators and competitors should cause the optimal size of particular species to differ from that of the entire taxon or species group (again, as identified by similar *baupläne* and trophic strategies).

A corollary of tenet 1, and one central to the explanation for the island rule presented here, is that relevant selective pressures vary in their importance in a predictable manner among species of different body size (figure 2.5; right-hand column of figure 2.6). Thus, insular populations of small species often increase in body size on ecologically simplified islands (i.e., in the absence of larger competitors and predators), converging back on the optimal body size for that species group (again, as determined by common *bauplän* and trophic strategy; figure 2.6). This trend toward gigantism in otherwise small species may also be reinforced by immigrant selection (selection for the larger, and consequently more vagile, phenotypes during active immigration), which should be most

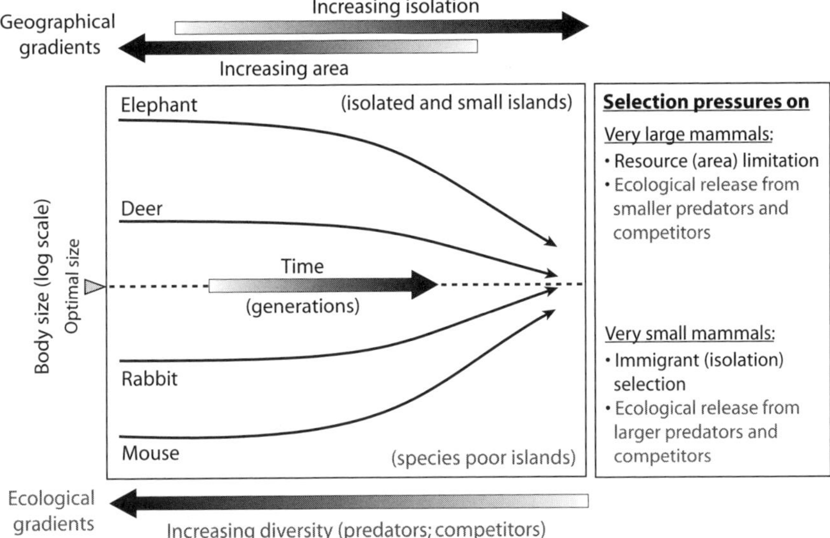

Figure 2.6. A general explanation for body size evolution in populations of insular animals. It is based on the assumptions that there is an optimal body size for a functional group of species (as defined by their *bauplän* and trophic strategy), that ecological interactions (in mainland and other species-rich communities) cause the optimal size for each species to be different from that of the group, and that selection pressures associated with geographic and ecological gradients (normal and gray italics font, respectively) vary in a predictable manner with body size of the species (right-hand column).

intense in the smaller species (see descriptions of "immigration selection," *sensu* Lomolino [1984, 1985, 1989]; and the equivalent phenomenon of selection for "thrifty genotypes" in Polynesians, *sensu* Bindon and Baker [1997]). Typically large species, on the other hand, are less challenged by the physiological demands of immigration, but more limited in their abilities to obtain adequate resources to maintain populations on all but the very large islands. In addition, large species are also influenced by ecological interactions in species-rich systems. Indeed, deer, hippos, elephants, and other large vertebrates may have originally evolved their massive size in response to intense ecological pressures of mainland communities (i.e., to outcompete smaller, more specialized competitors, and to escape predators by "outgrowing" them). Once these ecological pressures are removed, such as what occurs on species-poor islands, species of extreme size should tend to converge on the hypothesized optimum for that functionally defined group of species (shaded triangles in figures 2.5 and 2.6).

The conceptual model in figure 2.6 also provides novel sets of predictions regarding the island rule. First, it explicitly adds a temporal component to the island rule, by suggesting that the length of time a species is on an island will influence the degree of dwarfism or gigantism it has developed. To date, this has not been an important consideration in the study of the island rule because most focal species have presumably been present on islands long enough for their trajectory in body size to have either been completed or to be near completion. The situation has changed, however, because species under study now include those introduced onto oceanic islands during periods of historic colonizations by Europeans, imperiled species purposely translocated onto islands (e.g., the small off-shore islands of Australia and New Zealand), and species that persist within islandlike (i.e., heavily fragmented, smaller, and ecologically simplified) remnants of their native range. Indeed, recent evidence suggests that patterns consistent the island rule can manifest quickly (e.g., changes in body size of introduced mice on off-shore islets of New Zealand [King 2005] and by mammals and birds inhabiting heavily fragmented remnants of their native habitats in Denmark [Schmidt and Jensen 2003, 2005]). Thus, the temporal component of the island rule is likely to become a larger focus of research in the future (Lomolino et al. 2006). Second, the conceptual model (figure 2.6) predicts that the degree of change in body size attained by a focal organism (regardless of whether it is toward gigantism or dwarfism) is dependent on the geographic and ecological characteristics of the particular islands it inhabits (especially island area and isolation, and diversity of predators and competitors). Thus, much of the residual variation about the general trendline describing the island rule may be explicable once the characteristics of insular ecosystems are taken into account.

Third and most importantly, this conceptual model provides a general explanation for what seemed to be unrelated and sometimes contrary patterns (gigantism in some species, dwarfism in others), and across a broad range of functional groups and taxa (e.g., mammals, reptiles and invertebrates; terrestrial and aquatic species, both recent and extinct). We can, however, explain an even broader diversity of related patterns, including some apparent anomalies, if we overlay the causal models of body size evolution (figures 2.5 and 2.6) onto the geographic template of insular distributions (figure 2.2). As figure 2.7 reveals, once put in this context, the island rule emerges as not just one, but a set, of complementary patterns which vary depending on the species and the archipelagoes in question. Here, we generate insular distribution functions for three sets of species (small mammals, mesocarnivores, and large herbivorous mammals) by assuming a particular pattern of covariation

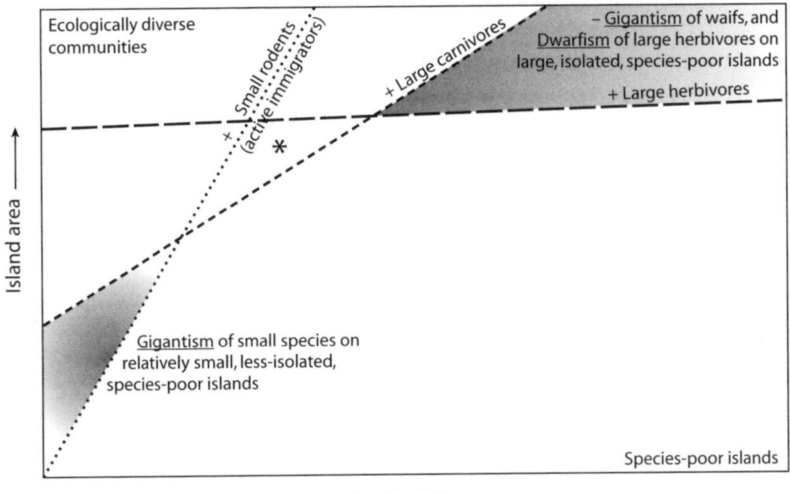

Figure 2.7. The geographic and ecological context of the island rule: body size evolution of insular animals placed within the context of an integrative theory of island biogeography. See figures 2.1, 2.2, 2.5, and 2.6; dashed lines=insular distribution functions of three sets of species—small rodents, meso- to large carnivores (e.g., large canids, felids, and ursids), and large herbivores (e.g., ungulates and proboscideans; species present ["+"] above these constraint lines). The region marked with the asterisk delineates islands that are likely to lack meso- to large carnivores because, although within their immigration capacities, these islands are probably too isolated or too small to support persistent populations of their prey (rodents and large mammals). Note: the effects of *in situ* speciation (which would be most important on the very large and isolated islands) are not included in this version of the model.

in fundamental capacities of the species (tenet 6). In this case, we are assuming that larger species will tend to have greater vagilities and greater resource requirements (translating into lower slopes but higher intercepts of their insular distribution functions), which is reasonable and well evidenced at least for actively immigrating mammals (and likely other vertebrates as well; see Lomolino 1989, 1999, 2000; see also Calder 1984, McNab 2002).

This geographically and ecologically more explicit, process- and species-based model (figure 2.7) explains why these microevolutionary trends are not universal, but should vary in a predictable manner among species (differing in their original body size and in their fundamental capacities) and among islands (varying in area, isolation, and other factors influencing accessibility, carrying capacity and diversity of competitors and predators). As figure 2.7 illustrates, species such as small mammals should exhibit gigantism only on islands that they can colonize (either as active immigrators or as waifs) and where larger competitors and predators are likely to be absent (i.e., on relatively small, near islands for the active immigrators, and on larger but relatively isolated islands for the waifs. On the other hand, typically large mammals (e.g., deer, hippos, and elephants) should exhibit dwarfism only on the very large and isolated islands, which lack the mesoherbivores and carnivores that likely contributed to selection for their large size on the mainland in the first place (see Palombo 2001, 2005, Palombo et al. 2005, Raia and Meiri 2006). Thus, isolated islands—the evolutionary arenas for both the titanic and the Lilliputian marvels—are often inhabited by a depauperate but predictable assemblage of species; frequently dominated by large rodents and relatively small deer, hippos, or elephants, but lacking carnivorous mammals.

The inferences from this model with respect to body size evolution of insular carnivores are especially interesting. Central to this explanation for the island rule (figures 2.5 and 2.6) is that insular populations of extreme size will undergo gigantism or dwarfism on ecologically simplified islands, converging on an intermediate and presumed optimal size. Given the requirements of being carnivorous, however, those mammals are less likely to be of extreme size and seldom should they be able to maintain their populations on ecologically depauperate islands (i.e., those that by definition lack persistent prey populations) for periods required for substantial evolutionary divergence in body size. Indeed, although predators may repeatedly colonize such islands, we expect that either their residence will be ephemeral (because their predation—unchecked in species-poor systems—often leads to predatory exclusion of their prey and, in turn, collapse of their own populations as well; see Lomolino 1984, Schoener et al. 2001, 2002) or their diets will shift toward prey more readily available in insular environments (e.g., sea birds, fish, shoreline invertebrates, and carcasses of marine mammals; see Goltsman et al., 2005, pp. 406, 412). Given this catch-22 of being an insular carnivore, it is surprising, at least in retrospect, that there actually is a signal consistent with the island rule for such species (figures 2.3a and 2.4; the inferred significance of statistical analyses of this pattern depends on which measure of body size is used [that of skulls or teeth], whether the data include carnivores of extreme size and populations inhabiting very large, mainlandlike islands [e.g., Borneo, Sumatra, Great Britain, and Java], and whether the results are evaluated under the constraints of a one-tailed or two-tailed test; see Lomolino's [2005, pp. 1684–85, figure 2] reanalysis of Meiri et al.'s [2004] data; see also Price and Phillimore 2007, Meiri 2007, Meiri et al. 2007). Meiri et al.'s (2008b) recent studies of body size of Borneo's

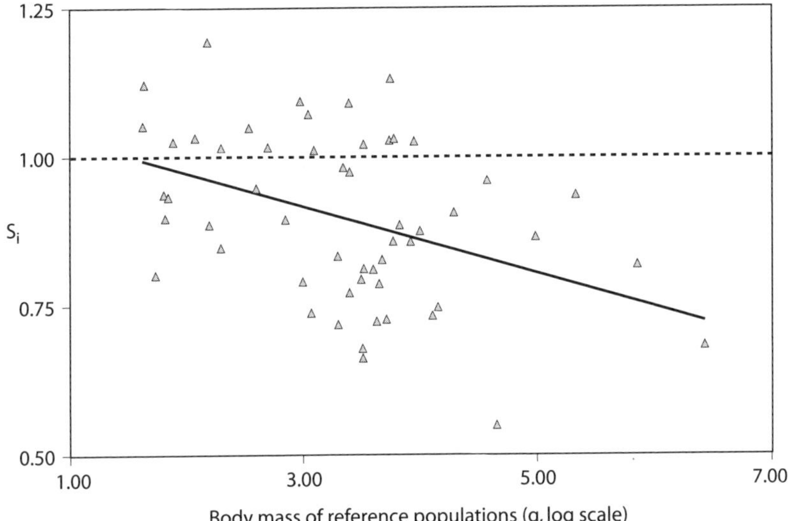

Figure 2.8. Body size trends of mammals from the island of Borneo. They exhibit a graded trend toward increased degree of dwarfism with increased ancestral body size, consistent with the island rule. S_i is body size of insular populations expressed as a proportion of that of their mainland relatives (expressed as mass equivalents by comparing cubed linear dimensions; comparing condylobasal length of skulls of insular forms to that of the largest skulls of that species in the region (data from Meiri et al. 2008b).

mammals are especially relevant to this hypothesis regarding the influence of large carnivores, resource requirements, and ecological release on body size evolution. They report a graded trend toward increased dwarfism in otherwise large (>100 g) Bornean mammals (figure 2.8), being consistent with the island rule and presumably a function of the absence of large predators (e.g., tigers [*Panthera tigris*], leopards [*P. pardus*], and saber-toothed cats [*Hemimachairodus zwierzyckii*]) on this island at least since the early Holocene.

Conclusion: The Way Forward

Just as immigration, evolution, and extinction produce reticulated histories of biotas (Brooks 2004, Lieberman 2004) that colonize new regions and diverge in isolation, only to suffer eventual range collapse and extinction or reinitiate the cycle by colonizing other regions (including those of their ancestors), the natural sciences develop in an analogous fashion. The reticulating phylogeny of island theory weaves a complex web of early discoveries and articulations of new theories, followed by expansions and contractions in their conceptual domains, replacements by competing theories, or synthesis and reintegration with innovations from other relevant disciplines. Thus, the developmental history of island biogeography, and in particular the equilibrium model, provides invaluable lessons; not just on how MacArthur and Wilson achieved their paradigmatic masterpiece, but on how today's biogeographers can once again transform the field. We are encouraged by the recent efforts of our colleagues, including the distinguished contributors to this volume, to provide such fundamental advances—not by derision of competing scientists and disproof of their ideas, but by genuine consilience and collaborative syntheses of complementary theories and insights to achieve a more comprehensive understanding of the ecological and evolutionary development of isolated biotas.

Acknowledgments

We thank Jonathan Losos and Robert Ricklefs for their invitation to participate in the symposium and contribute to this book, and we thoroughly enjoyed the opportunity to interact with other participants and colleagues in attendance. Jonathan Losos, Robert Ricklefs, Robert J. Whittaker, Michael Willig, and an anonymous reviewer provided numerous helpful comments, and Maria Rita Palombo provided data on body size of Mediterranean mammals during the Pleistocene.

Literature Cited

Agassiz, L. 1840. *Etudes sur les Glaciers/Studies of the Glaciers*. Translated into English and edited by A. V. Carozzi, *Studies on Glaciers, Preceded by the Discourse of Neuchâtel*. New York: Hafner, 1967.

Allen, A. P., J. H. Brown, and J. F. Gillooly. 2002. Global biodiversity, biochemical kinetics, and the energetic-equivalence rule. *Science* 297:1545–48.

Apanius, V., N. Yorinks, E. Bermingham, and R. E. Ricklefs. 2000. Island and taxon effects in the prevalence of blood parasites and activity of the immune system in Lesser Antillean birds. *Ecology* 81:1959–69.

Berry, R. J. 1984. *Evolution in the Galapagos Islands*. New York: Academic Press.

Brooks, D. R. 2004. Reticulations in historical biogeography: The triumph of time over space in evolution. In *Frontiers of Biogeography*, ed. M. V. Lomolino and L. R. Heaney, 123–44. Sunderland, MA: Sinauer Associates.

Brown, J. H. 1971. Mammals on mountaintops: Nonequilibrium insular biogeography. *American Naturalist* 105:467–78.

———. 1978. The theory of insular biogeography and the distribution of boreal birds and mammals. *Great Basin Naturalist Memoirs* 2:209–27.

———. 1981. Two decades of homage to Santa Rosalia: toward a general theory of diversity. *American Zoologist* 21:877–88.

———. 1995. *Macroecology*. Chicago: University of Chicago Press.

———. 2004. Concluding remarks. In *Frontiers of Biogeography*, ed. M. V. Lomolino and L. R. Heaney, 361–68. Sunderland, MA: Sinauer Associates.

———. 1995. *Macroecology*. Chicago: University of Chicago Press.

Brown, J. H., and A. Kodric-Brown. 1977. Turnover rates in insular biogeography: Effect of immigration on extinction. *Ecology* 58:445–49.

Brown, J. H., and M. V. Lomolino. 1989. On the nature of scientific revolutions: Independent discovery of the equilibrium theory of island biogeography. *Ecology* 70:1954–57.

———. 2000. Concluding remarks: Historical perspective and the future of island biogeography theory. *Global Ecology and Biogeography* 9:87–92.

Brown, J. H., J. F. Gillooly, A. P. Allen, M. Van Savage, and G. West. 2004b. Toward a metabolic theory of ecology. *Ecology* 85:1771–89.

Brown, P., T. Sutikna, M. Morwood, R. P. Soejono, Jatmiko, E. W. Saptomo, and R. A. Due. 2004a. A new small-bodied hominin from the late Pleistocene of Flores, Indonesia. *Nature* 431:1055–61.

Brown, W. L., and E. O. Wilson. 1956. Character displacement. *Systematic Zoology* 5:49–64.

Buffon, G. L. L., Comte de. 1761. *Histoire Naturelle, Generale et Particuliere*, vol. 9. Imprimerie Royale, Paris.

Calder, W. A., III. 1984. *Size, Function and Life History*. Cambridge, MA: Harvard University Press.

Candolle, A. P. de. 1820. *Essai Elementaire de Geographie Botanique*. De l'imprimerie de F. G. Levrault.

Cardillo, M. 1999. Latitude and rates of diversification in birds and butterflies. *Proceedings of the Royal Society of London, Series B* 266:1221–25.

Carlquist, S. 1974. *Island Biology*. New York: Columbia University Press.

Currie, D. J. 1991. Energy and large scale patterns of animal and plant species richness. *American Naturalist* 137:27–49.

Dammerman, K. W. 1922. The fauna of Krakatau, Verlaten Island and Sebesy. *Treubia* 3:61–112.

———. 1948. The fauna of Krakatau, 1883–1933. *Koninklïjke Nederlandsche Akademie Wetenschappen Verhandelingen* 44:1–594.

Darlington, P. J., Jr. 1938. The origin of the fauna of the Greater Antilles, with discussion of dispersal of animals over water and through the air. *Quarterly Review of Biology* 13:274–300.

———. 1943. Caribidae of mountains and islands: Data on the evolution of isolated faunas and on atrophy of wings. *Ecological Monographs* 13:37–61.

———. 1957. *Zoogeography: The Geographical Distribution of Animals*. New York: John Wiley & Sons.

Darwin, C. 1859. *On the Origin of Species by Means of Natural Selection or the Preservation of Favored Races in the Struggle for Life*. London: John Murray.

Darwin, C. 1860. *Journal of Researches into the Natural History and Geology of the Countries Visited During the Voyage of H.M.S. Beagle Round the World.* London: John Murray.

Diamond, J. M. 1975. Assembly of species communities. In *Ecology and Evolution of Communities*, ed. M. L. Cody and J. M. Diamond, 342–444. Cambridge, MA: Harvard University Press, Belknap Press.

———. 1977. Colonization cycles in man and beast. *World Archaeology* 8:249–61.

Docters van Leeuwen, W. M. 1936. Krakatau, 1883–1933. *Annales du Jardin Botanique de Buitenzorg* 46–47:1–506.

Ebach, M. C., and R. S. Tangney. 2007. *Biogeography in a Changing World.* New York: CRC Press.

Elton, C. 1927. *Animal Ecology.* New York: Macmillan.

Elton, C. S. 1958. *The Ecology of Invasions by Animals and Plants.* London: Methuen & Co.

Emerson, B. C., and N. Kolm. 2005. Species diversity can drive speciation. *Nature* 434:1015–17.

Erwin, T. C. 1981. Taxon pulses, vicariance, and dispersal: An evolutionary synthesis illustrated by carabid beetles. In *Vicariance Biogeography: A Critique,* ed. G. Nelson and D. E. Rosen, 159–96. New York: Columbia University Press.

Fallon, S. M., E. Bermingham, and R. E. Ricklefs. 2003. Island and taxon effects in parasitism revisited: Avian malaria in the Lesser Antilles. *Evolution* 57:606–15.

Fisher R. 1930. *The Genetical Theory of Natural Selection.* Oxford: Clarendon Press.

Forster, J. R. 1778. *Observations Made during a Voyage Round the World, on Physical Geography, Natural History and Ethic Philosophy.* London: G. Robinson.

Foster, J. B. 1964. Evolution of mammals on islands. *Nature* 202:234–35.

Francisco-Ortega1, J., I. Ventosa, R. Oviedo, F. Jiménez, P. Herrera, M. Maunder, and J. L. Panero. 2008. Caribbean Island Asteraceae: Systematics, molecules, and conservation on a biodiversity hotspot. *Botanical Review* 74:112–31.

Gaston, K. J., and T. M. Blackburn. 2000. *Pattern and Process in Macroecology.* Oxford: Blackwell Scientific Publications.

Gause, G. F. 1934. *The Struggle for Existence.* Baltimore: Williams and Wilkins.

Givnish, T. J. 1998. Adaptative plant evolution on islands. In *Evolution on Islands,* ed. P. R. Grant, 281–304. New York: Oxford University Press.

Goltsman, M., E. P. Kruchenkova, S. Sergeev, I. Volodin, and D. W. MacDonald. 2005. "Island syndrome" in a population of Arctic foxes (*Alopex lagopus*) from Mednyi Island. *Journal of Zoology* 267:405–13.

Grant, P. R. 1998. *Evolution on Islands.* New York: Oxford University Press.

Haila, Y. 1986. On the semiotic dimension of ecological theory: The case of island biogeography. *Biology and Philosophy* 1:377–87.

Hanski, I. 1986. Population dynamics of shrews on small islands accord with the equilibrium model. *Biological Journal of the Linnean Society* 28:23–36.

Hardy, G. H. 1908. Mendelian proportions in a mixed population. *Science* 28:49–50.

Hawkins, B. A., R. Field, H. V. Cornell, D. J. Currie, J.-F. Guegan, D. M. Kaufman, J. T. Kerr, G. G. Mittelbach, T. Oberdorff, E. M. O'Brien, E. E. Porter, and J.R.G. Turner. 2003. Energy, water, and broad-scale geographic patterns of species richness. *Ecology* 84:3105–17.

Heaney, L. R. 1978. Island area and body size of insular mammals: Evidence from the tri-colored squirrel (*Callosciurus prevosti*) of Southwest Africa. *Evolution* 32:29–44.

———. 2000. Dynamic disequilibrium: A long-term, large-scale perspective on the equilibrium model of island biogeography. *Global Ecology and Biogeography* 9:59–74.

Hooker, J. D. 1853. *The Botany of the Antarctic Voyage of H.M.S. Discovery Ships "Erebus" and "Terror" in the Years 1839–1843*. London: Lovell Reeve.

———. 1867. Lecture on Insular Floras. London: Delivered before the British Association for the Advancement of Science at Nottingham, August 27, 1866.

Humboldt, A. von. 1805. *Essai sur la Geographie des Plantes Accompagne d'un Tableau Physique des Regions Equinoxiales, Fonde sur des Mesures Executees, depuis le Dixieme Degre de Latitude Boreale jusqu'au Dixieme Degre de Latitude Australe, pendant les Annees 1799, 1800, 1801, 1802 et 1803*. Paris: Levrault Schoell.

Hutchinson, G. E. 1959. Hommage to Santa Rosalia, or why are there so many kinds of animals? *American Naturalist* 93:145–59.

Hutchinson, G. E., and R. H. MacArthur. 1959. A theoretical ecological model of size distributions among species of animals. *American Naturalist* 93:117–25.

Jianu, C. M., and Weishampel, D. B. 1999. The smallest of the largest: A new look at possible dwarfing in sauropod dinosaurs. *Geologie en Mijnbouw* 78: 335–43.

Kalmar, A., and D. J. Currie. 2006. A global model of island biogeography. *Global Ecology and Biogeography* 15:72–81.

King, C. M. 2005. *Handbook of the Mammals of New Zealand*, 2nd ed. Oxford: Oxford University Press.

Kuhn, T. S. 1996. *The Structure of Scientific Revolutions*, 3rd ed. Chicago: University of Chicago Press.

Lack, D. 1947. *Darwin's Finches*. Cambridge: Cambridge University Press.

———. 1976. *Island Biology Illustrated by the Land Birds of Jamaica*. Studies in Ecology vol. 3. Berkeley: University of California Press.

Levins, R., and R. C. Lewontin. 1980. Dialectics and reductionism in ecology. *Synthese* 43:47–78.

Lieberman, B. S. 2004. Range expansion, extinction and biogeographic congruence: A deep time perspective. In *Frontiers of Biogeography*, ed. M. V. Lomolino and L. R. Heaney, 111–24. Sunderland, MA: Sinauer Associates.

Linnaeus, C. 1781. On the increase of the habitable earth. *Amonitates Academicae* 2:17–27.

Lomolino, M. V. 1984. Immigrant selection, predatory exclusion and the distributions of *Microtus pennsylvanicus* and *Blarina brevicauda on* islands. *American Naturalist* 123:468–83.

———. 1985. Body size of mammals on islands: The island rule re-examined. *American Naturalist* 125:310–16.

Lomolino, M. V. 1986. Mammalian community structure on islands: immigration, extinction and interactive effects. *Biological Journal of the Linnaean Society* 28:1–21.

———. 1988. Winter immigration abilities and insular community structure of mammals in temperate archipelagoes. In *Biogeography of the Island Region of Western Lake Erie*, ed. J. F. Downhower, 185–96. Columbus: Ohio State University Press.

———. 1989. Bioenergetics of cross-ice movements of *Microtus pennsylvanicus, Peromyscus leucopus* and *Blarina brevicauda*. *Holarctic Ecology* 12:213–18.

———. 1990. The target area hypothesis: The influence of island area on immigration rates of non-volant mammals. *Oikos* 57:297–300.

———. 1993. Winter filtering, immigrant selection and species composition of insular mammals of Lake Huron. *Ecography* 16:24–30.

———. 1994. Species richness patterns of mammals inhabiting nearshore archipelagoes: Area, isolation and immigration filters. *Journal of Mammalogy* 75:39–49.

———. 1996. Investigating causality of nestedness of insular communities: Selective immigrations or extinctions? *Journal of Biogeography* 23:699–703.

———. 1999. A species-based, hierarchical model of island biogeography. In *Ecological Assembly Rules: Perspectives, Advances, Retreats*, ed. Evan A. Weiher and Paul A. Keddy, 272–310. New York: Cambridge University Press.

———. 2000. A species-based theory of insular zoogeography. *Global Ecology and Biogeography* 9:39–58.

———. 2005. Body size evolution in insular vertebrates: Generality of the island rule. *Journal of Biogeography* 32:1683–99.

Lomolino, M. V., and L. R. Heaney, eds. 2004. *Frontiers of Biogeography*. Sunderland, MA: Sinauer Associates.

Lomolino, M. V., B. R. Riddle, and James H. Brown. 2006a. *Biogeography*, 3rd ed. Sunderland, MA: Sinauer Associates.

Lomolino, M. V., D. F. Sax, and J. H. Brown, eds. 2004. *Foundations of Biogeography*. Chicago: University of Chicago Press.

Lomolino, M. V., D. F. Sax, B. R. Riddle, and J. H. Brown. 2006b. The island rule and a research agenda for studying ecogeographic patterns. *Journal of Biogeography* 33:1503–10.

Lomolino, M. V., and M. D. Weiser. 2001. Toward a more general species-area relationship: Diversity on all islands great and small. *Journal of Biogeography* 28:431–45.

Losos, J. B., and K. de Queiroz. 1997. Evolutionary consequences of ecological release in Caribbean *Anolis* lizards. *Biological Journal of the Linnaean Society* 61:459–83.

Losos, J. B., and D. Schluter. 2000. Analysis of an evolutionary species-area relationship. *Nature* 408:847–50.

Losos, J. B., and R. S. Thorpe. 2004. Evolutionary diversification of Caribbean *Anolis* lizards. In *Adaptative Speciation*, ed. U. Dieckmann, M. Doebeli, J.A.J. Metz, and D. Tautz, 322–44. Cambridge: Cambridge University Press.

Lotka, A. J. 1925. *Elements of Physical Biology*. Baltimore: Williams and Wilkins.

MacArthur, R. H., and T. H. Connell. 1966. *The Biology of Populations*. New York: John Wiley & Sons.

MacArthur, R. H., and R. Levins. 1964. Competition, habitat selection, and. character displacement in a patchy environment. *Proceedings of the National Academy of Sciences U.S.A.* 51:1207–10.

———. 1967. The limiting similarity, convergence and divergence of coexisting species. *American Naturalist* 101:377–85.

MacArthur, R. H., and E. O. Wilson. 1963. An equilibrium theory of insular zoogeography. *Evolution* 17:373–87.

———. 1967. *The Theory of Island Biogeography*. Monographs in Population Biology vol. 1. Princeton, NJ: Princeton University Press.

Matthew, W. D. 1915. Climate and evolution. *Annals of the New York Academy of Sciences* 24:171–318.

Mayr, E. 1942. *Systematics and the Origin of Species*. New York: Columbia University Press.

Mayr, E., and J. M. Diamond. 2001. *The Birds of Northern Melanesia: Species, Ecology and Biogeography*. New York: Oxford University Press.

McNab, B. K. 1994a. Energy conservation and the evolution of flightlessness in birds. *American Naturalist* 144:628–42.

———. 1994b. Resource use and the survival of land and freshwater vertebrates on oceanic islands. *American Naturalist* 144:643–60.

———. 2002. Minimizing energy expenditure facilitates vertebrate persistence on oceanic islands. *Ecology Letters* 5:693–704.

———. 2002. *The Physiological Ecology of Vertebrates: A View from Energetics*. Ithaca, NY: Cornell University Press.

Meiri, S. 2007. Size evolution in island lizards. *Global Ecology and Biogeography* 16:689–93.

Meiri, S., T. Dayan, and D. Simberloff. 2004. Body size of insular carnivores: little support for the island rule. *American Naturalist* 163:469–79.

———. 2007. Guild composition and mustelid morphology—character displacement but not character release. *Journal of Biogeography* 34:2148–58.

Meiri, S., N. Cooper, and A. Purvis. 2008a. The island rule: Made to be broken? *Proceedings of the Royal Society of London, Series B* 275:141–48.

Meiri, S., E. Meijaard, S. A. Wich, C. P. Groves, and K. M. Helgen. 2008b. Mammals of Borneo—small size on a large island. *Journal of Biogeography* 35:1087–94.

Merton, R. K. 1961. Singletons and multiples in scientific discovery: A chapter in the sociology of science. *Proceedings of the American Philosophical Society* 105:470–86.

Meyer, A. 1993. Phylogenetic relationships and evolutionary processes in East African cichlid fishes. *Trends in Ecology and Evolution* 8:279–84.

Millien, V. 2006. Morphological evolution is accelerated among island mammals. *PLoS Biology* 4(10):e321.

Millien, V., and J. Damuth. 2004. Climate change and size evolution in an island rodent species: New perspectives on the island rule. *Evolution* 58:1353–60.

Millien, V., S. K. Lyons, L. Olson, F. A. Smith, A. B. Wilson, and Y. Yom-Tov. 2006. Ecotypic variation in the context of global climate change: Revisiting the rules. *Ecological Letters* 9:853–69.

Morwood, M., R. P. Soejono, R. G. Roberts, T. Sutikna, C.S.M. Turney, K. E. Westaway, Rink, J.-x. Zhao, G. D. van den Bergh, Rokus Awe Due, D. R. Hobbs, M. W. Moore, M. I. Bird, and L. K. Fifield. 2004. Archaeology and age of a new hominin from Flores in eastern Indonesia. *Nature* 431:1087–91.

Morwood, M. J.; P. Brown, Jatmiko, T. Sutikna, E. Wahyu Saptomo, K. E. Westaway, Rokus Awe Due, R. G. Roberts, T. Maeda, S. Wasisto, and T. Djubiantono. 2005. Further evidence for small-bodied hominins from the Late Pleistocene of Flores, Indonesia. *Nature* 437:1012–17.

Munroe, E. G. 1948. The geographical distribution of butterflies in the West Indies. Ph.D. dissertation, Cornell University, Ithaca, NY.

———. 1953. The size of island faunas. In *Proceedings of the Seventh Pacific Science Congress of the Pacific Science Association (1949, New Zealand)*, vol. IV, *Zoology*, 52–53. Auckland: Whitcombe and Tombs.

———. 1963. Perspectives in biogeography. *The Canadian Entomologist* 95: 299–308.

Palombo, M. R. 2001. Dwarfing in insular mammals: The case of endemic elephants of Mediterranean islands. Sixth European Workshop on Vertebrate Palaeontology, Italy.

———. 2005. How could endemic proboscideans help us in understanding the "island rule"? In *Proceedings of the Second International Congress "The World of Elephants, The Mammoth Site of Hot Springs, South Dakota,"* ed. L. Agenbroada, G. Haynesb, E. Johnson, and M. Rita Palombo, 132–35.

Palombo, M. R., M. Mussi, P. Gioia, and G. Cavarretta. 2005. Studying Proboscideans: Knowledge, problems, and perspectives. *Quaternary International* 126–128:1–3.

Patterson, B. D., and W. Atmar. 1986. Nested subsets and the structure of insular mammalian faunas and archipelagoes. *Biological Journal of the Linnaean Society* 28:65–82.

Pearl, R. 1925. *The Biology of Population Growth*. New York: A. A. Knopf.

Pickett, T. A., J. Kolasa, and C. G. Jones. 2007. *Ecological Understanding: The Nature of Theory and the Theory of Nature,* 2nd ed. New York: Academic Press.

Preston, F. W. 1960. Time and space and the variation of species. *Ecology* 41:611–27.

Price, T. D., and A. B. Phillimore. 2007. Reduced major axis regression and the island rule. *Journal of Biogeography* 34:1998–99.

Raia, P., and S. Meiri. 2006. The island rule in large mammals: Paleontology meets ecology. *Evolution* 60:1731–42.

Raunkiaer, C. 1904. Om biologiske Typer, med Hensyn til Planternes Tilpasning til at overleve ugunstige Aarstider. *Botanisk Tidsskrift* 16:14.

———. 1934. *The Life Forms of Plants and Statistical Plant Geography*. Oxford: Clarendon Press.

Ricklefs, R. E., and G. W. Cox. 1972. Taxon cycles of the West Indian avifauna. *American Naturalist* 106:195–219.

———. 1978. State of taxon cycle, habitat distribution and population density in the avifauna of the West Indies. *American Naturalist* 112:875–95.

Ricklefs, R.E., and I. J. Lovette. 1999. The roles of island area *per se* and habitat diversity in the species-area relationships of four Lesser Antillean faunal groups. *Journal of Animal Ecology* 68:1142–60.

Riddle, B. R., and D. J. Hafner. 2004. The past and future roles of phylogeography in historical biogeography. In *Frontiers of Biogeography*, ed. M. V. Lomolino and L. R. Heaney, 93–110. Sunderland, MA: Sinauer Associates.

Rohde, K. 1992. Latitudinal gradients in species diversity: The search for the primary cause. *Oikos* 65:514–27.

Roughgarden, J., S. D. Gaines, and S. W. Pacala. 1987. Supply side ecology: The role of physical transport processes. In *Organization of Communities: Past and Present*, ed. J. H. R. Gee and P. S. Giller, 491–518. Oxford: Blackwell Science Publications.

Sander, M. P., O. Mateus, T. Laven, and N. Knötschke. 2006. Bone histology indicates insular dwarfism in a new Late Jurassic sauropod dinosaur. *Nature* 441:739–41.

Sax, D. F., S. D. Gaines, and J. H. Brown. 2002. Species invasions exceed extinctions on islands worldwide: A comparative study of plants and birds. *American Naturalist* 160:766–83.

Schmidt, N. M., and P. M. Jensen. 2003. Changes in mammalian body length over 175 years—adaptations to a fragmented landscape? *Conservation Ecology* 7:6. http://www.consecol.org/vol7/iss2/art6/.

———. 2005. Concomitant patterns in avian and mammalian body length changes in Denmark. *Ecology and Society* 10:5. http://www.ecologyandsociety.org/vol10/iss2/art5/.

Schoener, T. W., D. A. Spiller, and J. B. Losos. 2001. Predators increase the risk of catastrophic extinction of prey populations. *Nature* 412:183–86.

———. 2002. Predation on a common *Anolis* lizard: Can the food-web effects of a devastating predator be reversed? *Ecological Monographs* 72:383–407.

Sclater, P. L. 1858. On the general geographical distribution of the members of the class Aves. *Journal of the Linnean Society, Zoology* 2:130–45.

———. 1897. On the distribution of marine mammals. *Proceedings of the Zoological Society of London* 41:347–59.

Simpson, G. G. 1940. Mammals and land bridges. *Journal of the Washington Academy of Science* 30:137–63.

———. 1943. Mammals and the nature of continents. *American Journal of Science* 241:1–31.

———. 1944. *Tempo and Mode in Evolution*. New York: Columbia University Press.

———. 1956. Zoogeography of West Indian land mammals. *American Museum Novitates* no. 1759.

———. 1980. *Splendid Isolation: The Curious History of Mammals in South America*. New Haven, CT: Yale University Press.

Sismondo, S. 2000. Island biogeography and the multiple domains of models. *Biology and Philosophy* 15:239–58.

Steadman, D. W. 2006. *Extinction and Biogeography of Tropical Pacific Birds.* Chicago: University of Chicago Press.

Stuessy, T. F. 2007. Evolution of specific and genetic diversity during ontogeny of island floras: The importance of understanding process for interpreting island biogeographic patterns. In *Biogeography in a Changing World*, ed. M. C. Ebach and R. S. Tangney, 117–34. New York: CRC Press.

Thornton, I. 1996. *Krakatau: The Destruction and Reassembly of an Island Ecosystem.* Cambridge, MA: Harvard University Press.

Thornton, I.W.B. 1992. K. W. Dammerman: Forerunner of island theory? *Global Ecology and Biogeography Letters* 2:145–48.

Van Valen, L. 1973. A new evolutionary law. *Evolutionary Theory* 1:1–33.

Volterra, V. 1926. Variazioni flultuazioni del numero d'individui in specie convirenti. *Memorie Accademia dei Lincei* 2:31–113.

———. 1931. *Lecons sur la Theorie Mathematique de la Lutte pour la Vie.* Paris: Gauthier-Villars.

Wagner, W. L., and V. A. Funk. 1995. *Hawaiian Biogeography: Evolution on a Hot Spot Archipelago.* Washington, DC: Smithsonian Institution Press.

Wallace, A. R. 1857. On the natural history of the Aru Islands. *Annals and Magazine of Natural History*, Supplement to Volume 20, December.

———. 1869. *The Malay Archipelago: The Land of the Orangutan and the Bird of Paradise.* New York: Harper.

———. 1876. *The Geographical Distribution of Animals,* 2 vols. London: Macmillan.

Watson, J. D. 1968. *The Double Helix; A Personal Account of the Discovery of the Structure of DNA.* New York: Atheneum.

Wegener, A. 1912a. Die Entstehung der Kontinente. *Petermanns Geogr. Mitt.* 58:185–95, 253–56, 305–8.

———. 1912b. Die Entstehung der Kontinente. *Geologische Rundschau* 3:276–92.

———. 1915. *Die Entstehung der Kontinente und Ozeane.* Braunschweig: Vieweg. (Other editions 1920, 1922, 1924, 1929, 1936.)

Weinberg, W. 1908. Über den Nachweis der Vererbung beim Menschen. *Jahreshefte des Vereins für Vaterländische Naturkunde in Württemberg* 64: 368–82.

Whittaker, R. J. 1998. *Island Biogeography: Ecology, Evolution and Conservation.* New York: Oxford University Press.

Whittaker, R. J., M. B. Bush, and K. Richards. 1989. Plant recolonization and vegetation succession on the Krakatau Islands, Indonesia. *Ecological Monographs* 59:59–123.

Whittaker, R. J., and J. M. Fernandez-Palacios. 2007. *Island Biogeography: Ecology, Evolution and Conservation,* 2nd ed. New York: Oxford University Press.

Whittaker, R. J., K. A. Triantis, and R. J. Ladle. 2008. A general dynamic theory of oceanic island biogeography. *Journal of Biogeography* 35:977–94.

Willdenow, K. L. 1792 (translated into English 1805). *Grundriss de Kräuterkunde zu Vorlesungen (Principles of Botany).* Berlin: Haude und Spener.

Williams, C. B. 1964. *Patterns in the Balance of Nature.* London: Academic Press.

Willis, J. C. 1915. The endemic flora of Ceylon, with reference to geographical distribution and evolution in general. *Philosophical Transactions of the Royal Society of London, Series B* 206:307–42.

———. 1922. *Age and Area: A Study in Geographical Distribution and Origin of Species*. Cambridge: Cambridge University Press.

Wilson, E. O. 1959. Adaptive shift and dispersal in a tropical ant fauna. *Evolution* 13:122–44.

———. 1961. The nature of the taxon cycle in the Melanesian ant fauna. *American Naturalist* 95:169–93.

———. 1994. *Naturalist*. New York: Island Press

Wright, D. H. 1983. Species-energy theory: An extinction of species-area theory. *Oikos* 41:496–506.

Wright, S., J. Keeling, and L. Gillman. 2006. The road from Santa Rosalia: A faster tempo of evolution in tropical climates. *Proceedings of the National Academy of Sciences U.S.A.* 103:7718–22.

The MacArthur-Wilson Equilibrium Model

A CHRONICLE OF WHAT IT SAID AND HOW IT WAS TESTED

Thomas W. Schoener

THE DOMAIN OF THIS CHAPTER is the development and testing of the MacArthur-Wilson Species Equilibrium Model. Naturally, most testing (as well as theoretical extension) followed rather closely the initial presentation (MacArthur and Wilson 1963, 1967) of this exciting, innovative conceptualization. My objective in this chapter is to focus mainly on this earlier research. As I discuss at the end of this chapter, papers citing the MacArthur-Wilson book have become very numerous in recent years. For this reason, an exhaustive review of current work is beyond the scope of my chapter. Rather, I focus on how the main aspects of the model, as presented by MacArthur and Wilson, have been evaluated in what I consider to be the most notable papers, many of which come from the older literature. As certain other chapters in this volume attest, the MacArthur-Wilson Species Equilibrium Model continues to inspire new research ideas, some far removed from the original kernels planted in the 1960s; because of my historical emphasis, I leave it to these other chapters to chart such future directions.

Basic Features of the MacArthur-Wilson Species Equilibrium Model

The MacArthur-Wilson Species Equilibrium Model was first presented as a graph of gross extinction and immigration rates against the number of species present on an island (MacArthur and Wilson 1963, 1967). In its most general form it makes two assumptions (figure 3.1):

1. The rate of immigration of new species (those not yet on the island) decreases monotonically with increasing number of species already present. It reaches zero when all species in the source area (there are P of them) are on the island.

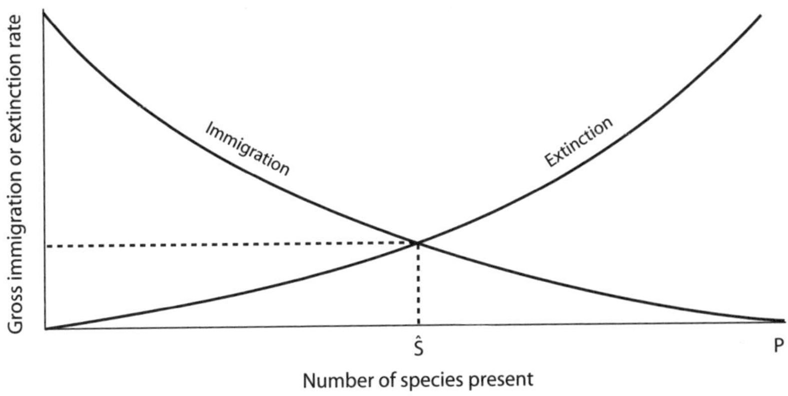

Figure 3.1. The graphical version of the MacArthur-Wilson Species Equilibrium Model. The model is for a particular island. $\hat{S}$ is the number of species at equilibrium (when gross immigration equals gross extinction), and *P* is the number of species in the source pool. Rate curves are monotonic but nonlinear. The intercept of the dashed line on the ordinate is the turnover rate at equilibrium.

2. The rate of extinction of species increases monotonically as the number of species increases (the more species there are, the more to go extinct).

These two assumptions imply that an equilibrium between immigration and extinction will eventually occur, at which time the immigration and extinction rates will have the same value, called the turnover rate at equilibrium.

Both of these model results, equilibrium and turnover, were predictions bold for their time, and as such drew substantial controversy. An equilibrium in numbers of species runs counter to a previous view that far islands would have fewer species than near islands because of lower dispersal rates, but that, given enough time, they would approach the number of species on near islands (both ultimately limited by the number of "available" species in source areas and perhaps by opportunities for *in situ* speciation). Species turnover was even more controversial: many lists and manuals giving the species of some taxon found on a particular island had been and were continuing to be published; how could the species on islands be dynamic, such that the very identities of catalogued species change from one survey to the next? The degree to which equilibrium and turnover in fact have been found by investigators will concern us shortly, but first I note a few niceties for the MacArthur-Wilson Species Equilibrium Model.

The graphical model was first presented with nonlinear species immigration and extinction curves. MacArthur and Wilson(1967) argued that the immigration curve should be concave, declining more rapidly at first because the better dispersers would be the first to arrive, leaving poorer and poorer dispersers as the only species not on the island and thereby reducing the absolute rate of decline of the species immigration curve. Their major argument for the concavity of the species extinction curves, as elaborated a bit later by Wilson (1969) in a "Brookhaven Symposia in Biology" volume, was completely different: the more species, the greater the likelihood of deleterious, i.e., extinction-producing, species interactions (as a first approximation this extinction rate would be proportional to the square of the number of species on the island). A fair bit later, Gilpin and Armstrong (1981) presented their species-by-species theory, showing that the same argument MacArthur and Wilson used to justify the concavity of immigration curves applied to the concavity of extinction curves—if all species possible (P of them) are present and one loses species, one will lose the most extinction-prone species first. Put another way, Gilpin and Armstrong showed that it is simply the variation in the individual-species extinction and immigration probabilities (rates) that can give concavity.

Despite the greater realism of the nonlinear model, a linear version, first presented in detail in Wilson's (1969) Brookhaven paper, gives us a feel for some of the important properties of this more limited version of an equilibrium model. He wrote the linear model as

$$dS/dt = \text{gross immigration} - \text{gross extinction, or}$$

$$dS/dt = \lambda_A(P - S) - \mu_A S \Rightarrow \hat{S} = P\left(\frac{\lambda_A}{\mu_A + \lambda_A}\right), \tag{3.1}$$

where S is the number of species on the island at time t, λ_A is the per species immigration rate, μ_A is the per species extinction rate, P is the number of species in the source pool, and $\hat{S}$ is the number of species at equilibrium. Figure 3.2 graphs this model.

The differential equation (3.1) can be solved for the colonization curve, or the curve relating number of species on the island to time since the colonization process began; it is a convex exponential (figure 3.3), i.e.,

$$S(t) = \hat{S}(1 - e^{-(\lambda_A + \mu_A)t}). \tag{3.2}$$

The convex form of the colonization curve is also a prediction that can be tested (the model above would not lead uniquely to this form, however). Inspection of equation (3.2) allows two additional insights. First, the rate of approach to equilibrium varies positively with both the immigration *and* extinction parameters (even though extinction diminishes

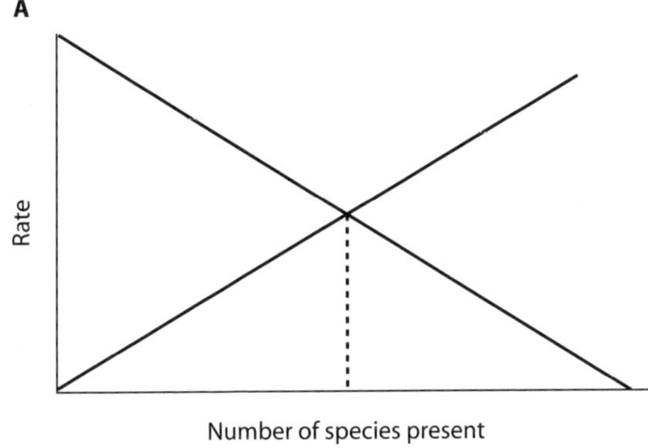

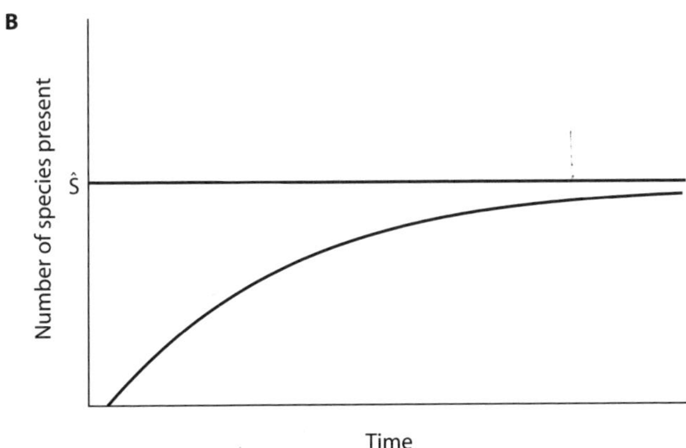

Figure 3.2. A. The linear version of the MacArthur-Wilson Equilibrium Model. Conventions as in figure 3.1. B. The colonization curve (species on an island versus time since beginning of the immigration/extinction process).

the number of species). Second, equilibrium is approached at a decreasing rate (the slope of the colonization curve diminishes with time). This implies that islands not at equilibrium yet not too far from equilibrium are going to be strongly influenced by the same factors (the immigration and extinction rates, and whatever affects those quantities) as are islands effectively at equilibrium.

Evidence for the Species Equilibrium

I now discuss the degree to which empirical tests supported the idea that islands are in a state of species equilibrium (turnover is considered in the

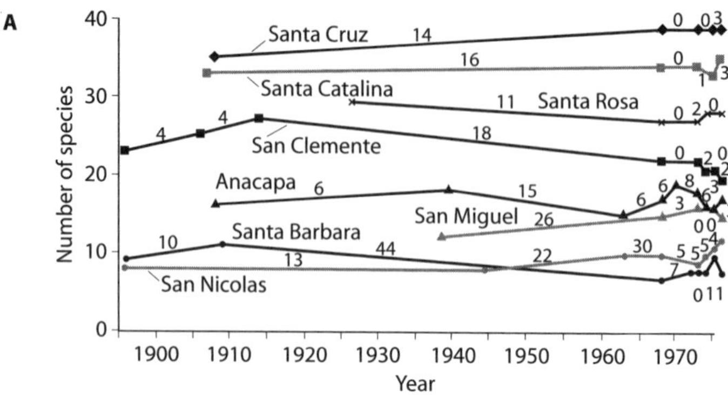
A
Number of species
Year
Santa Cruz
Santa Catalina
Santa Rosa
San Clemente
Anacapa
San Miguel
Santa Barbara
San Nicolas

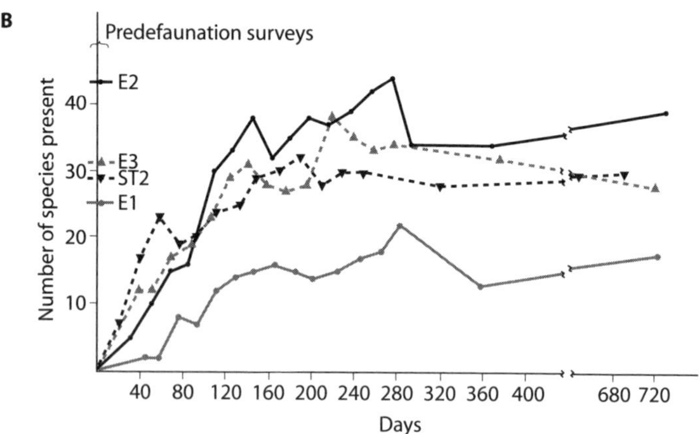
B
Predefaunation surveys
E2
E3
ST2
E1
Number of species present
Days

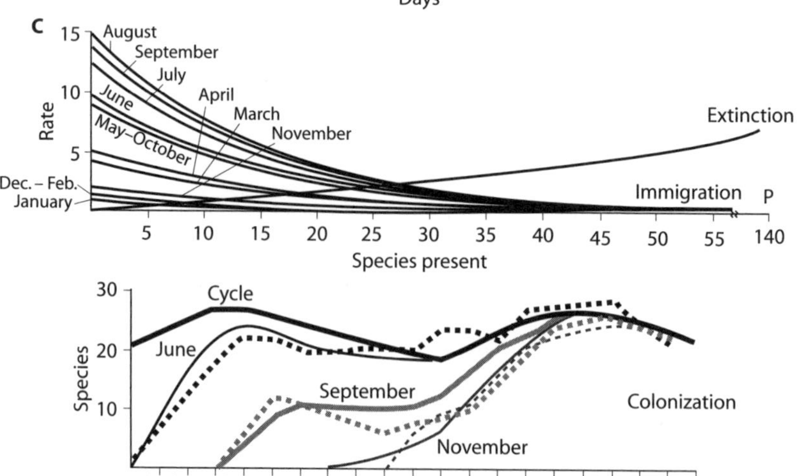
C
Rate
August
September
July
June
April
March
November
May–October
Dec. – Feb.
January
Extinction
Immigration
P
Species present
Species
Cycle
June
September
November
Colonization
Time (months)

next section). Although the discussion is grouped by system, it is arranged roughly in order of decreasing correspondence to the equilibrium prediction.

1. *Arthropods of red mangrove islands.* Shortly after publication of MacArthur and Wilson's book, Wilson and his student Simberloff performed a major test of the MacArthur-Wilson Species Equilibrium Model, reported in a set of papers entitled "Experimental zoogeography of islands." (Wilson and Simberloff 1969, Simberloff and Wilson 1969; the history of these experiments is recounted in Wilson's chapter in this volume). The mangrove *Rhizophora mangle* grows as isolated units in shallow marine waters; the areas of such "islands" can range from a few cm^2 (a single recently rooted propagule) to groups of many clustered trees. Wilson and Simberloff hired a pest-extermination company to place sheeting over a number of moderately sized such islands and gas the arthropods within; this "defaunation" killed nearly all of the arthropods inhabiting the islands, and then Simberloff and Wilson monitored the recolonization of the islands. The islands typically recovered to their predefaunation species numbers in something less than a year, although the most distant island had a slower approach, not fully achieving its previous value even after two years (Simberloff and Wilson 1970). This is expected from equation (3.3). The form of the colonization curve (species versus time) was convex (figure 3.3b), also in accordance with the theory (see also next section).

2. *Birds of the Channel Islands.* Breeding birds of nine islands off the California coast were first surveyed in 1917. Jones and Diamond (1976) performed a number of surveys on each island, beginning approximately fifty years later; these repeated surveys extended over a period of four years (figure 3.3a). The data showed a great deal of constancy in the number of species, although species identities were quite different (see next section).

3. *Birds of the islands in the Aegean Sea.* During 1988–92, Foufopoulos and Mayer (2007) resurveyed five islands that were first surveyed by

Figure 3.3. A. Birds of the Channel Islands, California (USA). Number of breeding species *S* for each island plotted against survey year. The number written over the line connecting each pair of points is the percent turnover between those surveys (Jones and Diamond 1976). B. Colonization curves of four mangrove islets, Florida (USA). E-2 is the nearest island and E-1 is the farthest island (Simberloff and Wilson 1970). C. Oscillating equilibrium in marine epifaunal invertebrates. Gross immigration and extinction rate curves (top) and colonization curves (bottom) (Osman 1978). The number of species changes seasonally as a result of corresponding changes in the immigration-rate curves.

Watson (1964) in 1954–61. No species count changed by more than one species, thereby providing strong evidence for an equilibrium species number.

4. *Birds and plants of Krakatau.* In 1883 a huge volcanic eruption destroyed two-thirds of the Indonesian island of Krakatau and buried its remnants and two neighboring islands under 30–60 meters of ash, with no apparent plants or animals surviving. Unlike for the mangrove islands discussed above, prior records of the bird or plant inhabitants of the islands were unavailable for this unanticipated "natural experiment." For birds, however, the recolonization (as documented by Dammermann [1948]) appeared to MacArthur and Wilson (1967) to be leveling off after only 25–36 years, a pattern that they interpreted as major support for their equilibrium theory. Subsequent studies, however, showed that the conclusion was premature, equilibrium perhaps not being quite attained even a century after the eruption (Bush and Whittaker 1991, Thornton et al. 1993). Plants, in contrast, showed a much slower rate of recovery, and in MacArthur and Wilson's (1967, figure 22A) illustration there was little if any indication of convexity in the colonization curve. The most recent censuses, about 100 years later (Bush and Whittaker 1993), show otherwise: equilibrium appears perhaps nearly attained for seed plants (figure 3.4, top left) and ferns (compiled by Thornton et al. 1993). For plants, MacArthur and Wilson (1967) predicted that extinction rates might actually decline at first during the period when initially arriving species facilitate the establishment and persistence of subsequent species; figure 3.4 (top right) reproduces the relevant figure from their book. In fact, nonmonotonic curves have been reported (Thornton et al. 1993), but for immigration rather than extinction (figure 3.4, bottom left). However, as Lomolino et al. (2005) pointed out, this discrepancy most likely reflects how immigration is defined in the two treatments (initial immigration [so *sensu stricto*], versus recent immigration plus establishment, the quantity available from censuses widely spaced in time). Interestingly, a similar nonmonotonicity is apparent, perhaps to a slightly lesser extent, for birds (figure 3.4, bottom right). In any event, recent species-by-species analysis of extinction among Krakatau plants (Whittaker et al. 2000) concludes that successional loss of habitat (as well as to a lesser extent other habitat disturbance or loss) largely accounts for the extinction of well-established species.

5. *Marine epifaunal invertebrates on rocks.* To simulate colonization of rocks, Osman (1978) set out artificial panels in the marine subtidal of Massachusetts. This experiment produced an oscillating equilibrium (figure 3.3C), in which the number of species increased toward a regular cyclical rise and fall of species. The low point of the cycle occurred in the

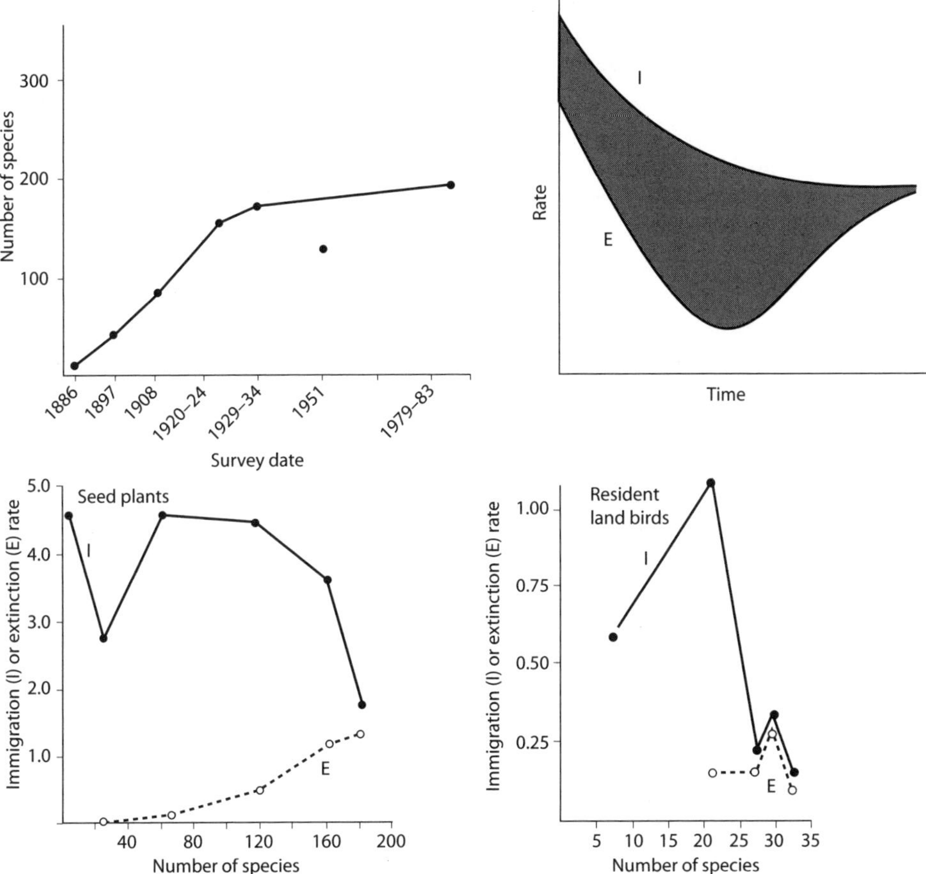

Figure 3.4. Top left. Species number versus time for seed plants of Krakatau (Rakata; Thornton et al. 1993). Top right. Immigration and extinction curves vs. time for a biota showing succession (MacArthur and Wilson 1967, fig. 23). Bottom left. Immigration and extinction curves versus number of species present for seed plants of Krakatau (Thornton et al. 1993). Bottom right. Immigration and extinction curves versus number of species present for resident land birds of Krakatau (Rakata; Thornton et. al. 1993).

winter months and the high point in the late summer, as would be expected in this highly seasonal environment.

6. *Birds on islands off Australia and New Zealand.* Censuses conducted over periods of 50–124 years for fifteen islands in the Australasian region showed that their number of passerine species did not fluctuate around an apparent equilibrium; rather, in fourteen of fifteen cases

the species counts increased, up to 900% of the original values. Abbott and Grant (1976), who compiled these data, argued that direct human changes were insufficient to account for these systematic increases. Rather, they suggested (somewhat presciently) that climatic warming might have been responsible. Abbott and Grant (1976) entitled their paper "Nonequilibrial bird faunas on islands," and these islands certainly stand in marked contrast to the Channel Islands discussed above.

7. *Plants and ants on islands of the Bahamas.* Surveys of both plants and ants spanning nearly two decades on approximately 200 Bahamian islands by Morrison (2002, 2003, in prep.) showed a similar "nonequilibrial" situation as for the birds of the previous example. However, the direction of change was the opposite: islands lost plant and ant species during the second decade of the study rather than gained them. Foliar cover of plant populations whose species did not completely go extinct showed a steady decline over that time. The relative abundance of ant populations that did not go extinct declined in the second decade of the study as well. Although several hurricanes struck the region (see no. 9 below), the direct impact of hurricanes did not appear to be the main cause. Morrison suggested that decreasing precipitation and increasing temperatures in the region, along with potential increased herbivory of plants due to hurricane and drought stress, could be contributing factors.

8. *Birds on Skokholm Island.* Abbott and Grant (1976) compiled data on numbers of bird species for a small island off the British mainland, recorded 1928–67, with time off for the war years. The species number fluctuated between 5 and 13, with substantial temporal autocorrelation (figure 3.5). These are large percentage changes, so they might be interpreted, as Abbott and Grant did, as evidence against equilibrium. However, MacArthur and Wilson's (1963, 1967) original theory went well beyond the simple deterministic, graphical or algebraic model presented above, including a stochastic version with per-unit-time probabilities of immigration and extinction rather than fixed rates. The implication for present purposes is that the "equilibrium" number of species is expected to vary around some mean, rather than be constant once an average equilibrium is attained. Box 3.1 reproduces and extends somewhat the MacArthur-Wilson mathematics to show that the variance/mean number of species will fall between 0 and 1. For the Skokholm data, the mean is 6.59 and the variance is 4.37, so that variance/mean ~ 2/3. This relatively high value is to be expected (box 3.1) from the high extinction rate that should characterize the low populations on this very small island (see also comments in the next section under no. 5).

9. *Hurricane effects on Bahamian lizards and spiders.* In 1996 the massive Hurricane Lili swept east to west across the Bahamas, bringing

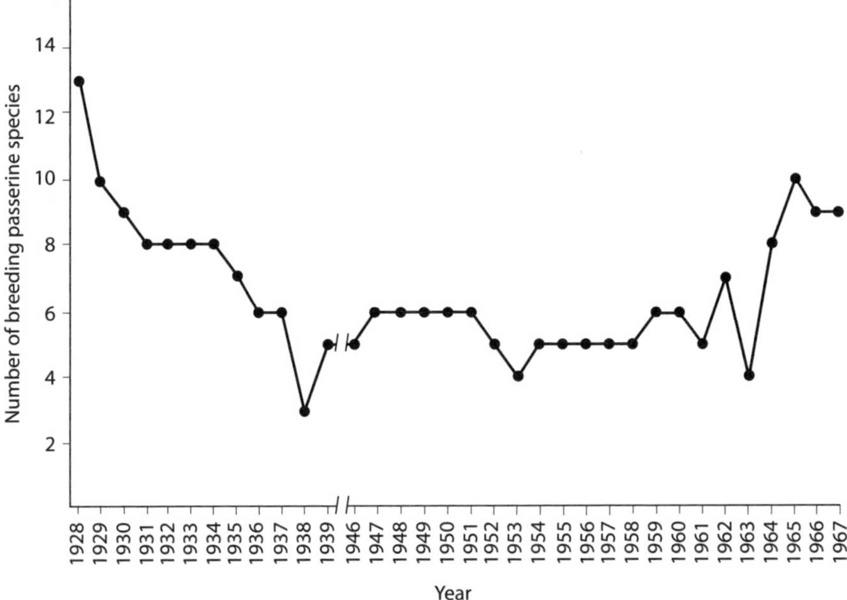

Figure 3.5. Number of breeding passerine bird species for Skokholm Island through time (figure drawn from data in Abbott and Grant 1974). Note the substantial variability in number of species over the time period studied.

with it a storm surge of up to ca. 5 meters (Spiller et al. 1998). Such an inundation was devastating for many small, very low islands of the region, including the eleven islands to the west of the main island of Great Exuma, directly in the path of Lili. As part of an introduction experiment, Losos and Spiller had been collecting faunal data for these eleven islands, as well as eight protected islands to the east of Great Exuma, up to the very moment the hurricane struck. After its passage, they retrieved their boat from a tree (so the story goes, anyway) and recensused the islands. On the exposed islands, every lizard and web-spider individual originally inhabiting the islands was gone. However, spiders on the devastated islands were not entirely absent: webs of a few individuals of a species never before found on the islands (*Metazygia bahama*) were found clinging to bare rock (subsequently this species completely disappeared). It is notable in this regard that the first colonist of Krakatau was also a spider (Thornton 1996)! On the protected islands, no lizard population became extinct as a result of the hurricane, and the likelihood of extinction for spider populations on those islands was negatively related to their population size.

BOX 3.1. Derivation of the limits of the variance/mean for number of species around equilibrium.

Note: λ and μ are MacArthur and Wilson's (1967) notation and refer to gross rates in this derivation.

MacArthur and Wilson show that

$$\frac{\text{var}}{\text{mean}} \leq \frac{d\mu/dS}{(d\mu/dS)-(d\lambda/dS)} = \frac{1}{1+\dfrac{|d\lambda/dS|}{|d\mu/dS|}}.$$

We can graph the gross immigration and extinction rate as follows (this corresponds to our figure 3.1)

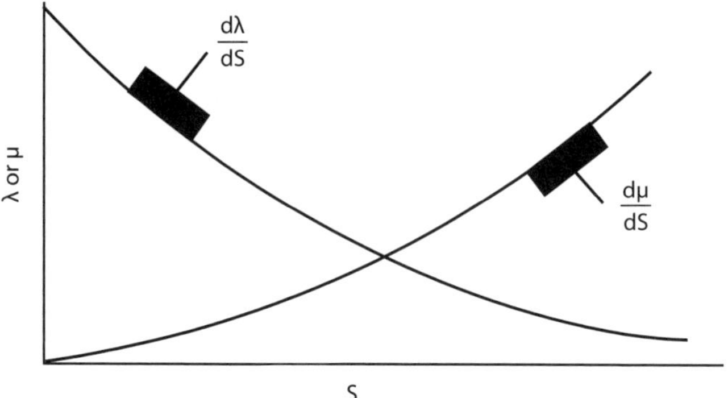

Then

$$\text{var} = \frac{\lambda_{\hat{s}} + \mu_{\hat{s}}}{2\left(\dfrac{d\mu}{dS} - \dfrac{d\lambda}{dS}\right)}$$

at equilibrium. Letting $\lambda_{\hat{s}} = \mu_{\hat{s}} = X$ (the common value of the two rates at equilibrium), it follows that (see figure)

(Continued)

(*Continued*)

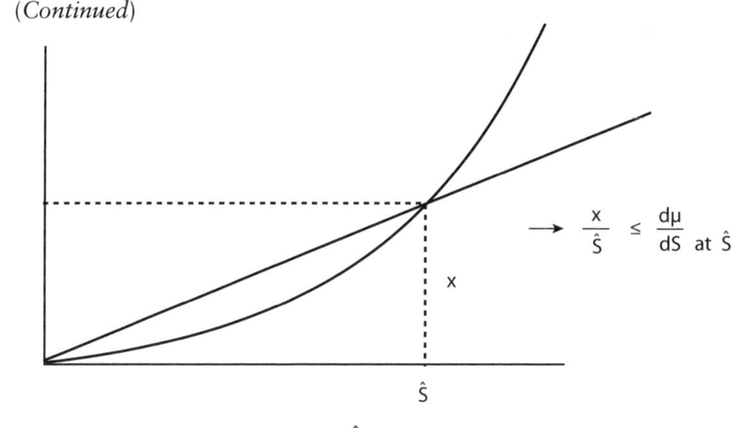

$$\frac{\text{var}}{\text{mean}} = \frac{X/\hat{S}}{\frac{d\mu}{dS} - \frac{d\lambda}{dS}} \Rightarrow \frac{X}{\hat{S}} \leq \frac{d\mu}{dS} \text{ at } \hat{S}.$$

Substituting into the previous equation, we get

$$\frac{\text{var}}{\text{mean}} \leq \frac{d\mu/dS}{\frac{d\mu}{dS} - \frac{d\lambda}{dS}}.$$

If $|d\mu/dS| \cong |d\lambda/dS|$, var/mean ≤ 1⁄2, the MacArthur-Wilson result, but in general var/mean falls between 0 and 1.

When was the original species equilibrium recovered for this natural defaunation? The answer depends on the organism. For spiders, recolonization was rapid and in one year, the number of species on average was the same as before the hurricane struck (figure 3.6; the islands were all less than15 km from the main island of Great Exuma). In complete contrast, the number of species of lizards was still at zero on the exposed islands, and at the last survey date (2001) only two of the islands had been colonized by lizards (in protected areas, three of five islands naturally having lizards were colonized, but none of the eight introduction islands was). Thus it appears that equilibrium depends on the organism: for highly vagile organisms like spiders, which disperse mainly by ballooning through the air, equilibrium can be recovered quickly, just as it was for Simberloff and Wilson's mangrove arthropods discussed above. For lizards, which have to disperse by rafting or floating (Schoener and Schoener 1984), attainment of equilibrium may take a very long time, indeed a

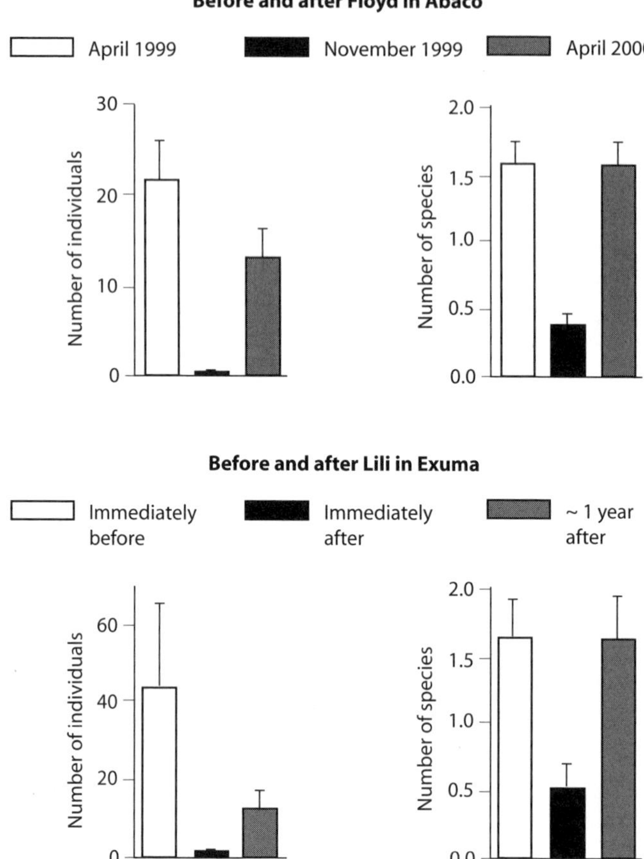

Figure 3.6. Mean number of individuals and of species (±one standard error) for web spiders on islands immediately before, immediately after, and one year after devastating hurricanes struck two regions of the Bahamas (Schoener and Spiller 2006). The patterns for two different hurricanes in two different regions are nearly identical.

longer time than the next devastating catastrophe, implying that lizards may never be at equilibrium. Certainly, spiders and lizards on the same set of islands differ in the likelihood that they will be at equilibrium at a randomly chosen point in time.

Two footnotes to these results are interesting. First, even though the number of species of spiders attained equilibrium after one year, the total number of spider individuals fell short of the value before Lili struck (figure 3.6). Second, the pattern in both number of species and number of individuals was repeated with nearly the same relative values after the storm surge of Hurricane Floyd in 1999 wiped out the spiders on a more northerly group of islands, those off the main island of Great Abaco (figure 3.5; Schoener and Spiller 2006). That two hurricanes would recently occur, for both of which predefaunation data were available, seems serendipitous, although the likelihood of further confirmation of these patterns is perhaps not small, given the increase in hurricane frequency presently characterizing the Caribbean.

10. *Arthropods in soybean fields.* An even more extreme example of draconian extinction being frequent relative to how quickly equilibrium is attained was described by Price (1976). Croplands are highly temporary habitats for which "defaunation" is a scheduled human activity; combined with seasonal variation, this results in catastrophic extinction followed by a period of little to no recolonization. Once the crop has been replanted and is growing again, arthropods begin to colonize it, but they do not have time to reach an equilibrium before the next catastrophic harvest.

Conclusions about Equilibrium

These examples allow us to make the following conclusions concerning the existence of equilibrium:

First, equilibrium can be steady (a constant number of species), cyclical (a regular fluctuation in number of species), or moving directionally (a slow, undirectional change in numbers of species brought on by a systematic change in immigration and/or extinction rates, e.g., due to climate change). Many examples of the first possibility were discussed above, and Osman's (1978) work on marine epifaunal communities illustrates the second. The third is not clearly demonstrated by any of the examples above, except possibly the birds of Australian and New Zealand islands and the plants and ants of Bahamian islands; however, one could certainly argue that those two examples are nonequilibrial, and indeed that is what their investigators have done.

Second, even for a steady average equilibrium, there is expected to be a variance according to the stochastic version of the MacArthur-Wilson equilibrium model (as well as for any other such type of model). In the case of Skokholm Island, the variance was large but within theoretical expectations.

Third, a system approaching equilibrium can have similar properties to one at equilibrium, e.g., with respect to area and distance effects (see below). Because number of species is expected to approach equilibrium at a decreasing rate (second derivative negative), as in equation (3.2), communities are expected to show qualitatively the same effects of factors affecting immigration and extinction rates, even if those communities are moderately far from the equilibrium number, and perhaps even over the majority of the colonization period. Hence it would not be fair to argue that, because an island community is not at equilibrium, a species-area effect as predicted by the equilibrium model (if island area is related to extinction rate; see below) will not occur.

Fourth, the more frequent the disturbance rate, the less likely equilibrium is to be attained. Disturbance, as in lizards of the Bahamas and arthropods of soybean fields, can wipe out a biota before, and sometimes well before, there is time to attain equilibrium.

Fifth, for a given rate of disturbance, equilibrium is more likely to be attained by organisms that are good dispersers (giving a higher immigration rate and thus a faster approach to equilibrium as in equation [3.2]). An example is lizards and spiders on the same Bahamian islands; the latter attain species equilibrium quickly after catastrophic hurricanes, whereas the former may never do so.

Evidence for Species Turnover

The second prediction of the MacArthur-Wilson Species Equilibrium Model, that species lists will vary in composition even after equilibrium is attained (as well as on the way to equilibrium) is even less intuitive than the equilibrium prediction itself; we now review evidence for species turnover. The most commonly used measure of this quantity is relative turnover, given as

$$\text{Turnover (relative) over a unit time interval } (t_1 - t_2) = \frac{\text{extinctions of species already present + immigrations of new species}}{\text{number of species at } t_1 + \text{number of species at } t_2} \times 100 \qquad (3.3)$$

(This equation can actually be viewed as having two averages, one in the numerator and the other in the denominator, so the 2's in each of these averages cancel out. A second kind of measure, absolute turnover, does not normalize by species counts but simply computes the average of the absolute numbers of species immigrating and becoming extinct over the time period.)

1. *Arthropods of red mangrove islands.* The Simberloff-Wilson colonization curves show a roughly monotonic approach to equilibrium (figure 3.3, top), and this is accompanied by a patchy record of individual-species presences and absences, with particular species immigrating and then going extinct, some repeatedly, during the colonization process. Moreover, once the old equilibrial number is regained, the composition

of arthropod species is substantially different from that determined shortly before the artificial defaunation.

2. *Birds of the Channel Islands.* Using a formula similar to equation (3.3), Diamond (1969) found very high turnover for birds of the Channel Islands separated by censuses 51 years apart. The conclusion was challenged by Lynch and Johnson (1974), who argued that, among other problems, species were missed during one or the other census, thereby artificially inflating the reported turnover rate—a phenomenon labeled "pseudoturnover" by Simberloff (1974). However, subsequent censuses by Jones and Diamond (1976) annually over a period of several years showed that in fact turnover was substantial, primarily because of entire missed sequences of immigration followed by extinction for particular species—"cryptoturnover" (Simberloff 1974). In fact, their year-by-year data showed turnover at 0.5–4.9%, whereas the two censuses in Diamond's (1969) original study gave 0.3–1.2%, if anything too small. Hence, if the original two censuses missed species, they were more than compensated for by entirely missed immigration/extinction sequences for particular species during the long interval between the censuses.

Shortly after the data were published, Diamond and May (1977) presented an elegant mathematical treatment of how measured ("apparent") turnover is expected to decline with increasing time between censuses (box 3.2). For the "island" treated by Diamond and May—the Farne archipelago (near Skokholm; see above)—predictions match data rather well (figure 3.7). The turnover rate per year $T(1)$ equals 0.13 or 13%. For intervals exceeding about ten years ($T \geq 10$), turnover is underestimated by about an order of magnitude. Note that the possible variety of species for this high-latitude site is limited by a rather low diversity of immigrants, so to some extent the same species wink in and out. Also note that most of the species are migrants and present in very small numbers, further contributing to a high turnover rate.

BOX 3.2. Derivation of the relation of apparent turnover T after time t (the period between two successive censuses) to t.

Let $I_i(t)$ and $E_i(t)$ for Species i be the probability of, respectively, being present at t yrs if initially absent and of of being absent at t yrs if initially present. The incidence (which gives the fraction of time periods occupied by a given species [or the fraction of islands at any time occupied by a given species]) is given by

(*Continued*)

(*Continued*)

$$p_i = \frac{\lambda_i}{\mu_i + \lambda_i},$$

$$I_i(t) = p_i(1 - (1 - \mu_i - \lambda_i)^t), \quad E_i(t) = (1 - p_i)(1 - \mu_i - \lambda_i)^t).$$

Equilibrium species number sums over the incidences:

$$S^* = \sum_{\bar{c}=1}^{S_T} p_{i,} \qquad S_T = P.$$

The apparent rates (those quantities measured by the investigator over a census period), $\Lambda(t)$ and $M(t)$, sum up the I_i's and E_i's for all species:

$$\Lambda(t) = \sum_i (1 - p_i) I_i(t),$$

$$M(t) = \sum_i p_i E_i(t).$$

Note that $\Lambda(t)$=gross immigration and $M(t)$=gross extinction. They are equal at equilibrium.

The apparent turnover $T(t)$ is calculated as

$$T(t) = \frac{\text{gross immigration} + \text{gross extinction}}{(S_1 + S_2)t},$$

which is our equation (3.1) divided by the length of the time interval. At equilibrium, $S_1=S_2=S^*$, so substituting from the above equations, we get the apparent turnover after time t as

$$T(t) = \frac{2\sum (p_i)(1 - p_i)(1 - (1 - \mu_i - \lambda_i)^t)}{2S^* t}$$

$$= \sum \frac{\lambda_i \mu_i}{(\mu_i + \lambda_i)^2} (1 - (1 - \mu_i - \lambda_i)^t)$$

3. *Birds of the Aegean islands.* Despite a very strong tendency toward equilibrium in species numbers, the five islands studied by Foufopoulos and Mayer (2007) showed a great deal of turnover over the same period, comparable to values for other temperate islands as reviewed above.

4. *Birds and plants of Krakatau.* MacArthur and Wilson's (1967) original estimates for extinction of birds in this archipelago are now

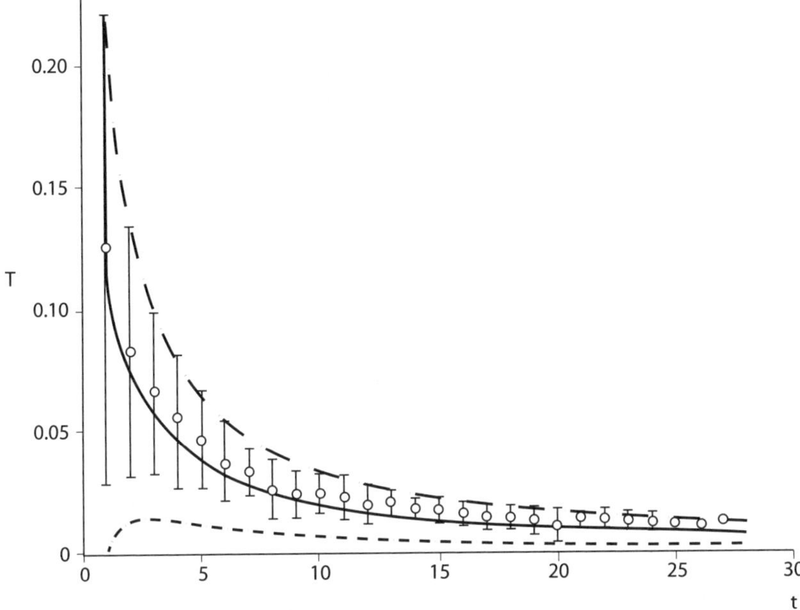

Figure 3.7. Apparent turnover rate (T) of breeding land bird species on the Farne Islands, expressed as the fraction of breeding species immigrating or becoming extinct per year and calculated from differences in the species list for pairs of censuses t years apart. Circles, mean observed T; vertical bars, observed mean$\pm$1 standard deviation. Solid curve, mean predicted T; dotted curves, predicted mean$\pm$1 standard deviation (Diamond and May 1977).

known to be much too high, perhaps by a factor of about 3 (Thornton et al. 1988): Their estimates are 0.5–1.6% per year, whereas recent estimates are 0.25–0.42% per year. Similarly, previous extinction rates for the plants of Krakatau are "significantly overestimated" (Whittaker et al. 2000): New data reduced the pseudoturnover contribution, and the extinctions that are thought to have occurred involved human introductions as well as rare or ephemeral species. As stated above, losses were mainly due to successional loss of habitat and to a lesser extent, other habitat disturbance or loss.

5. *Birds of tropical islands other than Krakatau.* Values of turnover for the Channel and Aegean Islands, which are temperate, are large. In contrast, certain tropical islands (those not subjected to recent disturbance) have much lower turnover. Abbott and Grant (1976) noted that, over a 72-year interval, the Tres Marías Islands off western Mexico had only two immigrations. Even more extreme, Slud's (1976) data show that the Neotropical Cocos Island had no turnover in 72

years, and One Tree Island in the Great Barrier Reef region had no turnover during six continuous years of observation (Heatwole et al. 1981). Most sensationally, a seven-year survey (1984–90) by Mayer and Chipley (1992; see this paper for additional references), with additional censuses in 1954 and 1976, found no immigrations and only one extinction for Guana Island in the Caribbean. This stability is also in contrast to the Australasian islands discussed in the equilibrium section (no. 6).

However, some tropical islands show higher turnover. In 1986 an extensive hydroelectric project flooded a huge area in the Caroni Valley of Venezuela, creating islands in Lago Guri that had formerly all been part of a single land mass. Surveys by Terborgh and colleagues (1997; see Terborgh's chapter in this volume) found that a new equilibrium was achieved in just seven years on the smaller islands, while the larger islands are still declining. Similar phenomena occurred in relation to the massive changes when the Panama Canal was constructed (review in Lomolino et al. 2005). Here, as in Lago Guri, turnover was somewhat lower the larger the island; it was also lower for far than near islands (except for the nearest three islands; Wright 1985). The general patterns are consistent with the MacArthur-Wilson Species Equilibrium Model (see next section) or a modified such model (see Wright [1985] for details). However, while turnover is substantial for these tropical islands, they are perhaps not comparable examples to those of the preceding paragraph, as the islands were in a recently very much disturbed state, being essentially young landbridge islands relaxing to a new equilibrium. Further, the islands studied were very close to the mainland, having indeed been recently a part of it. Finally, as pointed out in the previous section, the extinction component of turnover for the Krakatau archipelago is now known to be much smaller than was originally thought, despite the recently disturbed nature of that region.

Thus there may well be a difference for birds in turnover between the average temperate versus tropical island. Why might this occur? Mayer and Chipley (1992) suggested it is because tropical birds have lower immigration rates (they are locally more sedentary), lower mortality, and are nonmigratory.

6. *Spiders on Bahamian Islands*. What kinds of species show turnover? The question is easiest to answer for extinction, which shows a strong relation to population size when looked at empirically or theoretically (e.g., the above studies for Bahamian spiders, Channel Island birds; see the theoretical review in Schoener et al. 2003). This brings up the issue: How important, in terms of total population numbers of all individuals combined, are species showing turnover? Indeed, some-

thing of a contradiction runs through the various theoretical papers written by MacArthur: some papers assume a community of competitors that is commonly at population-size equilibrium (MacArthur 1968); other papers postulate that turnover, which involves the entire disappearance of species (to say nothing of changes in abundance) is commonplace.

An attempt to answer this question precisely was made for Bahamian spiders by Schoener and Spiller (1987), who calculated the percentage of all individuals combined belonging to populations becoming extinct over particular intervals, ranging from one to five years. Using one-year intervals, 2.8% belonged to populations becoming extinct. Using five-year intervals, still only 4.8% did so. Turnover, while quite large in terms of species number (about 35% per year), does not involve the most abundant species, those that should often have the greatest food-web effects and in any event are of most interest to ecosystem, as opposed to biodiversity, ecologists. In this system, often the same species become extinct and reimmigrate, much as portrayed in Hanski's (1982) core-satellite scheme. Population-persistence curves, which give the fraction of species populations remaining *n* years after a particular census, show this more precisely (figure 3.8). The curve for all web-spider species combined levels off quite sharply (even on a semilogarithmic scale). Interestingly, the individual species vary in the degree to which a leveling off occurs: *Gasteracantha cancriformis* has a practically exponential decline, i.e., a straight line on a semilog plot (produced by a constant per time probability of a population becoming extinct). In contrast, *Eustala cazieri* and *Metapeira datona* show a marked curvature even on a semilog scale, implying that many of their populations persist for long periods of time. The mostly ephemeral nature of the populations going extinct is similar to the situation for Krakatau plants (Whittaker et al. (2000).

Conclusions about Turnover

These examples allow us to make the following conclusions about species turnover:

First, complete turnover events (immigration followed by extinction of a particular species) are often missed in surveys, which typically are separated by substantial intervals. While it is possible that the opposite type of error will occur (designating a species absent that was in fact present because of an incomplete survey, thereby inflating the turnover estimate), for intermittent censuses, missed complete sequences are expected to be common enough so that turnover will typically be underestimated.

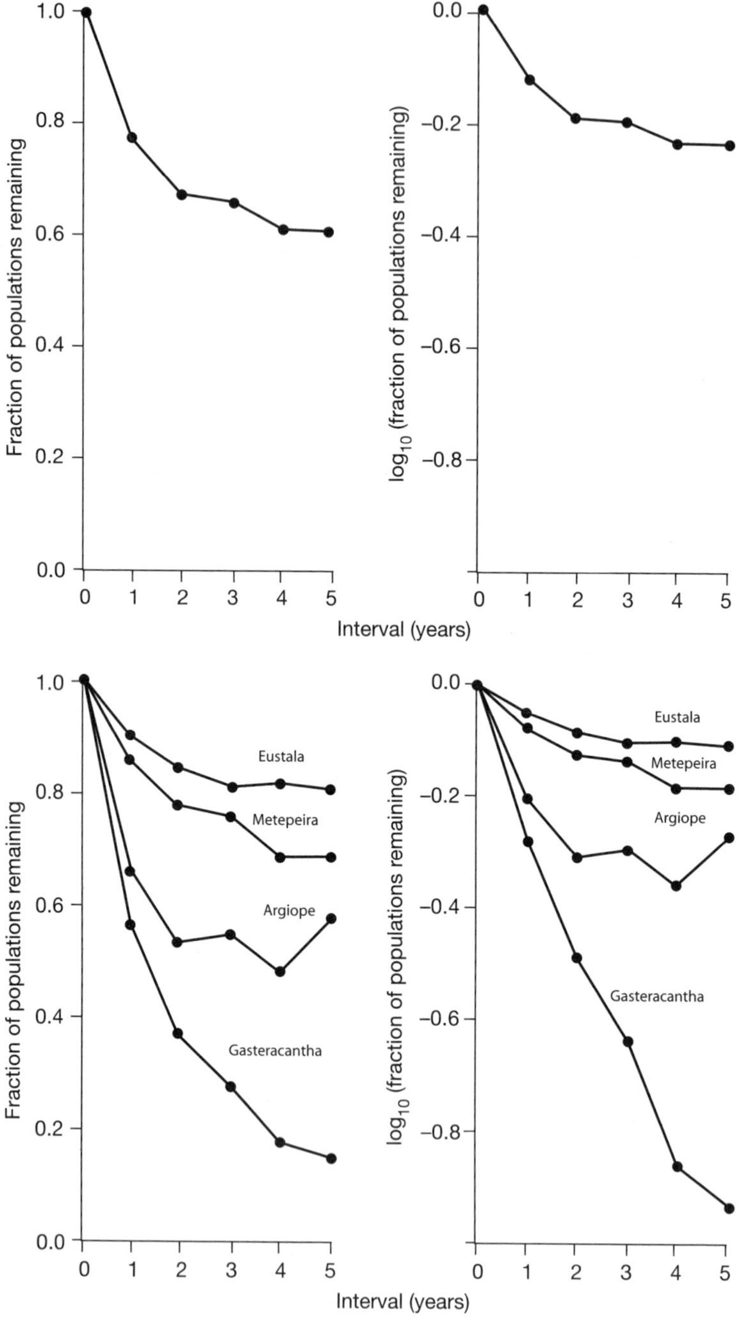
1.0
0.8
0.6
0.4
0.2
0.0
Fraction of populations remaining
0.0
−0.2
−0.4
−0.6
−0.8
log10 (fraction of populations remaining)
0 1 2 3 4 5
Interval (years)
Eustala
Metepeira
Argiope
Gasteracantha

Second, turnover tends to be greater for small islands and for far islands, in accordance with the MacArthur-Wilson Species Equilibrium Model (see next section).

Third, turnover can be very low for tropical islands, but for those recently disturbed or created, this is not necessarily the case.

Fourth, species turning over may comprise a low fraction of the total number of individuals in the biota—this results from the very strong relation between extinction rate (one of the two components of turnover) and population size. Such species can be important for species-diversity studies but would seem epiphenomenal for ecosystem studies.

Species-Distance and Species-Area Relations

The MacArthur-Wilson Species Equilibrium Model makes predictions about the effects of an island's distance from the source of immigrants and about an island's area, as follows. Assume

1. Near islands have higher immigration rates than far islands, for the same number of species present; and
2. Small islands have higher extinction rates higher than large islands, for the same number of species present. This is because average population size is smaller for the smaller islands, hence the per species extinction likelihood is greater—note that a decreasing relation between extinction likelihood and population size has been repeatedly demonstrated, e.g., Jones and Diamond (1976), Terborgh and Winter (1980), Schoener and Schoener (1983b), Schoener and Spiller (1987), Pimm et al. (l988), Laurance (1990).

These assumptions imply two results (figure 3.9). First, near islands (of the same area as far islands) have more species. Second, large islands (at the same distance as small islands) have more species. Both predictions are consistent with numerous examples from the literature (reviewed in Lomolino et al. 2005). Note that the graphs of figure 3.9 also imply that absolute turnover (intercept on the ordinate) is greater for near than far islands and greater for small than large islands (for relative turnover, equation [3.3], which can be different, see Williamson [1978]).

Figure 3.8. Population-persistence curves for web spiders on 108 islands of the Bahamas. Top. All species combined. Bottom. Individual species curves (Schoener and Spiller 1987). Note that, overall, while some species become extinct rather quickly, about the same percentage persist throughout the study period. The bottom panels show the four commonest species, which differ considerably among themselves, and sometimes in comparison to the overall pattern.

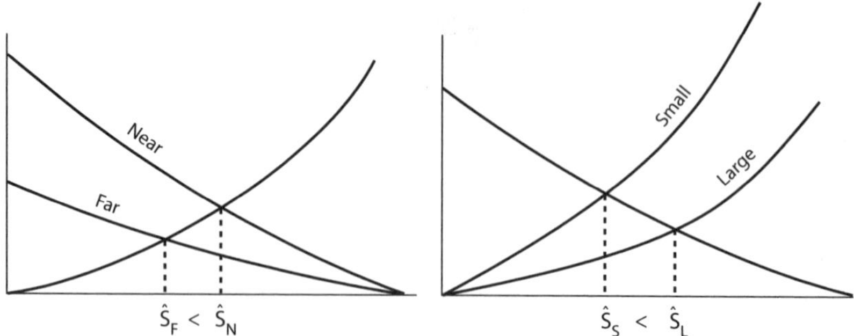

Figure 3.9. Left. The distance effect for the MacArthur-Wilson Equlibrium Model. Far islands have lower immigration rates than near islands, resulting in a smaller number of species present at equilibrium. Right. The area effect for the MacArthur-Wilson Equilibrium.Model. Large islands have lower extinction rates than near islands, resulting in a larger number of species present at equilibrium. Axes as in figure 3.1.

Species-distance relations have had a variety of explanations, only one of which follows from the original MacArthur-Wilson (1963, 1967) model.

First, far islands are less likely to be at species equilibrium than near islands because of their lower immigration rates, but given enough time will eventually achieve the same number of species as otherwise similar near islands. This is a nonequilibrium explanation for the common observation of biotic poverty on isolated islands.

Second, far islands have a less diverse range of habitats, thereby supporting fewer species that depend on those habitats. This explanation says that far and near islands are not "otherwise similar," but differ in the key feature of habitat complexity. Lack (1976) used this idea to explain the lower species diversity of birds on far islands. The explanation is somewhat circular for the entire biota, of course, as a lower habitat diversity for birds would probably imply a lower plant-species diversity, and that would in turn beg explanation.

Third, because of a lower immigration rate far islands may reach equilibrium at a smaller number of species than do near islands. This is the MacArthur-Wilson (1967) explanation, and it is a bit difficult to express without mathematics; the graphical model (figure 3.9 left) is more transparent: an island having a lower immigration rate will balance with its extinction-rate curve at a smaller number of species. It of course differs from the first explanation in that this predicted state of affairs is expected to last forever (at least until the immigration rates change).

Tests distinguishing the first from the third explanations are rare except for short-term experiments such as that of Simberloff and Wilson on mangrove arthropods discussed above (see Schoener [1988] for other examples). Schoener and Schoener (1983a) were able to distinguish the second from the third explanation for Bahamian resident birds and lizards, which showed distance (and area) effects. The fraction of vegetation in different height categories was used to construct habitat-diversity indices, and Lack was correct that far islands had a lower habitat diversity than near islands. However, accounting for that relation in partial correlation still resulted in significantly negative distance relations. This last result is certainly consistent with the MacArthur-Wilson Species Equilibrium Model, although some of the islands, as least for lizards, may not be at species equilibrium (see above).

The list of explanations for the species-area effect is even longer than that for the species-distance effect (Spiller and Schoener, in press):

First, some kind of random sampling could produce the effect, independently of a well-defined mechanistic process. For example, imagine only that large islands have more individuals of a given kind of organism than do small: draw (or allow to colonize; see "third" below) more individuals from the source's species-abundance distribution for large than small islands, and more species will result on large islands.

Second, populations are larger on larger islands, implying lower extinction rates there. This is the MacArthur-Wilson assumption, and like the distance effect is somewhat difficult to express without mathematics; note from the graph (figure 3.9, right) that an island having a lower extinction rate will balance with the immigration-rate curve at a greater number of species. The assumption relating population size to extinction likelihood is very well supported by data, as discussed above.

Third, interception area (or shoreline) is larger for larger islands, implying a greater immigration rate for larger islands (not just a smaller extinction rate). This so-called "target effect" has been shown for a variety of organisms (reviewed in Lomolino et al. 2005). These include the striking result of Buckley and Knedlhans (1986) in which species diversity of seaborne plant propagules is linearly related to shoreline length for islands off Australia, and the demonstration of Lomolino (1990) that immigration rates of mammals to islands in the St. Lawrence River were positively correlated with island area (see also Rey 1981, Schoener and Schoener 1981, Hanski and Peltonen 1988). It is possible, of course, to add this effect to MacArthur and Wilson's original model, giving a more complicated set of curves. The corresponding effect for area is the "rescue effect" of Brown and Kodric-Brown (1977), in which extinction rate varies with distance: the nearer the island, the more likely populations on that island will be "rescued" from extinction by numerical reinforcements from the mainland; the greater flow from the mainland also could enhance genetic diversity on the island and prevent inbreeding depression, again reducing the chance of extinction. Oddly, few demonstrations of this rescue effect seem to exist additional to the arthropods-on-thistle-head example in Brown and Kodric-Brown's seminal paper. Smith (1980) showed that talus-inhabiting pikas (*Ochotonia princeps*) had lower extinction rates on "islands" near to a source of immigrants (see also Wright 1985, Lawrance 1990). One system, Bahamian web spiders, shows all four possible relations—the traditional area and distance relations of MacArthur and Wilson, as well as the relations of immigration to island area and extinction to island distance (Toft and Schoener 1983).

Fourth, habitat diversity is higher on larger islands, leading to the ability to support a greater diversity of ecologically distinct species there. Perhaps even more than for distance, the relation of species number to habitat diversity to area is likely to hold; the altitudinally zoned, diverse vegetation characterizing higher islands, which tend to be larger, constitutes a good example. Indeed, sometimes the relation of species number to habitat diversity is stronger than that for area, e.g., the study by Watson (1964) of birds on the Aegean Islands; an overview is presented by Ricklefs and Lovette (1999).

Fifth, abiotic disturbance is larger on smaller islands, implying a greater extinction rate there. Evidence for this idea comes again from Bahamian lizards: larger islands, which tend to be higher, were less likely to lose their lizards as a result of the storm surge that accompanied Hurricane Floyd (Schoener et al. 2001): lizards could survive the inundation if on high enough ground. This example also illustrates a consequence of the correlation between two island traits—maximum altitude and area. Altitude was in fact more important than area in forestalling extinction (Schoener et al. 2001); however, when altitude was not taken into account in the statistical analysis, area was significant.

Sixth, within-island multiplication of species is greater for larger islands. This idea was demonstrated conclusively by Losos and Schluter (2000) for Caribbean *Anolis* lizards, and it is discussed elsewhere in this volume (Losos and Parent). It has also been measured and modeled for endemic land mammals by Heaney and colleagues (summary in Heaney 2004).

Plots of species-area relations are commonplace in the literature, and they fall into two general categories, a linear relation on a semilogarithmic scale (as implied by an exponential function)

$$S = c_1 + c_2 \log A \tag{3.4}$$

and a linear relation on a log-log scale (as implied by a power function)

$$\log S = \log c + z \log A \Rightarrow S = cA^z, \tag{3.5}$$

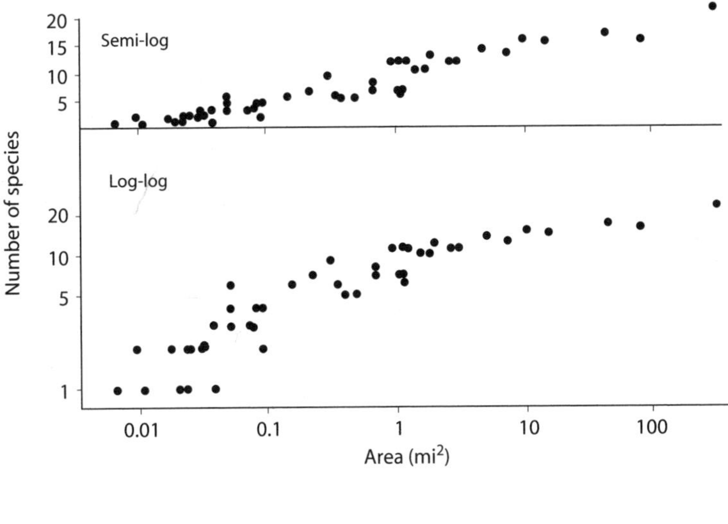

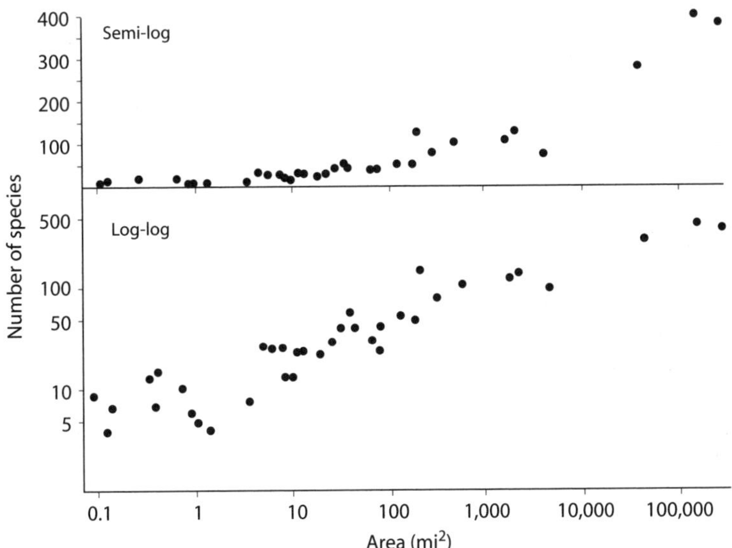

Figure 3.10. Species-area plots showing the semilog and log-log relation, top and bottom respectively. Top. Shetland land birds. Bottom. Malaysian faunal region land birds. (Schoener 1976)

where S is number of species, A is area, and c_1, c_2, c, and z are constants typically to be fitted to the data. Which description is better, equation (3.4) or (3.5)? Connor and McCoy (1979) interpret their review of 100 data sets to say that the two fit about equally. Clear examples of each of the two are given in figure 3.10, in which arithmetic bird species number increases linearly with log(area) for the Shetland islands, whereas logarithmic bird species number increases linearly with log(area) for the Malaysian region (note, incidentally, that the plot for Malaysian islands on a semilogarithmic scale is especially accelerating for the largest islands, perhaps due to within-island species multiplication).

What is the form of the species-area relation implied by the MacArthur-Wilson model? Using equation (3.1) above as a starting point, Schoener (1976) has shown that where abundances at equilibrium are complementary (defined as abundances summed over all species equaling ρA, where ρ is the density of all individuals combined and A is island area),

$$S = [\lambda_A \rho A / 2\mu_N]\,[-1 + \sqrt{1 + (4P_{\mu_N} / \lambda_A \rho A)}] \tag{3.6}$$

where λ_A, μ_A, and P are as in equation (3.1) and μ_N is proportional to μ_A. Equation (3.6) results from assuming (1) $\mu_A = \mu_N / \bar{N}$, where $\bar{N}$ is the average population size and (2) $\bar{N} = \rho A/S$ (other possible assumptions are in Schoener [1976]). Substituting these into equation (3.1) and solving the resulting quadratic in S gives equation (3.6). For this expression, unlike the descriptive power or exponential functions, the number of species asymptotes at P, the number in the source (note that within-island diversification by *in situ* speciation is not in the model). In other words, no matter what the area, there can be no more species on the island than that number available for colonization, a property that must be true for any MacArthur-Wilson-like model. Note also that the slope on a log-log plot ($z = d\log S/d\log A$) is not constant but goes from 0 to 0.5 in this model (a model in which individual species abundances are additive, not complementary, extends the range of z to 1.0; Schoener [1976]).

In the equilibrium species-area model (equation [3.6]), the greater the λ_A (per species immigration constant) the smaller $d\log S/d\log A$. Indeed z is smaller for less remote islands within a single archipelago (also z for an archipelago of habitat islands, where immigration is presumably very high, can be very small, e.g., Watling and Donnelly [2006]; see Holt, this volume, for review). But far archipelagoes have smaller z's than near archipelagos (figure 3.11; also see Connor and McCoy 1979). This is probably because of a differentially high λ (per species immigration rate) among birds that have been able to colonize such archipelagoes. To elaborate, for far archipelagos most immigration is from other islands within the archipelago; for this component, both P and the immigration rate (which var-

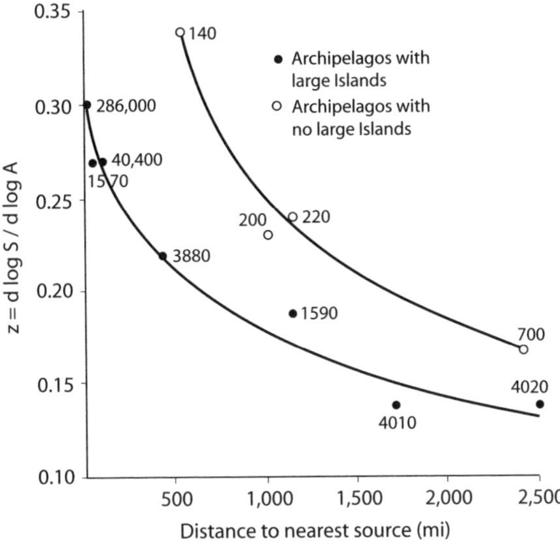

Figure 3.11. The species-area slope (log-log) or z versus distance to the nearest source (as measured from the edge of the archipelago to the nearest large land mass). Numbers give area of the largest island; clear circles are archipelagoes with only islands less than 710 mi^2; shaded circles are archipelagoes with largest island greater than 1500 mi^2 (details in Schoener 1976).

ies with P) are relatively small. Immigration from outside the archipelago (say from some large continental source) is minimal despite the large number of species in the pool, P, because of the much lower λ. For near archipelagos, most of the colonization is from sources external to the archipelago, and this gives an immigration curve with a large intercept on the rate axis as well as a large P. Figure 3.12 illustrates this argument. Various evidence additional to that just cited suggests that this model is on the right track. For example, the species-area slope for birds on islands of Burtside Lake, Minnesota (United States) is unusually high, but P is very large and the islands are very small (Rusterholz and Howe 1979).

The species-area representation of the MacArthur-Wilson model (equation [3.6]) also suggests a relation between the per species extinction rate μ_A and the steepness of the species area slope z: the greater the extinction rate, the greater the slope. Assuming that an increase in predation intensity can be represented by an increase in per species extinction likelihood, this implies that a biota subjected to predation should have a larger log species/log area slope (see also Holt 1996, Holt et al. 1999). However, Ryland and Chase (2007) used a different extension of the MacArthur-Wilson Species Equilibrium Model to get the opposite result: the greater

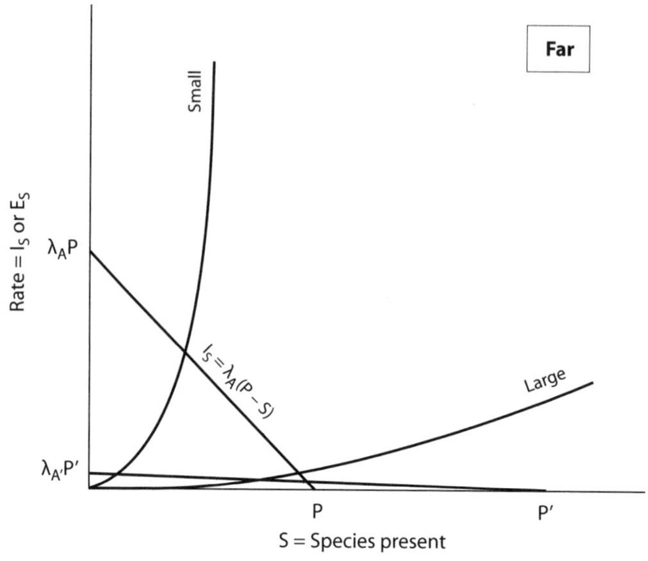

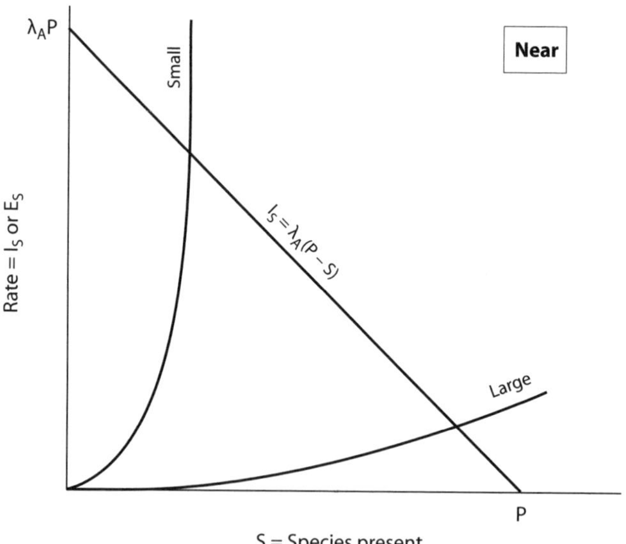

Figure 3.12. Equilibrium for near and far archipelagoes. See text for explanation and definition of symbols (Schoener 1976).

the predation intensity on a biota, the smaller the species-area slope. In the Ryland-Chase extension, the contribution that predators make to the per species extinction rate is assumed additive, not multiplicative as in equation (3.6), and this seemingly minor change in functional form reverses the direction of the prediction. In Holt's chapter in this volume, this analysis is generalized to allow for the extinction factor or addend to itself be a function of area, and in that case results can be more complicated. In neither approach, however, is there a mechanistic or biological justification for the functional form of the respective assumption about how predation affects extinction. Moreover, using a completely different approach, a continuous-time Markov model, Holt (1996; Holt et al. 1999) predicted that the higher the trophic rank, the larger the species-area slope. This result follows from the colonizing properties of predators and prey: higher trophic ranks cannot colonize unless a member of the next lowest rank is present. However, they won't always colonize even when that is true—this is a necessary but not sufficient condition. This leads to a larger (or at best equal) species-area slope, the higher the trophic rank in a given community. Finally, equation (3.6) suggests another way that predators might have larger species-area slopes than prey: the lower the population density ρ of the group in question, the larger the slope.

The preponderance of data collected so far (Hoyle 2004, Ryland and Chase 2007), including ten-year averages for web-spider data from 64 islands from the central Bahamas (near Staniel Cay; Spiller and Schoener [in press]), supports the prediction that predators should have steeper species-area slopes (z's) than prey. There is even a rough correspondence between (surmised) low population density and greater z among birds (Schoener 1976); raptors are in the low-density group. However, more northerly Bahamian spider data (from the Abaco region; Schoener and Spiller 2006) if anything suggest the opposite, supporting the prediction from equation (3.6).

A final form for the species-area relation has been suggested by Lomolino and Weisen (2001; see precursor ideas in Lomolino [2000] and Lomolino et al., this volume), one having essentially an S-shaped segment, i.e., a greater rate of increase for intermediate-sized islands than either for small or large islands; note that the low slope for the smallest islands (where variation in species number is expected to be greatest because of stochastic factors) is the feature of this concept that makes it very different from any of the species-area curves proposed so far, descriptive or mechanistic. Some evidence for such a slope was indeed given in MacArthur and Wilson's (1967) book for a particular case—plants on the Micronesian atoll of Kapingamarangi (Niering 1963). However, their explanation was quite different: freshwater lenses are absent on islets below a certain area, giving a very low and constant species number there. The upper, leveling-off portion of the "S" has the same explanation as that for equation (3.6) above: any system with an upper limit to the number of species available to immigrate to an island (*P* in this case) will have a species-area curve that will tend to level off in its upper portion. In their analysis of 102 insular data sets, Lomolino and Weiser (2001) showed that an increasing portion of the species-area curve is quite general: the initial flat portion of the species-area curve typically included a substantial portion of an archipelago's islands. The authors also point out that the final portion of the species-area curve, should there exist within-island species multiplication, may again accelerate.

Bibliology of the MacArthur-Wilson Species Equilibrium Model

I would like to close by reminding the reader of the word "chronicle" in the title of this chapter. This word attempts to bolster the legitimacy of my approach of dealing with the mostly older papers (see the introduction). At an early stage of preparing my presentation, I was concerned about the following question: Were most papers that dealt specifically with the MacArthur-Wilson Species Equilibrium Model in fact older? This would necessarily be true were most papers that *cited* the MacArthur-Wilson book and paper older. Optimistically, I went to the Science Citation Index to see what the more recent papers had to say, hoping to lace my presentation with a few appropriate citations. MacArthur and Wilson's 1963 paper has a reasonable number of new citations, showing a modest if mildly erratic rise to about twenty-five citing references per two-year period (figure 3.13). However, I was shocked to find that MacArthur and Wilson's 1967 book in recent years (2000–2007 inclusive) had over 2,000 citations, dashing any hope for an easy resolution of my question.

The pattern of citations itself is very interesting (figure 3.13). The number of citing references for the book increases sharply from 1967 until about 1985, at which point it levels off, showing an apparent "citation equilibrium." However, in 2000 the number of citations begins another steep climb that continues unabated to the present time. Does the recent pattern of increasing citations imply that the influence of the MacArthur-Wilson theory, at least *sensu lato*, is again on the rise? Or is it simply a by-product of a recent increase in the overall numbers of citations, no matter what the significance of the work?

To analyze further, one would like some measure of the increase in citations that might be expected simply from the increase in number of citing papers, perhaps a comparison to a work that could serve as a "citation standard." I was hard pressed to think of any such ecological work, given the ups and downs that so many ideas have received in this field. Then I

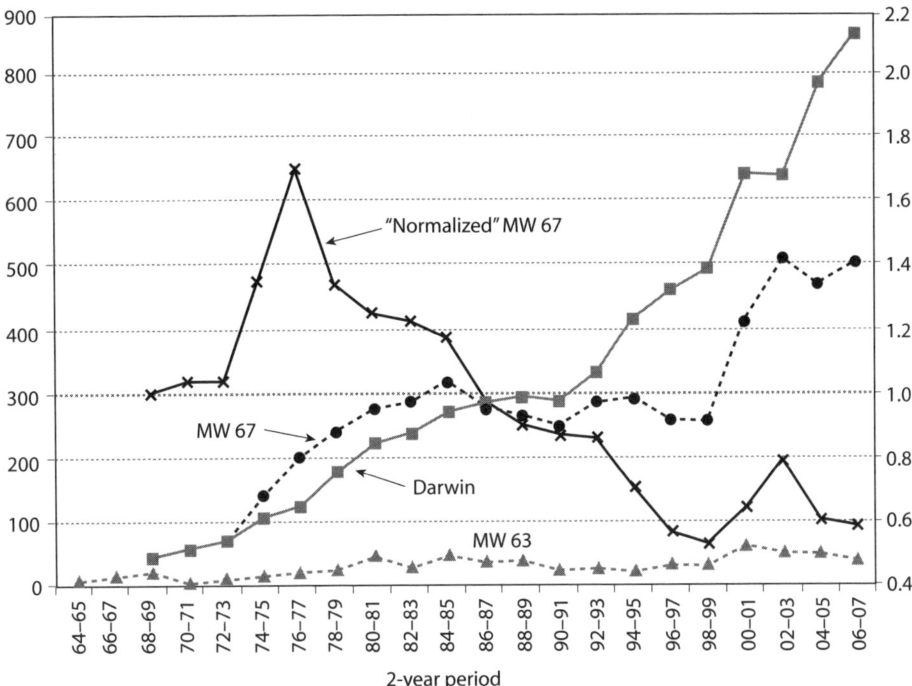

Figure 3.13. Absolute number of citations (left vertical axis) or normalized (by *On the Origin of Species)* number of citations (right vertical axis) per two-year period, against two-year interval. Triangles: Absolute number for MacArthur and Wilson's (1963) paper. Circles: Absolute number for MacArthur and Wilson's (1967) book. Squares: Absolute number for *On the Origin of Species* (all editions listed). Crosses: Normalized MacArthur-Wilson (1967) citations, i.e., circles divided by squares.

had an inspiration: surely Darwin's *On the Origin of Species* is a work that has not waned in influence and has had many years to achieve a constant citation rate per citing reference. All one had to do was normalize the MacArthur-Wilson numbers by dividing the latter by the number of citations of Darwin's enshrined work. Strikingly, using this measure we find a completely different result than just using the raw number of citations: the MacArthur-Wilson book reached the apogee of its influence in 1975, after which it underwent an almost linear decline. One has the nagging feeling, however, that something has gone wrong with the analysis, and this is reinforced by looking at the citation curve for *On the Origin of Species* alone. It steadily increases, exceeding the MacArthur-Wilson book at about 1986 (where the lines cross in figure 3.13) and then continuing upward, even at a slightly increasing rate.

So where does this leave us? Is the true phenomenon of importance the relentless rise of Darwin, rather than anything to do with the MacArthur-Wilson statistics? If so, how can we explain the increasing popularity of Darwin—is that just due to the increasing number of citing references, or is something more going on? No doubt this topic will be discussed at length in 2009 during the 150th anniversary of *On the Origin of Species.*

Acknowledgments

I thank R. Holt, M. Lomolino, J. Losos, R. Ricklefs, and an anonymous reviewer for insightful comments on previous versions of this chapter and NSF Grant No. DEB-0444763 for support.

Literature Cited

Abbott, I., and P. R. Grant. 1976. Nonequilibrial bird faunas on islands. *American Naturalist* 110:507–28.

Brown, J. H., and A. Kodric-Brown. 1977. Turnover rates in insular biogeography: Effect of immigration on extinction. *Ecology* 58:445–49.

Buckley, R. C., and S. B. Knedlhans. 1986. Beachcomber biogeography: Interception of dispersing propagules by islands. *Journal of Biogeography* 13:69–70.

Bush, M. B., and R. J. Whittaker. 1991. Krakatau: Colonization patterns and hierarchies. *Journal of Biogeography* 18:341–56.

Connor, E. F., and E. D. McCoy. 1979. The statistics and biology of the species–area relationship. *American Naturalist* 113:791–833.

Dammermann, K. W. 1948. The fauna of Krakatau 1883–1933. *Koninklijke Nederlandsche Akademie Wetenschappen Verhandelingen* 44:1–594.

Diamond, J. M. 1969. Avifaunal equilibria and species turnover rates on the Channel Islands of California. *Proceedings of the National Academy of Sciences U.S.A.* 64:57–63.

Diamond, J. M., and R. M. May. 1977. Species turnover on islands: Dependence on census interval. *Science* 197:266–70.

Foufopoulos, J., and G. C. Mayer. 2007. Turnover of passerine birds on islands in the Aegean Sea (Greece). *Journal of Biogeography* 34:1113–23.

Gilpin, M. E., and R. A. Armstrong. 1981. On the concavity of island biogeographic rate functions. *Theoretical Population Biology* 20:209–17.

Hanski, I. 1982. Dynamics of regional distribution: The core and satellite species hypothesis. *Oikos* 38:210–21.

Hanski, I., and A. Peltonen. 1988. Island colonization and peninsulas. *Oikos* 51:105–6.

Heaney, L. R. 2004. Conservation biogeography in oceanic archipelagos. In *Frontiers of Biogeography: New Directions in the Geography of Nature*, ed. M. V. Lomolino and L. R. Heaney, 345–60. Cambridge: Cambridge University Press.

Heatwole, H., T. Done, and E. Cameron. 1981. *Community Ecology of a Coral Cay*. The Hague: Junk.

Holt, R. D. 1996. Food webs in space: An island biogeographic perspective. In *Food Webs: Integration of Patterns and Dynamics,* ed. G. A. Polis and K. O. Winemiller, 313–23. London: Chapman and Hall.

Holt, R. D., J. H. Lawton, G. A. Polis, and N. D. Martinez. 1999. Trophic rank and the species-area relationship. *Ecology* 80:1495–504.

Hoyle, M. 2004. Causes of the species-area relationship by trophic level in a field-based microecosystem. *Proceedings of the Royal Society of London, Series B* 271:1159–64.

Jones, H. L., and J. M. Diamond. 1976. Short-time-base studies of turnover in breeding bird populations on the California Channel Islands. *Condor* 78: 526–49.

Lack, D. 1976. *Island Biology*. Oxford: Blackwell Scientific Publications.

Laurance, W. F. 1990. Comparative responses of five arboreal marsupials to tropical forest fragmentation. *Journal of Mammalogy* 71:641–53.

———. 1991. Ecological correlates of extinction proneness in Australian tropical rain forest mammals. *Conservation Biology* 5:79–89.

Lomolino, M. V. 1990. The target area hypothesis: the influence of island area on immigration rates of non-volant mammals. *Oikos* 57:297–300.

———. 2000. Ecology's most general, yet protean pattern: The species-area relationship. *Journal of Biogeography* 27:17–26.

Lomolino, M. V., B. R. Riddle, and J. H. Brown. 2005. *Biogeography*, 3rd ed. Sunderland, MA: Sinauer Associates.

Lomolino, M. V., and M. D. Weiser. 2001. Towards a more general species-area relationship. *Journal of Biogeography* 28:431–45.

Losos, J. B., and D. Schluter. 2000. Analysis of an evolutionary species-area relationship. *Nature* 408:847–50.

Lynch, J. F., and N. K. Johnson. 1974. Turnover and equilibria in insular avifaunas, with special reference to the California Channel Islands. *Condor* 76:370–84.

MacArthur, R. H. 1968. The theory of the niche. In *Population Biology and Evolution*, ed. R. C. Lewontin, 159–76. Syracuse, NY: Syracuse University Press.

MacArthur, R. H., and E. O. Wilson. 1963. An equilibrium theory of insular zoogeography. *Evolution* 17:373–87.

———. 1967. *The Theory of Island Biogeography*. Princeton, NJ: Princeton University Press.

Mayer, G. C., and R. M. Chipley. 1992. Turnover in the avifauna of Guana Island, British Virgin Islands. *Journal of Animal Ecology* 61:561–66.

Morrison, L. W. 2002. Determinants of plant species richness on small Bahamian islands. *Journal of Biogeography* 29: 931–41.

———. 2003. Plant species persistence and turnover on small Bahamian cays. *Oecologia* 136:51–62.

Niering, W. A. 1963. Terrestrial ecology of Kapingamarangi Atoll, Caroline Islands. *Ecological Monographs* 33:131–60.

Osman, R.W. 1978. The influence of seasonality and stability on the species equilibrium. *Ecology* 59:383–99.

Pimm, S. L., H. L. Jones, and J. Diamond. 1988. On the risk of extinction. *American Naturalist* 132:757–85.

Price, P. W. 1976. Colonization of crops by arthropods: Non-equilibrium communities in soybean fields. *Environmental Entomology* 5:605–11.

Ricklefs, R. E., and I. J. Lovette. 1999. The roles of island area per se and habitat diversity in the species-area relationships of four Lesser Antillean faunal groups. *Journal of Animal Ecology* 68:1142–60.

Rusterholz, K. A., and R. W. Howe. 1979. Species-area relations of birds on small islands in a Minnesota Lake. *Evolution* 33:468–77.

Ryberg, W. A., and J. M. Chase. 2007. Predation-dependent species area relationships. *American Naturalist* 170:636–42.

Schoener, A. 1988. Experimental island biogeography. In *Analytical Biogeography*, ed. A. A. Myers and P. S. Giller, 483–512. London: Chapman and Hall.

Schoener, A., and T. W. Schoener. 1984. Experiments on dispersal: Short-term floatation of insular anoles, with a review of similar abilities in other terrestrial animals. *Oecologia* 63:289–94.

Schoener, T. W. 1976. The species-area relationship within archipelagos. In *Proceedings of the Sixteenth International Ornithological Congress (1974)*, ed. H. J. Frith and J. H. Calaby, 629–42. Canberra: Australian Academy of Sciences.

Schoener, T. W., J. Clobert, S. Legendre, and D. A. Spiller. 2003. Life history models of extinction: A test with island spiders. *American Naturalist* 162:558–73.

Schoener, T. W., and A. Schoener. 1981. The dynamics of the species-area relation in marine fouling systems. *American Naturalist* 118:335–60.

———. 1983a. The time to extinction of a colonizing propagule increases with island area. *Nature* 302:732–34.

———. 1983b. Distribution of vertebrates in some very small islands. II. *Journal of Animal Ecology* 52:237–62.

Schoener, T. W., and D. A. Spiller. 1987. High population persistence in a system with high turnover. *Nature* 330:474–77.

———. 2006. Nonsynchronous recovery of community characteristics in island spiders after a catastrophic hurricane. *Proceedings of the National Academy of Sciences U.S.A.* 103:2220–25.

Schoener, T. W., D. A. Spiller, and J. B. Losos. 2001. Natural restoration of the species-area relation for a lizard after a hurricane. *Science* 294:1525–28.

Simberloff, D. S. 1974. Equilibrium theory of island biogeography and ecology. *Annual Review of Ecology and Systematics* 5:161–82.

Simberloff, D. S., and E. O. Wilson. 1969. Experimental zoogeography of islands. The colonization of empty islands. *Ecology* 50:278–96.

———. 1970. Experimental zoogeography of islands: A two-year record of colonization. *Ecology* 51:934–37.

Slud, P. 1976. Geographic and climatic relationships of avifaunas with special reference to comparative distribution in the neotropics. *Smithsonian Contributions to Zoology* 212:1–149.

Smith, A. T. 1980. Temporal changes in insular populations of the pika (*Ochotoma princeps*). *Ecology* 61:8–13.

Spiller, D. A., J. B. Losos, and T. W. Schoener. 1998. Impact of a catastrophic hurricane on island populations. *Science* 281:695–97.

Spiller, D. A., and T. W. Schoener. In press. Species-area. In *Encyclopedia of Islands*, ed. R. Gillespie and D. Clague. Berkeley: University of California Press.

Terborgh, J., and B. Winter. 1980. Some causes of extinction. In *Conservation Biology,* ed. M. E. Soulé and B. A. Wilcox, 119–34. Sunderland, MA: Sinauer Associates.

Terborgh, J., L. Lopez, and G. Tellos. 1997. Bird communities in transition: The Lago Guri Islands. *Ecology* 78:1494–501.

Thornton, I. 1996. *Krakatau.* Cambridge, MA: Harvard University Press.

Thornton, I., R. A. Zann, and P. A. Rawlinson. 1993. Colonization of Rakata (Krakatau Is.) by non-migrant land birds from 1883 to 1992 and implications for the value of island biogeography theory. *Journal of Biogeography* 20: 441–52.

Toft, C. A., and T. W. Schoener. 1983. Abundance and diversity of orb spiders on 106 Bahamian Islands: Biogeography at an intermediate trophic level. *Oikos* 41:411–26.

Watling, J. I., and M. A. Donnelly. 2006. Fragments as islands: A synthesis of faunal responses to habitat patchiness. *Conservation Biology* 20:1016–25.

Watson, G. E. 1964. Ecology and evolution of passerine birds in the island of the Aegean Sea. Ph.D. dissertation, Yale University, New Haven, CT.

Whittaker, R. J., R. Field, and T. Partomihardjo. 2000. How to go extinct: Lessons from the lost plants of Krakatau. *Journal of Biogeography* 27:1049–64.

Wilson, E. O. 1969. The species equilibrium. In *Diversity and Stability in Ecological Systems*, 38–47. Brookhaven Symposia in Biology no. 22. Upton, NY: Brookhaven National Laboratory.

Wilson, E. O., and D. S. Simberloff. 1969. Experimental zoogeography of islands: Defaunation and monitoring techniques. *Ecology* 50:267–78.

Wright, S. J. 1985. How isolation affects turnover of species. *Oikos* 44:331–40.

A General Dynamic Theory of Oceanic Island Biogeography: Extending the MacArthur-Wilson Theory to Accommodate the Rise and Fall of Volcanic Islands

Robert J. Whittaker, Kostas A. Triantis, and Richard J. Ladle

> *A theory attempts to identify the factors that determine a class of phenomena and to state the permissible relationships among the factors . . . substituting one theory for many facts. A good theory points to possible factors and relationships in the real world that would otherwise remain hidden and thus stimulates new forms of empirical research. . . . If it can also account for, say, 85% of the variation in some phenomenon of interest, it will have served its purpose well.*
>
> —*MacArthur and Wilson (1967, p. 5)*

MacArthur and Wilson's (1963, 1967) dynamic equilibrium theory of island biogeography has a clear claim to be the most influential body of theory within ecological biogeography. Central to its continuing influence, their model invokes fundamental dynamic processes operating on populations, in order to explain key emergent patterns of system species richness, turnover, and endemism. As they envisaged, their theory has found application (with varying success) to all types of insular system, from microcosms to oceanic islands, and from ponds to habitat islands of woodland in "seas" of human-transformed habitat (Whittaker and Fernández-Palacios 2007).

The aim embodied in the 1967 monograph was to promote a research agenda for island biogeography in which the particularities of historical narratives were set aside in the search for the general mechanisms, laws, and rules and their emergent outcomes, beginning at the population level. Within the better-known opening chapters, it can be considered a largely macroecological approach (*sensu* Brown 1995), whereas the later chapters develop the accompanying evolutionary theory concerning, for example, species radiation and the taxon cycle. There have been numerous attempts to link evolutionary and ecological dynamics building on the MacArthur-Wilson model (e.g., Wilson 1969, Diamond 1975, Heaney 1986, 2000, Peck 1990, Cowie 1995, Peck et al. 1999, Losos and Schluter 2000, Price 2004, Emerson and Kolm 2005a,b, Heaney et al. 2005), notwithstanding which, the model has been less successful and is arguably less complete when applied to oceanic island systems operating on evolutionary time scales than when applied to "ecological islands" (e.g., Haila 1990, Paulay 1994, Cowie 1995, Stuessy et al. 1998, Borges and Brown 1999, Heaney 2000, Whittaker and Fernández-Palacios 2007, Gillespie and Baldwin, this volume, Losos and Parent, this volume).

Recently, Heaney (2007) has called for the development of a comprehensive new model of oceanic island biogeography, reunifying ecological and evolutionary biogeography. Such a model should be based on the identification of general patterns, describe these patterns quantitatively, and capture the underlying mechanisms (Brown and Lomolino 2000). Here, we sketch out an extension to the MacArthur-Wilson dynamic model that combines their reasoning with a simplified model of the ontogeny of oceanic islands to derive a general dynamic theory for the biogeography of oceanic islands.

The MacArthur-Wilson Dynamic within the "Radiation Zone"

The MacArthur-Wilson model recognizes that, for a discrete and isolated biological system, the number of species at any point in time must be a function of the number previously occurring there plus those gained through immigration and/or speciation (specifically via cladogenesis[1]), minus those having gone locally extinct. Their theory proposes that these three fundamental processes should vary in a predictable fashion in response to time since system initiation, and in relation to two principal controlling geographical/environmental influences: isolation and area. Immigration rate (I, species immigrating to the island per unit time) should decline as a function of isolation (distance), and extinction rate (E, species being lost from the island per unit time) should decline as a function of increasing area (a general surrogate for island carrying capacity, K). Taking the case of a newly formed and barren island, I starts at its highest rate and declines as a hollow exponential curve as the proportion

[1]Anagenesis (the evolutionary change from a colonist species to a neo-endemic form) does not lead to an increased number of species on an island (although it does increase endemism). Thus, as they were primarily concerned with understanding variation in species richness, MacArthur and Wilson (1963, 1967) focused on evolutionary change giving rise to increased richness, i.e., cladogenesis (*sensu* Stuessy et al. 1990) when outlining their dynamic equilibrium model.

of species propagules arriving on an island that represent new species declines, while *E* gradually rises as the resource space is occupied. Expressed per unit time, *I* is shown as forming a concave falling curve, with *E* forming a convex rising curve (MacArthur and Wilson 1967, figure 20) and, in time, these rates intersect to provide a dynamic equilibrium, a condition at which *I* and *E* are in balance, with a continual turnover (*T*) of species occurring thereafter.

MacArthur and Wilson (1963, 1967) recognized that, on the more remote islands, the pace of immigration is sufficiently slow that increasing proportions of the biota on such islands are the result of *in situ* evolutionary change, with species gain via speciation (again, in this context they were mostly focused on net expansion of lineages), most pronounced on larger islands towards the outer limits of the distributional reach of a taxon: which they denoted the "radiation zone." Hence, they argued that species gain through *in situ* speciation increased with island/archipelago remoteness and with island area.

The Implications of the Limited Life Span of Oceanic Islands

The simplest classification of types of islands found within seas and oceans divides them into three classes: continental shelf islands (many of which have been joined to continents at Pleistocene sea-level minima, i.e. they are land-bridge islands), continental fragments (ancient continental islands), and oceanic islands (Wallace 1902, Whittaker and Fernández-Palacios 2007). Our focus herein is on the last of these groups, the true oceanic islands. They are formed in varied tectonic circumstances but are largely volcanic in origin, building from the oceanic crust to form land masses isolated from mainland source pools by open stretches of ocean. While those formed in arcs associated with subduction zones can be renewed over extended periods of tectonism, many remote oceanic islands (e.g., in hot-spot archipelagoes, fracture zones, etc.) are formed by volcanic activity of limited duration, and once formed experience subsidence and erosion, resulting in their eventual demise, or persistence in tropical waters only as low-lying atolls, sustained by coral growth. Thus, with some well-known exceptions, remote islands forming volcanically over oceanic crust are typically short-lived. The significance of the island life cycle of these oceanic islands has been recognized by a number of authors (e.g., Paulay 1994, Stuessy et al. 1998, Stuessy 2007), most presciently by Peck (1990, p. 375), who wrote that "A relationship [of numbers of eyeless terrestrial cryptozoans] with island age should be expected, but it would not be a straight line. . . . Rather the relationship should be a curve which rises fast at first, reaches a peak or plateau, and then decreases as erosion destroys the island."

In two recent papers, we have developed this line of reasoning more fully, suggesting that common elements in the ontogeny of oceanic islands should produce common emergent trends in diversity (Whittaker et al. 2007, 2008). Similar to the simple core model at the heart of the MacArthur-Wilson island theory, which focused principally on species richness (a metric indicative of "ecological dynamics"), we focus first on some simple metrics of "evolutionary dynamics," in particular on numbers and proportions of species restricted to single islands (i.e., single-island endemics, SIEs). SIE data arguably provide only crude metrics but have been used in a number of recent studies as indicators of evolutionary dynamics (e.g., Peck et al. 1999, Emerson and Kolm 2005a, Triantis et al. 2008).

It should be noted that we use the term "evolutionary dynamics" in a broad sense, to encompass biotic and abiotic processes occurring over evolutionary time scales that determine emergent outcomes of species numbers, endemism, and phylogeography. While there is evidence indicating long-term persistence of many island biotas in the absence of catastrophic disturbance, erosion, and subsidence (i.e., where islands are fairly stable and persistent) (Ricklefs and Bermingham 2002), it is not possible to assume when examining the phylogeny of an island clade that all species that have formed within an island or archipelago have persisted to the present day. Hence, estimates of evolutionary rates available in the literature should be regarded as diversification rates, i.e., meaning rates of speciation minus extinction. This recognizes that within radiating archipelagic lineages some species may have formed and long ago gone extinct, something that must, for example, have happened repeatedly during the 32 Ma history (Price and Clague 2002) of the Hawaiian Chain. So, whereas in the model developed herein we invoke trends in speciation and extinction rates through time, in practice, when it comes to evaluating the phylogenetic evidence, we have to accept that even when looking at apparent evidence of speciation rates on young islands like Hawaii (the Big Island), it is more proper to consider them *diversification* rates (i.e., $S-E$). Moreover, when examining numbers of single-island endemics, each of speciation, extinction, and interisland colonization has a role. The limitations of using SIEs as metrics are discussed below.

The Premises and Properties of the General Dynamic Theory

Premises of the General Dynamic Theory

The general dynamic model (GDM) rests on three key premises as stated in table 4.1. The first two premises derive directly from MacArthur and Wilson (1967), and encapsulate both (1) their immigration/speciation-extinction dynamics, and (2) the argument that speciation

Table 4.1
The Three Premises Underlying the General Dynamic Model of Oceanic Island Biogeography

Premise	*Support for the premise*
Biological processes:	
The MacArthur-Wilson model is an essentially correct summation of the key biological processes, i.e., island biotas are a function of rates of immigration, extinction and speciation, which lead toward a biotic equilibrium broadly as they envisaged.	A large body of literature supports the importance of these processes, but evidence of attainment of equilibrium for distant oceanic archipelagoes remains equivocal as progress toward equilibrium is very slow (e.g., Cowie 1995, Whittaker and Fernández-Palacios 2007).
Evolutionary response:	
Diversification within island lineages is typically greatest on larger islands that are remote (i.e. where interactions with closely-related fellow colonists is least) and where lineage persistence for non-trivial periods of time is permitted.	1. Island systems near the effective dispersal limits of a higher taxon, where few lineages colonize, typically show the greatest diversification per colonist lineage (the "radiation zone" of MacArthur and Wilson, 1967). 2. Within oceanic island archipelagos, single island endemics (SIEs) have a far larger minimum area threshold and increase disproportionately with increasing area relative to native species of the taxon (Peck et al. 1999, Triantis et al. 2008).
Geological progression:	
Oceanic islands are formed volcanically and typically have short life spans; in the simplest scenarios an island builds relatively speedily to maximum area and altitudinal range in its youth, next becomes increasingly dissected as it erodes, and then gradually subsides/erodes to disappear back into the sea or persist as a low-lying atoll.	Geological dating of oceanic islands indicates much support for this, especially for the Hawaiian hot-spot chain of islands (Price and Clague 2002), although not all volcanic islands follow such a simple developmental sequence (reviewed in Whittaker and Fernández-Palacios 2007).

Source: After Whittaker et al. 2008.

and diversification in insular habitats are encouraged through the ecological opportunity signified by the concept of "empty niche space," intertwined with the geographical opportunity provided by isolation (e.g., Lack 1947, Peck et al. 1999, Heaney 2000, Gillespie, 2004, Levin 2004). The final premise recognizes (3) that oceanic islands have a typical developmental life cycle from youth, to maturity, to old age and eventual loss (e.g., Nunn 1994, Price and Clague, 2002), and, crucially, that this life cycle plays itself out at a temporal scale resonant with and strongly influencing the evolutionary dynamics shaping the biota of oceanic island archipelagoes and basins (Peck 1990, Peck et al. 1999, Price and Clague 2002, Stuessy et al. 2005, Whittaker and Fernández-Palacios 2007).

Properties of the General Dynamic Theory

In this section we set out the general properties of the GDM through a series of graphical representations, inspired by MacArthur and Wilson's (1963, 1967) familiar dynamic model. We begin with Heaney's (2000) representation of the radiation zone concept, figure 4.1, showing how, for a given taxon, declining frequency of colonization translates into decreasing richness combined with increased absolute and relative importance of *in situ* cladogenesis.

In figure 4.2 we set out a general representation of the life history of an oceanic island, assuming the simplest of oceanic island histories, from initial appearance as a new volcanic island, building to a high cone-shaped form, of maximal area and height, and then becoming increasingly dissected and eroded. In time, such islands typically both subside (some rapidly and substantially, e.g., Moore and Clague 1992) and erode (aerially and through marine action), resulting in loss of both elevational range and area, until they disappear back into the sea, or persist in tropical seas as atolls—coralline islands of low elevation (Nunn 1994, Stuessy et al. 1998, Price and Clague 2002). Maximum topographic complexity will typically occur some time after the maximal elevation and area have been reached and passed.

In reality, most oceanic islands have rather more complicated histories than depicted, sometimes involving separate islands fusing to become one, and often involving catastrophic episodes of volcanism (tailing off with age) and slope failures (sometimes massive) (Price and Clague 2002, Whelan and Kelletat 2003, Le Friant et al. 2004); while Pleistocene climate change and sea-level fluctuations have also left detectable imprints on their biogeography (e.g., Peck 1990, Price and Elliott-Fisk 2004, Carine 2005). Furthermore, those oceanic islands that have formed within island arcs in association with plate margins can experience yet more

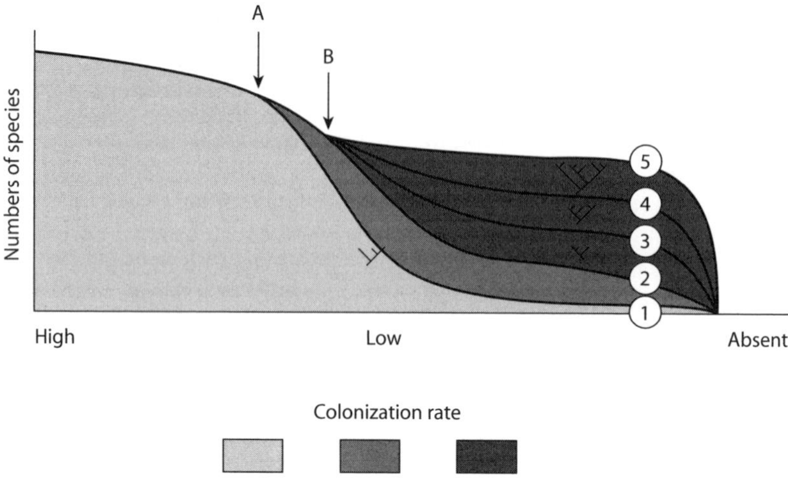

Figure 4.1. L. R. Heaney's (2000) model of the development of species richness on large islands or archipelagos that experience varying rates of colonization due to varying degrees of isolation. According to Heaney, on islands near a species-rich source, high rates of gene flow will inhibit speciation. As the average rate of gene flow drops below approximately the level of one individual per generation, anagenesis will begin to take place and endemic species will develop. These endemic species (between lines 1 and 2) will have their sister-taxon in the source area, not on the island/archipelago. As colonization becomes still less frequent, and as time passes, phylogenesis will produce endemic clades diversified within the island/archipelago (species between lines 2 and 3). Over time, the oldest clades will become progressively more species rich (between lines 3 and 5).

complex histories, involving both vertical and lateral displacement (e.g., Buskirk 1985, Keast and Miller 1996) and can be more persistent than assumed herein (Paulay 1994). Hence, the simplified ontogenetic argument presented here is most applicable to hot-spot archipelagoes, and while it should, in principle, apply to other volcanic oceanic island archipelagoes, some modification will be necessary to accommodate alternative and more complex geological scenarios.

Considering the simplified scenario in figure 4.2, the model implies that (1) the maximum carrying capacity K of an island, in terms of biomass and number of individuals across all species, will be reached roughly coincident with maximum area and elevational range (figure 4.3a), with (2) the maximum heterogeneity of environment, and thus maximum opportunity for within-island allopatry, occurring somewhat later, but still within the "middle age" of the island (figure 4.4).

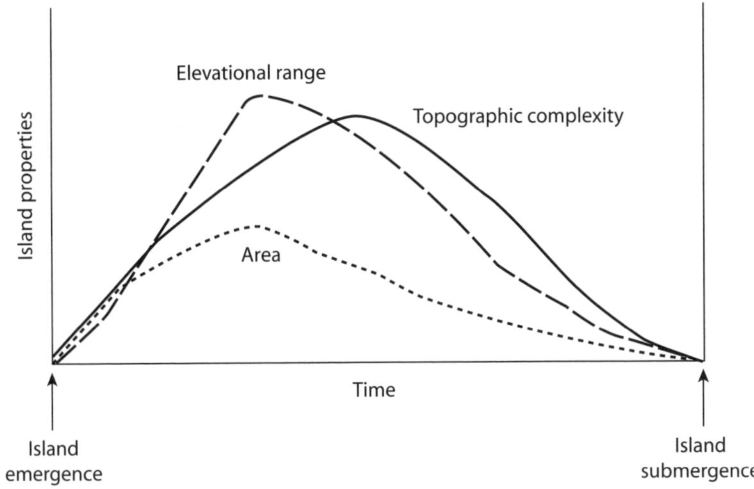

Figure 4.2. Idealized relationships between the age (x-axis, time) and area (dotted line), elevational range (dashed line), and topographic complexity (solid line), of a hypothetical oceanic island. Island maximum altitude and area both peak before maximum topographic complexity, but all three are expected to show a humped pattern. As the period of growth is typically shorter than the period of decline, time may best be considered a logarithmic function. From Whittaker et al. (2008).

Implications and Predictions of the General Dynamic Theory

These arguments allow us to extend the MacArthur-Wilson theory to incorporate the implications of both an extended preequilibrium phase and an extended postequilibrium phase where K is declining and $E>(I+S)$. Figures 4.3a and 4.3b combine these arguments to provide a graphical model of the dynamic processes involved in the developmental cycle of an island within an oceanic archipelago. The period from island emergence to maximal carrying capacity is typically far shorter than the period of decline (consider, e.g., Stuessy et al. 1998, Carracedo and Tilling 2003, Le Friant et al. 2004), such that the time axis should be represented as some form of logarithmic or power function.

With regard to evolutionary dynamics, the key propositions in relation to the generalized life cycle of an island are:

1. in *youth,* initially most species can be attributed directly to immigration, typically from older islands in the archipelago;
2. during *immaturity,* speciation rates (and rates of cladogenesis) peak relatively early on, when there are enough lineages present to "seed" the process (see

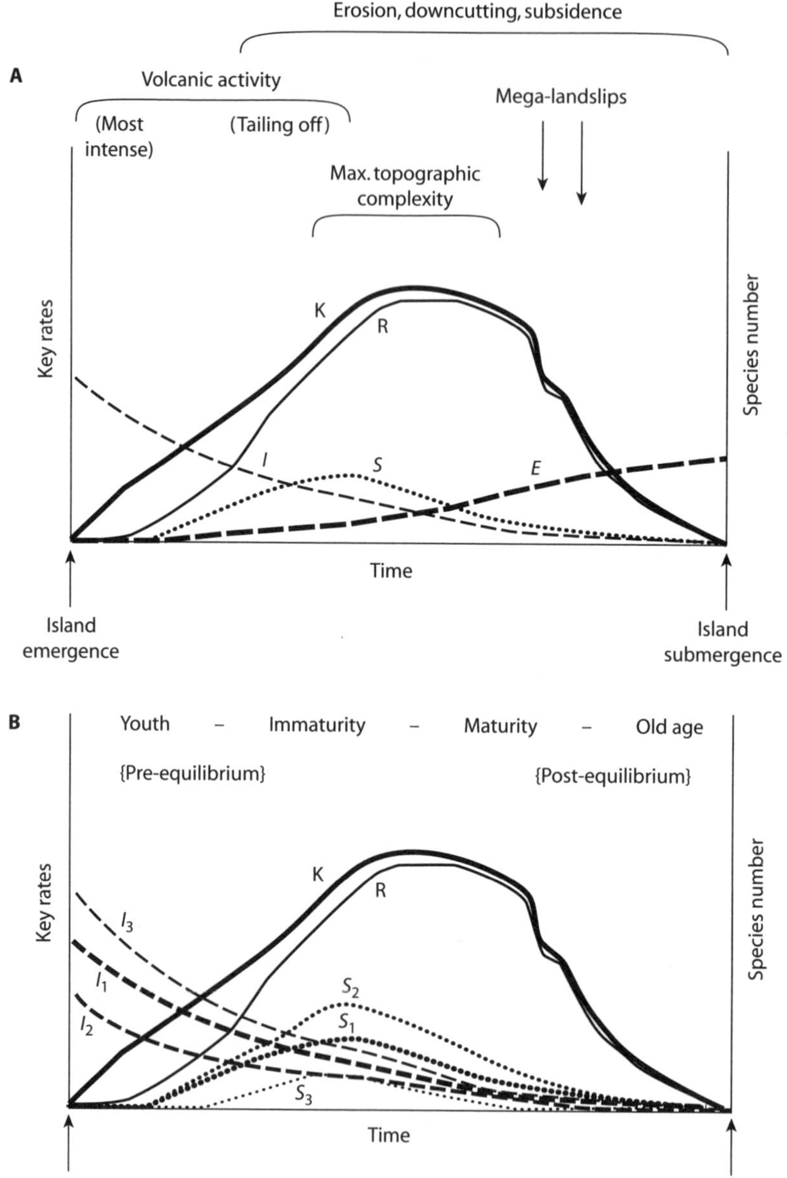

Figure 4.3. Graphical representation of the key rates and properties of the general dynamic model (GDM) of oceanic island biogeography. Island building being typically much more rapid than decline, time should be considered as some form

Percy et al. 2008), but when there is also plenty of adaptive opportunity in the form of empty niche space;
3. in *maturity*, species richness peaks, while speciation continues to add new species, partly due to the increasingly dissected topography, which generates increased opportunities for within-island allopatry;
4. in *old age* speciation declines to a low relative and absolute rate in tandem with reduced *K* and increased *E* (and thus reduced richness) as islands decline in elevation, topographic relief, area, and habitat diversity in old age; and
5. *finally*, all is lost, the island founders.

It is worth noting that the form of the *I, E,* and *S* curves for a series of islands should be expected to vary in relation to not only the usually considered parameters of area and isolation of islands, but also the temporal resolution of the analysis. This is probably of greatest significance when considering the early phase of island emergence and biotic colonization. An illustration of this comes from empirical analyses of the recolonization of the Krakatau Islands following their sterilization by volcanism in 1883, and of the interrupted colonization of the emerging island of Anak Krakatau from the 1930s onward (Bush and Whittaker 1991, Thornton 1996, Whittaker and Fernández-Palacios 2007). These studies found departures from MacArthur and Wilson's (1967) smoothly falling *I* and smoothly rising *E* rates for several taxa, as a result of factors such as (a) initially hostile environmental conditions preventing widespread colonization early on, (b) accelerated phases of colonization as successional thresholds were passed (the formation of the first woodlands, etc.), (c) episodes of extinction linked to (b), and (d) bursts of extinction and immigration linked to further disruptive volcanic activity.

---

of log function, as also the case for figures 4.2 and 4.4. A. Showing the postulated relationships between the biological characteristics and the ontogeny of a single island, where, for key rates: *I* is immigration rate, *S* is speciation rate, and *E* is extinction rate (each rate referring to number of species per unit time); and for species number: *K* is the potential carrying capacity, and *R* is realized species richness. For islands showing sudden extensive loss of territory due to landslips (as suggested by the kinks in the *K* and *R* curves) the extinction rate curve would require modification. B. Modification of *I* and *S* curves in relation to distance between islands or mobility of the taxa concerned. The amplitude of the *S* curve will vary between archipelagoes and major taxa as a function of the size of the available species pool/ease of dispersal. This variation in accessibility is signified by the variation between *I*1, *I*2, and *I*3 curves, corresponding respectively to *S*1, *S*2, and *S*3 curves. Note that a suite of modified *R* curves should also be shown, to match the variations in the balance of rates of immigration, speciation, and extinction, but have been omitted to reduce clutter. From Whittaker et al. (2008).

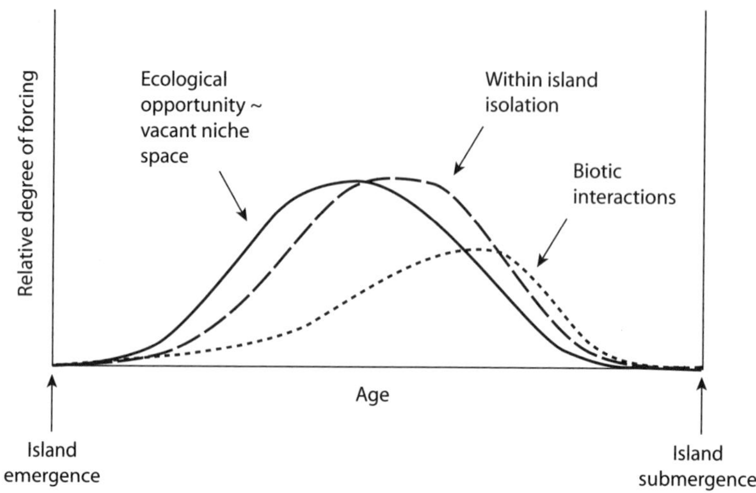

Figure 4.4. Schematic representation of relative roles of different forcing factors through the life cycle of the island. Considering figures 4.2 and 4.3, we can derive the prediction that the greatest opportunities for adaptive radiation (solid line, first peak) will occur earlier than those for non-adaptive processes linked to within-island isolation (dashed line, second peak). Biotic interactions within and across trophic levels may be expected to become more important in the later stage of the island life cycle (dotted line, third peak), past the point of maximum carrying capacity and where extinction rate is climbing with island erosion/ subsidence. Such biotic/competitive mechanisms may produce species involved in tight mutualisms, or fine subdivisions of resources sympatrically, but not at a rate sufficient to prevent the eventual decline in the proportion of SIEs. From Whittaker et al. (2008).

Such complexities are evident over time periods of years up to several decades. However, as we are here concerned with systems running over several million years, we can think in terms of a temporal resolution of analysis of hundreds to thousands of years, in which Krakatau-like successional dynamics will be largely undetectable. Hence, our model shows smoothly falling I and rising E rates essentially from time zero, ignoring the likelihood that when analyzed at a very fine temporal resolution we might expect to see a more complex early development pattern.

Although true oceanic islands arise in varied geological circumstances, they are frequently clustered together in space, forming distinct archipelagoes within which the timing of formation of each island varies significantly (e.g., Nunn 1994, Carracedo and Tilling 2003). Thus, as each island goes through its own life cycle, an archipelago develops in which a wide array of island ages/stages is available at any single time. Hence, a young island is supplied by colonists from nearby older islands, and in time supplies colonists to the next island(s) to form. Therefore, archipelagoes such as the Canaries or Hawaii can be conceived of as consisting of a series of terrestrial platforms each going through the sequences shown in figures 4.2–4.4, but each at a different point along the time axis.

Considering a single island forming within an existing archipelago, developing to maximum size, and elevational range, then becoming increasingly dissected through erosion, and finally entering a long phase of decline in area, elevation, and environmental complexity, we expect a general hump-shaped trend in potential carrying capacity (K) and similar trends in species richness (R), and in speciation rate (S) (figure 4.3). Extinction of species can occur at any stage, but will be driven by differing processes at different stages of an island's life cycle. During the building/maturity phase, high-magnitude catastrophes (large volcanic eruptions, mega-landslides) will be more important—if highly unpredictable—while the more gradual erosion and subsidence processes associated with older islands will eventually force the background extinction rate to rise above the combined processes of addition (speciation and immigration), inexorably driving species number toward zero for islands that founder beneath the waves, completing the cycle.

We may also derive a general prediction (table 4.2) for the trend in the proportion of single-island endemic species (pSPIE) during the ontogeny of a particular focal island. Initially, as the island ecosystems are seeded (colonized through successional processes) from the nearby older islands in the archipelago, most species are not SIEs, although they may well include archipelago-level endemics, so the pSIE will be low. However, as the available propagule pool is relatively limited, and ecological space is initially unsaturated, speciation rate picks up, often generating significant radiations within single genera (e.g., Gillespie and Baldwin, this volume), thus increasing the proportions of SIEs and simultaneously generating an increased species-to-genus ratio. As the process continues, some part of this diversification process may be attributable to the arrival of "keystone species" such as *Metrosideros* in the Hawaiian system, providing stimulus to diversification in interacting animal lineages (Percy et al. 2008, and cf. Emerson and Kolm 2005). However, as the island ages and declines, it follows that a point is reached at which $E>(I+S)$, and so species richness and the number of SIEs (nSIE) will each decline.

A further prediction follows, that the proportion of SIEs on our focal island should also decline, for the following reasons: (1) the area threshold for SIEs is on average larger than for non-SIE native species (Triantis et al. 2008), partly as the latter may persist even as fairly small populations if reinforced by occasional propagule flow from other islands;

TABLE 4.2
Predictions Derivable from the General Dynamic Model

1. Island species number and the number of SIEs should be a humped function of island age and, when examining snapshot data across an archipelago, this will be combined with a positive linear relationship with area.
2. The amplitudes of the curves shown in figure 4.3a should vary in relation to the size of the island at maturity, with higher peak richness and SIE numbers on islands that attain greatest size (area and elevation) at maturity.
3. The relative amplitudes of the immigration and speciation rate curves should vary in relation to the effective isolation of islands, i.e., in relation either to distance between islands and their sources or to the mobility of the taxon, as shown in figure 4.3b.
4. Lineage radiation (leading to multiple SIEs on individual islands) should be most prevalent after the initial colonization phase, in the period leading up to island maturity, coinciding with maximal carrying capacity (K) and the development of maximal topographic complexity.
5. Montane representatives on old, declining islands should gradually be lost because of loss of habitat, meaning that surviving montane forms are increasingly likely to be relatively old (i.e., basal) forms in relation to other members of an archipelagic radiation.
6. The proportion of SIE should also be a humped function of island age, as islands that decline to small size and carrying capacity should lose SIEs in accordance with the second premise of the GDM (and see also: prediction 8).
7. SIE per genus should be higher on younger islands; intermediate-aged islands will have more lineages showing speciation than do young or old islands; SIE per genus should decline on older islands so that as islands lose SIE, there is a tendency towards monotypic genera, preserving maximal ecological spacing in the remaining endemics.
8. As islands age, some of their SIE species should colonise a younger island, so that they become multi-island species instead. Hence, the GDM also predicts that the progression rule should be a common/dominant phylogeographical pattern within an archipelago.
9. Using Stuessy et al.'s (1990, 2006) approach to classifying speciation modes, there should be a tendency on old, submerging islands for anagenesis to be an increasingly prominent speciation signal. Note: This assumes that where SIEs are the only member of their genus the explanation is in situ speciation. In practice we expect that on the oldest islands "anagenesis" will often be a misnomer, as there will be a trend towards survival of single relicts from former radiations.

(*Continued*)

10. Adaptive radiation will be the dominant process on islands where the maximum elevational range occurs, as it generates greatest richness of habitats (major ecosystem types), including novel ones few colonists have experienced. Nonadaptive radiation will become relatively more important on slightly older islands, past their peak elevation, due to increased topographical complexity promoting intra-island allopatry (figure 4.4). Similarly, composite islands (e.g. Tenerife, formed from three precursors), should have provided more opportunity than islands of simpler history for within-island allopatry, producing sister-species that lack clear adaptive separation (e.g., Gruner 2007).

Source: From Whittaker et al. 2008.

(2) the loss of habitat diversity (e.g., upland habitats, lava tubes [Borges and Hortal 2009]), and corresponding increase in habitat similarity with the coastal lowlands of other islands in the group, results in the collapse of radiations of neo-endemics (including many habitat specialists) on the focal island, while widespread coastal generalists would be anticipated to persist best; and (3) as the focal island supplies colonists to the next island to form, some of the SIE species of the focal island colonize the new island (in accordance with the progression rule [Funk and Wagner 1995]) and lose their status as SIEs. This last mechanism will apply most strongly in hot-spot archipelagoes involving a clear age progression; it may not be so evident in more complex island arc systems, and would not be anticipated at all in, e.g., poorly dispersing sightless troglodytes.

The GDM thus allows us to derive several predictions (table 4.2) about the emergent properties of the biota: (a) of a single oceanic island through time; and (b) of the islands of an oceanic archipelago at a single point in time. Given the extended time period (millions of years) over which data would ideally be required to fully explore the generality of the assumptions and predictions, we have to make use principally of predictions about temporal "snapshot" patterns in order to assess support for the GDM. This requires the selection of oceanic archipelagoes in which a meaningful portion of the life cycle shown ultimately by a single island is available for study in the form of separate islands of widely different age/stage. The key problem in doing this is that the islands within an archipelago do not all attain identical properties at maturity, and in particular they may vary significantly in maximum attained area and elevational range: properties of key importance (table 4.1, figure 4.3a). To deal with this analytically we need to include a term for island size, assuming that all islands within a group follow the same general trajectory, but that the amplitude of the curves will vary in relation to the maximum area attained.

Evaluation

Macroecological Analysis of Diversity Data

The postulated humped trends of particular diversity attributes/metrics in relation to island age (table 4.2) constitute a particularly distinguishing and testable feature of the GDM. Whittaker et al. (2008) therefore began the empirical evaluation of the GDM by using data from five oceanic island archipelagoes (the Canaries, Galápagos, Marquesas, Azores, and the Hawaiian Islands) satisfying two criteria: (1) they provide a good span of island ages (maximum island ages were used in the analyses); and (2) fairly comprehensive survey work and compendia were available for particular taxa. Details of data sets, modeling approaches, and specific aspects of island histories, etc., are provided in Whittaker et al. (2008).

Tests of the GDM factoring in both island age and area take the form Diversity $(D) = a + b(\text{Time}) + c(\text{Time}^2) + d(\log\text{Area})$, where the use of a logarithmic function of area follows standard practice, empirically derived in numerous published analyses, and where the expectation is for positive exponents for Area and Time but a negative exponent for Time^2 to reflect a humped relationship between diversity and island age. We term these fitted regression models ATT^2 (i.e., Area+Time+Time^2) models to distinguish them from the theoretical GDM. These models were compared with the semilogarithmic and power models for island area (the most commonly favored in the literature), plus a semilogarithmic island age model and a parabolic age model (i.e., $D = b_1 + b_2 \times \text{Age} + b_3 \times \text{Age}^2$) to explore the fits derivable from area or age alone. The diversity metrics used were species richness (SR), number of SIEs (nSIE), proportion of SIEs (pSIE), and a simple diversification index (DI), which is the ratio of nSIE to the number of genera containing SIEs (where nSIE = 0, DI was also set to 0).

The ATT^2 models describing species richness were statistically significant for each of the fourteen data sets, with a mean R^2 value of 0.85 ± 0.08 (SD) and in each case the relationship with island age was humped in form (table 4.3). Similar findings pertained for each of thirteen tests for each of the three SIE-based metrics, which were again significant in all cases. The island age component was humped except in four cases, namely, nSIE and pSIE for Azorean snails, and pSIE and DI for Galápagos beetles. The ATT^2 model (with humped age relationship) provided the best model (based on adjusted R^2 values) in between eight and ten cases for each metric (table 4.3). The four alternative models are each simpler compared to the ATT^2 models, being two-parameter $(T + T^2)$ or one-parameter models. The two conventional area models each provide higher adjusted R^2 values than the ATT^2 model for between one and four cases (depending on the metric used) but, unlike this model, neither pro-

TABLE 4.3
Summary of Tests of the General Dynamic Model Using Diversity Metrics from Five Archipelagoes

Island group	*Taxon*	*No. of islands*	*% endemism*	*% SIE*	*SR*	*nSIE*	*pSIE*	*DI*
Canary	Arthropods	7	40%	22%	0.93**	0.88**	0.82**	0.77**
Canary	Plants	7	40%	15%	0.91**	0.90**	0.90**	0.99**
Canary	Snails	7	91%	84%	0.87**	0.84**	0.88**	0.90**
Hawaii	Arthropods	10	99%	72%	0.83	0.74	0.71**	0.90*
Hawaii	Coleoptera	10	99%	83%	0.84*	0.77	0.93**	0.93**
Hawaii	Flowering plants	10	90%	54%	0.94**	0.83**	0.73**	0.79**
Hawaii	Snails	10	99%	88%	0.67	0.61	0.96**	0.74
Galapagos	Insects	13	66%	29%	0.80**	0.65**	0.55**	0.52**
Galapagos	Insects (small orders)	13	62%	30%	0.76**	0.48**	0.28	0.34**
Galapagos	Beetles	13	70%	28%	0.82**	0.73**	0.70**†	0.47
Galapagos	Plants	13	30%	5%	0.84**	0.80**	0.73**	0.81**
Marquesas	Plants	10	46%	23%	0.95**	0.63	0.68	0.85**
Azores	Plants	9	7.2%	<1%	0.83*	—	—	—
Azores	Snails	9	51%	31%	0.90**	0.90†	0.94*†	0.66

Source: Compiled from Whittaker et al. 2008.

Notes: The table shows number of islands considered, the overall proportion of endemism in the archipelago, the percentage of single island endemics (SIEs) and the unadjusted R^2 values for the ATT^2 model Diversity = $a + b$(Time) + c(Time^2) + d(logArea). Diversity metrics: SR = species richness of native species, nSIE = number of SIE, pSIE = proportion of SIE, DI = a simple diversification index [the ratio of nSIE to the number of genera containing SIEs (where nSIE = 0, DI was also set to 0)]. All regression models were significant at $P < 0.05$. Asterisks indicate model performance of the ATT^2 model compared with the following alternative models: the semilogarithmic area model, the power model, a semilogaeithmic time model, and a parabolic time model, using adjusted R^2 values, which penalize more complex models in comparison to simpler ones.** indicates that the ATT^2 model was the best model (highest adjusted R^2 value),* indicates cases where the ATT^2 model had equivalent adjusted R^2 values (+/–0.2), and no asterisk indicates that one of the alternative models had higher explanatory power. † indicates a humped time relationship was not observed.

vides significant fits to all data sets, with nonsignificant fits most evident for the three Canarian taxa (i.e., standard species-area models are inadequate in this archipelago; see also Triantis et al. 2008). The time-only models generally performed poorly in comparison to the ATT^2 models, with one exception, the Azorean snail data, for which, contrary to the expectations of the GDM, the relationship with time is not humped. This particular result can be accounted for within the GDM reasoning if it is accepted that the maximum geological age for some islands differs substantially from the effective age of the island in biological terms; although some might consider this special pleading (see details in Whittaker et al. [2008] and see analyses for other Azorean groups by Borges and Hortal [2009]). In summary, the analyses demonstrate that the ATT^2 model provides a generally good fit with data from a range of plant and invertebrate taxa from five oceanic island archipelagoes, both for numbers of native species (SR) and for metrics more directly indicative of evolutionary dynamics (nSIE, pSIE, DI). It is worth emphasizing that in the majority of the cases studied the relationship between the diversity metrics used and island age, when included in a model with island area, was hump shaped, despite the fact that the modeling approach did not impose such a relationship (see table 4.3).

The effectiveness of the ATT^2 model in fitting data for particular archipelagoes and taxa is expected to depend on the effective isolation of the archipelago (figure 4.3b) and on the extent to which the archipelago provides a full range of island developmental stages. For example, for archipelagoes providing only young (and/or rejuvenated) islands, it would be consistent with the GDM for a simpler "log(area)+linear time" model to provide a better fit than the full ATT^2 model (Borges and Hortal 2009). However, across the data sets evaluated, comparison with the alternative models provides confirmation that the ATT^2 model, while not the simplest model (and not necessary in all cases), has greater generality than the traditional diversity-area models, or time-only models.

There are a variety of limitations to these tests: (1) the biological data are undoubtedly incomplete, (2) the islands have had more complex histories of formation than we assume, and (3) Pleistocene sea-level fluctuations have altered island areas and repeatedly joined and divided some islands. In addition, it is important to recognize that species may acquire and lose SIEs in several ways, e.g., (1) some current SIE species may have originated on another island (or land mass), from which they subsequently became extinct; (2) some species that evolved *in situ* as an SIE may have gone extinct and so are not around to be counted; (3) some former SIE species may have colonized another island(s) to become multi-island endemics (MIEs); (4) some MIEs occur on islands that were formerly connected at times of lowered sea level, indicating that their current disjunct distribution may derive from localized vicariance. Hence, we emphasize that the three metrics based on SIE data should be regarded as evolutionary dynamics metrics rather than either diversification or speciation indices. Nonetheless, we hold that *in situ* speciation will typically be the main driver of change in each of the three evolutionary metrics (nSIE, pSIE, DI) in the lengthy period leading up to the establishment of a dynamic evolutionary equilibrium (*sensu* Wilson 1969), whereas within-archipelago migration and within-island extinction become more important influences on numbers and proportions of SIEs during the even longer period of island "senescence."

General Evaluation of the GDM

> *As volcanism continually requires the founding of new local populations, genetic shifts and/or other episodic evolutionary change would be expected to accelerate during the growth phase of each successive Hawaiian volcano. These influences, however, would decline as each volcano completes its active phase and becomes dormant. . . . We suggest that the youngest island at any one time has always been Hawaii's major evolutionary crucible.*
> —*Carson et al. (1990, p. 7057)*

It is intrinsically difficult to obtain evidence of changes in rates of the vital processes (i.e., migration/immigration, speciation, and extinction) through time and in relation to other island attributes (spacing, overall archipelago isolation, Quaternary climate change, etc.). This is especially the case for the biotas of remote oceanic islands, many of which can be accounted for by mean colonization rates of one species every few thousand years (e.g., Wagner and Funk 1995, Peck et al. 1999). Similarly, attributing evolutionary outcomes to nonadaptive versus adaptive processes (prediction 10, table 4.2) is challenging (but see Barrett 1996, Cameron et al. 1996, Price and Wagner 2004), suggesting that testing some of the predictions in table 4.2 will be rather difficult to accomplish. Hence, while the indices of evolutionary dynamics evaluated in table 4.3 are crude, we have followed other recent authors (e.g., Peck et al. 1999, Emerson and Kolm 2005a) in adopting the rationale that SIE data are a good starting point and are likely to be indicative of trends and patterns in other metrics of evolutionary dynamics. In support of this, tallies of data for the overall number of Canarian endemic plants across the seven main islands of the archipelago (reproduced in Whittaker and Fernández-Palacios 2007) show that, at least in this case, the pattern for the number of Canarian endemics is strongly correlated with the nSIE and again shows a humped relationship with island age.

Several of the predictions derived from the GDM (specifically 4, 5, 7–10: table 4.2) concern the mode (figure 4.4) and pattern of lineage

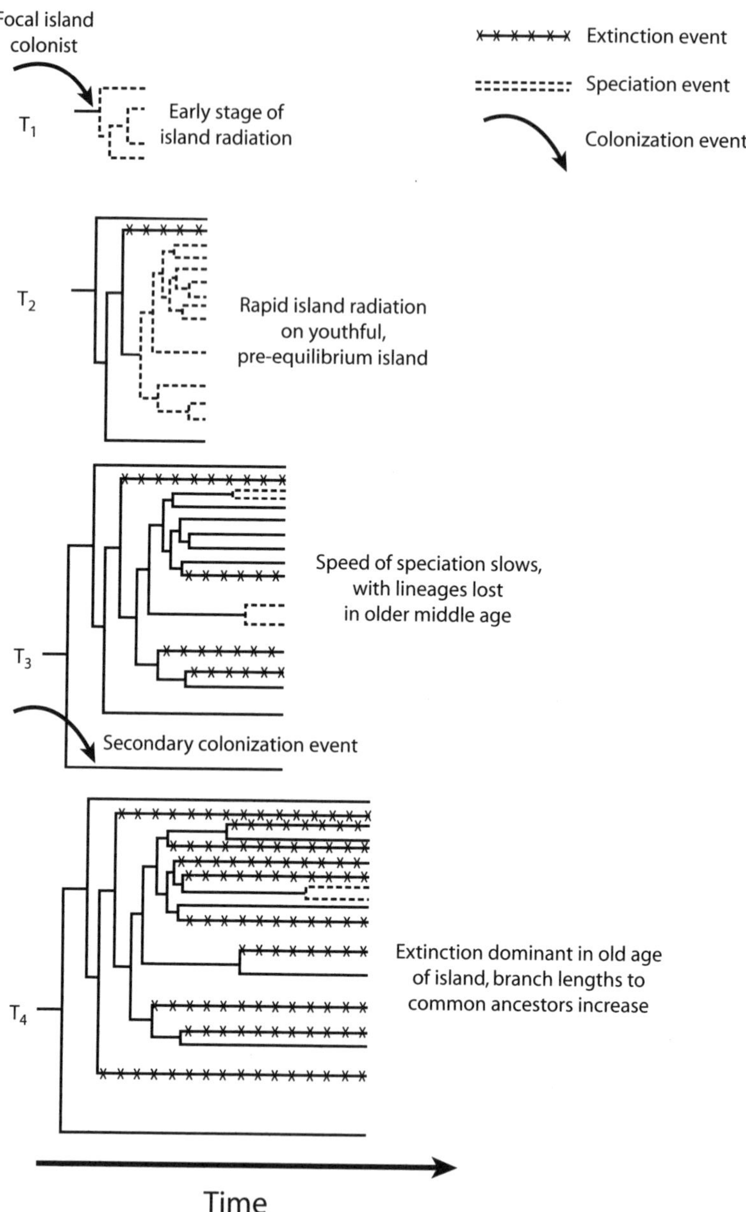

Figure 4.5. A hypothetical island lineage conforming to the general dynamic model, examined at four points in time (*T1* to *T4*). *T1*, a single colonization event to our focal island early in its life cycle leads to rapid onset of radiation, exploiting the relatively uncontested niche space. *T2*, during the period leading up to island maturity, a full array of habitats is available, and opportunities for within-

development. As an aid to visualizing the latter, figure 4.5 shows how a typical lineage might look at different points in time on an island progressing from youth to old age. Although we are not yet able to evaluate these model predictions systematically, we can begin to explore these aspects of the model with reference to existing literature from island systems.

Silvertown (2004) notes that large endemic taxa within the Canarian endemic flora are typically monophyletic (e.g., 63 species of Crassulaceae, and 37 species of *Echium*), i.e., they typically derive from single-colonization events. Silvertown suggests that this may be indicative of the operation of niche preemption by early-colonizing lineages that may have inhibited the success of later-arriving mainland relatives and also have spread out across the archipelago as new islands formed, frequently radiating into new habitats. These interpretations are broadly consistent with the GDM, and particularly the notion of greatest lineage radiation occurring on relatively young islands (e.g., Cowie 1995, Carine et al. 2004, Silvertown et al. 2005).

Turning to Hawaii, Gillespie and Baldwin (this volume) identify three basic categories of Hawaiian taxa in respect to speciation rapidity: (a) groups that diversify based on sexual selection speciate rapidly and in cases attain highest diversity very quickly on the youngest island (e.g., *Laupala* crickets and some *Drosophila*); (b) groups that predominantly diversify ecologically (many animal, some plant lineages) may reach their highest diversity after a somewhat longer period of time, on a youthful but not perhaps on the youngest island; and (c) groups that appear to have diversified mostly in allopatry (or in parapatry) (e.g., *Orsonwelles* spiders, many plant groups) show a progressive increase in species numbers with island age, implying that this mode of speciation tends to be rather slower and that equilibrium may not have been reached within the approximately five million year span provided between Hawaii and

---

island allopatry also gradually increase, with both circumstances encouraging speciation and diversification (many, short branches in the tree). *T*3, the frequency of formation of new species is expected to slow and to increasingly be balanced by losses as island erosion and subsidence reduce the available habitat space. With the passage of time, secondary colonization events from an older island following the progression rule, or sometimes backwards colonization events are possible. Thus, the clade is becoming more diverse and paraphyletic (ancestral) on our focal island compared to the next youngest island to form in the chain (or to *T*1 of the focal island). *T*4, speciation rate declines in tandem with reduced *K*, and extinction increasingly weeds out the tree, nevertheless, while the number of branches/species may be reduced the genetic diversity may remain high (compared to *T*1 or *T*2) due to the possession of older endemic lineages (longer branch lengths in a pruned tree).

Kauai. These findings were based on phylogenetic analyses of a range of taxa that were established on the Hawaiian Islands at the time that Kauai was the youngest island in the chain (if not earlier). They appear to be broadly supportive of a number of the predictions arising from the GDM (see table 4.2, predictions 4, 6, and 7) but at the same time highlight that the GDM is capable of further refinement.

Several other phylogenetic analyses also indicate that younger islands are particularly active arenas for genetic differentiation and speciation (although strictly the evidence is generally for diversification rates; see above) (e.g., Carson et al. 1990, Kaneshiro et al. 1995, Barrier et al. 2001, Levin 2004, Percy et al. 2008). On the Hawaiian Islands, Levin (2004) reports that the estimated "speciation rate" for plants is a negative function of island age, varying from 0.20 species per lineage per million years (Myr) on Kauai (5.7 million years old) to 2.1 species per lineage per million years on Hawaii (0.5 Myr). Studies from the flora of the Juan Fernández Islands also support the idea of high *initial* rates of radiation, with faster rates evident on the younger island (Levin 2004, and see Crawford et al. 1992).

We find additional support for the likelihood that relatively high speciation rates can account for "explosive early" patterns of lineage diversification in recent simulation modeling by Rabosky and Lovette (2008), in a paper providing a method for distinguishing the signal of speciation from extinction in molecular phylogenies. Further analyses of island radiations using this approach hold promise for the evaluation of the ideas presented herein. However, from the data currently available, it has to be allowed that apparently faster evolutionary rates on younger islands could, at least in cases, be the outcome of the effects of erosion and subsidence on older islands reducing the persistence of neoendemic lineages within the older islands (as in figure 4.5, and see Peck et al. 1999, Stuessy 2007). Such extinctions are always going to be hard to quantify from traditional forms of data as we are highly unlikely to find comprehensive fossil evidence for species lost as a result of island erosion and subsidence. There are, however, numerous cases where island phylogenies point to the past existence and extinction of ancestral species that once occurred on land areas that no longer exist, i.e., former uplands and lost islands (those now submerged) (e.g., Wagner and Funk 1995, Keast and Miller 1996, Price and Clague 2002, Butaud et al. 2005, Emerson and Oromí 2005, Pulvers and Colgan 2007), providing general exemplification of the point that island decline forces extinctions and in time a net reduction in diversity.

Phylogeographic analyses of island lineages provide further evidence of the processes of movement and evolution across archipelagoes. One commonly supported pattern involves taxa showing a pattern of movements from older to younger islands within an archipelago, with speciation occurring on newly colonized islands (see figure 4.5). This *progression rule* pattern (Funk and Wagner 1995) is particularly evident in archipelagoes showing a clear linear age sequence of islands, consistent with our general theory (table 4.2, prediction 8). Examples drawn from many that provide support for this rule include, from Hawaii, *Drosophila, Hesperomannia, Hibiscadelphus, Kokia, Orsonwelles, Remya, Metrosideros,* and *Tetragnatha*; from Macaronesia, *Olea, Gallotia, Gonopteryx, Hegeter, Pimellia,* and possibly *Dysdera*; from Galápagos, scarabs and weevils; and from the Austral Islands, *Misumenops rapaensis* (original references in Whittaker et al. 2008, and see Gillespie and Baldwin, this volume, Percy et al. 2008).

We acknowledge that various other phylogeographical patterns (or no resolvable pattern) have been detected from these and other oceanic archipelagoes. In some cases, e.g., Galápagos birds, evolutionary scenarios involve multiple phases of island hopping and of alternating periods of allopatry and sympatry within a single radiation (Lack 1947, Grant and Grant, 1996). Moreover, data for some lineages are most parsimoniously explained by a sequence of colonization in contradiction to the age sequence (e.g., Kvist et al. 2005, Sanmartín et al. 2008). So it should be understood that the progression rule is not without exceptions (see Funk and Wagner 1995, Gillespie and Roderick 2002). However, based on the GDM, it should be expected to be a dominant pattern, followed by many taxa in archipelagoes showing a strong island age sequence, and especially so in taxa which happen to colonize early in the developmental history of an archipelago, yet which also exhibit sufficient dispersal limitation to speciate within the islands of that archipelago.

We are under no illusions that the general dynamic model described herein provides a complete theory of oceanic island biogeography and evolution, but we do consider that it provides an analytically tractable framework that is largely consistent with the larger body of theoretical ideas we have discussed herein. Modification will be necessary for those classes of island that conform poorly to our ontogenetic model, including many island arc archipelagoes and islands of mixed continental/oceanic origins showing complex histories of horizontal and vertical movement, erosion, and rebuilding (e.g., Buskirk 1985, Keast and Miller 1996). For those oceanic islands that do conform to the simple ontogenetic model, perhaps one of the most important omissions from the framework is the role of Pleistocene climate change and accompanying variation in the configuration of islands (e.g., Nunn 1994, Carine 2005, Whittaker and Fernández-Palacios 2007, Ávila et al. 2008). Global environmental change in the Pleistocene altered not only the number, area, and elevational range of islands in these archipelagoes, but also their relationship with source pools. For instance, Carine (2005) argues that the evolutionary pattern in Macaronesian *Convolvulus* is suggestive of discrete waves of colonization, which he explains through the "colonization window" hypothesis. This postulates that colonization opportunities have varied through time as a function of both the geotectonic mechanisms discussed herein (island formation, island sterilization/disturbance) and periods of climate change. Thus, low sea-level stands during the Pleistocene saw the emergence of stepping-stone islands, aiding dispersal among the more persistent islands of Macaronesia, and between them and the mainland. Similar arguments have been invoked elsewhere, and the notions that dispersal distances and directionality of dispersal related to major current systems can change through time, provide additional components that require integration into a comprehensive general theory of oceanic island biogeography (Cook and Crisp 2005, Cowie and Holland 2006).

Conclusion

In this paper we have outlined a general dynamic theory for the biogeography of oceanic islands, which explicitly places MacArthur and Wilson's (1963, 1967) dynamic equilibrium model into the geological and evolutionary context of oceanic archipelagoes. The GDM is a deliberately simplified representation of diversity dynamics on oceanic islands. Our aim was to capture the few major factors that drive diversity patterns on oceanic islands of different sizes and ages, not to produce a precise predictive model. The main intended advantage of the GDM is not the better fit of the ATT^2 models (which are directly derived from the GDM), since other higher-order models can have this property too, but that it may offer an improved theoretical framework for describing and understanding the evolutionary biogeography of oceanic islands. We envisage that the GDM is capable of further theoretical and empirical development, for example (1) modification to incorporate alternative repeated geological scenarios, (2) tests of genetic/functional trait variation at subspecies level for multi-island native species/endemics, (3) extension to take account of principles of community assembly on oceanic islands (see Gillespie and Baldwin, this volume), (4) analysis of the fit of the model for non-native species, and (5) translation of the current graphical models into a more precise mathematical format. Thus, although a more complete, formal treatment awaits further development, we hope the GDM can offer the foundation for a newly expanded theory of island biogeography, unifying ecological and evolutionary biogeography.

Acknowledgments

We are grateful to Henning Adsersen, Paulo Borges, Mark Carine, Brent Emerson, Larry Heaney, Joaquín Hortal, Jonathan Losos, José María Fernández-Palacios, Aris Parmakelis, Carsten Rahbek, Robert Ricklefs, Spyros Sfenthourakis, Tod Stuessy, Kathy Willis, and attendees of the Harvard symposium for discussion and/or comments on this and/or our 2008 *Journal of Biogeography* paper (on which this chapter is based). RJW is grateful to the organizers for the invitation to participate, and for financial support to attend the meeting. KAT was supported in this work by a Marie Curie Intra-European Fellowship Program (project "SPAR," No. 041095).

Literature Cited

Ávila, S. P., P. Madeira, N. Mendes, A. Rebelo, A. Medeiros, C. Gomes, F. García-Talavera, C.M. da Silva, M. Cachão, C. Hillaire-Marcel, and A. M. de Frias Martins. 2008. Mass extinctions in the Azores during the last glaciation: Fact or myth? *Journal of Biogeography* 35:1123–29.

Barrett, S.C.H. 1996. The reproductive biology and genetics of island plants. *Philosophical Transactions of the Royal Society of London, Series B* 351:725–33.

Barrier, M., R. H. Robichaux, and M. D. Purugganan. 2001. Accelerated regulatory gene evolution in an adaptive radiation. *Proceedings of the National Academy of Sciences U.S.A.* 98:10208–3.

Borges, P.A.V., and V. K. Brown. 1999. Effect of island geological age on the arthropod species richness of Azorean pastures. *Biological Journal of the Linnean Society* 66:373–410.

Borges, P.A.V., and J. Hortal. 2000. Time, area and isolation: Factors driving the diversification of Azorean arthropods. *Journal of Biogeography* 36:178–91.

Brown, J. H. 1995. *Macroecology*. Chicago: University of Chicago Press.

Brown, J. H., and M. V. Lomolino. 2000. Concluding remarks: Historical perspective and the future of island biogeography theory. *Global Ecology and Biogeography* 9:87–92.

Buskirk, R. E. 1985. Zoogeographic patterns and tectonic history of Jamaica and the northern Caribbean. *Journal of Biogeography* 12:445–61.

Butaud, J-F., R. Rives, D. Verhaegen, and J-M. Bouvet. 2005. Phylogeography of Eastern Polynesian sandalwood (*Santalum insulare*), an endangered tree species from the Pacific: a study based on chloroplast microsatellites. *Journal of Biogeography* 32:1763–74.

Cameron, R.A.D., L. M. Cook, and J. D. Hallows. 1996. Land snails on Porto Santo: adaptive and non-adaptive radiation. *Philosophical Transactions of the Royal Society of London, Series B* 351:309–27.

Carine, M. A. 2005. Spatio-temporal relationships of the Macaronesian endemic flora: A relictual series or window of opportunity? *Taxon* 54:895–903.

Carine, M. A., S. J. Russell, A. Santos-Guerra, and J. Francisco-Ortega. 2004. Relationships of the Macaronesian and Mediterranean floras: Molecular evidence for multiple colonizations into Macaronesia and back-colonization of the continent in *Convolvulus* (Convolvulaceae). *American Journal of Botany* 91:1070–85.

Carracedo, J. C., and R. I. Tilling. 2003. *Geología y Volcanología de Islas Oceánicas. Canarias–Hawai.* Santa Cruz de Tenerife, Spain: CajaCanarias–Gobierno de Canarias.

Carson, H. L., J. P. Lockwood, and E. M. Craddock. 1990. Extinction and recolonization of local populations on a growing shield volcano. *Proceedings of the National Academy of Sciences U.S.A.* 87:7055–57.

Cook, L. G., and M. D. Crisp. 2005. Directional asymmetry of long-distance dispersal and colonization could mislead reconstructions of biogeography. *Journal of Biogeography* 32:741–54.

Cowie, R. H. 1995. Variation in species diversity and shell shape in Hawaiian land snails: In situ speciation and ecological relationships. *Evolution* 49:1191–202.

Cowie, R. H., and B. S. Holland. 2006. Dispersal is fundamental to biogeography and the evolution of biodiversity on oceanic islands. *Journal of Biogeography* 33:193–98.

Crawford, D. J., T. F. Stuessy, D. W. Haines, M. B. Cosner, M. O. Silva, and P. Lopez. 1992. Allozyme diversity within and divergence among four species of *Robinsonia* (Asteraceae: Senecioneae), a genus endemic to the Juan Fernandez Islands, Chile. *American Journal of Botany* 79:962–66.

Diamond, J. M. 1975. Assembly of species communities. In *Ecology and Evolution of Communities*, ed. M. L. Cody and J. M. Diamond, 342–444. Cambridge, MA: Harvard University Press.

Emerson, B. C., and N. Kolm. 2005a. Species diversity can drive speciation. *Nature* 434:1015–17.

———. 2005b. Emerson and Kolm reply. *Nature* DOI: 10.1038/nature04309.

Emerson, B. C., and P. Oromí. 2005. Diversification of the forest beetle genus *Tarphius* in the Canary Islands, and the evolutionary origins of island endemics. *Evolution* 59:586–98.

Funk, V. A., and W. L.Wagner. 1995. Biogeographic patterns in the Hawaiian Islands. In *Hawaiian Biogeography: Evolution on a Hot Spot Archipelago, ed.* W. L. Wagner and V. A. Funk, 379–419. Washington, DC: Smithsonian Institution Press.

Gillespie, R. G. 2004. Community assembly through adaptive radiation in Hawaiian spiders. *Science* 303:356–59.

Gillespie, R. G., and G. K. Roderick. 2002. Arthropods on islands: colonization, speciation, and conservation. *Annual Review of Entomology* 47:595–632.

Grant, P. R., and B. R. Grant. 1996. Speciation and hybridization in island birds. *Philosophical Transactions of the Royal Society of London, Series B* 351: 765–72.

Gruner, D. S. 2007. Geological age, ecosystem development, and local resource constraints on arthropod community structure in the Hawaiian Islands. *Biological Journal of the Linnean Society* 90:551–70.

Haila, Y. 1990. Towards an ecological definition of an island: A northwest European perspective. *Journal of Biogeography* 17:561–68.

Heaney, L. R. 1986. Biogeography of mammals in SE Asia: Estimates of rates of colonization, extinction and speciation. *Biological Journal of the Linnean Society* 28:127–65.

———. 2000. Dynamic disequilibrium: A long-term, large-scale perspective on the equilibrium model of island biogeography. *Global Ecology and Biogeography* 9:59–74.

———. 2007. Is a new paradigm emerging for oceanic island biogeography? *Journal of Biogeography* 34:753–57.

Heaney, L. R., J. S. Walsh, Jr., and A. T. Peterson. 2005. The roles of geological history and colonization abilities in genetic differentiation between mammalian populations in the Philippine archipelago. *Journal of Biogeography* 32: 229–47.

Kaneshiro, K. Y., R. G. Gillespie, and H. L. Carson. 1995. Chromosomes and male genitalia of Hawaiian *Drosophila*: Tools for interpreting phylogeny and geography. In *Hawaiian Biogeography: Evolution on a Hot Spot Archipelago*, ed. W. L. Wagner and V. A. Funk, 55–71. Washington, DC: Smithsonian Institution Press.

Keast, A., and S. E. Miller, editors. 1996. The origin and evolution of Pacific Island Biotas, New Guinea to Eastern Polynesia: Patterns and processes. Amsterdam: SPB Academic.

Kvist, L., J. Broggi, J. C. Illera, and K. Koivula. 2005. Colonisation and diversification of the blue tits (*Parus caeruleus teneriffae*-group) in the Canary Islands. *Molecular Phylogenetics and Evolution* 34:501–11.

Lack, D. 1947. *Darwin's Finches: An Essay on the General Biological Theory of Evolution*. Cambridge: Cambridge University Press.

Le Friant, A., C. L. Harford, C. Deplus, G. Boudon, R.S.J. Sparks, R. A. Herd, and J. C. Komorowski. 2004. Geomorphological evolution of Montserrat (West Indies): Importance of flank collapse and erosional processes. *Journal of the Geological Society* 161:147–60.

Levin, D. A. 2004. The ecological transition in speciation. *New Phytologist* 161:91–96.

Losos, J. B., and D. Schluter. 2000. Analysis of an evolutionary species-area relationship. *Nature* 408:847–50.

MacArthur, R. H., and E. O. Wilson. 1963. An equilibrium theory of insular zoogeography. *Evolution* 17:373–87.

———. 1967. *The Theory of Island Biogeography*. Princeton, NJ: Princeton University Press.

Moore, J. G., and D. A. Clague. 1992. Volcano growth and evolution of the island of Hawaii. *Bulletin of the Geological Society of America* 104:1471–84.

Nunn, P. D. 1994. *Oceanic Islands*. Oxford: Blackwell.

Paulay, G. 1994. Biodiversity on oceanic islands: Its origin and extinction. *American Zoologist* 34:134–44.

Peck, S. B. 1990. Eyeless arthropods of the Galapagos Islands, Ecuador: Composition and origin of the cryptozoic fauna of a young, tropical, oceanic archipelago. *Biotropica* 22:366–81.

Peck, S. B., P. Wigfull, and G. Nishida. 1999. Physical correlates of insular species diversity: the insects of the Hawaiian Islands. *Annals of the Entomological Society of America* 92:529–36.

Percy, D. M., A. M. Garver, W. L. Wagner, H. F. James, C. W. Cunningham, S. E. Miller, and R. C. Fleischer. 2008. *Proceedings of the Royal Society of London, Series B* DOI:10.1098/rspb.2008.0191.

Price, J. P. 2004. Floristic biogeography of the Hawaiian Islands: Influences of area, environment and paleogeography. *Journal of Biogeography* 31:487–500.

Price, J. P., and D.A. Clague. 2002. How old is the Hawaiian biota? Geology and phylogeny suggest recent divergence. *Proceedings of the Royal Society of London, Series B* 269:2429–35.

Price, J. P., and D. Elliott-Fisk. 2004. Topographic history of the Maui Nui Complex, Hawai'i, and its implications for Biogeography. *Pacific Science* 58:27–45.

Price, J. P., and H. L. Wagner. 2004. Speciation in Hawaiian angiosperm lineages: Cause, consequence, and mode. *Ecology* 58:2185–2200.

Pulvers, J. M., and D. J. Colgan. 2007. Molecular phylogeography of the fruit bat genus *Melonycteris* in northern Melanesia. *Journal of Biogeography* 34:713–23.

Rabosky, D. L., and I. L. Lovette. 2008. Explosive evolutionary radiations: Decreasing speciation or increasing extinction through time? *Evolution* 62: 1866–75.

Sanmartín, I., P. van der Mark, and F. Ronquist. 2008. Inferring dispersal: A Bayesian approach to phylogeny-based island biogeography, with special reference to the Canary Islands. *Journal of Biogeography* 35:428–49.

Silvertown, J. 2004. The ghost of competition past in the phylogeny of island endemic plants. *Journal of Ecology* 92:168–73.

Silvertown, J., J. Francisco-Ortega, and M. Carine. 2005. The monophyly of island radiations: An evaluation of niche pre-emption and some alternative explanations. *Journal of Ecology* 93:653–57.

Stuessy, T. F. 2007. Evolution of specific and genetic diversity during ontogeny of island floras: The importance of understanding process for interpreting island biogeographic patterns. In *Biogeography in a Changing World*, ed. M. C. Ebach and R. S. Tangney, 117–33. Boca Raton, FL: CRC Press.

Stuessy, T. F., D. J. Crawford, and C. Marticorena. 1990. Patterns of phylogeny in the endemic vascular flora of the Juan Fernandez Islands, Chile. *Systematic Botany* 15:338–46.

Stuessy, T. F., D. J. Crawford, C. Marticorena, and R. Rodríguez. 1998. Island biogeography of angiosperms of the Juan Fernandez archipelago. In *Evolution and Speciation of Island Plants,* ed. T. F. Stuessy and M. Ono, 121–38. Cambridge: Cambridge University Press.

Stuessy, T., J. Greimler, and T. Dirnböck. 2005. Landscape modification and impact on specific and genetic diversity in oceanic islands. *Biologiske Skrifter* 55:89–101.

Stuessy, T. F., G. Jakubowsky, R. Salguero Gómez, M. Pfosser, P. M. Schlüter, T. Fer, B.-Y. Sun, and H. Kato. 2006. Anagenetic evolution in island plants. *Journal of Biogeography* 33:1259–65.

Triantis, K. A., M. Mylonas, and R. J. Whittaker. (2008) Evolutionary species-area curves as revealed by single-island endemics: Insights for the inter-provincial species-area relationship. *Ecography* 31:401–7.

Wagner, W. L., and V. A. Funk, eds. 1995. *Hawaiian Biogeography: Evolution on a Hot Spot Archipelago*. Washington, DC: Smithsonian Institution Press.

Wallace, A. R. 1902. *Island Life*, 3rd ed. London: Macmillan.

Whelan, F., and D. Kelletat. 2003. Submarine slides on volcanic islands—a source for mega-tsunamis in the Quaternary. *Progress in Physical Geography* 27:198–216.

Whittaker, R. J., and J. M. Fernández-Palacios. 2007. *Island Biogeography: Ecology, Evolution, and Conservation*, 2nd ed. Oxford: Oxford University Press.

Whittaker, R. J., R. J. Ladle, M. B. Araújo, J. M. Fernández-Palacios, J. Delgado, and J. R. Arévalo. 2007. The island immaturity–speciation pulse model of island evolution: An alternative to the "diversity begets diversity" model. *Ecography* 30:321–27.

Whittaker, R. J., K. Triantis, and R. J. Ladle. 2008. A general dynamic theory of oceanic island biogeography. *Journal of Biogeography* 35:977–94.

Wilson, E. O. 1969. The species equilibrium. *Brookhaven Symposia in Biology* 22:38–47.

The Trophic Cascade on Islands

John Terborgh

One of the bits of conventional wisdom about islands most of us accept implicitly is that island vegetation is relatively defenseless against introduced herbivores (Carlquist 1974, Bowen and van Vuren 1997). Scores of anecdotal accounts of denudation of islands by goats, rabbits, pigs, and other introduced herbivores lie behind this conventional wisdom. The reports are so numerous and consistent that one cannot doubt their collective veracity (Coblentz 1978, Courchamp et al. 1999). But the simplistic conclusion to be drawn from these anecdotes—that island floras typically evolve reduced defenses against herbivores—may be understating a more complex and interesting reality.

A less often remarked upon generality is that essentially all islands support herbivores, be they insects, crustaceans, lizards, tortoises, birds, or even mammals. We are thus presented with a paradox: if most islands support native herbivores, then why are island floras so vulnerable to introduced herbivores, especially mammals?

At least two reasons come to mind. There very well may be more. The principal herbivores of remote islands are arthropods, but arthropod herbivores may be mismatched with respect to food plants since plants and arthropods are likely to colonize independently (Janzen 1973a, 1975). Plants generally arrive as seeds transported via wind or in the guts of birds or bats, whereas arthropods can be carried on the wind or in the plumage of birds, or rafted in driftwood. Thus colonizing arthropod herbivores will rarely find their preferred host plants on a given island and will consequently either fail to survive or be obliged to subsist on less preferred plant species on which larvae will develop slowly and in reduced numbers. Mismatching of plants and herbivores could result in reduced herbivore pressure and evolved relaxation of defenses.

There is some support for this idea. Back in the 1970s, two investigations independently reported that sweep net samples of arthropods from Caribbean islands contained conspicuously fewer species and individuals than samples from equivalent sites on the Netoropical mainland (Allen et al. 1973, Janzen 1973b). In keeping with this observation, it was noted shortly afterward that the bird communities of several Antillean islands are consistently deficient in the specialized insectivores that dominate the avifaunas of the mainland (Terborgh and Faaborg 1980). More than 85% of the individual birds captured in standard mist-netted samples at low-elevation sites on either the South or North American mainland were strict insectivores, whereas fewer than 20% of those captured in the Antilles were. The remaining 80% of the Antillean birds were omnivores, nectarivores, frugivores, and granivores, species living at lower trophic levels whose livelihoods were derived in part or in full from plants. This result pointed to something distinctive and fundamental about the organization of island avifaunas, but to my knowledge, no one has pursued it further.

A second reason island floras may be relatively lacking in antiherbivore defenses is that many of the nonarthropod herbivores of islands are terrestrial and therefore unable to access arboreal foliage (Carlquist 1965). One can point to the land iguanas and tortoises of the Galapagos, the flightless geese of Hawaii and other Pacific islands (James and Burney 1997, Steadman 2006), the land crabs of many midoceanic islands, and the outsized chuckwallas of the Sea of Cortez. In such a setting, a plant has only to grow to a meter or so to escape all but arthropod herbivores. The latter are likely to be controlled by predators—birds, lizards, spiders, and the like (Spiller and Schoener 1990). Reduced herbivory should translate rapidly into reduced investment in antiherbivore defenses, given that tannins and other antiherbivore compounds can constitute up to 35% of the dry weight of foliage (Coley et al. 1985). Thus, before we are tempted to draw broad generalizations about reduced antiherbivore defenses in island vegetation, it would be wise to investigate the specific context of the island(s) in question.

Theory

In pursuing this further, it would be helpful to refer to a theoretical framework. There is, in fact, a theory that can allow us to make predictions about levels of herbivory on islands, although the theory was not constructed with islands in mind. Proposed in 1981 by Oksanen, Fretwell, and others, it was termed "the exploitation ecosystems hypothesis" (EEH). A refined statement of it appears in Oksanen and Oksanen (2000). The theory, like most useful theories in ecology, is quite simple in outline. In essence, it follows Hairston, Smith, and Slobodkin (1960) in assuming three trophic regimes in terrestrial ecosystems (figure 5.1). The key variable is

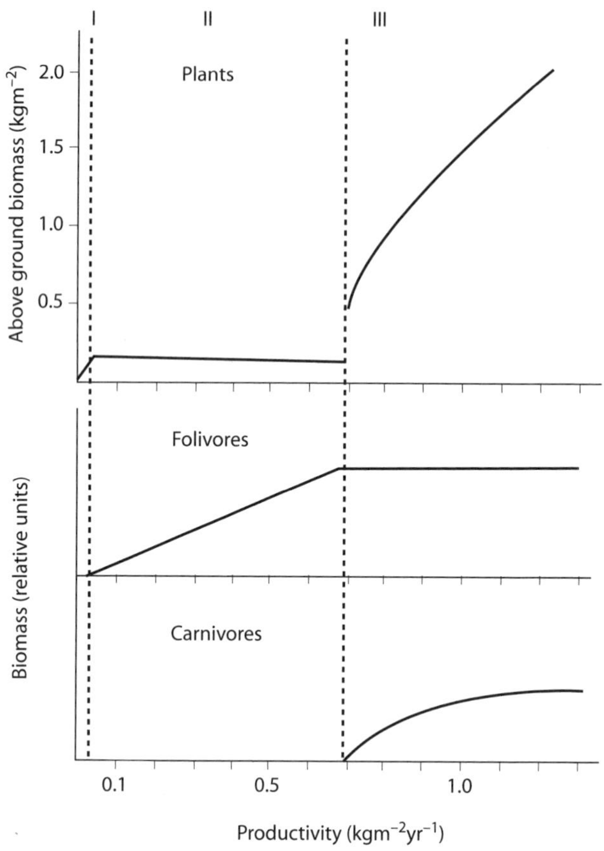

Figure 5.1. Exploitation Ecosystem Hypothesis. Trophic levels are added in stepwise fashion as ecosystem productivity increases (from Oksanen and Oksanen 2000).

productivity. At the lowest productivity levels, barely above zero, there are only producers. Such type-I ecosystems are found only in the most extreme deserts and in the high Arctic or Antarctic (examples in Oksanen et al. 1981, Oksanen and Oksanen 2000). We would expect plants living under such circumstances to allocate relatively little of their meager resources to herbivore defenses (Blossey and Nötzold 1995).

At slightly higher productivity levels, the amount of energy transformed by the ecosystem becomes sufficient to support a consumer trophic level. We shall call these type-II ecosystems. Since some arthropods can subsist on quantities of resources that are almost invisible to humans, we would expect arthropods to enter at lower productivity levels than vertebrates.

The EEH presumes that, as productivity increases beyond the herbivory threshold, herbivory increases apace, maintaining the plant biomass at a roughly constant level.

At some point (again, probably sooner for arthropods than for vertebrates) productivity crosses a second threshold, and a third trophic level—predators—enters the picture in type-III ecosystems. With still further increases in productivity, predators are presumed to maintain consumers at more or less constant levels, just as the consumers maintained the plant biomass at nearly constant levels in type-II ecosystems. This being so, edible (nonwoody) plant biomass increases with further gains in productivity up to a maximum determined by the physical environment. The EEH thus incorporates both bottom-up and top-down forcing.

Now, what has this to do with islands? It has a lot to do with islands if we make a simple substitution of parameters. The most informative variable of island biogeography—island size—is an excellent surrogate for productivity (other factors, climate, soils, etc., being equal). The substitution of area for productivity was pioneered by Schoener (1989) and is known as the productivity-space hypothesis. Biogeographical arguments can also link island area to the length of food chains (Holt 1996). Applying this logic, the smallest islands should support only producers, somewhat larger islands should support producers and consumers, and so forth.

Our focus for the remainder of this inquiry will be type-II islands, those supporting producers and consumers, but not predators of a dominant herbivore. I shall consider type-II islands originating in two distinct ways, via contraction and via colonization, and show that their herbivore communities display some convergent properties independent of the taxa involved. We shall also see that type-II ecosystems are unlike any we ever encounter in our normal travels. Natural type-II ecosystems have become extremely rare and one has to go, quite literally, to the ends of the earth to find them, at least in the tropics.

Results: Lago Guri

The first case I shall present involves a type-II ecosystem created by the contraction of a type-III ecosystem to an area (i.e., productivity level) too small to support predators of vertebrates and some invertebrates. In the case in question, the area contraction took place when the Caroní Valley in Venezuela was flooded in 1986 by the huge (4,300 km^2) Guri hydroelectric impoundment (Morales and Gorzula 1986). Flooding fragmented the formerly continuous dry forest of the mainland, creating hundreds of islands ranging from tiny specks of <<1 ha to >760 ha.

Our first surveys of some of these islands in 1990 indicated that three-quarters or more of all vertebrates present on the nearby mainland had already disappeared from islands of < 12 ha, leaving strongly imbalanced animal communities. Some functional groups were underrepresented (e.g., pollinators, seed dispersers) whereas others were entirely absent (predators of vertebrates). Nearly all persistent species exhibited hyper-abundance, that is, their local population densities on islands were elevated far above their densities on the mainland (Terborgh et al. 1997a,b). Persistent hyperabundant groups included birds, some lizards and amphibians, spiders, small rodents, and several generalist herbivores: red-footed tortoise (*Geochelone carbonaria*), common iguana (*Iguana iguana*), red howler monkey (*Alouatta seniculus*), and leaf-cutter ants (*Atta* spp., *Acromyrmex* spp.) (Terborgh et al. 1997b, Lambert et al. 2003, Rao et al. 2001, Aponte et al. 2003, Orihuela et al. 2005).

Since many of our results from the Lago Guri island system have been published elsewhere, I shall provide only a brief summary here, focusing particularly on herbivory. We studied herbivory indirectly via assessments of plant demography at sites supporting high, medium, and low densities of generalist herbivores. Herbivore abundance varied inversely with island size so that "small" islands (below 1.5 ha) supported the highest herbivore densities, "medium" islands (between 3 and 12 ha) supported intermediate densities, and "large" landmasses (88 and 190 ha, mainland) supported low densities. To assess the effects of herbivore density on plant demography, we followed the fates of 3030 small saplings (≥ 1 m tall and < 1 cm diameter at breast height [dbh]), 3997 large saplings (≥ 1 cm, < 10 cm dbh), and 4771 adult trees (≥ 10 cm dbh) for 5 years at 12 sites (table 5.1).

The mortality of small and large saplings was elevated on both small and medium islands, but the differences were not always statistically significant. Far more pronounced were the decreases of recruitment into both stem size classes. Recruitment into the adult tree class (≥ 10 cm dbh) did not differ in relation to landmass size. In sum, demographic effects associated with hyperbundant herbivores were greater for recruitment than mortality and restricted to small stem size classes.

Given that common iguanas and red howler monkeys confine most or all of their feeding activities to the canopy, and that tortoises were not found on small islands, leaf-cutter ants emerged as the herbivore most likely responsible for the low recruitment rate of saplings (Lopez and Terborgh 2007). We obtained further evidence implicating leaf-cutter ants and perhaps other arthropods by setting out tree seedlings under fine wire mesh cages. Seedling survival was high under cages, even at sites supporting *Atta* densities 100 times greater than observed on the mainland (Lopez and Terborgh 2007). In some cases, uncaged seedlings were defoliated during

TABLE 5.1
Demography of Small and Large Saplings on Small, Medium, and Large Landmasses at Lago Guri, Venezuela, 1997–2002

Landmass size	*Relative no. stems/225 m²* *(1997)*	*Relative proportion died 1997–2002*	*Relative proportion recruited 1997–2002*	*Relative no. stems/225 m²* *(2002)*
		Small saplings		
Small	0.36	1.53	0.19	0.25
Medium	0.79	1.31	0.33	0.64
		Large saplings		
Small	1.24	2.07	0.32	1.04
Medium	1.57	1.60	0.39	1.47

Source: Modified from Terborgh et al. 2006, p.257, table2.

Note: Values given are relative to those observed on the large landmasses that served as controls.

the first night of exposure, whereas seedlings survived up to 3 years under cages (figure 5.2).

We found that hyperabundant leaf-cutter ants were relatively unselective in their choice of foliage compared to ants living in widely separated colonies on large landmasses (Rao et al. 2001). Similar observations were made on red howler monkeys (Orihuela et al. 2005). The observation of decreased selectivity under hyperabundance carries important implications.

First, it shows that plant defenses conferring low preference status under "normal" circumstances act in a conditional fashion, being effective only at low herbivore densities. We found that most plant species become vulnerable at high herbivore densities, as indicated by the fact that mortality of saplings exceeded recruitment in nearly every species present on small and medium islands. Relaxed defenses in response to insularity was not a factor in this situation because all plants stranded on Guri islands carried genotypes evolved under mainland conditions. "Edge effects" and exposure to prevailing winds had no discernible effect on the mortality or recruitment of any size class of stems (Terborgh et al. 2006).

Second, the facultative ability of leaf-cutter ants, howler monkeys, and presumably other generalist herbivores to subsist on species of foliage that are ordinarily rejected allows their numbers to increase as much as an

Figure 5.2. Top: Dry forest understory of a large landmass control site at Lago Guri, Venezuela. Bottom: Understory of a small island supporting a hyperdense population of leaf-cutter ants.

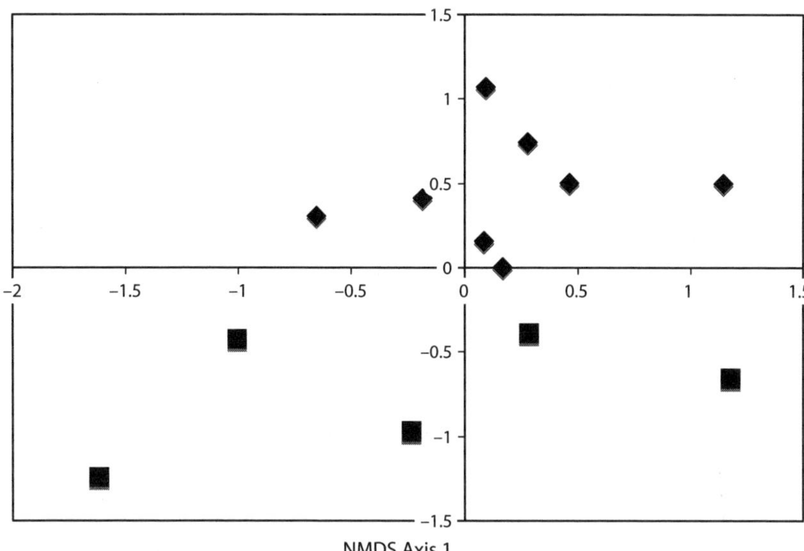

Figure 5.3. Two dimensional NMDS ordination of stems ≥1, <10 cm dbh found in 225 m² sampling plots located on *Atta* colonies (squares) and away from *Atta* colonies (diamonds) on medium islands in Lago Guri, Venezuela. The two sets of points are distinct by multiresponse permutation, p=0.001.

order of magnitude above those considered "normal." Thus the "carrying capacity" for generalist herbivores released from top-down control is many times greater than normal density, at least as a transient condition (Beschta and Ripple 2008).

Third, community-wide suppression of plant recruitment by hyper-abundant herbivores leads to collapse of the characteristic dry forest vegetation of the Caroní Valley and its replacement by an entirely novel plant community never before documented.

We were not able to quantify the plant species composition of the vegetation that would emerge under steady-state type-II conditions because transformation of the vegetation of the islands we studied was still in mid-course when the project ended in 2003. We did, however, obtain some hints of what might be in store by inventorying saplings growing on top of five *Atta* colonies on four medium islands (figure 5.3). The figure shows a nonmetric multidimensional scaling ordination of stems ≥1 cm and <10 cm dbh growing in 225 m² plots centered on *Atta* colonies and at sites beyond the foraging radius of existing *Atta* colonies on the same islands. In each case, points representing *Atta* colony samples fall near the periphery of the ordination space and far from the corresponding off-*Atta*-colony samples, indicating marked compositional divergence. Just how marked the divergence was can be judged by a pair of examples. The 3 most abundant species growing on *Atta* colonies on the island of Ambar, representing 258 out of 419 stems (62%), were not represented in 302 stems from 2 off-colony sites on the same island. Conversely, none of the 3 most abundant species in off-colony samples was contained in the 90 stems growing on an *Atta* colony on the island of Panorama. Interestingly, there was no consistent direction of divergence of the various *Atta* colony samples in ordination space, in keeping with the fact that different plants tended to dominate at different sites.

Plants able to survive and even increase at *Atta* colony sites included both common and rare elements of the local dry forest vegetation. The five colony sites supported from 90 to 275 saplings of 14 to 38 species, a majority of which can be presumed to be survivors from precolony times rather than newly established individuals (table 5.2). Each site was dominated by a small number of species, from 1 to 5, that made up 50% or more of the stems. The great majority of species were represented by only 1, 2, or 3 stems at each site. The collection of dominant species is taxonomically diverse, yet most of them were exceptional in possessing coriaceous evergreen leaves, an uncommon feature in the semideciduous dry forest vegetation of the Caroní Valley. Another characteristic that may have deterred *Atta* herbivory, found in two legumes (*Acacia sp.*, *Calliandra laxa*), was the possession of compound leaves with finely divided leaflets that were individually much smaller than the usual load carried by *Atta* workers.

Another noteworthy feature of the results is that the lists of species that dominated on each island show little overlap. Here we appear to have a good example of what Hurtt and Pacala (1995) have termed "winner by default." Any given island will carry only a sample of the regional floristic diversity and a given site within an island will offer an even more limited diversity. Thus, the "best competitor" in the regional species pool will not always be on hand to "win" in a given situation and other species will succeed instead. In an open competition run over many generations in the presence of hyperabundant herbivores, the winners might be further pared down to an even smaller group of species than we observed on the four islands.

The species listed in table 5.2 appear to be the vanguard of a drastically altered vegetation adapted to a type-II world of hyperabundant herbivores. One can anticipate that most of the less common species still surviving on *Atta* colonies at the time of our census will eventually die out, leaving only the most resistant species. One can further anticipate that a huge loss in plant diversity will accompany the winnowing process. Speculating even further, one could anticipate that a type-II world at equilibrium would be characterized by a low diversity

Table 5.2
Numbers of the Five Most Abundant Sapling Species Found in Five 225 m^2 Plots Centered on *Atta* Colonies on Four Medium Islands in Lago Guri, Venezuela

Species	*Chotacabra*	*Panorama*	*Lomo*	*Ambar no. 1*	*Ambar no.2*	*Total*
Protium sagotianum				59	77	136
Hymenaea courbaril			77		14	91
Eugenia punicifolia	39	11	33			83
Gustavia sp.				54	11	65
Brownea coccinea	22			37		59
Hirtella paniculada				43	13	56
Ocotea glomerata	56					56
Cupania sp.			27			27
Guatteria schomburkii	24					24
Myrtaceae 'rusty twigs'			16			16
Coccoloba falax				14		14
Maytenus guianensis			14			14
Casearia silvestris		13				13
Coursetia ferruginea	13					13
Bunchosia mollis		10				10
Calliandra laxa		10				10
Talisia heterodoxa					10	10
Acacia paniculata		8				8

of highly defended plant species and, accordingly, reduced densities of herbivores.

Results: Primary Type-II Islands

Is this merely wild speculation, or can we find real-world examples of equilibrial type-II ecosystems with which to test the idea? The answer is yes, though well-documented examples are few. Before humans transformed the ecology of the world's islands, the oceans undoubtedly contained hundreds or perhaps thousands of islands supporting type-II ecosystems. Many islands of the Pacific and the Indonesian archipelago would have qualified, as would many of the Philippines and West Indies. But human conquest of the world's islands was accompanied by habitat destruction, introductions of domestic and commensal animals, and consequent extinctions that have forever altered the ecology of the vast majority of the world's islands. Introduced rats, rabbits, cats and other human commensals have fundamentally disrupted the ecology of even remote subantarctic islands like Macquarie, Kerguelen, Crozet and the Tristan da Cunha group (Courchamp et al. 1999). But fortunately, a few extremely isolated islands have survived more or less intact, and it is to these we must go to find the answer to our question.

In pondering this issue, and pursuing it in the literature, I found three cases that are supported by sound natural history data. Two are isolated islands in the Indian Ocean: Christmas Island and the Aldabra Atoll, and the third is East Plana Cay in the Bahamas. Each of these islands supports a generalist herbivore in the absence of predators, and in each case, the herbivore belongs to a different taxonomic class or phylum. On Christmas Island the herbivore is a land crab, *Becarcoidea natalis*; on Aldabara it is a tortoise, *Geochelone gigantea*; and on East Plana Cay, it is a mammal, the Bahamian hutia, *Geocapromys ingrahami* (table 5.3).

In all three cases, the herbivores maintain population densities and biomasses greatly exceeding those of equivalent herbivores in the presence of predators (Coe et al. 1976, Iverson 1982). We shall see that these three cases, disparate as they are in geography and taxonomy, have much in common with each other and with the case of the Lago Guri islands already considered.

All three islands are small, isolated from other islands and remote from the mainland, suggesting low turnover (MacArthur and Wilson 1967). We can thus safely presume that the type-II ecosystems they support are ancient and that their extraordinary herbivores and the plants upon which they subsist have been evolving together for millennia. Research

TABLE 5.3
Generalist Herbivores of Three Remote Oceanic Islands: Their Population Densities and Biomasses

Island	*Location*	*Generalist herbivore*	*Body mass*	*Population density/km²*	*Biomass kg per km²*
Christmas	10° 29′S, 105° 38′E	*Becarcoidea natalis*	≤500 g	1,300,000	145,000
Aldabra	8° 25′S, 48° 20′E	*Geochelone gigantea*	≤250 kg	2,700	58,300
East Plana	22° 23′N, 73° 30′W	*Geocapromys ingrahami*	755 g (m) 660 g (f)	3,000	2,100

conducted on each of the three islands offers distinct insights into the nature and operation of type-II ecosystems.

Christmas Island

Christmas Island lies 360 km south of Java in the Indian Ocean and supports only one macroherbivore, the red crab, *Becarcoidea natalis*. The crabs, weighing up to 500 g, live in burrows on the forest floor at densities estimated at 1.3/m² (Green 1997). The crabs consume leaf litter and any other edible plant parts that fall to the ground. Crabs as a dominant herbivore are not unusual. Related species occupy scores of islands in the Pacific Ocean and the mangrove zone of tropical shorelines around the world (Sherman 2002).

The crabs of Christmas Island have recently come under threat, but in a way that initiated a fortuitous experiment. In a tragic but typical inadvertency, the notoriously destructive yellow crazy ant, *Anoplolepis gracilipes*, arrived on Christmas Island over 70 years ago. For decades it remained at low density until 1989, when huge, multiqueened, "supercolonies" were noticed. Since then, the ant has been spreading in a front across the island with worker densities reaching thousands/m² (O'Dowd et al. 2003). Crabs have no defense against the ants and are killed by them so that ant-occupied zones have become crabless. The slow spread of the ant across the island allowed investigators to compare tracts of forest with and without crabs.

Removal of the island's dominant herbivore has resulted in a stunning transformation of the vegetation (O'Dowd et al. 2003: figure 5.4). All three trophic levels present on the island have been affected: consumers,

Figure 5.4. Understory of forest on Christmas Island, Indian Ocean: Top: Natural state with red crabs. Bottom: Without red crabs after invasion of the yellow crazy ant (*Anoplolepis gracilipes*) (from O'Dowd et al. 2003, p. 815).

producers, and decomposers. In the natural state of the island, crabs consumed most plant matter falling from the canopy: leaves, flowers, and fruits (Green et al. 1999). Seedlings of many species are also consumed (O'Dowd and Lake 1990, Green et al. 1997). Crab foraging thus maintains the forest floor in a condition strikingly reminiscent of that of small Lago Guri islands, bare of leaf litter and most regenerating plants (compare figures 5.2 and 5.4). Extirpation or exclusion of the crab released seeds and seedlings from predation, whereupon the understory quickly became crowded with tree saplings (Green et al. 1997). Seedling diversity jumped from 6 to 22 species per 80 m^2 (O'Dowd and Lake 2003). Leaf litter that had previously been consumed by crabs now lay on the forest floor to decompose slowly, as in mainland forests. Portions of Christmas Island that have been invaded by the ant are undergoing a catastrophic shift in vegetation, perhaps as profound as the one we documented on islands in Lago Guri, with the distinction that the change is in response to a release from herbivore pressure rather than the opposite.

Aldabra

The Aldabra Atoll supports the Aldabra giant tortoise, one of three surviving members of a once-extensive radiation in the western Indian Ocean of up to eight species of tortoises (Gerlach 2004, 2005). Approximately 150,000 tortoises weighting up to 250 kg each occupy the 155 km^2 Aldabra Atoll. The atoll consists of several discrete islands, some of which lack surface water and, consequently, tortoises. Occupied portions of the island support tortoise densities of up to 2,700 per km^2 (Coe et al. 1979; table 5.3).

The principal islands of the western Indian Ocean, Madagascar, Mauritius, Reunion, and Rodrigues, all harbored giant tortoises that were quickly exterminated, along with the elephant bird, dodo, solitaire, and other species, after humans discovered the islands. Nevertheless, the legacy of the extinct tortoises lives on in the native vegetation as indicated by the presence of many plant species possessing the unusual trait of heterophylly (figure 5.5).

The juvenile leaves of these plants are mostly small and grasslike, not at all resembling the adult leaves. Recently, a team of researchers conducted leaf choice experiments with captive Aldabra tortoises. The tortoises overwhelmingly selected adult over juvenile leaves (figure 5.5) despite greater natural accessibility of the latter (Eskildsen et al. 2004). Moreover, they showed that the transition from juvenile to adult leaf morphology takes place at a height equivalent to the reach of a foraging tortoise (figure 5.6).

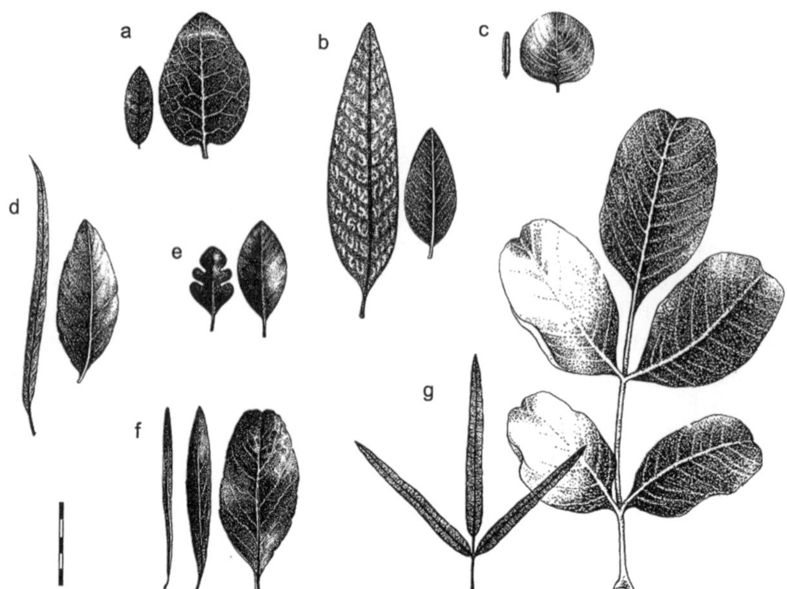

Figure 5.5. Heterophylly in some plants of the Mascarene Islands (Mauritius Reunion, and Rodrigues) western Indian Ocean. (from Eskildsen et al. (2004). Juvenile leaves are on the left: a) *Diospyros egrettarum*, b) *Tarenna borbonica*, c) *Eugenia lucida*, d) *Cassine orientalis*, e) *Turraea casimiriana*, f) *Maytenus pyria*, g) *Gastonia mauritiana*.

East Plana Cay

The last of the three cases concerns the hutias of East Plana Cay. The Bahamian hutia was thought possibly to be extinct until Garrett Clough confirmed its presence in 1966 on East Plana Cay, a 450 ha island lying to the windward of other Bahamian islands (Clough 1969). Perhaps its small size and windward position served to protect it from invasion by rats (*Rattus* spp.), for humans, rats, cats, dogs, etc., had long since exterminated the hutia populations of all other Bahamian islands.

The vegetation of East Plana Cay is low, shrubby, and relatively undiverse. The diet of hutias is comprised principally of the foliage, and doubtless other parts, of six common plant species belonging to the following genera: *Strumpfia*, *Conocarpus*, *Foresteria*, *Phyllanthus*, *Croton,* and *Tournefortia*. These include members of families, e.g., Boraginaceae, Combretaceae, Euphorbiaceae, that produce potent antiherbivore defenses, so one can surmise that the vegetation of East Plana Cay is comprised of a selection of the most resistant species from the Bahamian flora (Clough 1972).

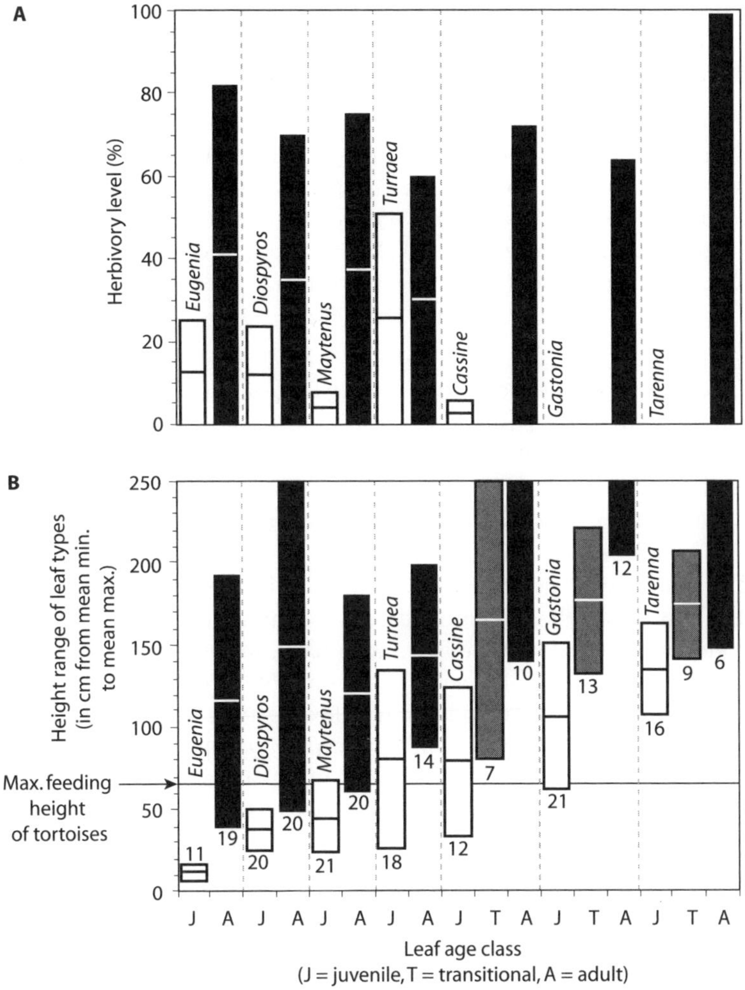

Figure 5.6. A. Proportions of adult (black bars) versus juvenile (open bars) leaves of seven heterophyllous plant speces eaten by Aldabra tortoises. B. Vertical ranges of juvenile (white bars) and adult (black bars) foliage of seven heterophyllous plant species. Checkered bars indicate foliage showing transitional morphology. Numbers below the bars refer to sample sizes. The horizontal line represents the browse line for Aldabra tortoises (from Eskildsen et al. 2004).

Persistence of the hutia on only one small island made it highly vulnerable to extinction, prompting Clough and others to establish an additional population by releasing 11 hutias (6 males and 5 females) on Little Wax Cay (24° 53′ N, 76° 47′ W), a small island in the Exuma group, some 300 km to the northwest of East Plana Cay (Campbell et al. 1991). That was in 1973. Twelve years later, in 1985, another investigator estimated the number of hutias on Little Wax Cay at 1200. Four years after that, a third party led by David Campbell returned to the island in April, 1989, to conduct vegetation analysis (Campbell et al. 1991).

> Even as one approached Little Wax Cay from the sea, it is obvious that the vegetation of the cay had been massively perturbed. Large areas of the island were bald, without closed, living canopy, in sharp contrast to neighboring cays, which do not have hutias. Many of the trees and shrubs were recently killed and remained as gaunt skeletons, which had not yet decomposed. Closer examination of the cay revealed that large areas were paved with hutia fecal pellets. (Campbell et al. 1991, p. 538)

Campbell et al. go on to state that they found no evidence of seven plant species documented by Russell in a 1958 survey of Little Wax Cay undertaken prior to the introduction of hutias. They conclude that "as the edible plants of Little Wax Cay are being destroyed by hutias, the vegetation of the Cay is likely to become dominated by toxic plants, and it is inevitable that the population of hutias on the Cay will soon begin to fall" (Campbell et al 1991).

The results of Campbell et al. clearly indicate that the vegetation of Little Wax Cay was lacking in defenses against herbivory prior to the introduction of hutias. Whether hutias had ever previously been on the island is not known, but they had presumably been absent for at least 100 years prior to the introduction, allowing time for the vegetation to adjust to type-I conditions. Similar uncertainty applies to the history of East Plana Cay, as well. The Bahamas once supported a large owl that might have controlled hutias, but the owl has been extinct for several thousand years since the Bahamas were colonized by humans (Steadman et al. 2007).

Discussion and Conclusions

Plants of type-II insular ecosystems do carry anti-herbivore defenses—but only against native herbivores. Defenses found in the vegetation of type-II islands are various, depending on the accessibility of propagules and/or foliage to native herbivores. On Christmas Island, where terrestrial crabs are the herbivore, defenses are expressed at the propagule (seed and seedling) stage (Green et al. 1997); on Aldabra and other islands of the Western Indian Ocean, where tortoises were the principal herbivore, it is at the stage of juvenile leaves; and on East Plana Cay, where a mammal capable of climbing is the selective agent, conventional chemical defenses are expressed in mature foliage (Campbell et al. 1991). Given that native herbivores of type-II islands are often earthbound, like crabs and tortoises, they might select for height-limited defenses that would prove ineffective against introduced mammals like goats or cattle. Height-limited defenses are also found in African acacias, though the height at which thorns cease to be produced is the height of a giraffe (Archibald and Bond 2003).

Herbivore densities in type-II ecosystems are consistently high multiples of those observed in type-III systems on continental mainlands. This was true both for the secondary type-II systems of Lago Guri islands and the three primary type-II systems described just above. Hyperabundant herbivores thus appear to be characteristic of type-II systems. Transitions from type-II to type-I or from type-III to type-II ecosystems may entail what Scheffer et al. (2001) have termed "catastrophic regime shifts" involving major changes in plant species composition.

The intense herbivore pressure that prevails in type-II systems could be expected to drive plant-herbivore arms races. To this point there is little evidence, though consistently high herbivore densities suggest that the herbivores "win." Plant investment in antiherbivore defenses necessarily entails trade-offs with growth and reproduction and must therefore be self-limiting (Coley et al. 1985). Animals subsisting on heavily defended plant material may themselves experience decrements in growth and reproductive performance, but such decrements may not be strongly disadvantageous in the context of predator-free type-II islands. In the language of foraging ecology, the herbivores of type-II systems become energy maximizers instead of time minimizers (MacArthur and Pianka 1966).

Any plants that were fully resistant to a resident herbivore could take over an island like Aldabra or East Plana Cay and shut out the herbivores, but that does not appear to happen. Plant diversity on type-II islands appears to be low, but it is far from zero. Hyperabundant herbivores thus fail to eliminate plant diversity and persist on type-II islands, presumably for millennia. This could be understood if selection favored herbivore genotypes that could tolerate the defenses of the most common plant species. Such frequency-dependent selection would prevent monopolization of the vegetation by any one plant species and would help stabilize plant diversity, though perhaps at a low level compared to type-III systems.

The evolution of plant defenses is usually considered in relation to the feeding preferences of herbivores, but defenses can also serve as a currency of interspecific competition between plants (Blossey and Nötzold 1995). Fast-growing, weakly defended plants should predominate under low herbivory, such as in type-I systems. Where predators regulate herbivore densities, herbivore pressure is likely to fluctuate in both space and time, establishing a regime of lottery competition (Chesson and Warner 1981). Plants sharing a common herbivore could display reciprocal demography, just as do prey species sharing a common predator (Holt 1977). Thus, a regime of low, patchy herbivory (type III) could be expected to maintain higher overall levels of plant diversity than one without herbivory (type I) or continuously high herbivory (type II). In the absence of herbivory, interspecific competition between plant species would limit diversity, whereas under intense herbivory, only species with strong defenses could persist (Lubchenco 1978). An analogy to the intermediate disturbance hypothesis seems apt here (Connell 1978, Molino and Sabatier 2001). If herbivore pressure proves to be a strong regulator of plant diversity on islands, then the presence/absence of generalist herbivores could act as a major biotic filter for plant species composition superimposed on the traditional geographic filters of area, isolation, and elevation.

How does the EEH intersect with classical island biogeography? Perhaps the intersection is broader than we currently imagine. Productivity and herbivory have not been major issues in island biogeography. Investigators have most often focused on the number of species of birds or lizards or, less commonly, other groups, such as bats, ants, and beetles. Inspired by MacArthur and Wilson (1967), investigators have overwhelmingly fixated on the physical parameters of area, isolation, and elevation, while remaining largely blind to the potential of interisland variation in biotic conditions to contribute to explanations of biogeographic patterns. An outstanding exception to this statement is found in the prescient work of Schoener and his colleagues (see their chapter in this volume).

Development of a more holistic view of island biogeography, one that takes into account both physical and biotic variables, has been hindered by the lack of a biotic complement to the MacArthur-Wilson theory. Here I suggest that the EEH, and modifications thereof, can provide the missing biotic complement. I'm not suggesting that the EEH, or anything like it, can substitute for MacArthur-Wilson. The success of MacArthur-Wilson is outstanding and beyond debate. What I am suggesting is that the biotic conditions of an island can, and undoubtedly do, contribute to explaining such biogeographic features as the presence or absence of individual species and the species richness of a particular taxon.

To support this contention, I offer four highly abbreviated examples. (1) MacArthur himself was puzzled by a phenomenon he termed "density overcompensation" (MacArthur et al. 1972). The term refers to the oft-repeated finding of greater total bird densities on islands than in similar habitat on the corresponding mainland, notwithstanding greater species diversity on the latter. We observed density overcompensation in birds on Lago Guri islands and obtained evidence pointing to bottom-up (productivity) effects associated with the presence of howler monkeys at hyperabundant densities and a concomitant acceleration of nutrient cycling (Feeley and Terborgh 2005, 2006, 2008). Top-down effects (reduced predation) could also help to explain density overcompensation. (2) Diamond's (1975) famous "checkerboard" distributions represent a biotic mechanism (competitive exclusion) that operates to regulate the presence/absence of individual species on particular islands (see Simberloff and Collins, this volume). (3) Schoener and Spiller (1996) have shown that spider diversity on tiny Bahamian islets is strongly regulated from the top down by the presence or absence of the lizard *Anolis sagrei*, an important predator of spiders. (4) Exogenous inputs, such as nutrients withdrawn from the sea and transported to seabird nesting islands as fish and manure, can transform the vegetation of entire islands in a bottom-up effect (Croll et al. 2005).

It is likely that one could find many more examples to add to these if one searched the literature. Suffice it to say that biotic interactions of various kinds, including bottom-up and top-down effects, can contribute to a more complete understanding of island biology.

These speculations lead us to reconsider the nature of island vegetation in relation to the exploitation ecosystem hypothesis. The smallest islands should support type-I ecosystems. The relevant range of island areas has not been determined, but the presence of crabs and/or reptilian herbivores on islands of less than 1 km^2 suggests that most tropical type-I islands must be tiny (Burness et al. 2001). Even mammals can persist on some very small islands. East Plana Cay is only 4.5 km^2 and Little Swan Island, which supported an endemic hutia until domestic cats were released onto it in the 1960s, is only 2.5 km^2 (Morgan and Woods 1986). Islands supporting type-II ecosystems were probably once numerous in the world's oceans in all but the most remote (and perhaps high-latitude) locations. Plant species native to such islands must have carried defenses against resident herbivores, but, as practically all such islands are now inhabited by man and his commensals, the ecosystems of extremely few survive intact. Predators enter the picture on much larger islands where they maintain herbivores at the low densities typical of type-III ecosystems (Burness et al. 2001).

Finally, the world's largest islands (e.g., Madagascar, New Guinea, New Zealand) once carried complete ecosystems, replete with top carnivores and megaherbivores (here defined operationally as herbivorous animals large enough to escape predation as adults; Burness et al. 2001). Megaherbivores, like the hyperabundant herbivores of type-II ecosystems, are capable of overriding all but the most assertive antiherbivore defenses, so we could expect that relatively undefended plant species would be relegated to fugitive status as ephemerals or gap colonists, or confined to rock faces or other inaccessible sites, as is the case of a number of highly endangered plants of the Hawaiian archipelago (Carlquist 1970).

Megaherbivores have roamed the continental landmasses of the earth since the early Mesozoic, with only a temporary hiatus after the end-Cretaceous extinctions. As recently as the late Pleistocene, proboscidians (elephants) of several genera were found on all continents except Australia and Antarctica. Judging from the known distribution of elephants in Africa today, proboscidians were ubiquitous generalists, ranging essentially everywhere between the extremes of rainfall, temperature, and elevation gradients. Even now, African elephants occur from the edge of the Sahara to the Cape of Good Hope, from the Indian Ocean to the Atlantic, and from the lowlands of the Congo Basin to above timberline on Mt. Kilimanjaro and Mt. Kenya (Coe 1967, Owen-Smith 1988). The ubiquity of proboscidians in Africa, and their former presence elsewhere in the world, including the high Arctic, underscores the extreme implausibility of climate change as the factor responsible for the disappearance of proboscidians and other megafauna from all parts of the world except Africa and southern Asia (Barnosky et al. 2004).

Unfortunately, the EEH does not consider megaherbivores, an oversight that exemplifies the shifting baseline of our anthropocentric society. Nevertheless, the EEH can be extended quite simply by adding a type-IV regime to accommodate megaherbivores, but there remain some questions about the range of productivity levels that would support type-II, -III, and -IV ecosystems.

It stands to reason that, if type-IV ecosystems once occupied all but the most extreme situations within continents, type-III ecosystems would have occupied very limited areas. Indeed, given the prehistoric ubiquity of megaherbivores and their island counterparts, such as the elephant bird, giant tortoises, and moas, it is reasonable to wonder whether Type-III ecosystems ever existed other than on islands. Today, elephants are found in areas of extremely low productivity in the Namibian desert where rainfall is less than 100 mm/yr (Viljoen 1989). Referring back to figure 5.1, that would place the threshold to type-IV ecosystems at the far left of the diagram at a level of productivity around 0.1 kg/m^2yr^{-1}.

We can thus surmise that type-IV ecosystems occupied more than 90% of the unglaciated, nondesert habitat of the planet since the Mesozoic (extinction crises and their aftermaths excepted). Type-I, -II, and -III ecosystems would have been relegated primarily to islands where water barriers filtered the colonization of large vertebrates (Holt 1996). The type-II and -III ecosystems that now occupy most of the more-or-less "natural" habitat remaining on the continents are therefore of recent anthropogenic origin.

To summarize, I propose that the four ecosystem states, I, II, III, and IV, comprise a trophic cascade in herbivory (table 5.4). As in more

Table 5.4
The Trophic Cascade in Herbivory

Ecosystem type	*Trophic levels*	*Herbivore pressure*	*Plant defenses*	*Plant diversity*
I	producers only	low	low	low
II	producers+consumers	high	high	low
III	producers+consumers+ predators	low	variable	high
IV	producers+consumers+ predators+megaherbivores	high	high	low

conventional top-down trophic cascades, successive states are characterized by alternating, high (types II and IV) and low (types I and III) levels of herbivory (Paine 1980, Scheffer et al. 2001). Plant defenses should adapt to herbivore pressure through natural selection, induced responses, and/or species selection based on constitutive properties. Plant diversity should be low in the absence of herbivory (type I; pure bottom-up forcing) and in the presence of hyperabundant herbivores or megafauna (types II and IV; strong top-down forcing); it should be high in the presence of predators that cause a moderate level of herbivory to fluctuate in space and time (type III; mixed top-down and bottom-up forcing).

I grant that some of this is unabashed speculation, but everything I propose can be supported or refuted by appropriate empirical tests. Those desiring to conduct such tests should not delay. Already, more than 90% of the earth's ice-free terrain has been fundamentally altered. Continental areas were generally type-IV until human-mediated overkill liquidated megaherbivores nearly everywhere. Now, type-IV ecosystems remain only in small and shrinking portions of Africa and southern/southeastern Asia. The remainder of continental earth has relaxed to type-III conditions (lacking megaherbivores but retaining large carnivores such as wolves and jaguars) or type-II conditions (large carnivores eliminated and native herbivores replaced by livestock; Valone et al. 2001). The implications for conservation of this trophic downgrading of the earth's ecosystems are largely unexplored. The best chances for finding examples of type-I, -II, and -III ecosystems that have arisen naturally and are still undegraded must remain among the world's islands. Sadly, very few islands remain anywhere that have not undergone anthropogenic shifts in state. Documenting the ecology of these last remaining intact islands before alien species arrive and transform them should be a research goal of the highest priority.

Acknowledgments

I wish to express my deep gratitude to Lauri Oksanen for his friendship and for the inspiration his ideas have given me, for they have opened my eyes to a diversity of island ecosystems I had never previously imagined. I am grateful to Luis Balbas and to EDELCA (Electrificación del Caroní) for long-standing support of the Lago Guri project. Financial support from the MacArthur Foundation and National Science Foundation is gratefully acknowledged (DEB-9707281, DEB-0108107). I also thank two reviewers for insightful and helpful comments.

Literature Cited

Allen, J. D., W. Barnthouse, R. A. Prestbye, and D. R. Strong. 1973. On foliage arthropod communities of Puerto Rican second growth vegetation. *Ecology* 54:628–37.

Aponte, C., G. R. Barreto, and J. Terborgh. 2003. Consequences of habitat fragmentation on age structure and life history in a tortoise population. *Biotropica* 35:550–55.

Archibald, S., and W. J. Bond. 2003. Growing tall vs. growing wide: tree architecture and allometry of *Acacia karroo* in forest, savanna, and arid environments. *Oikos* 102:3–14.

Barnosky, A. D., P. L. Koch, R. S. Feranec, S. L. Wing, and A. B. Shabel. 2004. Assessing the causes of Late Pleistocene extinctions on the continents. *Science* 306:70–75.

Beschta, R. L., and W. J. Ripple. 2008. Wolves, trophic cascades, and rivers in the Olympic National Park, USA. *Ecohydrology* 1:118–30.

Blossey, B., and R. Nötzold. 1995. Evolution of increased competitive ability in invasive nonindigenous plants: A hypothesis. *Journal of Ecology* 83:887–89.

Bowen, L., and D. van Vuren. 1997. Insular endemic plants lack defenses against herbivores. *Conservation Biology* 11:1249–54.

Burness, G. P., J. M. Diamond, and T. Flannery. 2001. Dinosaurs, dragons, and dwarfs: the evolution of maximal body size. *Proceedings of the National Academy of Sciences U.S.A.* 98:14518–23.

Campbell, D. G., K. S. Lowell, and M. E. Lightbourn. 1991. The effect of introduced hutias (*Geocapromys ingrahami*) on the woody vegetation of Little Wax Cay, Bahamas. *Conservation Biology* 5:536–41.

Carlquist, S. 1965. *Island Life*. Garden City, NY: The Natural History Press.

———. 1970. *Hawaii: A Natural History*. Garden City, NY: The Natural History Press.

———. 1974. *Island Biology*. New York: Columbia University Press.

Chesson, P. L., and R. R. Warner. 1981. Environmental variability promotes coexistence in lottery competitive systems. *American Naturalist* 117:123–43.

Clough, G. C. 1969. The Bahamian hutia: A rodent refound. *Oryx* 10:106–9.

———. 1972. Biology of the Bahamian hutia, *Geocapromys ingrahami*. *Journal of Mammalogy* 55:670–72.

Coblentz, B. E. 1978. The effects of feral goats (*Capra hircus*) on island ecosystems. *Biological Conservation* 13:279–86.

Coe, M. J. 1967. *The Ecology of the Alpine Zone of Mount Kenya*. Monographiae Biologicae No. XVII. The Hague: W. Junk.

Coe, M. J., D. Bourn, and I. R. Swingland. 1979. The biomass, production and carrying capacity of giant tortoises on Aldabra. *Philosophical Transactions of the Royal Society of London, Series B* 286:163–76.

Coe, M. J., D. H. Cummings, and J. Phillipson. 1976. Biomass and production of African large herbivores in relation to rainfall and primary production. *Oecologia* 22:341–54.

Coley, P. D., J. P. Bryant, and F. S. Chapin III. 1985. Resource availability and plant anti-herbivore defense. *Science* 230:895–99.

Connell, J. H. 1978. Diversity in tropical rain forests and coral reefs. *Science* 199:1302–10.

Courchamp, F., M. Langlais, and G. Sugihara. 1999. Control of rabbits to protect island birds from cat predation. *Biological Conservation* 89:219–25.

Croll, D. A., J. L. Maron, J. A. Estes, E. M. Danner, and G. V. Byrd. 2005. Introduced predators transform subarctic islands from grassland to tundra. *Science* 307:1959–61.

Diamond, J. M. 1975. Assembly of species communities. In *Ecology and Evolution of Communities*, ed. M. L. Cody and J. M. Diamond, 342–444. Cambridge, MA: Harvard University Press.

Eskildsen, L. I., J. M. Olesen, and C. G. Jones. 2004. Feeding response of the Aldabra giant tortoise (*Geochelone gigantea*) to island plants showing heterophylly. *Journal of Biogeography* 31:1785–90.

Feeley, K. J., and J. W. Terborgh. 2005. The effects of herbivore density on soil nutrients and tree growth in tropical forest fragments. *Ecology* 86:116–24.

———. 2006. Habitat fragmentation and effects of herbivore (howler monkey) abundances on bird species richness. *Ecology* 87:144–50.

———. 2008. Trophic drivers of species loss from forest fragments. *Animal Conservation* 11:366–68.

Gerlach, J. 2004. *Giant Tortoises of the Indian Ocean*. Frankfurt am Main: Edition Chimaira.

———. 2005. Interpreting morphological and molecular data on Indian Ocean giant tortoises. In *African Biodiversity: Molecules, Organisms, Ecosystems*, ed. B. A. Huber, B. J. Sinclair, and K-H. Lamp, 213–19. New York: Springer.

Green, P. T. 1997. Red crabs in rain forest on Christmas Island, Indian Ocean: Activity patterns, density and biomass. *Journal of Tropical Ecology* 13:17–38.

Green, P. T., P. S. Lake, and D. J. O'Dowd. 1999. Monopolization of litter processing by a dominant land crab on a tropical oceanic island. *Oecologia* 199: 435–44.

Green, P. T., D. J. O'Dowd, and P. S. Lake. 1997. Control of seedling recruitment by land crabs in rain forest on a remote oceanic island. *Ecology* 78:2474–86.

Hairston, N. G., F. E. Smith, and L. B. Slobodkin. 1960. Community structure, population control, and competition. *American Naturalist* 94:421–44.

Holt, R. D. 1977. Predation, apparent competition and the structure of prey communities. *Theoretical Population Biology* 12:197–229.

———. 1996. Food webs in space: An island biogeographic perspective. In *Food Webs: Integration of Pattern and Process*, ed. G. A. Polis and K. O. Winemiller, 313–23. New York: Chapman and Hall.

Hurtt, G. C., and S. W. Pacala. 1995. The consequences of recruitment limitation: Reconciling chance, history, and competitive differences between plants. *Journal of Theoretical Biology* 176:1–12.

Iverson, J. B. 1982. Biomass in turtle populations: A neglected subject. *Oecologia* 55:69–76.

James, H. F., and D. A. Burney. 1997. The diet and ecology of Hawaii's extinct flightless waterfowl: Evidence from coprolites. *Biological Journal of the Linnean Society* 62:279–97.

Janzen, D. H. 1973a. Dissolution of mutualism between *Cecropia* and its *Azteca* ants. *Biotropica* 5:15–28.

———. 1973b. Sweep samples of tropical foliage insects: Effects of seasons, vegetation types, elevation, time of day, and insularity. *Ecology* 54:687–708.

———. 1975. Behavior of *Hymenaea courbaril* when its predispersal seed predator is absent. *Science* 189:145–47.

Lambert, T. D., G. H. Adler, C. M. Riveros, L. Lopez, R. Ascanio, and J. Terborgh. 2003. Rodents on tropical land-bridge islands. *Journal of Zoology (London)* 260:179–87.

Lopez, L., and J. Terborgh. 2007. Seed predation and seedling herbivory as factors in tree recruitment failure on predator-free forested islands. *Journal of Tropical Ecology* 23:129–37.

Lubchenco, J. 1978. Plant species diversity in a marine intertidal community: Importance of herbivore food preference and algal competitive abilities. *American Naturalist* 112:23–39.

MacArthur, R. H., J. M. Diamond, and J. R. Karr. 1972. Density compensation in island faunas. *Ecology* 53:330–42.

MacArthur, R. H., and E. Pianka. 1966. On optimal use of a patchy environment. *American Naturalist* 100:603–9.

MacArthur, R. H., and E. O. Wilson. 1967. *The Theory of Island Biogeography*. Princeton, NJ: Princeton University Press.

Molino, J-F., and D. Sabatier. 2001. Tree diversity in tropical rain forests: A validation of the intermediate disturbance hypothesis. *Science* 294:1702–4.

Morales, L. C., and S. Gorzula. 1986. The interrelations of the Caroní River Basin ecosystems and hydroelectric power projects. *Interciencia* 11:272–7.

Morgan, G. S., and C. A. Woods. 1986. Extinction and the zoogeography of West Indian land mammals. Biological Journal of the Linnean Society 28: 167–203.

O'Dowd, D. J., P. T. Green, and P. S. Lake. 2003. Invasional "meltdown" on an oceanic island. *Ecology Letters* 6:812–17.

O'Dowd, D. J., and P. S. Lake. 1990. Red crabs in rainforest, Christmas Island: differential herbivory of seedlings. *Oikos* 58:289–92.

Oksanen, L., S. D. Fretwell, J. Arruda, and P. Niemelä. 1981. Exploitation ecosystems in gradients of primary productivity. *American Naturalist* 118: 240–61.

Oksanen, L., and T. Oksanen. 2000. Logic and realism of exploitation ecosystems hypothesis. *American Naturalist* 155:703–23.

Orihuela Lopez, G., J. Terborgh, and N. Ceballos. 2005. Food selection by a hyperdense population of red howler monkeys (*Alouatta seniculus*). *Journal of Tropical Ecology* 21:445–50.

Owen-Smith, R. N. 1988. Megaherbivores: The influence of very large body size on ecology. Cambridge: Cambridge University Press.

Paine, R. T. 1980. Food webs: Linkage, interaction strength, and community structure. *Journal of Animal Ecology* 49:667–85.

Rao, M., J. Terborgh, and P. Nuñez. 2001. Increased herbivory in forest isolates: Implications for plant community structure and composition. *Conservation Biology* 15:624–33.

Scheffer, M., S. Carpenter, J. A. Foley, and B. Walker. 2001. Catastrophic shifts in ecosystems. *Nature* 413:591–96.

Schoener, T. W. 1989. Food webs from the small to the large. *Ecology* 70: 1559–89.

Schoener, T. W., and D. A. Spiller. 1996. Devastation of prey diversity by experimentally introduced predators in the field. *Nature* 381:691–94.

Sherman, P. M. 2002. Effects of land crabs on seedling densities and distributions in a mainland neotropical rain forest. *Journal of Tropical Ecology* 18: 67–89.

Spiller, D. A., and T. W. Schoener. 1990. A terrestrial field experiment showing impact of eliminating top predators on foliage damage. *Nature* 347:469–72.

Steadman, D. W. 2006. Extinction and biogeography of tropical Pacific birds. Chicago: University of Chicago Press.

Steadman, D. W., R. Franz, G. S. Morgan, N. A. Albury, B. Kakuk, K. Broad, S. E. Franz, K. Tinker, M. P. Pateman, T. A. Lott, D. M. Jarzen, and D. L. Dilcher. 2007. Exceptionally well preserved late Quarternary plant and vertebrate fossils from a blue hole on Abaco, The Bahamas. *Proceedings of the National Academy of Science U.S.A.* 104:19897–902.

Terborgh, J., and J. Faaborg. 1980. Saturation of bird communities in the West Indies. *American Naturalist* 116:178–95.

Terborgh, J., K. Feeley, M. Silman, P. Nuñez, and B. Balukjian. 2006. Vegetation dynamics of predator-free land-bridge islands. *Journal of Ecology* 94:253–63.

Terborgh, J., L. Lopez, and J. Tello. 1997a. Bird communities in transition: The Lago Guri Islands. *Ecology* 78:1494–501.

Terborgh, J., L. Lopez, J. Tello, D. Yu, and A. R. Bruni. 1997b. Transitory states in relaxing land bridge islands. In *Tropical Forest Remnants: Ecology, Management, and Conservation of Fragmented Communities*, ed. W. F. Laurance and R. O. Bierregaard, Jr., 256–74. Chicago: University of Chicago Press.

Valone, T. J., M. Meyer, J. H. Brown, and R. M. Chew. 2001. Timescale of perennial grass recovery in desertified arid grasslands following livestock removal. *Conservation Biology* 16:995–1002.

Viljoen, P. J. 1989. Spatial distribution and movements of elephants (*Loxodonta africana*) in the northern Namib Desert region of the Kaokoveld, South West Africa/Namibia. *Journal of Zoology* 219:1–19.

Toward a Trophic Island Biogeography

REFLECTIONS ON THE INTERFACE OF ISLAND BIOGEOGRAPHY AND FOOD WEB ECOLOGY

Robert D. Holt

IN THIS ESSAY, I explore the interplay of two of the most important conceptual frameworks in community ecology—island biogeography and food web ecology (figure 6.1). My goal is to lay out steps toward their synthesis—with the ultimate objective being to stimulate the fuller development of what we might call "trophic island biogeography." I start by sketching key insights at the heart of each paradigm, and point out ways they were already related (albeit for the most part implicitly, or sketchily) in the famed 1967 monograph by Robert MacArthur and and E.O. Wilson, *The Theory of Island Biogeography*. I then use simple modifications of the canonical model of colonization and extinction on an island presented in that monograph to consider questions such as top-down effects of predators on the species-area relationships of prey, and bottom-up effects of prey on food chain length and predator species-area relationships. Next, I consider a number of interesting complications which arise when bottom-up and top-down effects occur simultaneously, and in particular emphasize the potential importance of island area as a moderator of intrinsically unstable trophic interactions. To round off the paper, I briefly discuss a number of areas of active inquiry in community ecology that will be important for a fully developed trophic island biogeography, and then conclude by reflecting on how trophic interactions in fragmented landscapes in some ways resemble, and in other ways radically differ from, those in isolated oceanic islands.

Island Biogeography Theory

A central question posed in the opening chapters of MacArthur and Wilson's monograph was: *What factors govern variation in the number of species found on islands, as a function of island area and distance from*

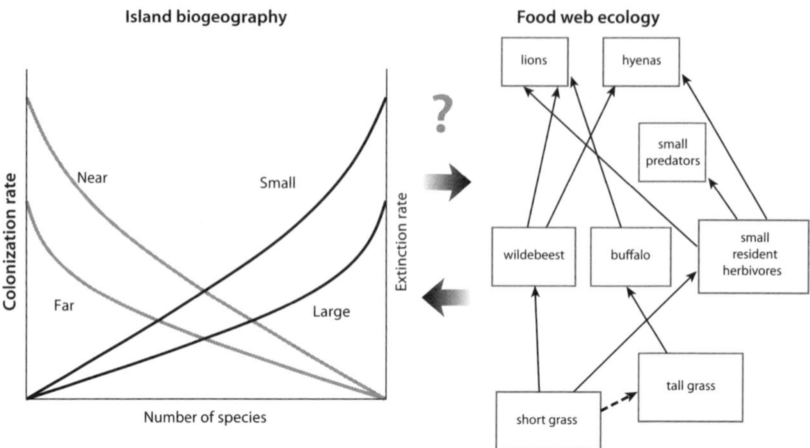

Figure 6.1. Two of the most important conceptual paradigms in community ecology—island biogeography and food web ecology. The text explores the question of how these paradigms are related. The right panel is a simplified food web of the Serengeti ecosystem (Holt et al. 2008).

continental source pools? Their answer, the "equilibrium theory," as portrayed in the model on the left side of figure 6.1, focused on colonization and extinction (Schoener, this volume). This theory embodies two crucial insights that go well beyond island biogeography. First, *communities at all spatial scales are dynamic.* Viewed over the grand span of earth history, local communities ("local" denotes the spatial scale where individuals potentially interact, for instance by competition) assemble via colonization from external sources (augmented by occasional *in situ* speciation) and are depleted by extinctions (Graham et al. 1996). MacArthur and Wilson (1967) argued that a similar dynamism occurs even over shorter time scales. Subsequent literature has often focused on the celebrated, and indeed controversial, hypothesis by MacArthur and Wilson (1967) that communities are at or near equilibrium, so the number of species remains roughly constant in the face of continual turnover in composition. But the deeper message that communities are dynamic does not depend on the assumption of equilibrium. Long-term censuses on both islands and continents often reveal extinctions and recolonizations over short time scales (Williamson 1981, Schoener, this volume). Extinctions can be deterministic—due to disturbance, succession, interspecific interactions, or shifts in climate—or simply the stochastic winking in and out of rare community members. Unraveling the mechanics of community assembly and disassembly mandates a close focus on colonization and extinction, which are thus essential for understanding all communities, whether or not they reach equilibrium.

Second, *space matters*. Most ecology textbooks show how the curves in figure 6.1 (left) vary with island area and distance. Colonization should reflect an island's distance from sources of colonists and the ability of species to traverse dispersal barriers. This insight was not new to MacArthur and Wilson (1967), but they did elegantly articulate the logic of demographic influences on colonization, as well as stepping stones and other determinants of colonization rates, using quantitative approaches that set a high standard for subsequent ecological theory. Within continents, spillover of species among habitats can boost local diversity; the absence of such spillover may lead to lower diversity on islands than on comparably sized mainland areas (MacArthur and Wilson 1967, pp. 16 and 115; Holt 1993, Rosenzweig 1995). Second, the area of an island influences extinction rates. This is partly simply because larger areas harbor more individuals—a "pure area" effect—and partly because larger areas contain more distinct habitats, which can buffer extinctions and sustain specialized niches—an "environmental diversity" effect. The pure area effect can reflect two processes. If a species' density is constant, its absolute numbers will scale with island area; smaller populations face larger dangers of extinction from demographic risk and other factors (Schoener, this volume). Moreover, if colonization is analogous to random sampling from a continental fauna, as small islands have few total individuals they in effect are a small sample and so could contain few species by chance alone (Schoener, personal communication). The emphasis on space was a fundamental insight provided by the theory of island biogeography that still resonates throughout both basic ecology and applied arenas such as conservation biology (Laurance, this volume).

Food Web Theory

The second canonical paradigm in figure 6.1—the food web—goes back at least to Charles Elton, with an intellectual lineage running through Lindeman, Hutchinson, Cohen, Pimm, and many others up to the present. The powerful metaphor of communities as interactive webs has stimulated an enormous amount of creative work. For instance, one can view webs as abstract networks of connections and focus on efficient descriptors describing those patterns (e.g., Martinez 1992). Or one can attach dynamical equations to each node (e.g., Yodzis 1998) and explore the implications of web structure for issues such as the relationship between stability and complexity (e.g., McCann 2000, Kondoh 2003), the vulnerability of webs to disturbance, invasion, and the extinction of resident species (e.g., Dunne et al. 2002), and the relative strength of top-down and bottom-up forces.

What is the relationship between these two ecological paradigms? Until recently, very little. Classical studies of food webs paid scant attention to the influence of spatial processes on food web structure and dynamics. The excellent monograph on food webs by Stuart Pimm (1982), for instance, deals with space only with respect to how distinct habitats can lead to food web compartmentalization. Tom Schoener (1989) in an important paper did provide an insightful discussion of how food chain length might be influenced by island size, and his paper helped stimulate some growth in this area (for reviews see Holt and Hoopes 2005, Polis et al. 2004). But until quite recently (Amaresakare 2008), analyses of spatial patterns and processes have overall been a rather minor theme in the food web literature.

Conversely, I think it is fair to say that classic island biogeography theory (and its modern descendant, metacommunity theory [Hubbell 2001, Holyoak et al. 2005]) largely emphasized the "horizontal" structure of communities, such as potential competition between members of a guild or taxon, with little attention given to food webs per se. Yet although MacArthur and Wilson (1967) do not directly discuss food webs, it should be noted that they do state that the extinction curve should be concave because of "interference" among species; interference might well include predation, as well as exploitative and interference competition (the concavity in the extinction curve may also arise because of variation in species-specific rates; see Schoener, this volume). Moreover, they do touch upon trophic interactions in two short, but telling, passages. In chapter 5, "Invasibility and the variable niche," the section titled "The closed community" comments on how predators influence coexistence. "Each of the conditions for reduction of diversity—competitors too similar, species too rare, predators too rare (or too common)—can prevent invaders from colonizing." This statement suggests that *local food web interactions can govern colonization*. In chapter 6, "Evolutionary changes following colonization," one reads "impoverishment of diversity often leads to lack of effective predators. This is because the *K* of predators is considerably lower than that of their prey, so they are precariously rare even on large islands." One way to parse this passage is that trophic structure (and in particular trophic rank) influences extinction. The second sentence in this quotation implies the first, in the sense that, if predators are differentially vulnerable to extinction, then communities with low diversities on islands are particularly likely to lack predators. Sampling effects could also play a role; effective predators may be absent in species-poor assemblages by chance alone. An alternative interpretation of the first sentence is that the impoverishment of prey diversity *itself* leads to a lack of *effective* predators. One mechanism leading to this is the increase of predator abundance with prey species richness, permitting predators to more effectively limit any particular prey population. This is apparent competition (Holt 1977, Holt and Lawton 1994), an indirect interaction among alternative prey species arising from a predator's numerical response to the entire suite of prey in its diet. The basic idea hinted at in chapter 6 of *The Theory of Island Biogeography* is thus that *trophic structure and rank can influence extinction rates.*

Hence, food web interactions may govern the two basic processes of island biogeography theory—colonization and extinction. Conversely, local food web structure itself should reflect these same processes. All local food webs are assembled by colonization, and depleted by extinction, both of which are spatially mediated processes. A recognition of the interplay of these two paradigms suggests that the time is ripe for their fusion into a "trophic island biogeography." As a start toward such a theory, it is useful to take the simplest version of the MacArthur-Wilson equilibrial theory, and ask how a consideration of trophic position influences its predictions for broad categories such as "predators and prey," or "specialist and generalist predators." The next sections present several complementary approaches to this theme.

Trophic Status as a Predictor Variable in Island Biogeography

As a simple start, with a food web in hand, by using various protocols (e.g., counting links up from the base, or using stable isotopes; Post and Takimoto 2007), one can assign a trophic rank to each species and then contrast "predators" (a set of high-ranked species) to "prey" (a set of low-ranked species). There could be systematic population-level attributes correlated with trophic rank that directly influence colonizing ability or extinction risk. For instance, predators are often rarer than their prey (Spencer 2000), and thus, *ceteris paribus*, more likely to go extinct on small islands due to demographic and environmental stochasticity. Figure 6.2 shows how these considerations influence a noninteractive model of island communities. The model is

$$\frac{dS_i}{dt} = I_i(K_i - S_i) - E_i S_i \tag{6.1}$$

where S_i is the number of species in a given trophic set i (for now, respectively, predator or prey), K_i is the number of species in this set i in the mainland species pool, I_i is the colonization rate per species, and E_i

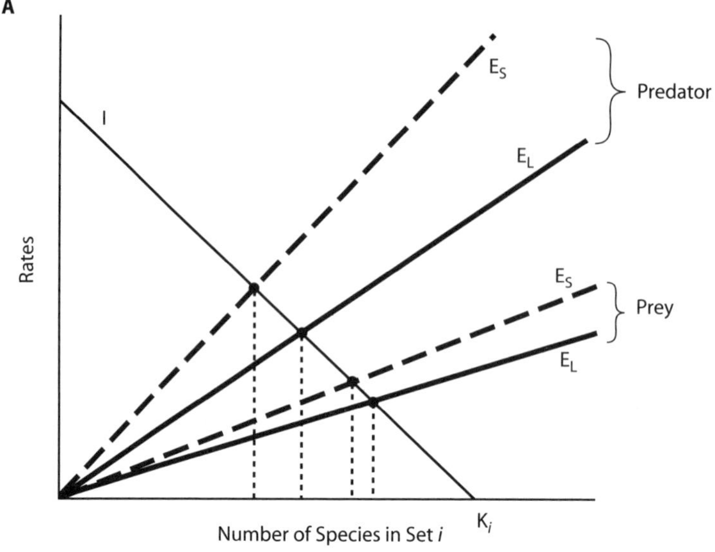

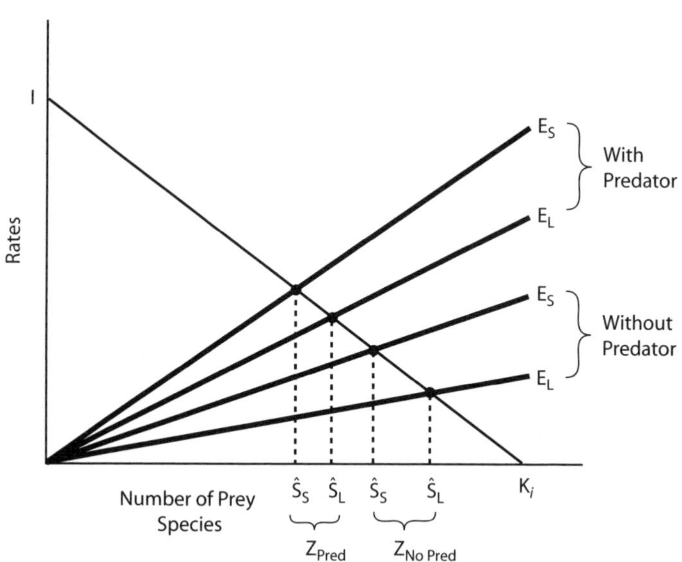

Figure 6.2. The MacArthur-Wilson equilibrial model applied to predators and prey. A. As explained in the text, as a deliberately oversimplified starting point, we assume a non-interactive community in which we have taxonomic or functional grounds to separate "predators" from "prey." For simplicity, we assume immigration rates are equivalent for these two classes. If predators are typically less dense than prey, this may not affect extinction rates on a large island much, but would make predators much more sensitive than their prey to reduced island size. B. Predators are present on both large and small islands. In the example shown, increased extinction due to predation reduces the effect of island area upon prey species richness (after Rydberg and Chase 2007).

is an extinction rate. (For simplicity, I assume that colonization and extinction rates are linear.) We assume extinction declines with the logarithm of island area A, i.e., $dE_i/d\log(A) < 0$. The equilibrial species richness in trophic set i is

$$S_i^* = \frac{I_i K_i}{I_i + E_i(a)} \tag{6.2}$$

[$a = \log(A)$]. If the strength of the species-area relationship for trophic set i is

$$z_i = d\log(S_i)/d\log(A) \tag{6.3}$$

(apt for any relationship that is roughly a power law, $S = cA^z$), after a little manipulation we have

$$z_i = \frac{|dE_i/da|}{I_i + E_i(a)} \tag{6.4}$$

(the vertical lines denote absolute value; log here refers to natural log). The numerator measures the sensitivity of extinction rates to island area. Start with islands large enough that all species have low extinction rates. As island size decreases, it may be reasonable to expect extinction rates for predators to increase more sharply than for prey, simply because predators tend to be relatively rare. Due to demographic stochasticity, a decline in a predator from 1000 to 100 individuals should increase extinction risk much more than a proportional decline in its prey from 10,000 to 1000, and so the numerator of (6.4) should be larger for predators. As indicated in figure 6.2A, this leads to the very simple prediction that there should be a stronger species-area relationship for predators than for prey. This prediction is not watertight, for z also depends on the rates in the denominator of (6.4). If extinction rates are high, few species will be present, and z-values will all be low, so there would only be minor, nearly undetectable differences between predators and prey. Equation (3.6) in Schoener (this volume; see also Schoener 1976a) relates the number of species present on an island to their aggregate density. If the total density is independent of species richness (a zero-sum assumption), this equation predicts that, over a given range of island areas, among taxa with comparable colonization and extinction rates and source pool diversities, those taxa with lower aggregate densities will show stronger species-area relationships than do taxa with higher aggregate densities. Consistent with this prediction, Schoener (1976a) notes that. in general, birds with relatively low z have relatively high summed population densities and vice versa; in particular, raptors have relatively high z.

The effect expected for distance is less clear. Predators are often larger than their prey and might behaviorally avoid physical transport processes that could take them across water gaps; this reduces colonization. A low immigration rate in the denominator of (6.3) inflates the impact of area sensitivity on extinction, and so increases z for predators. By a comparable argument, one expects a stronger species-distance relationship for predators than for prey. Some evidence matches this prediction. Shulman and Chase (2007) showed in experimental mesocosms that the ratio of predator to prey species declined with distance from a source pond (figure 6.3). Yet some predators are highly mobile, readily crossing barriers that impede prey. Greater mobility at higher trophic ranks should weaken species-area and species-distance relationships for predators, compared with their prey. The whole issue of how trophic rank influences colonization cries out for more empirical study and mechanistic modeling.

The model of figure 6.2 is a reasonable place to start, but it blatantly ignores the fact that the fates of predator and prey are closely intertwined. Comparable arguments pertain to any grouping of species into sets that differ in colonization and extinction rates (e.g., large- vs. small-body species in the same trophic level; species near the edges of their climatically defined geographical ranges vs. species near their range centers). The next section presents a first step toward incorporating trophic interdependencies.

Top-Down Effects in Island Biogeography

Sometimes, predators may be distributed largely independently of island/patch area and distance. Humans, for instance, deliberately or inadvertently introduce predators onto islands, or into islandlike habitats (e.g., trout have been introduced into isolated glacial lakes in New Zealand). The distribution of these predators should then be largely independent of prey species. How does such extrinsically determined predation modify prey colonization and extinction dynamics? The incidence and abundance of prey on islands can be strongly influenced by predation. This is particularly dramatic for introduced alien predators (Salo et al. 2007), but also occurs for predators and prey with a shared evolutionary history. Adler and Levins (1994) note that rodent numbers often increase with decreasing island area, and suggest that this reflects predator presence and abundance. An excellent example comes from islands in the Thousand Island Region of the St. Lawrence River, where occupancy and density of the short-tailed shrew (*Blarina brevicauda*) decline with

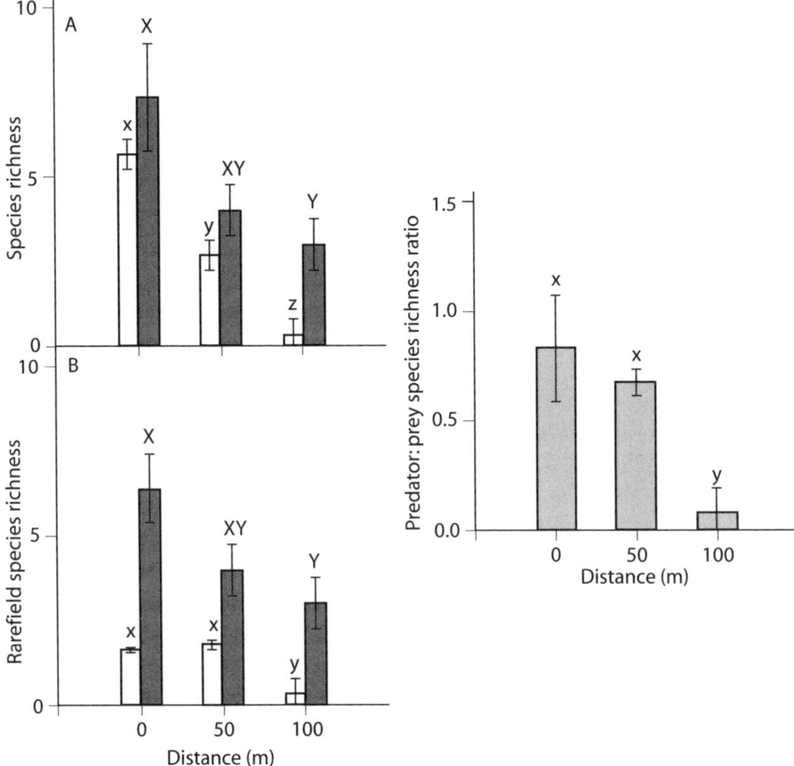

Figure 6.3. Predator-prey ratios vary systematically with distance from a source. Aquatic mesocosms (plastic tubs) were placed at varying distances from a pond, a source for aquatic insect colonists. In the left column, open and closed columns are respectively predators and prey; the top is raw data, the bottom, rarefied data. The right column is the ratio of predator to prey species There is a strong signal of distance from the pond on the trophic composition of the mesocosms, with predator species richness declining relatively more strongly at large distances (from Shulman and Chase 2007).

distance from the mainland, and conversely occupancy and density of the meadow vole (*Microtus pennsylvanicus*) increase (Lomolino 1984). *Blarina* disperses poorly across open water and ice; this explains its absence on distant, small islands. *Blarina* is also a voracious generalist predator, so given that it can colonize, its persistence may largely be independent of the vole. Conversely, when the shrew is present, it can limit or even eliminate *Microtus*. Thus, the vole exhibits ecological release on islands when freed of *Blarina* predation (Lomolino 1984). Likewise,

Nordstrom and Korpimaki (2004) showed in Fennoscandia that introduced minks are constrained to islands close to sources, and that mink predation in turn leads to a positive relationship between island bird species richness and distance. The presence of predators may act synergistically with disturbance to elevate prey extinction risks (Schoener et al. 2001). Experiments also show that predators can substantially reduce prey colonization success (Schoener and Spiller 1995, Kotiaho and Sulkava 2007).

Several authors have modified the basic MacArthur-Wilson (1967) model by adding top-down impacts of consumers onto prey extinction and colonization rates. Olff and Ritchie (1998) examined how herbivory influences plant species richness, where the presence of the herbivore is governed by extrinsic factors (e.g., as in livestock husbandry). They used a graphical model comparable to figure 6.2 to illustrate how grazing alters species richness by shifting colonization and extinction curves. For instance, by disturbing soil, herbivores open sites for germination, thus potentially boosting colonization. When grazers selectively attack competitive dominants, they may relax competition and reduce local extinctions (Harper 1969). Conversely, if grazers are unselective and grazing pressure is sufficiently intense, or competitively dominant plants can tolerate grazing better than can competitively inferior species, herbivores can boost extinction rates (Lubchenco 1978). Increases in extinction due to predation are likely common. For instance, Schoener and Spiller (1996) showed experimentally that predatory lizards directly depress spider prey species richness by elevating extinctions.

Ryberg and Chase (2007) recently modified the simple noninteractive model given by equation (6.1) by assuming that predators elevate extinction rates of prey by a constant additive amount, independent of island area. Here, I generalize their approach, allowing both intrinsic extinctions and extinctions from predation to vary with island area, as follows:

$$\frac{dS_i}{dt} = I_i(K_i - S_i) - (E_i(a) + E_i(a)')S_i. \qquad (6.5)$$

The equilibrial species richness is

$$S_i^* = \frac{I_i K_i}{I_i + E_i(a) + E_i'(a)}. \qquad (6.6)$$

Ryberg and Chase (2007) predict that, if predators uniformly and additively increase per species extinction rates of prey, islands with predators will have a more shallow species-area relationship than islands without predators. Manipulation of (6.6) shows that

$$z_i = \frac{\left|dE_i/da + dE_i'/da\right|}{I_i + E_i(a) + E_i'(a)}. \tag{6.7}$$

If predators elevate extinction uniformly across all islands, the second term in the numerator is zero, and there is an additional positive term in the denominator. This implies a lower z-value due to predation (figure 6.2B). If predation-driven extinctions increase with island size, the species-area relationship of the prey will be even weaker; decreased extinctions permitted by increasing island size will tend to be canceled out by increased extinctions from predation. Conversely, if extinction rates from predation are magnified on small islands, the effect of island size on species richness may be enhanced.

Equation (6.6) assumes that the most natural way to represent the impact of predation upon prey extinction is via an additive term. This is mathematically convenient, but does not as yet follow from any more microscopic derivation. Alternatively, one could assume that predators alter extinction rates multiplicatively by x, so that the extinction rate of the prey is $x(a)E_i(a)$ (T. Schoener, personal communication). After substitution, and manipulation, we find that

$$z_i = \frac{\partial \log(S_i)}{\partial a} = \frac{-1}{I + xE}\left(x\frac{\partial E}{\partial a} + E\frac{\partial x}{\partial a}\right). \tag{6.7'}$$

If the impact of predation upon prey extinction is independent of island area, $x > 1$ implies that predation *increases* the strength of the species-area relationship in prey.

Further study is required to determine whether (6.7) or (6.7′) provides the most "natural" or parsimonious representation of predation impacts upon prey extinction. But empirically there is support in the literature for the effects of predators on prey z-values going in both directions. Support for the prediction that predation flattens the species-area relationship comes from Ryberg and Chase (2007), who examined distributional patterns in two island-like habitats: orthopteran richness in Ozark glades (open rocky outcrops within a forest matrix), with and without the insectivorous collared lizard *Crotaphytus collaris*; and man-made ponds, with and without fish as predators on zooplankton. In both cases, for larger patch sizes, islands without predators clearly contained a greater richness of prey species than did islands with predators, and the former also had higher z-values. At low ranges of areas in both study systems, however, contrary to the model predictions (and as noted by Ryberg and Chase), the species-area relationships converged, suggesting minimal or no impact of predation upon prey species richness on small islands, or even possibly a slight positive effect. An area dependence in the impact of predation could reflect several factors. One such factor is that, among islands occupied by predators, their densities may decline sharply with decreasing island size (as shown in Lomolino [1984] for *Blarina*). For generalist predators like collared lizards and shrews, the reduced prey species richness expected on smaller islands may translate to a lower carrying capacity. If total mortality inflicted by predators on prey scales with predator density, the contribution of predation to extinctions in a focal group of prey species may be less important on smaller islands, because predators, even if present, tend to be rare.

But in other cases the impacts of predators on prey on small islands, compared to on large islands or continents, may be severe. Schoener and Spiller (1999) used removal experiments in the Bahamas to show that lizard predators much more strongly reduce spider density and species richness on small islands than on large islands. Several distinct mechanisms could be at play (and Schoener and Spiller [1999] suggest still others). Resources available for the prey themselves may be limited on small islands. If so, prey cannot tolerate as much predation and still persist, and even if they do persist it may only be at a lower abundance. Reduction to low densities by predation aggravates the risk of stochastic extinctions, just because absolute abundances are low on small islands. Fewer refuges may be available on small islands, making prey more vulnerable to exclusion from persistent generalist predators. Finally, generalist predators may be able to persist on just a few prey species, which permits the predators to drive other prey species extinct. Thus, top-down effects could amplify the species-area relationship in a prey guild.

Bottom-Up Effects in Island Biogeography

Now, I reverse the assumptions of the previous section. A food web at the very least describes bottom-up asymmetrical resource dependencies among species. For now we will assume the distribution of predators depends upon that of their prey, and for simplicity (relaxed below) assume also that, by contrast, prey distributions are independent of predation. I start by sketching the classic problem of the determinants of food chain length, focusing on specialist food chains, and then turn to the influence of trophic rank on the strength of the species-area relationship.

Understanding what limits food chain length is a long-standing puzzle in ecology. Ecological communities vary much more in species richness than in food chain length. But why? Traditional explanations are nicely summarized in Pimm (1982) and Post (2002), and these hypotheses have implications for how island size and distance might influence food chain length. For instance, energetic constraints suggest that longer food chains are expected in more productive habitats. Schoener (1989) generalized this observation and provided one way to link space to food web theory by pointing out that the total energy production of an island is productivity (energy/unit time/unit area) times area. He suggested that instead of productivity, *per se*, the total production contained within an island might govern the food chain length it can support—the "productive space" hypothesis. Schoener described this hypothesis as follows: "maximum food-chain lengths are determined by the amount of productive space required to allow critical component species populations [namely, ones at the top of the food web] to persist with some high probability." The hypothesis rests on a population-size argument. Consider a continental community with a classic "pyramid of numbers," so that density declines with increasing trophic rank in a food chain. Absolute population size is of course density times area. If we consider islands which have identical environmental conditions, but differ in area, a null model is that population size (total numbers, not density) for each species will be proportional to area. If there is a critical population size below which extinction is certain, the area at which this threshold will be reached will be larger for species at higher trophic ranks. This implies shorter food chains on smaller islands. Alternatively, assume that we compare these islands with another set of islands, which have a uniformly higher primary productivity. If this increase in production translates into a comparable increase in density at each trophic level, working through the same argument, one predicts that, with higher productivity, there is a lower critical island size below which the top predator dips below its critical abundance, than is observed on islands with lower productivity.

The productive space hypothesis is appealing, and is surely part of the story, but the jury is still out on the degree to which it entirely explains variation in food chain length among communities. Production does seem to be related to the decline in species diversity with increasing trophic rank (Rosenzweig 1995, Havens 1992, Duffy 2002), but the evidence to date suggests that it does not fully account for area effects on food chain length (Post 2002). One complication is that increased primary production may not translate neatly into proportional increases in abundance at each trophic level. For instance, shifts in species composition at lower trophic levels toward inedible species can lower the amount of production passing through to higher trophic levels. Satiation or interference competition may constrain predator numerical responses to increased food supplies. Increased production can destabilize predator-prey interactions; excursions to low densities may then aggravate extinction risks (the classic "paradox of enrichment"), particularly on small islands.

Finally, spatial subsidies on small islands can elevate the food base for predators above that expected from *in situ* productivity (Anderson and Wait, 2001; Schoener, this volume).

An alternative way for island area (and distance) to influence food chain length involves the consideration of trophic dependencies among species, in their own right. Introducing trophic dependencies into colonization-extinction dynamics can lead to the expectation that food chain length will increase with island area. I here summarize models exploring this idea presented earlier (Holt 1993, 1996, 1997a,b, 2002; see also Schoener et al. 1995) and weave in new thoughts and examples.

All species need resources and to some degree have specialized diets. If a species arrives on an island lacking its required resources, it cannot persist. On a continent, recurrent immigration can sustain "sink" populations at sites without resources, but if the distance between the mainland and island is sufficiently great, such sink populations will be absent or vanishingly rare. Consider an unbranched food chain of "stacked specialists." Species i has trophic rank i and feeds on species $i-1$. A useful descriptor of island distributions is the *incidence function* (Diamond 1975), which gives the percentage of islands occupied by species i, $p(i)$, as a function of island area, or distance to the mainland, or other island traits. In a food chain of stacked specialists, at equilibrium the incidence of species i is constrained by the incidence of all lower-ranked species on which it directly or indirectly depends. This leads to nested spatial distributions; islands without species $i-1$ are guaranteed not to harbor species i, but the converse need not hold.

We now define a *conditional incidence function* $p(i \mid i-1)$ to be the conditional probability that species of rank i is present, given that its required resource, species $i-1$, is present. Often, conditional incidence will increase with island area. Specialist herbivores, for instance, are often more likely present on larger populations of their host plants (Otway et al. 2005). The unconditional incidence function for species i is a product of conditional incidence functions, up the food chain:

$$p(i) = p(1)\prod_{j=2}^{i} p(j|j-1). \tag{6.8}$$

With this expression, and some simple assumptions, we can draw conclusions about how food chain length should vary with area and distance. The expected food chain length is simply the sum of incidence functions, up the chain:

$$E[L] = \sum_{i=1}^{n} p(i). \tag{6.9}$$

Assume that the incidence function for the basal species and the conditional incidence function for each higher-ranked species all increase with island area and decrease with increasing distance from the mainland. By application of the chain rule, we find that the expected food chain length also increases with area, and decreases with distance. As an example, Komonen et al. (2000) report that, following forest fragmentation, a specialist food chain supported by a bracken fungus was truncated on small forest fragments. So, with almost no biology at all, other than assuming trophic specialization and the garden variety expectation that island area and distance affect the likelihood that a species will be present, we can predict effects of island area and distance on food chain length.

As noted above, a principal motivation of MacArthur and Wilson's monograph was to understand how species richness covaried with island area and distance. Instead of a single food chain, assume the mainland community has m "stacked specialist" chains. What is the effect of trophic rank on z? For simplicity, assume all species of rank i have the same conditional incidence function. The expected number of species of rank i is simply $S_i = mp\ (i)$. The strength of the species-area relationship on a log-log plot is

$$z_i = \frac{d\log(S_i)}{d\log(A)} = \frac{d\log(p(i))}{d\log(A)} = z_{i-1} + \frac{1}{p(i|i-1)}\frac{d\,(p(i|i-1))}{d\log(A)}. \tag{6.10}$$

If conditional incidence increases with area, this expression implies that

$$z_1 < z_2 < z_3 < \ldots. \tag{6.11}$$

The strength of the species-area relationship should thus increase with trophic rank.

Trophic Island Biogeography: Steps Toward Generality

"Stacked specialist" food chain models are a sensible starting point for the development of a theory of trophic island biogeography. But such trophic specialization does not typify most food webs, which contain a mix of tight specialists and highly generalized consumers. Developing models of multispecies webs which pay attention to the detailed pattern of trophic interactions, and how these change during community assembly to feed back onto colonization-extinction dynamics, is a significant challenge. One approach is to craft detailed community assembly models that specify rules for the explicit distribution of trophic specialization and generalization in source food webs, and then use these to assemble island communities. Here I focus instead on an alternative approach to trophic island biogeography. I ignore the details of the web of interactions and instead make broad qualitative assumptions about how diversity in one trophic level influences rates of colonization and extinction in another, using a somewhat simpler and extended version of a model presented in Holt and Hoopes (2005). The goal is to craft qualitative theoretical predictions describing how species richness scales with area, contrasting generalists with specialists, and predators with their prey.

We first assume donor control, so predators do not influence prey colonization-extinction dynamics. The prey follow model (6.1) above and show island area and distance effects. Colonization-extinction dynamics in the predators is controlled in a bottom-up fashion by the number of prey species present on an island, S, as well as by island area and distance. It is well known that there can be a codependency in species richness among trophic levels. For instance, the composition of local arthropod herbivore communities is strongly affected by plant community composition (Siemann et al. 1999, Schaffers et al. 2008). So a reasonable rule of thumb is that a more diverse prey base should be able to support a more diverse assemblage of consumers.

The number of predator species on an island is P, which can change by colonization or extinction. This is assumed to given by an expression like (6.1) above. I use a prime to denote predator immigration and extinction rates. The immigration rate of the predator guild I' is assumed to increase with the number of prey species present on the island. Likewise, we assume the extinction rate E' decreases with increasing island area, for a fixed number of prey species, and also decreases with an increasing number of prey species, for a fixed island area. Taking logarithms of (6.2), as before, after some manipulation it can be shown that the z-value of the predators is related to the z-value of their prey by the following compact expression:

$$z_{pred} = z_{prey} Q + \frac{1}{I' + E'} \left| \frac{dE'}{d\log(A)} \right|. \tag{6.12}$$

where

$$Q = \frac{E'}{I'(E' + I')} \frac{dI'}{d\log(S)} + \frac{1}{E' + I'} \left| \frac{dE'}{d\log(S)} \right|. \tag{6.13}$$

The first term on the right side of expression (6.12) describes the indirect effect of area upon predator log(species richness), mediated through the species richness of the prey. The second term describes the direct effect of area upon predator extinction, controlling for prey species richness. With these expressions in hand, we can now address several qualitative issues in trophic island biogeography.

How should the z-values for specialists differ from those for generalist predators? Consider colonization. It is often reasonable to expect predator colonization to increase with prey species richness. For a specialist, colonization requires the prior presence of its required prey. On small, species-poor islands, there is a high probability that any particular prey species will be absent, precluding colonization by specialists that need it. Colonization by specialists should be more likely, the more prey species are present. For generalists, colonization may also depend positively upon prey species richness. For instance, an increased number of prey species may increase the total food supply and permit a higher initial rate of increase. If different prey species provide distinct limiting nutrients (called "obligate generalism" in Holt et al. 1999), colonizing predators may require multiple prey species to enjoy positive growth rates at all. But more usually, a generalist should be able to colonize communities containing many different subsets of the mainland prey community. If so, there may be a relatively weak effect of prey species richness upon colonization by generalists, compared to specialists.

It also seems reasonable that predator extinction rates should decrease with an increase in prey species richness. Ritchie (1999) provides a nice empirical example for prairie dog colony extinction rates, which decline with increasing plant species richness. But again, this effect may be stronger for specialists than for generalists. For specialist predators, their extinction rates can be no less than those of their required prey types—when a given prey species goes extinct, it drags all its specialist consumers with it. Generalists, by contrast, may subsist on other prey species, and so a reduction in prey species richness could imply a more modest increase in extinction rates. This should imply a lower Q for generalists, compared to specialists.

The final term in (6.12) is the direct effect of area upon predator extinction rates, controlling for prey species richness. Two factors are at play here. First, all else being equal, a decrease in area will proportionally shrink absolute population sizes. A systematic difference in the average densities of generalist vs. specialist predators would then imply a comparable difference in area sensitivity. I know of no data that directly address systematic differences in abundance as a function of degree of trophic specialization. Second, specialist predator-prey interactions are prone to unstable dynamics, with recurrent phases at low densities. Predators face a differential risk of extinction in these phases, a risk that is magnified on small islands. Moreover, as discussed below, small islands may lack spatial mechanisms that stabilize specialist predator-prey dynamics, further aggravating extinction risks of specialist predators versus generalists. It is thus plausible to hypothesize that extinction rates of specialists will be more sensitive to area, than will be the case for generalists.

These observations lead to the prediction that for a given trophic level $z_{specialist} > z_{generalist}$. When will predators have a steeper species-area relationship than their prey, i.e., $z_{pred} > z_{prey}$? It is sufficient that $Q > 1$, which is more likely if both predator immigration and extinction rates vary strongly with prey species richness. Direct area effects on the predator can also make it possible for predator z-values to exceed those of their prey, even if $Q < 1$.

If one accepts the above arguments, it is overall more likely for the z-values of predators to exceed those of their prey, when predators are relatively specialized in their diets; when overall immigration rates of predators are low, relative to extinction; and, when there are additional effects of area upon predator extinction rates, arising for reasons other than the effect of area upon prey species richness.

Empirical studies of the relationship between trophic rank and the species-area relationship, where comparison is made among taxa within a given set of islands or habitat patches, reveal patterns broadly consistent with these theoretical expectations. In a nice study of how trophic specialization influences the species-area relationship, Steffan-Dewenter and Tscharntke (2000) showed that the predicted effect of trophic generalization on the magnitude of z is found in butterflies differing in dietary breadth and distributed across habitat fragments; z-values increase monotonically from butterflies which are extreme generalists, to oligophages, to tight specialists on a single host plant (figure 6.4). Trophic generalists had lower z-values (between 0.05 and 0.1) than their host plants (0.13), whereas oligophages and monophages had higher values (0.16 and 0.21). This pattern matches the above theoretical predictions. Kruess and Tscharntke (2000) report species-area relationships for herbivorous insects, and their relatively specialized parasitoids, in meadows of red clover and vetch in central Europe, and demonstrate that z is considerably higher for the parasitoids than for their hosts (figure 6.5). Holt et al. (1999) review other examples. In assemblages dominated by trophic specialists, stronger species-area relationships (higher z) typically are seen at higher trophic ranks. But generalists reveal a mix. Some examples fit, but others do not. Even generalists can show strong area effects. For instance, Spencer et al. (1999) studied effects on predator extinction in arthropod communities in temporary ponds in Israel, and found the proportion of the community comprised of generalist predators to increase

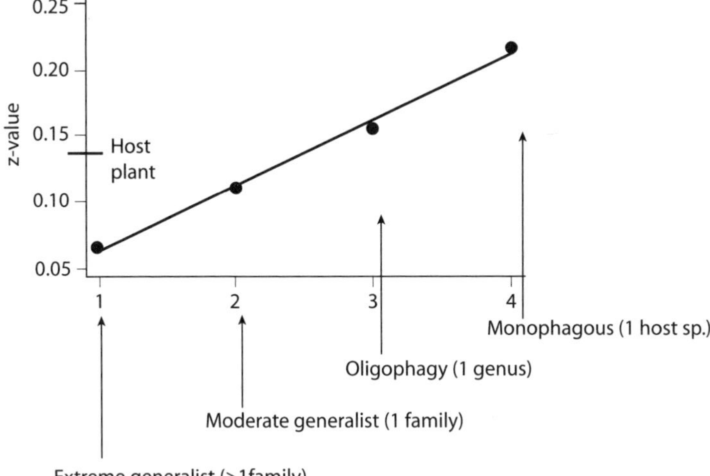

Figure 6.4. In butterflies of central Europe, there is a systematic relationship between the value of z, and the degree of trophic specialization (from Steffan-Dewenter and Tscharntke 2000).

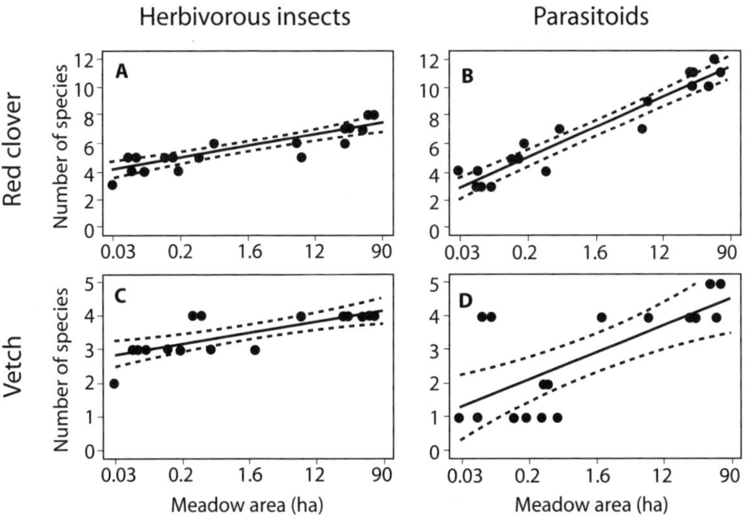

Figure 6.5. An example of stronger species-area relationships for specialist natural enemies (parasitoids) than their prey (host insects), in meadows in central Europes (from Kruess and Tscharntke 2000).

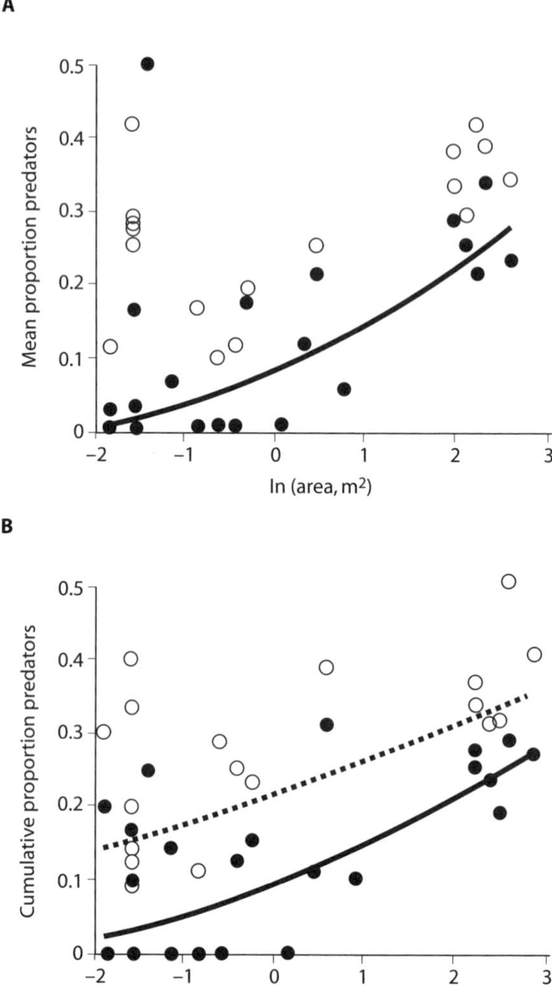

Figure 6.6. An example of stronger species-area relationships for generalist predators than for their prey (aquatic organisms in temporary ponds) (from Spencer et al. 1999, and unpublished data provided by Leon Blaustein). Closed circles: macroscopic predators; open circles: all predators.

strongly with log(area) (figure 6.6). John Glasser (1982) reanalyzed the classic Simberloff-Wilson (1969) study of arthropod communities on mangrove islets and found a suggestion of successional patterns in web structure. He classified species into three trophic groups: herbivores, predators, and parasites, and then plotted their colonization curves. One result (his figure 7) reveals a pronounced area effect on trophic organization: at the end of the study, the large islands E7 and E9 had a larger predator species-to-herbivore species ratio than did the small islands E1 and E2. But invertebrate predators on islands in the Gulf of California do not show a systematic increase in z-values with trophic rank (Holt et al. 1999; G.A. Polis, personal communication); consumers such as scorpions are highly generalized and have lower z-values than do lower-ranked trophic levels on the same islands (e.g., plants).

Piechnik et al. (2008) have recently analyzed the Simberloff-Wilson dataset in more detail, and conclude that there is a succession in niche breadth among consumers, with generalists colonizing before specialists. It is plausible, as Piechnik et al. suggest, that this reflects the sequential dependence of colonization expected for specialist consumers, which have to wait for establishment of their required resources before colonizing, as assumed in the theory sketched above. As Montoya et al. (2006) note, some community patterns may best be explained by an assembly "process whereby species sequentially partition resources as they invade an ecological community. Rare, trophically specialized species enter the community later than do generalists."

An alternative, complementary explanation for the higher z-values shown by specialists may be that generalists are good colonists for reasons other than their ability to exploit a variety of prey. From (6.12) and (6.13), if immigration rates are higher for generalists, then even with comparable area dependencies in the rate constants, the z-values for generalists will be lower. Why might this be a reasonable expectation? Model (6.1) (*et seq.*) assumes a noninteractive community. If we consider competition among predators for a moment, the question that arises is what permits the coexistence of specialist and generalist consumers? Given trade-offs in exploitative ability, as a broad rule of thumb (albeit with exceptions) one expects guilds of specialists, each with skills honed to their own particular prey, to outcompete generalists. Generalists could nonetheless persist in a metacommunity, given a trade-off between competitive abilities and colonizing abilities, so that generalists arrive before specialists, say, following local disturbances. This might preadapt generalist consumers to be among the earlier colonizers onto isolated islands.

Putting the Pieces Together: Some First Steps

In reality, communities emerge from the interplay of both bottom-up and top-down forces, as well as "horizontal" forces (competition, mutualism). This leads to a wide range of complex and interesting issues in spatial community ecology (Amaresakare 2008), and below I explore some that must be considered en route to a fully fleshed-out theory of trophic island biogeography.

One way to proceed is to develop models that explicitly describe colonization and extinction by each species. For a moment, consider again a food chain of stacked specialists. Schoener (Schoener et al. 1995, Appendix) and I (Holt 1996, 1997) independently developed Markov chain patch occupancy models that, in the spirit of MacArthur and Wilson (1967), track colonization and extinction at each trophic level. With this model, we relax the assumption of donor control. I will not repeat the analyses here but instead summarize results. The "state" of each island is the length of its food chain. For simplicity, we assume the basal species in the chain to be an effective colonizer, i.e., its incidence is unity. A fraction of islands, P_1, have just the basal species (e.g., a plant), a fraction P_2 have that species and a prey species that utilizes it (e.g., an herbivore), and the remaining fraction P_3 have the full food chain. The predator can colonize only after the prey has become established. If the prey species goes extinct, so does the predator; in addition, the predator might go extinct on its own. A model based on these assumptions is

$$\frac{dP_2}{dt} = c_{12}(1 - P_2 - P_3) - e_{21}P_2 - c_{23}P_2 + e_{32}P_3$$
$$\frac{dP_3}{dt} = c_{23}P_2 - e_{32}P_3 - e_{31}P_3 \qquad (6.14)$$

(The subscript "*ij*" denotes "transition from state *i* to state *j*.") At equilibrium, we can solve to examine how occupancies depend on area. We assume that extinction rates decline with increasing area, and consider three basic possibilities:

1. $e_{21}=e_{31}$. Prey extinction is not affected by the predator. We might call this "biogeographic donor control." In food web ecology, donor control denotes situations in which resource recruitment is independent of consumption by a consumer. If a predator does not alter prey extinctions, then even if predation is biologically significant (e.g., causes decreased local prey abundance), this will not be reflected in occupancy.
2. $e_{21} > e_{31}$. The prey extinction rate is reduced by the predator. This seems counterintuitive, but the effect is well grounded in theory and empirical examples are known. May (1972), for instance, showed in a model of a three-link food chain that a top predator attacking an herbivore could stabilize plant-herbivore dynamics if the top predator experiences direct density dependence (e.g., from territoriality), and the herbivore on its own has weak direct density dependence and is easily saturated by its own resource. On small islands in the Baltic, for instance, voles in

the absence of predation explode to high numbers and overgraze their food resources to the point of local extinction, whereas numbers stay steady and bounded away from zero when predators are present (Banks et al. 2004).

3. $e_{21} < e_{31}$. The final possibility is for the predator to increase prey extinctions. This may be the most likely of the three logical possibilities.

For the two first possibilities, larger island area implies a longer equilibrial food chain length. In the third, food chain length can *decrease* with increasing island area, or proximity to the source. This paradoxical effect can arise if an increase in area strongly decreases extinctions by the predator alone. An intuitive explanation goes as follows. If the predator colonizes small islands, by assumption it goes extinct rapidly, leaving the prey behind. But on a large island, the predator may persist and grow, drive its prey extinct, and then itself go extinct, reinitializing the island with just the basal species. Averaging over food chains on all islands of a given size, one might find shorter chains on larger islands, because these are precisely the arenas where predators persist long enough to exterminate their prey. This effect is particularly likely if predators have alternative resources which permit them to persist, at least for a while, in the absence of the focal prey species.

Fundamental features of predator-prey ecology suggest that there should be strong dependencies on island area of extinction rates in food webs. Classical predator-prey theory predicts that, if predators effectively limit their prey, unstable dynamics arise with periods at low densities. On a small island there will be recurrent periods of low absolute abundances, hence elevated extinction risks. All else being equal, unstable predator-prey interactions should be more persistent on large islands. This is a pure area effect.

Another pure area effect arises because the larger the island, the less likely it will contain well-mixed populations. Many taxa are relatively sluggish, and with limited within-island dispersal, partially independent populations are likely to emerge within large islands (Holt 2002). One active area of research in community ecology at present is metacommunity ecology (Holyoak et al. 2005), which is an intellectual descendant of island biogeographic theory. A "metacommunity" is a set of local communities, connected by dispersal. In a metacommunity, colonization into a focal patch comes from other occupied patches, rather than a fixed external source. Even if one is primarily interested in islands, there are good reasons to consider the implications of metacommunity dynamics for understanding within-island processes. When a species first colonizes an island, it rarely immediately occupies the entirety of the island, but establishes a beachhead, from which it expands. If dispersal is limited *within* islands, one can view island area as being a proxy for the number of local sites potentially connected by within-island dispersal (Holt 1992). Larger islands in effect are larger meta-communities, comprised of more such local sites. Area effects on extinction rates reflect the diverse ways island area influences internal metacommunity dynamics.

Predator-prey models incorporating space, dispersal, and localized interactions in metacommunities are often more stable than nonspatial models (Holt 1984, Hosseini 2003), due to several distinct stabilizing mechanisms that emerge in spatially distributed systems. All these mechanisms should be sensitive to area, and so could contribute to systematic effects of island area on food web structure. There are several recent reviews of the influence of space on the persistence and stability of predator-prey and food web interactions (Hassell 2000, Briggs and Hoopes 2004, Holt and Hoopes 2005), and here I summarize key insights that seem particularly germane to island biogeography.

Even in homogeneous areas, localized interactions, limited dispersal, and stochastic variation generate heterogeneities in population abundance and interaction strengths that are broadly stabilizing (Hassell 2000, Briggs and Hoopes 2004). A large area can contain many local populations that become asynchronous in their dynamics, given limited within-island dispersal, permitting persistence of locally unstable predator-prey interactions. Experimental studies suggest that with localized predator-prey interactions, persistence is enhanced with increasing size of the arena containing the interaction (Huffaker 1958, Holyoak and Lawler 1996, McCauley et al. 2000, Ellner et al. 2001). Theoretical models predict that spatial patterns such as traveling waves emerge at scales larger than the local population, but smaller than the whole system, and these patterns can contribute to stability (Hassell et al. 1991). But these emergent spatially patterned interactions have characteristic spatial scales (Donalson and Nisbet 1999, Gurney and Veitch 2000), and so cannot be sustained on small islands (Hassell et al. 1991). Wilson et al. (1998) considered a food chain of a hyperparasitoid, a primary parasitoid, and a basal host, all interacting on a lattice, in effect an island with local dispersal and highly unstable local interactions. This theoretical study revealed that food chain persistence was strongly sensitive to lattice size. An order-of-magnitude larger lattice was needed to sustain the full tritrophic interaction, compared to the host-parasitoid interaction (figure 6.7). Larger islands also often contain internal hetereogeneities (e.g., distinct habitats) leading to spatial variation in parameters such as attack rates and intrinsic growth rates. In general, such environmental heterogeneities can stabilize predator-prey systems (Holt 1984, Hassell 2000, Schreiber et al. 2006).

Developing patch occupancy models for more complex multispecies assemblages is a challenging task, because of the proliferating number of

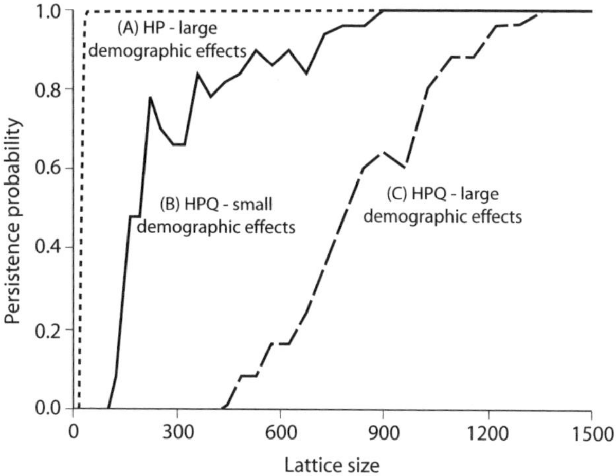

Figure 6.7. Larger lattices (a surrogate for island area) are more likely to retain strongly interacting food chains of hyperparasitoids, parasitoids, and hosts (from Wilson et al. 1998).

possible states and transitions (see Holt 1997a, 2002 for complexities arising even for simple food chains in a metacommunity context). The above models just blithely ignored all the reticulate detail of the structure of the web of interactions among species. As one example of the importance of such details, food chains may in some cases be longer on larger islands not because of the sequential additions of species at increasingly higher trophic ranks, as assumed above. A given predator species may be found across all islands, but be at a realized higher trophic rank on larger islands because those islands are also occupied by additional species at various intermediate ranks (Post and Takimoto 2007). For instance, in the Midwest United States the lake trout is the top predator across a wide range of lake volumes, but it is at a higher realized trophic rank in larger lakes, which compared to small lakes have many additional species of zooplankton and smaller fish providing long chains linking phytoplankton to the trout (Post et al. 2000b).

In general, larger areas may permit the persistence of otherwise unstable multispecies trophic interactions. When two prey species share a common enemy, one can indirectly exclude the other locally via the numerical response of their shared enemy. But this strong apparent competition (Holt and Lawton 1994, Hamback et al. 2006) may not cause extinction in spatially extended systems, if the prey inferior at withstanding predation more effectively colonizes empty patches, or if the predator prefers the prey with faster growth (King and Hastings 2003). Bonsall et al.

(2005) showed this experimentally for a parasitoid attacking two species of bruchid hosts; coexistence was prolonged when the interaction played out in a larger spatial arena. In a field study, Cronin (2007) showed experimentally that one plant hopper (*Delphacoides schlochoa*) strongly suppressed to the point of local extinction another plant hopper (*Prokelisia crocea*) (even though the two hosts occupied distinct habitats) due to numerical responses of shared parasitoids that straddled these habitats. He argued that coexistence occurred regionally because the species superior at withstanding the shared parasitoids was a poorer disperser. Such coexistence mechanisms are ineffective on small islands.

Moving to entire food webs, in his celebrated book on species invasions, Charles Elton (1958) argued that islands are prone to unstable dynamics and vulnerable to invasion, because of reduced species richness. McCann et al. (2005a,b) observe that this pattern (assuming it is true) could instead reflect the fact that island food webs are spatially constrained and so not buffered by the stabilizing mechanisms that emerge from interspecific interactions played out in expansive spatial arenas. Spencer and Warren (1996) carried out experiments on multispecies webs in small aquatic microcosms, where they compared the productive space hypothesis of Schoener (1989) with the effects of area, per se. They concluded that their results did not fit the productive space hypothesis very well, but "that spatial effects on the persistence of unstable food webs may be important." Spatial heterogeneity permits many mechanisms to operate—predator switching among habitats, source-sink relationships, and transient refuges—stabilizing even complex food webs (Holt 1984, Post et al. 2000a, Kondoh 2003, Eveleigh et al. 2007, Goodwin et al. 2005, Gripenberg and Roslin 2007). Conversely, on small islands the inherent instability of strong trophic interactions can be unleashed and cause extinctions. On large islands, within-island metacommunity processes may help counter the many ways species-rich webs have of being locally unstable.

Future Directions in Linking Food Webs to Island Theory

It is useful to provide pointers to some of the interesting and challenging complexities that need to be addressed in a mature trophic island biogeography. Many of these reflect important intellectual currents in contemporary community ecology.

Interaction modifications. There are behavioral effects by which predators can indirectly influence prey persistence in metacommunities. For example, the presence of predators in a patch can induce prey to emigrate, enhancing colonization rates into empty patches (Gilliam and

Fraser 2001, Prakash and de Roos 2002), thus facilitating prey persistence. This behavioral effect needs considerable space to operate effectively, and so might help further explain why strong local predator-prey interactions can persist on large islands, but not on small islands. Many other kinds of interaction modification (Abrams 1983) could modulate colonization-extinction dynamics. For instance, nonprey can interfere with the ability of a predator to capture its prey (Vos et al. 2001, Kratina et al. 2007, van Veen 2005); this is called "associational resistance" in plant-herbivore interactions (Atsatt and O'Dowd 1976, Hamback et al. 2000, Aquilino et al. 2005, Callaway et al. 2005). Such facilitation among prey has several consequences for trophic island biogeography. As overall prey species richness increases with island area, the stability of a specialist predator-prey interaction could be enhanced, relative to a monoculture, because predators are less able to overexploit their prey. So extinction rates of specialist predators and their prey may decline on larger islands. Countering this effect, however, successful colonization by specialist predators may be inhibited in richer prey communities. Colonization rates by specialists might actually peak at intermediate island sizes, then decline on larger islands.

Moreover, predator diversity can have diverse effects on the overall consumption of prey (Casula et al. 2006). Such diversity can augment predation pressure on prey (van Ruijven et al. 2005, Snyder et al. 2006), for instance because prey have fewer places to hide or modes of behavior that permit predator avoidance. Or, predators may interfere with each other, relaxing predation on their shared prey. If predator diversity increases with prey species richness, which effect predominates will govern how prey colonization and extinction rates change with predator species richness, which can then feed back onto colonization by the predators themselves.

Ecosystem dimensions of trophic island biogeography. Flows of materials between marine and terrestrial ecosystems can profoundly impact island communities. On unproductive islands, a regular influx of subsidies from marine sources can sustain terrestrial consumers even on very small islands (Anderson and Wait 2001), which can then exploit resident prey more effectively. Conversely, pulsed subsidies can lead to periods of relaxed predation upon resident island prey (Schoener, in press). Predators can limit the abundance of species (e.g., seabirds) that are key conduits of nutrients between islands and marine environments (Maron et al. 2006).

Transients. Oceanic island communities are likely to assemble one species at a time. After a species colonizes a food web, there is often a phase of pronounced transient dynamics, where abundances deviate very sharply from long-term equilibrial values, possibly for long periods of time (Hastings 2004). For instance, when a resident predator and prey are present, and a second prey species which does not compete directly with the resident is introduced, large-amplitude cycles in all species result enroute to a long-term stable equilibrium (Holt and Hochberg 2001). Though in the long run all species mathematically persist in this deterministic model, in biological practice extinctions may occur when species pass through transient low-density troughs. Noonburg and Abrams (2005) show that in a standard model of keystone predation—where a top predator facilitates coexistence of competitors by feeding preferentially on the dominant prey—invasion by one prey species into a community with the other species initially present and at equilibrium leads to very low densities, which in practice would likely preclude realistic coexistence. All these newly recognized effects of transient dynamics should be particularly important in small oceanic islands, where absolute abundances are in any case low. By contrast, on a continental island, the initial community is carved out of the original mainland biota, and such transient dynamics emerging during assembly should be less important in determining current community structure.

Cyclic assembly processes. Theoretical and experimental studies of food web assembly reveal that local communities receiving immigrants from an external source can go through cycles, from state A to B, and back again, or from A to B to C to D . . . and finally back to A. Cyclic compositional changes are common in theory (e.g., Morton and Law 1997, Steiner and Leibold 2004). Warren et al. (2003) in a microcosm study with protists found that the community could exist in two states, which we dub A and B. Predatory species could invade A, and transform it into B, and then themselves go extinct. After B had settled down, the predators could reinvade, and take the community back to A, and again the predators went extinct. For this process (and indeed any cyclical dynamics in composition) to be maintained, there needs to be a supply from external sources (either a continent or metacommunity) for one or more species.

A plausible cyclic assembly scenario emerges from considering the implications of garden variety, uncontroversial community ecology played out on islands, which can be understood (I hope) even without equations (see figure 6.8). Consider a source where two predators share two biotic, noncompeting resource populations and stably coexist (upper left corner of the figure). Predator coexistence requires niche partitioning, which we assume suffices for coexistence but is incomplete (i.e., there is dietary overlap). Assume predator 1 has a rate of exploitation of prey 1 (denoted by α) higher than on prey 2 (α'). Reciprocally, assume predator 2 is better at exploiting prey 2 at rate α but also exploits prey 1 at rate α'. In simple cases, for instance if the two predators have linear functional and

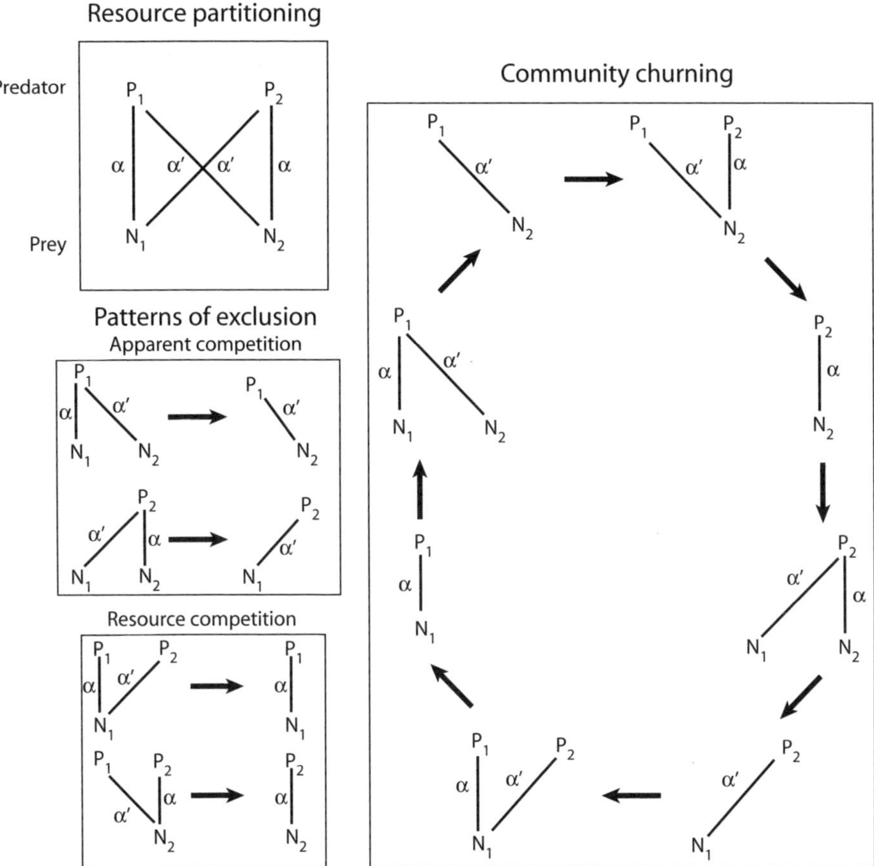

Figure 6.8. Community churning. Two predators persist on two prey species, because of resource partitioning on a mainland. Because of the reciprocal forces of resource competition and apparent competition in each three-species module, a series of colonizations and extinctions can be observed on islands, leading to a perpetual cycle in island community composition. See main text for details.

numerical responses to their prey, and the prey have logistic growth, it can be shown that an equilibrium with all species exists and is locally stable. We assume this food web module persists on the mainland, and that the two predators are effective at limiting prey abundance below carrying capacity. Despite the local stability of this four-species module at equilibrium, species losses can lead to a cascade of additional extinctions. If we consider the three-species subwebs within this four-species module, it is clear why instability looms, should a species be lost. Say a prey species is missing. If both predators are still present, we expect competitive exclusion; the predator better at utilizing that prey supplants the other. If instead one of the predators is absent, given our assumption about effective predation, exclusion due to apparent competitive advantage can occur; the prey species experiencing the lower predation rate indirectly supplants the more vulnerable prey species, mediated through the shared predator's numerical response.

When an island community is assembled, an interesting phenomenon emerges. If colonizations are rare, in any given time period only one species is likely to colonize. We start with an empty island. The two prey species colonize first, then a predator. This predator overexploits the prey species to which it is best adapted (as measured by the attack rate), leaving it sustained by the prey to which it has the lower attack rate. When the other predator colonizes, the first predator is now competively excluded. But the resulting two-species configuration is now open for colonization by the alternative prey species (which experiences a lower attack rate than the resident), after which the resident prey species is supplanted. This in turn permits the original predator to colonize again, restarting the cycle. These alternative shifts in species composition, driven by reciprocal shifts in the relative importance of resource competition and apparent competition, can lead to a constant, if leisurely, churning in island species composition, with colonists drawn from a stable mainland community. Variability in species composition among comparable islands may reflect not just the chance vicissitudes of colonization, but emergent heterogeneities due to inherent community instabilities.

More complex webs, and parasites. The theories presented above have assumed simple patterns of trophic organization, such as simple food chains, or discrete predator and prey trophic levels, as well as species with fixed properties. Realistic food webs are often very complex in their organization, with reticulate feeding relationships among large numbers of species, and on top of this complexity, the properties of food webs also should reflect the long-term imprint of coevolution among species, as well as speciation. One important class of trophic interactions that is still poorly understood in the context of food web ecology is host-parasite interactions, but it is increasingly clear that such interactions are ubiquitous and dynamically important (Lafferty et al. 2008). There are many potential implications of parasitisim for trophic island biogeography. For instance, many host-specific pathogens have strong area effects in incidence, with an increasing probability of being present on larger patches containing more of their hosts (e.g., the smut fungus *Ustilago scorzonerae* on its asteraceous host, *Scorzonera humilis*; Colling 2004). Although rather poorly documented, in some systems it is clear that the parasite load is less on distant islands; for instance, *Anolis* lizards in the northern Lesser Antilles have depauperate parasite faunas (Dobson et al. 1992). The presence of parasites can have surprising effects on predator-prey interactions. For instance, if prey sustain a pathogen, selective predation on infected prey can at times increase prey numbers and also prevent devastating epidemics (Packer et al. 2003). The strength of the effect of the predator on prey numbers may be greater on small islands, because pathogens are missing there. There may even be profound evolutionary effects from depauperate parasite communities on islands. Ricklefs and Bermingham (2007) suggest that one reason the Lesser Antilles had a more modest avian radiation than either the Galapagos or the Hawaiian Islands is that the latter two archipelagoes have relatively few pathogens, which when present can prevent secondary sympatry of budding species due to disease-mediated competition.

Food Webs in Fragmented Habitats

Why did *The Theory of Island Biogeography* resonate so thoroughly? Many biologists find islands intrinsically fascinating, and the interplay of empirical patterns and mathematical theory in MacArthur and Wilson (1967) presented a new paradigm for ecological studies. But beyond this, the late 1960s and 1970s were a time of increasing concern among environmentalists and scientists about the serious environmental problems caused by humanity around the globe, most notably extinction threats caused by habitat destruction and fragmentation. Indeed, Macarthur and Wilson (1967) opens on this theme: their first figure is the celebrated diagram by Curtis of forest fragmentation from a section of land in Wisconsin. Scenes of tropical deforestation—rich forests replaced by depauperate cattle pastures or miles upon endless deadeningly dull miles of oil palm plantations—are depressingly familiar to any well-traveled biologist. Fragmentation creates land-locked "islands" of habitat. The conceptual perspective provided by the island metaphor sparked an explosion of work on habitat fragmentation (Harris 1984), including observational studies, theory development, and long-term landscape experiments. Examples include the ongoing Biological Dynamics of Forest Fragments Project near Manaus, Brazil, 1979–present (Bierregard et al. 2001), and my own project on secondary succession in a fragmented landscape, near Lawrence, Kansas, 1983–present (Robinson et al. 1992, Cook et al. 2005). Laurance (this volume) provides an overview of the value—and limitations—of island theory for understanding habitat fragmentation.

A ubiquitous implication of habitat fragmentation is the disruption, elimination, or magnification of preexisting trophic interactions (see Terborgh chapter). Theoretical studies (Holt 1993, Holt et al. 1999, Bascompte and Sole 1998, Sole and Bascompte 2006, Sole and Montoya 2006) suggest that species (especially specialists) at higher trophic ranks may be differentially vulnerable to fragmentation. I mentioned above several studies of species-area relationships from fragmented habitats, consistent with these predictions. Empirical studies show that parasitoids (which are often relatively specialized) are more extinction-prone than their hosts (e.g., Cronin 2004), leading to reduced parasitism on smaller or more isolated habitat fragments (Kruess and Tscharntke 1994, Elzinga et al. 2005, Steffan-Dewenter and Tscharntke 2002, Tscharntke et al. 2002, Tscharntke and Brandl 2004, Valladores et al. 2006, van Nouyys 2005). At the community level, this differential susceptibility to fragmentation can lead to reduced predator-to-prey ratios with decreasing patch area (Didham et al. 1998, Ryall and Fahrig 2006), to trophic cascades (more intense herbivory on smaller patches where prey are freed from predation; Terborgh et al. 2001), and to steeper species-area relationships for predators than prey among fragments (Hoyle 2004). So some fragmentation effects do seem to match the above predictions of trophic island biogeography about food chain length and trophic influences on the strength of the species-area relationship.

However, although island biogeography continues to provide a powerful metaphor for thinking about habitat fragmentation, with the maturation of conservation biology it has become widely recognized that this metaphor can be limited, and at times misleading. Habitat fragments in some ways are like islands, but in some ways are radically different (Ewers and Didham 2006, Watling and Donnelly 2006, Laurance chapter). Edge effects can penetrate deep into fragments (Ewers and Didham 2006). The area separating fragments is not an empty sea, a mere barrier to dispersal, but sustains communities which often utilize the fragmented habitats to some extent. Coupling of distinct habitats by consumer or resource movement is a ubiquitous landscape process (Polis et al. 1997). Even as specialist predators become less important on small fragments (as predicted by trophic island biogeographic theory), generalist predators may become ever more present. For example, Robinson et al. (1995) showed that in the Midwestern United States, nest predation on forest birds by generalist predators increased strongly with fragmentation. Rand and Louda (2006) likewise showed that insect herbivores in remnant prairie patches in Nebraska experienced more intense predation due to generalist coccinellids sustained across a broader agricultural landscape, and comparable effects emerge in a wide range of landscape studies (Ryall and Fahrig 2006, Tscharntke et al. 2005, Rand and Tscharntke 2007). So even if top-down effects on small or distant oceanic islands are arguably unimportant, they may be very strong in small or isolated habitat patches embedded in anthropogenically modified landscapes, leading to strongly synergistic effects of predation with fragmentation (Davies et al. 2004, Rand et al. 2006). Moreover, transient dynamics are a key aspect of habitat fragmentation when landscapes shift rapidly. Holt and Hochberg (2001) conjectured that habitat destruction could lead to transient spikes in natural enemy impacts in remnant patches, as mobile predators crowd into remaining suitable areas. Thies et al. (2008) empirically demonstrated this effect; reductions in the area of rape crop cultivation led to a large short-term increase in mortality imposed by parasitoids on hosts in the remnant crop patches and elevated extinction risks, because parasitoids produced over a larger area surged into these areas.

So habitat fragments are not just islands. But it is clear that the island biogeograpic perspective has played a crucial historical role in stimulating analyses of habitat fragmentation (Laurance, this volume). Moreover, as humans continue to degrade the matrix habitat separating fragments, the long-term outcome may be island-like reserves, separated by a wasteland not all that different from a sterile ocean.

Coda

The Theory of Island Biogeography was a harbinger of the current rising tide of interest in spatial patterns and processes throughout the basic and applied ecological sciences, including food web ecology. Rather than end this essay by trying to summarize the ideas presented above, I would like to conclude on a more personal note. The *Theory of Island Biogeography* appeared in 1967. In 1970, I had the exceptional good fortune as a sophomore at Princeton of taking "Biogeography," taught by Robert MacArthur and Ed Fischer. Due to an improbable series of events, MacArthur became my advisor in a special university program, and he graciously took me along on his last lengthy field trip to Arizona in 1971, where I helped him carry out some of his foliage-profile measurements—he would stand in an opening in the chaparral, while I would disappear, thrashing along a randomly chosen direction he had picked into the thick, clothes-ripping grip of the scrub, carrying a checkerboard. The goal was for me to hold it up at different distances, until half the squares were hidden from his view. It was physically challenging, but I did manage to stumble across a Flammulated Owl, a few feet away from one of my sampling points—still the only one I have ever seen. In conversations over the campfire, and then in his office later, MacArthur gently guided my thinking toward an academic career in ecology (rather than physics, my major). On his sickbed in 1972, he handwrote letters of recommendation for me to deliver to Ed Wilson at Harvard, and elsewhere. I have no doubt this was instrumental in my getting into fine graduate schools. How lucky can a clueless young man from Tennessee be!

Acknowledgments

I have profited from many teachers, mentors, friends, and colleagues over the years, learning from them much related to the themes of my essay. As noted above, Robert MacArthur and E. O. Wilson had an impact on my own life, for which I am eternally grateful. In my undergraduate years, John Terborgh had (and continues to have) many influences on how I think about the world, and I recall with fondness my interactions with Henry Horn, John Bonner, and Tom Givnish. In graduate school, I was very fortunate to interact with Tom Schoener and Joel Cohen as teachers, and to have Bill Stubblefield and Russ Lande as friends. Since then, I have had the good fortune to have as professional collaborators thinking about these themes, and as friends, some of the finest scientists in the world: Gary Polis, John Lawton, Mike Hassell, Neo Martinez, David Post, Andrew Gonzalez, Stuart Pimm, Scott Robinson, Ilkka Hanski, David Steadman, George Robinson, Wendy Anderson, Scott Robinson, Manojit Roy, Rico Holdo, and Mike Barfield, among others. To all of you—thanks for the ride. I also thank the organizers for their invitation to contribute to this volume, and the University of Florida Foundation for its continued support.

Literature Cited

Abrams, P. A. 1983. Arguments in favor of higher-order interactions. *American Naturalist* 121:887–91.

Adler, G. H., and R. Levins. 1994. The island syndrome in rodent populations. *Quarterly Review of Biology* 69:473–90.

Amarasekare, P. 2008. Spatial dynamics of food webs. *Annual Review of Ecology, Evolution, and Systematics* 39:49–500.

Anderson, W. B., and D. A. Wait. 2001. Subsidized island biogeography hypothesis: Another twist on an old theory. *Ecology Letters* 4:289–91.

Aquilino, K. M., B. J. Cardinale, and A. R. Ives. 2005. Reciprocal effects of host plant and natural enemy diversity on herbivore suppression: An empirical study of a model tritrophic system. *Oikos* 108:275–82.

Atsatt, P. R., and D. O'Dowd. 1976. Plant defense guilds. *Science* 193:24–29.

Banks, P. B., K. Norrdahl, M. Nordström, and E. Korpimäki. 2004. Dynamic impacts of feral mink predation on vole metapopulations in the outer archipelago of the Baltic Sea. *Oikos* 105:79–88.

Bascompte, J., and R. V. Sole. 1998. Effects of habitat destruction in a prey-predator metapopulation model. *Journal of Theoretical Biology* 195:383–93.

Berlow, E. L. 1999. Strong effects of weak interactions in ecological communities. *Nature* 398:330–34.

Bierregaard, R., C. Gascon, T. E. Lovejoy, and R. Mesquita, eds. 2001. *Lessons from Amazonia: The Ecology and Conservation of a Fragmented Forest.* New Haven, CT: Yale University Press.

Bonsall, M. B., J. C. Bull, N. J. Pickup, and M. P. Hassell. 2005. Indirect effects and spatial scaling affect the persistence of multispecies metapopulations. *Proceedings of the Royal Society of London, Series B* 272:1465–71.

Bonsall, M. B., D. R. French, and M. P. Hassell. 2002. Metapopulation structures affect persistence of predator-prey interactions. *Journal of Animal Ecology* 71:1075–84.

Briggs, C. J., and M. F. Hoopes. 2004. Stabilizing effects in spatial parasitoid-host and predator-prey models: A review. *Theoretical Population Biology* 65:299–315.

Callaway, R. M., D. Kikodze, M. Chiboshvili, and L. Khetsuriani. 2005. Unpalatable plants protect neighbors from grazing and increase plant community diversity. *Ecology* 86:1856–62.

Cantrell, R. S., C. Cosner, and W. F. Fagan. 2001. How predator incursions affect critical patch size: The role of the functional response. *American Naturalist* 158:368–75.

Casula, P., A. Wilby, and M. B. Thomas. 2006. Understanding biodiversity effects on prey in multi-enemy systems. *Ecology Letters* 9:995–1004.

Chaneton, E. J., and M. B. Bonsall. 2000. Enemy-mediated apparent competition: empirical patterns and the evidence. *Oikos* 88:380–94.

Colling, G., and D. Matthies. 2004. The effects of plant population size on the interactions between the endangered plant *Scrozonera humilis* (Asteraceae), a specialized herbivore, and a phytopathogenic fungus. *Oikos*.105:71–78.

Cook, W. M., J. Yao, B. L. Foster, R. D. Holt, and L. B. Patrick. 2005. Secondary succession in an experimentally fragmented landscape: Community patterns across space and time. *Ecology* 86:1267–79.

Cronin, J. T. 2004. Host-parasitoid extinction and colonization in a fragmented prairie landscape. *Oecologia* 139:503–14.

———. 2007. Shared parasitoids in a metacommunity: Indirect interactions inhibit herbivore membership in local communities. *Ecology* 88:2977–90.

Davies, K. F., C. R. Margules, and J. F. Lawrence. 2004. A synergistic effect puts rare, specialized species at greater risk of extinction. *Ecology* 85:265–71.

Diamond, J. M. 1975. Assembly of species communities. In *Ecology and Evolution of Communities*, ed. M. L. Cody and J. M. Diamond, 342–444. Cambridge, MA: Harvard University Press.

Didham, R. K., J. H. Lawton, P. M. Hammond, and P. Eggleton. 1998. Trophic structure stability and extinction dynamics of beetles (Coleoptera) in tropical forest fragments. *Philosophical Transactions of the Royal Society of London, Series B* 353:437–51.

Dobson, A. P., S. V. Pacala, J. D. Roughgarden, E. R. Carper, and E. A. Harris. 1992. The parasites of Anolis lizards in the northern Lesser Antilles. *Oecologia* 91:110–17.

Donalson, D. D., and R. M. Nisbet. 1999. Population dynamics and spatial scale: effects of system size on population persistence. *Ecology* 80:2492–507.

Duffy, J. E. 2002. Biodiversity and ecosystem function: The consumer connection. *Oikos* 99:201–19.

Dunne, J. A., R. J. Williams, and N. D. Martinez. 2002. Network structure and biodiversity loss in food webs: Robustness increases with connectance. *Ecology Letters* 5:558–67.

Ellner, S. P., E. McCauley, B. E. Kendall, C. J. Briggs, P. R. Hosseini, S. Wood, A. Janssen, M. W. Sabelis, P. Turchin, R. M. Nisbet, and W. W. Murdoch. 2001. Habitat structure and population persistence in an experimental community. *Nature* 412:538–43.

Elton, C. S. 1958. *The Ecology of Invasions by Animals and Plants.* London: Methuen and Co.

Elzinga, J. A., H. Turin, J.M. van Damme, and A. Biere. 2005. Plant population size and isolation affect herbivory of *Silene latifolia* by the specialist herbivore *Hadena bicrurusi* and parasitism of the herbivore by parasitoids. *Oecologia* 144:416–26.

Eveleigh, E. S., K. S. McCann, P. C. McCarthy, S. J. Pollock, C. J. Lucarotti, B. Morin, G. A. McDougall, D. B. Strongman, J. T. Huber, J. Umbanhowar, and L.D.B. Faria. 2007. Fluctuations in density of an outbreak species drive diversity cascades in food webs. *Proceedings of the National Academy of Sciences U.S.A.* 104:16976–81.

Ewers, R. M., and R. K. Didham. 2006. Confounding factors in the detection of species responses to habitat fragmentation. *Biological Reviews* 81:117–42.

———. 2008. Pervasive impact of large-scale edge effects on a beetle community. *Proceedings of the National Academy of Sciences U.S.A.* 105:5426–29.

Gilliam, J. F., and D. F. Fraser. 2001. Movement in corridors: Enhancement by predation threat, disturbance, and habitat structure. *Ecology* 82(1):258–73.

Glasser, J. W. 1982. On the causes of temporal change in communities: Modification of the biotic environment. *American Naturalist* 119:375–90.

Goodwin, B. J., C. G. Jones, E. M. Schauber, and R. S. Ostfeld. 2005. Limited dispersal and heterogeneous predation risk synergistically enhance persistence of rare prey. *Ecology* 86:3139–48.

Graham, R. W., E. L. Lundelius, M. A. Graham, E. K. Schroeder, R. S. Toomey, E. Anderson, A. D. Barnosky, J. A. Burns, C. S. Churcher, D. K. Grayson, R. D. Guthrie, C. R. Harington, G. T. Jefferson, L. D. Martin, H. G. McDonald, R. E. Morlan, H. A. Semken, S. D. Webb, L. Werdelin, and M. C. Wilson. 1996. Spatial response of mammals to Late Quaternary environmental fluctuations. *Science* 272:1601–6.

Gripenberg, S., and T. Roslin. 2007. Up or down in space? Uniting the bottom-up versus top-down paradigm and spatial ecology. *Oikos* 116:181–88.

Gurney, W.S.C., and A. R. Veitch. 2000. Self-organization, scale, and stability in a spatial predator-prey interaction. *Bulletin of Mathematical Biology* 62:61–86.

Hamback, P. A., J. Agren, and L. Ericson. 2000. Associational resistance: Insect damage to purple loosestrife reduced in thickets of sweet gale. *Ecology* 81:1784–94.

Hamback, P. A. , J. A. Stenberg, and L. Ericson. 2006. Asymmetric indirect interactions mediated by a shared parasitoid: Connecting species traits and local distribution patterns for two chrysomelid beetles. *Oecologia* 148:475–81.

Harper, J. L. 1969. The role of predation in vegetational diversity. *Brookhaven Symposium in Quantitative Biology* 22:48–62.

Harris, L. D. 1984. *The Fragmented Forest: Island Biogeography and the Preservation of Biotic Diversity*. Chicago: University of Chicago Press.

Hassell, M. P. 2000. *The Spatial and Temporal Dynamics of Host-Parasitoid Interactions*. Oxford: Oxford University Press.

Hassell, M. P., H. N. Comins, and R. M. May. 1991. Spatial structure and chaos in insect population dynamics. *Nature* 353:255–58.

Hastings, A. 2004. Transients: The key to long-term ecological understanding? *Trends in Ecology and Evolution* 19:39–45.

Havens, K. 1992. Scale and structure in natural food webs. *Science* 257:1107–10.

Holt, R. D. 1977. Predation, apparent competition, and the structure of prey communities. *Theoretical Population Biology* 12:197–229.

———. 1984. Spatial heterogeneity, indirect interactions, and the coexistence of prey species. *American Naturalist* 124:377–406.

———. 1992. A neglected facet of island biogeography: The role of internal spatial dynamics in area effects. *Theoretical Population Biology* 41:354–71.

———. 1993. Ecology at the mesoscale: The influence of regional processes on local communities. In *Species Diversity in Ecological Communities,* ed. R. Ricklefs and D. Schluter, 77–88. Chicago: University of Chicago Press.

———. 1996. Food webs in space: An island biogeographic perspective. In *Food Webs: Contemporary Perspectives,* ed. G. Polis and K. Winemiller, 313–23. London: Chapman and Hall.

———. 1997a. From metapopulation dynamics to community structure: Some consequences of spatial heterogeneity. In *Metapopulation Biology*, ed. I. Hanski and M. Gilpin, 149–64. New York: Academic Press.

———. 1997b. Community modules. In *Multitropihc Interactions in Terrestrial Ecosystems*, ed. A. C. Gange, and V. K. Brown, 333–49. Oxford: Blackwell Science.

———. 2002. Food webs in space: On the interplay of dynamic instability and spatial processes. *Ecological Research* 17:261–73.

Holt, R. D., P. A. Abrams, J. M. Fryxell and T. Kimbrell. 2008. Reticulate food webs in space and time: Messages from the Serengeti. In *Serengeti III: Human Impacts on Ecosystem Dynamics,* ed. A.R.E. Sinclair, C. Packer, S.A.R. Mduma, and J. M. Fryxell, 241–76. Chicago: University of Chicago Press.

Holt, R. D., and M. E. Hochberg. 2001. Indirect interactions, community modules, and biological control: A theoretical perspective. In *Evaluating Indirect Ecological Effects of Biological Control,* ed. E. Wajnberg, J. K. Scott, and P. C. Quimby, 12–37. New York: CABI International.

Holt, R. D., and M. F. Hoopes. 2005. Food web dynamics in a metacommunity: Modules and beyond. In *Metacommuniites: Spatial Dynamics and Ecological Communities,* ed. M. Holyoak, M. Leibold, and R. D. Holt, 68–93. Chicago: University of Chicago Press.

Holt, R. D., and J. H. Lawton. 1994. The ecological consequences of shared natural enemies. *Annual Review of Ecology and Systematics* 25:495–520.

Holt, R. D., J. H. Lawton, G. A. Polis, and N. D. Martinez. 1999. Trophic rank and the species-area relationship. *Ecology* 80:1495–504.

Holyoak, M., and S. P. Lawler. 1996. Persistence of an extinction-prone predator-prey interaction through metapopulation dynamics. *Ecology* 77:1867–79.

Holyoak, M., M. A. Leibold, and R. D. Holt, eds. 2005. *Metacommunities: Spatial Dynamics and Ecological Communities.* Chicago: University of Chicago Press.

Hosseini, P. R. 2003. How localized consumption stabilizes predator-prey systems with finite frequency of mixing. *American Naturalist* 161:567–85.

Hoyle, M. 2004. Causes of the species-area relationship by trophic level in a field-based microecosystem. *Proceedings of the Royal Society of London, Series B* 271:1159–64.

Hubbell, S. P. 2001. *The Unified Neutral Theory of Biodiversity and Biogeography.* Princeton, NJ: Princeton University Press.

Huffaker, C. B. 1958. Experimental studies on predation: dispersion factors and predator-prey oscillations. *Hilgardia* 27:343–83.

King, A. A., and A. Hastings. 2003. Spatial mechanisms for coexistence of species sharing a common natural enemy. *Theoretical Population Biology* 64:431–38.

Komonen, A, R. Penttila, M. Lindgren, and I. Hanski. 2000. Forest fragmentation truncates a food chain based on an old-growth forest bracket fungus. *Oikos* 90:119–26.

Kondoh, M. 2003. Foraging adaptation and the relationship between food-web complexity and stability. *Science* 299:1388–91.

Kotiaho, J. S., and P. Sulkava. 2007. Effects of isolation, area, and predators on invasion: A field experiment with artificial islands. *Applied Soil Ecology* 35:256–59.

Kratina, P., M. Vos, and B. R. Anholt. 2007. Species diversity modulates predation. *Ecology* 88:1917–23.

Kruess, A., and T. Tscharntke. 1994. Habitat fragmentation, species loss, and biological control. *Science* 264:1581–84.

———. 2000. Effects of habitat fragmentation on plant-insect communities. In *Interchanges of Insects between Agricultural and Surrounding Landscapes*, ed. B. Ekbom, M. E. Irwin, and Y. Robert, 53–70. Dordrecht: Kluwer Academic Publishers.

Lafferty, K. D., S. Allesina, M. Arim, C. J. Briggs, G. De Leo, A. P. Dobson, J. A. Dunne, P.T.J. Johnson, A. M. Kuris, D. J. Marcogliese, N. D. Martinez, J. Memmott, P. A. Marquet, J. P. McLaughlin, E. A. Mordecai, M. Pascual, R. Poulin, and D. W. Thieltges. 2008. Parasites in food webs: The ultimate missing links. *Ecology Letters* 11:533–46.

Lomolino, M. V. 1984. Immigrant selection, predation, and the distributions of *Microtus pennsylvanicus* and *Blarina brevicauda* on islands. *American Naturalist* 123:468–83.

Lubchenco, J. 1978. Plant species-diversity in a marine intertidal community: Importance of herbivore food preference and algal competitive abilities. *American Naturalist* 112:23–39.

MacArthur, R. H., and E. O. Wilson. 1967. *The Theory of Island Biogeography*. Princeton, NJ: Princeton University Press.

Maron, J. L., J. A. Estes, D. A. Croll, E. M. Danner, S. C. Elmendorf, and S. L. Buckelew. 2006. An introduced predator alters Aleutian island plant communities by thwarting nutrient subsidies. *Ecological Monographs* 76:3–24.

Martinez, N. 1992. Constant connectance in community food webs. *American Naturalist* 139:1208–18.

May, R. M. 1972. Time-delay versus stability in population models with two and three trophic levels. *Ecology* 54:315–25.

McCann, K. 2000. The diversity-stability debate. *Nature* 405:228–33.

McCann, K., J. Rasmussen, and J. Umbanhowar. 2005. The dynamics of spatially coupled webs. *Ecology Letters* 8:513–23.

McCann, K., J. Rasmussen, J. Ubanhowar, and M. Humphries. 2005. The role of space, time, and variability in food web dynamics. In *Dynamic Food Webs*, ed. P. C. de Ruiter, V. Wolters, and J. C. Moore, 56–70. New York: Academic Press.

McCauley, E., B. E. Kendall, A. Janssen, S. Wood, W. W. Murdoch, P. Hosseini, C. J. Briggs, S. P. Ellner, R. M. Nisbet, M. W. Sabelis, and P. Turchin. 2000. Inferring colonization processes from population dynamics in spatially structured predator-prey systems. *Ecology* 2000:3350–61.

Montoya, J. M., S. L. Pimm, and R. V. Sole. 2006. Ecological networks and their fragility. *Nature* 442:259–64.

Morton, R. D., and R. Law. 1997. Regional species pools and the assembly of local ecological communities. *Journal of Theoretical Biology* 187:321–31.

Noonburg, E. G., and P. A. Abrams. 2005. Transient dynamics limit the effectiveness of keystone predation in bringing about coexistence. *American Naturalist* 165:322–35.

Nordstrom, M., and E. Korpimakki. 2004. Effects of island isolation and feral mink removal on bird communities on small islands in the Baltic Sea. *Journal of Animal Ecology* 73:424–33.

Olff, H., and M. E. Ritchie. 1998. Effects of herbivores on grassland plant diversity. *Trends in Ecology and Evolution* 13:261–65.

Otway, S. J., A. Hector, and J. H. Lawton. 2005. Resource dilution effects on specialist insect herbivores in a grassland biodiversity experiment. *Journal of Animal Ecology* 74:234–40.

Packer, C., R. D. Holt, P. J. Hudson, K. D. Lafferty, and A. P. Dobson. 2003. Keeping the herds healthy and alert: Implications of predator control for infectious disease. *Ecology Letters* 6:797–802.

Piechnik, D. A., S. P. Lawler, and N. D. Martinez. 2008. Food-web assembly during a classic biogeographic study: Species' "trophic breadth" corresponds to colonization order. *Oikos* 117:665–74.

Pimm, S. L. 1982. *Food Webs*, reprinted with a new Foreword, 2002. Chicago: University of Chicago Press.

Polis, G. A., W. B. Anderson, and R. D. Holt. 1997. Toward an integration of landscape and food web ecology: The dynamics of spatially subsidized food webs. *Annual Review of Ecology and Systematics* 28:289–316.

Polis, G. A., M. E. Power, and G. R. Huxel. 2004. *Food Webs at the Landscape Level*. Chicago: University of Chicago Press.

Post, D. M. 2002. The long and short of food-chain length. *Trends in Ecology and Evolution* 17:269–77.

Post, D. M., M. L. Pace, and N. G. Hairston, Jr. 2000. Ecosystem size determines food-chain length in lakes. *Nature* 405:1047–49.

Post, D. M., and G. Takimoto. 2007. Proximate structural mechanisms for variation in food-chain length. *Oikos* 116:775–82.

Prakash, S., and A. M. De Roos. 2002. Habitat destruction in a simple predator-prey patch model: How predators enhance prey persistence and abundance. *Theoretical Population Biology* 62:231–49.

Rand, T. A., and S. M. Louda. 2006. Spillover of agriculturally subsidized predators as a potential threat to native insect herbivores in fragmented landscapes. *Conservation Biology* 20:1720–29.

Rand, T. A., and T. Tscharntke. 2007. Contrasting effects of natural habitat loss on generalist and specialist aphid natural enemies. *Oikos* 116:1353–62.

Rand, T. A., J. M. Tylianakis, and T. Tscharntke. 2006. Spillover edge effects: The dispersal of agriculturally subsidized insect natural enemies into adjacent natural habitats. *Ecology Letters* 9:603–14.

Ricklefs, R. E., and E. Bermingham. 2007. The causes of evolutionary radiations in archipelagoes: Passerine birds in the Lesser Antilles. *American Naturalist* 169:285–97.

Ritchie, M. E. 1999. Biodiversity and reduced extinction risks in spatially isolated rodent populations. *Ecology Letters* 2:11–13.

Robinson, G. R., R. D. Holt, M. S. Gaines, S. P. Hamburg, M. L. Johnson, H. S. Fitch, and E. A. Martinko. 1992. Diverse and contrasting effects of habitat fragmentation. *Science* 257:524–26.

Robinson, S. K., F. R. Thompson, T. M. Donovan, D. R. Whitehead, and J. Faaborg. 1995. Regional forest fragmentation and the nesting success of migratory birds. *Science* 267:1987–90.

Rosenzweig, M. L. 1995. *Species Diversity in Space and Time*. Cambridge: Cambridge University Press.

Russell, F. L., S. M. Louda, T. A. Rand, and S. D. Kachman. 2007. Variation in herbivore-mediated indirect effects of an invasive plant on a native plant. *Ecology* 88:413–23.

Ryall, K. L., and L. Fahrig. 2006. Response of predators to loss and fragmentation of prey habitat: A review of theory. *Ecology* 87: 1086–93.

Rydberg, W. A., and J. M. Chase. 2007. Predator-dependent species-area relationships. *American Naturalist* 170:636–42.

Salo, P., E. Korpimaki, P. B. Banks, M. Nordstrom, and C. R. Dickman. 2007. Alien predators are more dangerous than native predators to prey populations. *Proceedings of the Royal Society of London, Series B* 274:1237–43.

Schaffers, A. P., I. P. Raemakers, K. V. Sykora and C.J.F. ter Braak. 2008. Arthropod assemblages are best predicted by plant species composition. *Ecology* 89:782–94.

Schoener, T. W. 1976. The species-area relationship within archipelagos. In *Proceedings of the 16th International Ornithological Congress (1974)*, ed. H. J. Frith and J. H. Calaby, 629–42. Canberra: Australian Academy of Science.

———. 1989. Food webs from the small to the large. *Ecology* 70:1559–89.

Schoener, T. W., and D. A. Spiller. 1995. Effect of predators and area on invasion: An experiment with island spiders. *Science* 267:1811–13.

———. 1996. Devastation of prey diversity by experimentally introduced predators in the field. *Nature* 381:691–94.

Schoener, T. W., and D. A. Spiller. 1999b. Variation in the magnitude of a predator's effect from small to large islands. In *Monografies de la societat d'historia natural de les balears 6: Ecologia de les illes*, 35–66. Palma, Spain: Grafiques Miramar, S. A.

Schoener, T. W., D. A. Spiller, and J. B. Losos. 2001. Predators increase the risk of catastrophic extinction of prey populations. *Nature* 412:183–86.

Schoener, T. W., D. A. Spiller, and L. W. Morrison. 1995. Variation in the hymenopteran parasitoid fraction on Bahamian islands. *Acta Oecologica* 16:130–21.

Schreiber, S. J., R. N. Lipcius, R. D. Seitz, and W. C. Long. 2006. Dancing between the devil and deep blue sea: The stabilizing effect of enemy-free and victimless sinks. *Oikos* 113:67–81.

Shulman, R. S., and J. M. Chase. 2007. Increasing isolation reduces predator-prey species richness in aquatic food webs. *Oikos* 116:1581–87.

Siemann, E., J. Haarstad, and D. Tilman. 1999. Dynamics of plant and arthropod diversity during old field succession. *Ecography* 22:406–14.

Simberloff, D. S., and E. O. Wilson. 1969. Experimental zoogeography of islands: The colonization of empty islands. *Ecology* 50:278–96.

Snyder, W. E., G. B. Snyder, D. L. Finke, and C. S. Straub. 2006. Predator biodiversity strengthens herbivore suppression. *Ecology Letters* 9:789–96.

Sole, R. V., and J. Bascompte. 2006. *Self-Organization in Complex Ecosystems.* Princeton, NJ: Princeton University Press.

Sole, R. V., and J. M. Montoya. 2006. Ecological network meltdown from habitat loss and fragmentation. In *Ecological Networks; Linking Structure to Dynamics in Food Webs*, ed. M. Pascual and J. A. Dunne, 305–47. Oxford: Oxford University Press.

Spencer, M. 2000. Are predators rare? *Oikos* 89:115–22.

Spencer, M., L. Blaustein, and J. E. Cohen. 1999. Species richness and the proportion of predatory animal species in temporary pools: Relationships with habitat size and permanence. *Ecology Letters* 2:157–66.

Spencer, M., and P. H. Warren. 1996. The effects of habitat size and productivity on food web structure in small aquatic microcosms. *Oikos* 75:419–30.

Steiner, C. F., and M. A. Leibold. 2004. Cyclic assembly trajectories and scale-dependent productivity-diversity relationships. *Ecology* 85:107–13.

Steffan-Dewenter, I., and T. Tscharntke. 2000. Butterfly community structure in fragmented habitats. *Ecology Letters* 3:449–56.

———. 2002. Insect communities and biotic interactions on fragmented calcareous grasslands—a mini review. *Biological Conservation* 104:275–84.

Swihart, R. K., Z. Feng, N. A. Slade, D. M. Mason, and T. M. Gehring. 2001. Effects of habitat destruction and resource supplementation in a predator-prey metapopulation mode. *Journal of Theoretical Biology* 210:287–303.

Terborgh, J., L. Lopez, P. Nuñez, M. Rao, G. Shahabuddin, G. Orihuela, M. Riveros, R. Ascanio, G. H. Adler, T. D. Lambert, L. Balbas. 2001. Ecological meltdown in predator-free forest fragments. *Science* 294:1923–26.

Thies, C., I. Steffan-Dewenter, and T. Tscharntke. 2008. Interannual landscape changes influence plant-herbivore-parasitoid interactions. *Agriculture, Ecosystems, and Environment* 125:266–68.

Tscharntke, T., and R. Brandl. 2004. Plant-insect interactions in fragmented landscapes. *Annual Review of Entomology* 49:405–30.

Tscharntke, T., T. A. Rand, and F.J.J.A. Bianchi. 2005. The landscape context of trophic interactions: Insect spillover across the crop-noncrop interface. *Annales Zoologici Fennici* 42: 421–32.

Tscharntke, T., I. Steffan-Dewenter, A. Kruess, and C. Thies. 2002. Characteristics of insect populations on habitat fragments: A mini review. *Ecological Research* 17:229–39.

Valladores, G., A. Salvo, and L. Cagnolo. 2006. Habitat fragmentation effects on trophic processes of insect-plant food webs. *Conservation Biology* 20:212–17.

Valone, T. J., and M. R. Schutzenhofer. 2007. Reduced rodent biodiversity destabilizes plant populations. *Ecology* 88:26–31.

Van Nouhuys, S. 2005. Effects of habitat fragmentation at different trophic levels in insect communities. *Annales Zoologici Fennici* 42:433–47.

van Ruijven, J., G. B. De Deyn, C. E. Raaijmakers, F. Berendes, and W. H. van der Putten. 2005. Interactions between spatially separated herbivores indirectly alter plant diversity. *Ecology Letters* 8:30–37.

van Veen, F.J.F., P. D. Van Holland, and H.C.J. Godfray. 2005. Stable coexistence in insect communities due to density- and trait-mediated indirect effects. *Ecology* 86:1382–89.

Vos, M., S. Moreno Berrocal, F. Karamaouna, L. Hemerik, and L.E.M. Vet. 2001. Plant-mediated indirect effects and the persistence of parasitoid-herbivore communities. *Ecology Letters* 4:38–45.

Warren, P. H., R. Law, and A. J. Weatherby. 2003. Mapping the assembly of protist communities in microcosms. *Ecology* 84:1001–11.

Watling, J. I., and M. A. Donnelly. 2006. Fragments as islands: A synthesis of faunal responses to habitat patchiness. *Conservation Biology* 20:1016–25.

Williamson, M. 1981. *Island Populations.* Oxford: Oxford University Press.

Wilson, H. B., M. P. Hassell, and R. D. Holt. 1998. Persistence and area effects in a stochastic tritrophic model. *American Naturalist* 151:587–95.

Winemiller, K.O. 1990. Spatial and temporal variation in tropical fish trophic networks. *Ecological Monographs* 60:331–67.

Yodzis, P. 1998. Local trophodynamics and the interaction of marine mammals and fisheriers in the Benguela ecosystem. *Journal of Animal Ecology* 67:635–58.

The Theories of Island Biogeography and Metapopulation Dynamics

SCIENCE MARCHES FORWARD, BUT THE LEGACY OF GOOD IDEAS LASTS FOR A LONG TIME

Ilkka Hanski

Two related notions about natural populations featured prominently in the writings of several ecologists in the 1950s. These authors realized that populations have a spatial structure, in the sense that a "population" in the wider landscape often consists of more or less distinct local populations. And secondly, these local populations may have more or less independent demographic fates, which has consequences for the dynamics of the regional population as a whole. Explaining their ideas at length in *The Distribution and Abundance of Animals* (1954), the Australian ecologists H. G. (Herbert) Andrewartha and L. Charles Birch put an especially strong emphasis on small-scale spatial structure of populations. They argued that local populations are often characterized by high rates of extinction and reestablishment, a viewpoint that contrasted with the then prevailing paradigm of stable populations regulated by density-dependent processes (reviewed by, e.g., Sinclair 1989). John Curtis, Professor of Botany in the University of Wisconsin, understood clearly the consequences of human-caused habitat loss and fragmentation on population processes and the spatial distribution of species. He wrote:

> Within the remnant forest stands, a number of changes of possible importance may take place. The small size and increased isolation of the stands tend to prevent the easy exchange of members from one stand to another. Various accidental happenings in any given stand over a period of years may eliminate one or more species from the community. Such a local catastrophe under natural conditions would be quickly healed by migration of new individuals from adjacent unaffected areas. . . . In the isolated stands, however, opportunities for inward migration are small or nonexistent. As a result, the stands gradually lose some of their species, and those remaining achieve unusual positions of relative abundance. (Curtis 1956, p. 729)

Not only does this paragraph describe the processes of local extinction and recolonization, but it also contains a vision of the extinction threshold. In the next paragraph on the same page Curtis commented on microevolutionary changes that are likely to take place in response to changing population structure due to habitat fragmentation. Quite a page! Carl Huffaker (1958), building upon the earlier theoretical work of the Australian Alexander J. Nicholson (1933), investigated in a fascinating experimental study the consequences of small-scale spatial structure of habitat for the dynamics and stability of predator-prey interaction. Mention should also be made of the "island model" in theoretical population genetics, already established by Sewall Wright in 1940.

The theories of island biogeography and metapopulation dynamics were introduced, respectively, by Robert MacArthur and Edward O. Wilson (1963, 1967) and by Richard Levins (1969, 1970) in the 1960s. From our present perspective, it is surprising that the island theory and metapopulation theory appear to have had their own independent origins, and origins that were independent of the work done on spatial population structures in the 1950s and earlier. In the case of MacArthur and Wilson (1963), the origin was their attempt to explain why large islands tend to have more species than small ones, while Levins's (1969, 1970) primary concerns were some demographic and evolutionary consequences of extinction-colonization dynamics. Of the papers and books that I cited in the first paragraph, MacArthur and Wilson (1967) referred only to Curtis (1956), by reproducing a figure illustrating the human-caused reduction in the total area and increase in the degree of fragmentation of woodland in the Cadiz Township in Wisconsin from 1831 until 1950 (reproduced here as figure 7.1). It is curious that, having included Curtis's fragmentation maps as the very first illustration in their book, MacArthur and Wilson made no real attempt to apply their model of island biogeography to fragmented landscapes without a mainland. I say more about this in the following sections; here it suffices to recapitulate that the written papers, chapters, and books suggest that there were several independent origins in the middle of the last century for the general idea that natural populations in larger regions consist of discrete local populations, and that this spatial structure of regional populations may have important consequences for their dynamics and long-term viability.

In the following two sections, my purpose is to show that MacArthur and Wilson's model of island biogeography and Levins's model of metapopulation dynamics are in fact special cases of a more general model, which can also accommodate the earlier descriptions of spatial population structure by Andrewartha and Birch and by Curtis. In this framework, the island model is a straightforward extension of the single-species

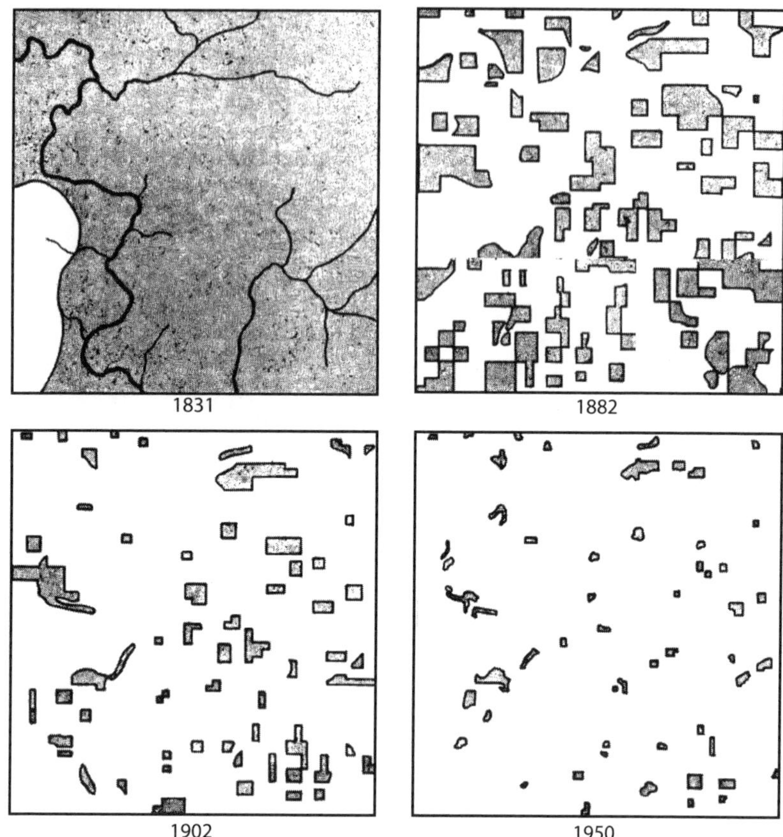

Figure 7.1. Reduction in the area and fragmentation of the woodland in the Cadiz township in Wisconsin from 1831 until 1950 (Curtis 1956). This figure was reproduced in MacArthur and Wilson (1967), p. 4. Curtis (1956) pioneered landscape ecology by calculating for the four maps the total area of woodland, the number of separate woodlots, the average size of woodlots, the length of the woodland periphery, and the periphery/area ratio. Incidentally, a look at this area today, with the help of GoogleEarth, reveals that some further fragmentation has occurred in the past 50 years, though the bigger woodland fragments in 1950 are still there (42°32′54.45″N, 89°45′52.06″W).

metapopulation model to many co-occurring but dynamically independent species. The reasons for laboring this point, which is rather obvious when you come to think about it, are twofold. It is of historical interest to ponder why the connection was not made explicitly early on. And secondly, the unified model, bringing together the key innovations in the respective models of MacArthur and Wilson and of Levins, has substantial power to predict the distribution of species in fragmented landscapes, and it leads to new insights about familiar patterns in the large-scale occurrence of species. Concerning the latter, I examine in this chapter how the species-area relationship, the feature of island communities that so much stimulated the work of MacArthur and Wilson (see the introduction to their 1963 paper), can be derived from the single-species metapopulation model, and I point out how intimately the species-area relationship is related to another well-established pattern in the occurrence of species, the distribution-abundance relationship.

Before moving on, I add a personal note. I am one of the many ecologists whose research has been greatly influenced by the works of MacArthur, Wilson, and Levins; it has been a privilege and source of enjoyment to write this chapter. I have taken the liberty of addressing selectively a few topics that stem directly from the classic models of island biogeography and metapopulation dynamics and to which I have attempted to make contributions over a prolonged period of time. This chapter is not a review of the literature, partly for lack of space but also because my particular purpose is to focus on the core concepts of MacArthur and Wilson and of Levins, and to highlight their role in the subsequent development of metapopulation models and theory. The simple MacArthur-Wilson island model and the Levins metapopulation model are by now largely history and replaced by many more specific models, and by a range of more general models of spatial dynamics, but these simple models splendidly exemplify the motto of this chapter: science marches forward, but the legacy of good ideas lasts for a long time.

The MacArthur-Wilson and Levins Models

As is well known, the setting of MacArthur and Wilson's island model involves a large mainland area, which is true mainland in the case of islands off the mainland but more generally a very large expanse of habitat, where P species have stable populations. Outside the mainland, there are islands, or more generally fragments of habitat, with dissimilar areas and with dissimilar distances (isolation) from the mainland. Migrants that originate from the mainland may establish new populations on the islands, and the island populations have a smaller or greater risk of local extinction. Migration among the islands is ignored; hence the essential dynamics of the model can be understood by considering just the mainland and one island. The MacArthur-Wilson model, in spite of its simplicity, is potentially a good description of the long-term dynamics of species occurring on true islands that are rather sparsely distributed off the mainland, making migration among the islands unlikely.

The core idea of the model is formulated in the following differential equation:

$$\frac{dS}{dt} = I(P - S) - ES, \tag{7.1}$$

which specifies the rate of change in S, the number of species present on a particular island. The number of species increases due to colonizations: each species in the mainland pool of P species that is not yet on the island (there are P–S such species) has the same probability of colonization, which translates into a constant colonization rate parameter I in the continuous-time model. The number of species on the island decreases due to extinctions: all species have the same extinction risk, and hence the total extinction rate is given by the constant extinction rate parameter E times the current number of species. At equilibrium,

$$\widehat{S} = IP(I + E). \tag{7.2}$$

Turning to Levins's metapopulation model, it is appropriate for highly fragmented landscapes such as shown in figure 7.1d: a large network of small or relatively small habitat fragments (patches) without any large expanse of habitat (mainland). To construct his model, Levins made the simplifying assumption that all patches are of the same size and that migration is global, equally likely among any pair of populations and patches (this is the island model assumption made in Sewall Wright's 1940 model, which, however, assumed stable populations). The set of local populations inhabiting the network of patches is called the metapopulation, a term that Levins (1970) coined, the size of which is given by the fraction of patches occupied, denoted by p.

Levins formulated the core idea of classic metapopulation dynamics with the following differential equation:

$$\frac{dp}{dt} = cp(1 - p) - ep. \tag{7.3}$$

Here c and e are the colonization and extinction rate parameters, describing the colonization capacity and the extinction-proneness of the species. Because colonization rate is proportional to just the fraction of occupied patches, which are the sources of migration, the Levins model does not contain any description of the landscape structure and it best applies to species for which the spatial configuration of habitat makes little difference due to frequent long-range migration.

It is of interest to ask why MacArthur and Wilson and Levins did not refer to each other's work in their respective publications, to say nothing about why they did not explore the conceptual and theoretical similarities in their models. The reason is not that they did not know about each other. They did, they met (see figure 1.3 in Wilson, this volume), and they even coauthored papers in the mid 1960s on the coexistence of competitors and maintenance of genetic polymorphism in heterogeneous environments (MacArthur and Levins 1964, Levins and MacArthur 1996). And as a matter of fact, Levins actually derived in a little-known paper published in 1963 (Levins and Heatwole 1963) the expression $\frac{ND}{M+D}$ for the equilibrium number of species on an island, which is the same as equation (7.2), though Levins used the inverse of extinction and colonization rates, the expected time to extinction D and the expected time to next colonization M (N is the number of species in the mainland pool). Ironically, MacArthur and Wilson did not give this simple equilibrium result in their 1963 paper, in which they first developed much of their theory, though it outlines many of the more advanced results subsequently discussed at length in 1967.

Turning to MacArthur and Wilson (1967), they discussed in their book "habitat islands on the mainland" (pp. 114–15), such as shown in figure 7.1, but rather than working in the direction of Levins's description of a network of local populations, they emphasized how habitat islands are different from true islands in being surrounded by other habitats that might harbor competitors of the focal species present in the habitat islands. They went on to describe what we would now call source-sink dynamics (in the sense of Pulliam 1988), and they discussed the implications of such dynamics for interspecific competition. Apparently, MacArthur and Wilson were so focused on what happens in a particular island, whether a true or a habitat island, that they did not attempt to extend their model formally to networks of local populations in fragmented landscapes—in spite of the very first figure in their book (figure 7.1). They were interested in communities of species—how does species number vary with the area or isolation of an island—rather than in single species, which would have facilitated the development of models for habitat networks. Finally, MacArthur and Wilson did not construct a measure of isolation that would have been applicable to islands in a network, in which colonization does not occur from the mainland but from multiple other populations in the neighborhood of the focal island.

Subsequent research has attempted to merge the conceptual frameworks of the island theory and the classic metapopulation theory in two major ways: first, by developing single-species metapopulation models that take from the island model the explicit description of landscape structure in terms of the areas and isolations of habitat patches (I describe this line of research in the next section) and second, by developing multispecies models by making use of Levins's description of habitat patch networks. This latter approach has led in the past few years to various models of metacommunity dynamics (reviewed in many chapters in Holyoak et al. 2005). There is a clear need for developing theory and models for metacommunities, but the task is difficult and the field is still searching for its basic concepts. Most of the current metacommunity models are not formally related to MacArthur and Wilson's island model nor to Levins's metapopulation model, for which reason they are not examined more closely in this chapter. One exception is the extension of Levins's model to two or more competing species, which I comment on in the final section of this chapter.

Spatially Realistic Metapopulation Models

Here I turn to models that mix assumptions from the island biogeographic model and the Levins metapopulation model. The qualifier "spatially realistic" indicates that the models take into account the actual spatial configuration of the habitat: how many patches are there in a network, how large are they, and how far apart are they located from each other? I show that the MacArthur-Wilson and Levins models are special cases of a spatially realistic metapopulation model.

The origin of these models is in Jared Diamond's (1975) incidence functions, which are based on a straightforward idea. Consider the occurrence of a species on a set of islands with dissimilar areas. Diamond grouped the islands in classes of similar areas, for instance islands from 1 to 10 ha, from 11 to 100 ha, and so forth. He then calculated the proportion of islands in a particular area class on which a particular species had been detected during a survey. The incidence function describes how the proportion of occupied islands changes with area—usually the incidence increases with area. The islands could equally well be classified based on some other property, such as the number of species present, and the incidence function would be constructed in a similar manner. More generally, we may not group the islands at all but define the incidence function $p(A)$ as the probability that the species is present on an island with area A.

In the case of mainland-island metapopulations, in which all migrants originate from the mainland, and assuming time-constant probabilities of extinction E and colonization C, the long-term probability of a species being present on an island is given by

$$p = \frac{C}{C + E}, \tag{7.4}$$

as already noted by Levins and Heatwole (1963) in the island biogeographic context (this result is a property of the Markov chain defined by the model assumptions). The incidence function is now obtained by making assumptions about how the colonization and extinction probabilities C and E depend on the area or some other property of islands (in continuous-time models the probabilities become rates).

The incidence functions played some role, though not a very big one, in the vigorous debate that broke out in the 1980s about the factors that influence the assembly of island communities—or factors that do not influence community assembly, as many participants found that "null models," which were presumed to involve no interspecific interactions, explained well the occurrences of species on islands. The volume edited by Strong et al. (1984) has many chapters on these issues (see Simberloff and Collins, this volume). At the same time, I was studying the dynamics of shrews and other small mammals on small islands. Stimulated by the work of Diamond and intrigued by the possibility of extracting some information about extinctions and colonizations from patterns of island occupancy, I constructed an incidence function by assuming that the annual extinction probability on island i is an inverse function of island area, $E_i = \frac{\mu}{A_i^{\zeta_{ext}}}$, and that the annual colonization probability declines exponentially with d_i, the isolation (distance) from the mainland, $C_i = \beta e^{-\alpha d_i}$, where μ, ζ_{ext}, α, and β are model parameters (Hanski 1993). Assuming further that the colonization probability approaches 1 when isolation approaches zero, we have $\beta=1$. The incidence function model is then given by

$$p_i = \frac{1}{1 + \frac{\mu e^{-ad_i}}{A_i^{\zeta_{ext}}}}. \tag{7.5}$$

Using data on the occurrence of *Sorex cinereus* on a set of 40 islands studied by Crowell (1986) and Lomolino (1993) in North America, I estimated the values of the model parameters (Hanski 1993). The figure in box 7.1 depicts how the predicted probability of occurrence (the incidence) depends on island area and isolation. Naturally, one could make some other structural assumptions about how colonization and extinction probabilities depend on island area and isolation than what was made above. Some assumptions lead to incidence functions in which several parameters occur as a product and hence their values cannot be estimated independently without making extra assumptions.

BOX 7.1. Measurement of connectivity in metapopulations without a mainland

In island biogeographic models with all migrants originating from the mainland, isolation of an island is given by its distance to the mainland. In metapopulations without a mainland, migrants to a particular habitat patch *i* originate from existing local populations in the surrounding habitat patches. A measure of connectivity, which reflects lack of isolation, may be constructed by summing up the contributions from all possible source populations *j*. These contributions are weighted by three factors (see the illustration). First, the area of the source patch *j*, which reflects the numbers of potential emigrants from that patch. To gain further flexibility, the area may be raised to power ζ_{em}, which reflects both the scaling of population size with patch area and the scaling of emigration with patch area. Second, the distance of the source patch *j* from the focal patch *i*, which influences the likelihood of individuals leaving patch *j* ever arriving at patch *i*. This likelihood is often assumed to be an exponential function of the distance d_{ij}, but some other distribution ("dispersal kernel") could be used instead. Parameter α gives the rate of decline in the exponential distribution of migration distances from population *j*. Third, the contribution of patch *j* depends on the probability of patch *j* being occupied. In reality, only patches that are currently occupied may send out migrants, but in the mean-field model the contribution of a patch is weighted by its probability of occupancy (the mean-field concept is discussed below). Finally, connectivity of patch *i* may depend on its own area, possibly raised to the power ζ_{im} to account for the scaling of immigration with patch area.

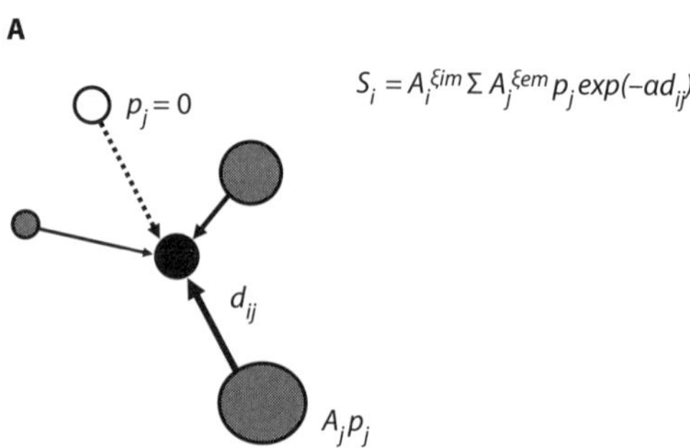

(*Continued*)

(*Continued*)

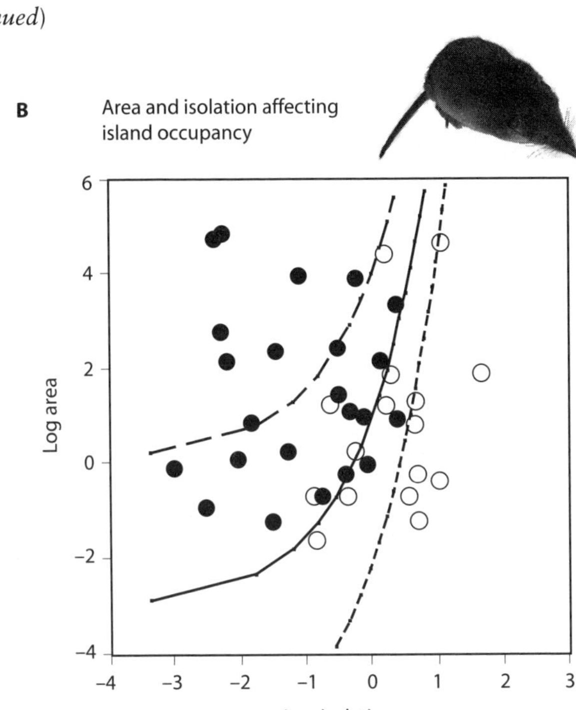

The two graphs illustrate how the pattern of patch occupancy depends in an analogous manner on isolation from the mainland in the case of islands off the mainland and on the above-described measure of connectivity in a metapopulation without a mainland. Black dots represent occupied, open circles unoccupied islands or habitat patches at the time of sampling. (B) Occurrence of the shrew *Sorex cinereus* on islands off the mainland. Isolation is here measured by distance to the mainland. The lines indicate the combinations of area and isolation for which the predicted probability of occupancy is greater than 0.1, 0.5, and 0.9, respectively (from Hanski 1993; data from Crowell 1986, Lomolino 1993). (C) Classic metapopulation of the silver-spotted skipper butterfly (*Hesperia comma*) on dry meadows in southern England. The line indicates the combinations of area and connectivity above which the predicted incidence of occupancy is greater than 0.5 (from Hanski 1994; data from Thomas and Jones 1993).

(*Continued*)

(*Continued*)

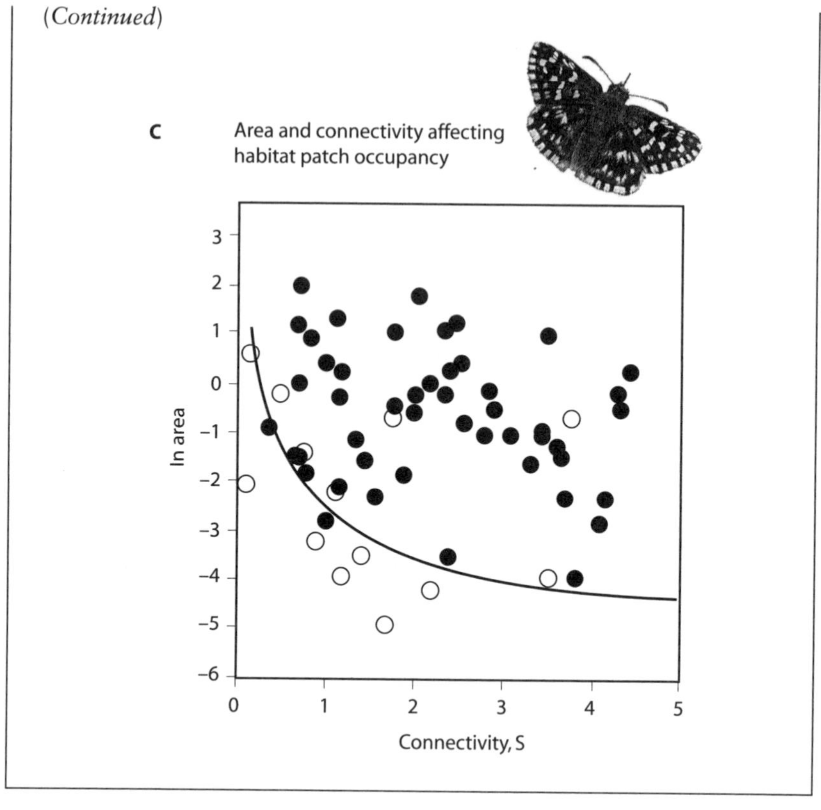

A brief digression is in place here. The incidence function exemplifies what is called the inverse approach in modeling (Tarantola 2005; for ecological applications see Wiegand et al. 2003, Ovaskainen and Crone 2009). Rather than estimating the parameters of ecological processes directly to predict patterns, here we use the pattern to estimate the parameters. The pattern consists of the probabilities of occupancy on a set of islands, the p_i values, which in practice are often approximated by just a single snapshot of presence-absence data. It is better if data are available for several years (Etienne et al. 2004), but even a single snapshot has much information if presence-absence data are available for many islands. And for systems with low rate of population turnover, extinctions and recolonizations, it would not help much to have data for many years, because most islands would stay in the state (occupied or not) in which they were observed in the first year. It is exactly for such systems, for which the direct measurement of the processes of extinction and recolonization would be difficult or impossible because of low rates, that the "pattern-oriented" approach represented by the incidence functions is potentially most helpful.

Though difficult to estimate directly, the rates of extinction and recolonization are of self-evident importance to population ecologists and conservation biologists. In my own work on three species of *Sorex* shrews inhabiting small islands in lakes in Finland, I examined how differences in body size among the three species affect their foraging behavior and life histories, and how these effects might be reflected in population dynamics. One approach was based on incidence functions, with which I estimated for each species the scaling of extinction risk with island area and hence with the carrying capacity: parameter ζ_{ext} in equation (7.5) (Hanski 1992). I found that, while extinction risk decreased very rapidly with increasing island area for the largest species, the scaling was shallow for the smallest species, consistent with the hypothesis that environmental stochasticity plays a bigger role in the dynamics and hence also in the extinction of small-bodied than large-bodied vertebrates (Pimm 1991, Hanski 1998a). I shall return to this observation in the next section while discussing the species-area relationship. Here it remains to note an important caveat to all this modeling: equation (7.5) assumes that enough time has elapsed without any major environmental changes so that the focal species occurs on the islands in a stochastic quasi-equilibrium between recurrent extinctions and recolonizations. This assumption has to be considered case by case.

Let us then turn to metapopulations without a mainland. The essential difference from the mainland-island situation just discussed is that now isolation has to be measured in a different manner, as in metapopulations without a mainland recolonization is the result of migration from any one of several possible source populations in the neighborhood of the focal habitat patch. Box 7.1 describes a measure of connectivity that can be used in this context; connectivity is the reverse of isolation, measuring lack of isolation. The apparent complication that arises in comparison with the measure of isolation from the mainland is that the value of connectivity changes in time, with a changing pattern of occupancy and population sizes in the source populations. In stochastic models that keep track of which particular habitat patches are occupied this is not a problem, but such models are difficult to analyze (Ovaskainen 2001, Ovaskainen and Hanski 2004) or one is forced to rely on numerical simulations. An alternative is to use a trick called the mean-field approximation: connectivity of patch *i* depends not on which particular other patches happen to be occupied at a particular time but instead on the probabilities of occupancy of the other patches, the p_i values. This may appear to be no solution at all, because surely the probability of occupancy is more difficult to determine than whether a patch is occupied or not. This is true for field studies, but for models the p_i values are very convenient. Now our model consists of a set of equations like equation (7.5), in which p_i for patch i depends on the corresponding p values for all the other patches in the network apart from i. This set of equations may be iterated until an equilibrium is reached, the set of p_i values that satisfies all the equations simultaneously (Hanski 1994). This will not work if there is no equilibrium, but single-species patch occupancy metapopulation models typically converge to a unique equilibrium (Ovaskainen and Hanski 2001). Another issue is how good the mean-field approximation is. I return to this question in the discussion, but note already here that, as far as the prediction of the equilibrium state is concerned (quasi-equilibrium in stochastic models), the mean-field approximation works rather well for heterogeneous patch networks, in which the habitat patches have dissimilar areas and dissimilar connectivities (for transient dynamics, see Ovaskainen and Hanski 2002). Luckily for this line of modeling, the real networks are always heterogeneous.

Working together with Otso Ovaskainen, I have constructed and analyzed a range of spatially realistic metapopulation models, including both stochastic models and their deterministic approximations (for reviews, see Hanski 2001, 2005, Hanski and Ovaskainen 2003, Ovaskainen and Hanski 2004). Of particular relevance here is a general equation for the deterministic rate of change in the incidence of occupancy of patch i, because this model has the MacArthur-Wilson model and the Levins model as two special cases. The spatially realistic model is given by

$$\frac{dp_i}{dt} = C_i(1 - p_i) - E_i p_i, \tag{7.6}$$

where C_i depends on the connectivity of patch i (see box 7.1). Assuming a mainland pool of P identical and independent species and constant colonization and extinction rate parameters for island i, the equilibrium incidence is given by $\hat{p} = c/(c + e)$, from which the basic MacArthur-Wilson model (equation [7.2]) follows by multiplying by P to obtain the equilibrium number of species. On the other hand, assuming a network of equally connected and equally large habitat patches, that colonization rate is proportional to the fraction of occupied patches (which is the same, at equilibrium, as the probability of any one patch being occupied, $C_i = cp_i$), and further assuming constant colonization and extinction rate parameters, we arrive at the Levins model, equation (7.3), with the equilibrium $\hat{p} = 1 - e/c$.

An attractive feature of the spatially realistic metapopulation models is that they can be parameterized with empirical data, as I showed in the case of a mainland-island model for shrews (box 7.1). The same applies to models that do not have a mainland. Methods of parameter estimation have been reviewed by Etienne et al. (2004) and many applications to real metapopulations have been discussed by Hanski (2005). Box 7.2 gives an extended example on the Glanville fritillary butterfly.

BOX 7.2. The Glanville fritillary metapopulation in the Åland Islands in Finland and extinction threshold

The Glanville fritillary butterfly *(Melitaea cinxia)* has a classic metapopulation in a large network of about 4,000 habitat patches in the Åland Islands, southwest Finland, within an area of 50 by 70 km^2 (map; Hanski 1999, Nieminen et al. 2004). The habitat patches are dry meadows with an average area of only 0.15 ha and never larger than a few ha (photograph). There is a high rate of population turnover, with around 100 local populations going extinct every year for various reasons (Hanski 1998b) and about the same number of new populations being established. The extinction rate declines with increasing patch area, and the colonization rate increases with connectivity (graphs on the left; data on annual extinction and colonization events have been binned in patch area and connectivity classes and only the average values are shown here; Ovaskainen and Hanski 2004). The graph on the right shows the size of the metapopulation as a function of the metapopulation capacity λ_M (Hanski and Ovaskainen 2000) in 25 habitat patch networks (these networks were delimited as clusters of patches in the entire large network shown in the map). The vertical axis shows the size of the metapopulation based on a survey of habitat patch occupancy in one year. The empirical data have been fitted by a spatially realistic model. The result provides a clear-cut example of the extinction threshold (from Hanski and Ovaskainen 2000).

(*Continued*)

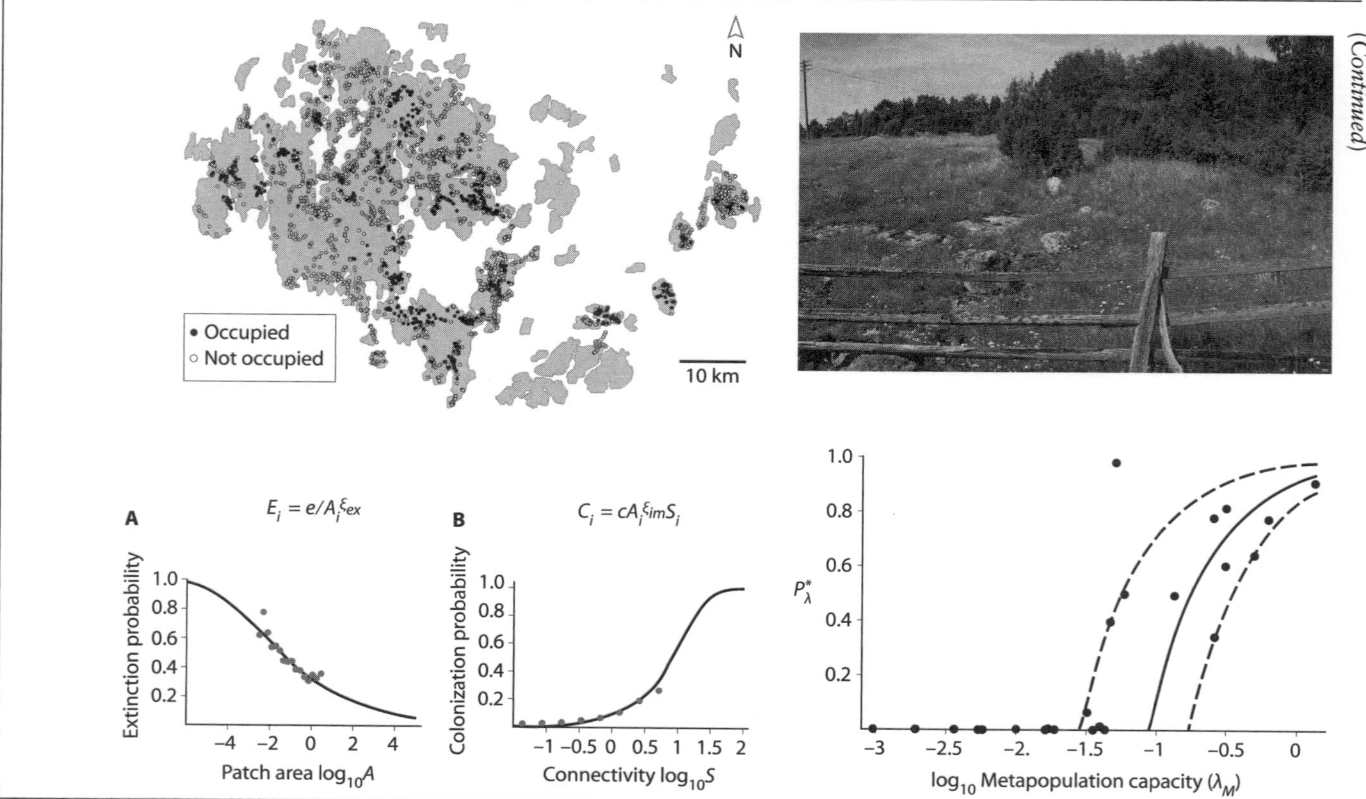

N
10 km
• Occupied
○ Not occupied
A
$E_i = e/A_i^{\xi_{ex}}$
Extinction probability
1.0
0.8
0.6
0.4
0.2
−4
−2
0
2
4
Patch area $\log_{10}A$
B
$C_i = cA_i^{\xi_{im}}S_i$
Colonization probability
1.0
0.8
0.6
0.4
0.2
−1
−0.5
0
0.5
1
1.5
2
Connectivity $\log_{10}S$
P^*_λ
1.0
0.8
0.6
0.4
0.2
−3
−2.5
−2.
−1.5
−1
−0.5
0
$\log_{10}$ Metapopulation capacity (λ_M)

(*Continued*)

The Species-Area Relationship Derived from Incidence Functions

MacArthur and Wilson (1963) originally developed their theory of island biogeography to explain a general pattern in the occurrence of species on islands: the species-area relationship. A couple of different functional forms had been suggested to describe the increasing number of species with increasing island area (e.g., Rosenzweig 1995), but the most common form is the one due to the Swedish ecologist Olof Arrhenius (1921) and used by MacArthur and Wilson, the power function species-area relationship, $S=kA^z$, where S is the number of species on an island, or within an area delimited more arbitrarily, A is the area, and k and z are two parameters. This relationship can be linearized by taking logarithms, and the parameter z then gives the slope of the logarithm of S against the logarithm of island area.

At the level of single species, the incidence function describes how the probability of occurrence of a particular species changes (usually increases) with increasing island area. For instance, in the case of equation (7.5), the logit of p_i, $\ln\left(\frac{p_i}{1-p_i}\right)$, increases linearly with the logarithm of island area, with the slope given by parameter ζ_{ext}. Clearly, there must be some relation between the incidence functions for individual species and the species-area relationship for the community of species, especially if the species have independent dynamics on the islands as assumed in the basic island model, equation (7.1).

Starting from equation (7.4) and observing that $S(A)=\sum_i p_i(A)$, Ovaskainen and Hanski (2003) calculated the slope of the power function species-area relationship as

$$z = \frac{\sum_i p_i(A)\left[1-p_i(A)\right]x_i(A)}{\sum_i p_i(A)},$$

where

$$x_i(A) = \frac{-d\log\left[E_i(A)/C_i(A)\right]}{d\log A}.$$

Assuming further that extinction and recolonization rates scale with island area as $E_i = e_i/A^{\xi_{ext}}$ and $C_i = c_i A^{\xi_{col}}$, $x_i(A)$ is independent of island area A, and it is convenient to describe an incidence function with two quantities, the "critical" island area A_i^* at which $p_i(A)=0.5$, and the slope of the incidence function at A_i^*, which is proportional to x_i (figure 7.2).

To actually predict the species-area relationship for a community of species based on their incidence functions, we need to know the distributions

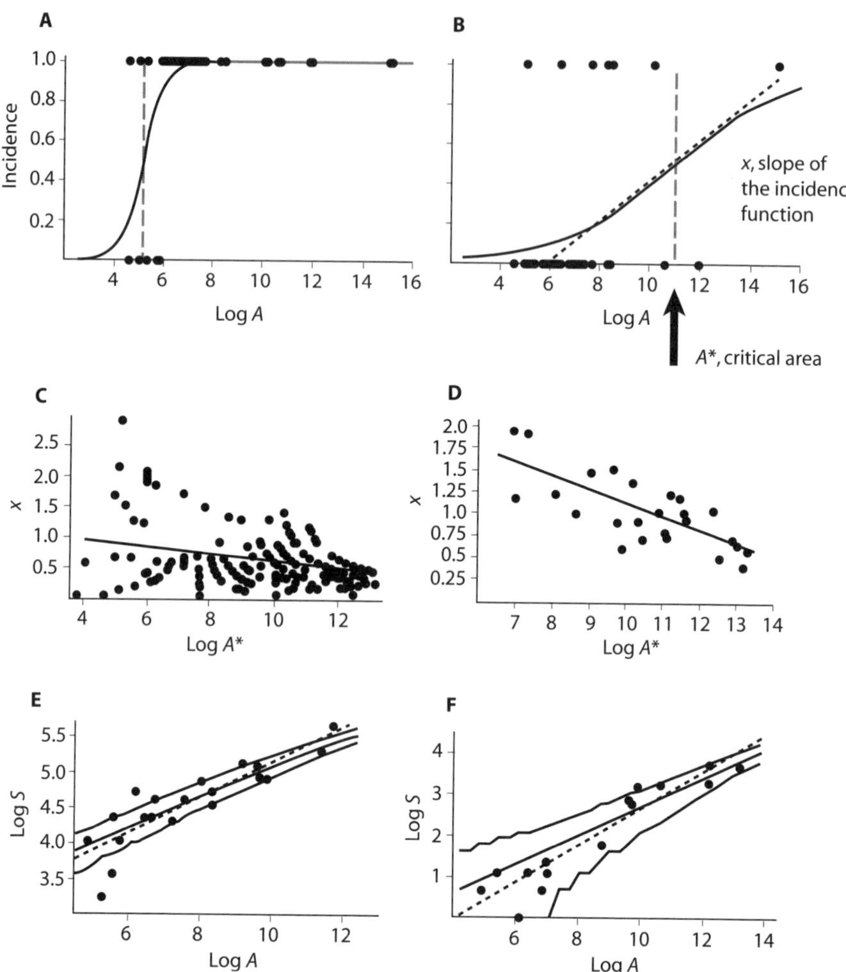

Figure 7.2. Two examples of incidence functions for the birds *Troglodytes troglodytes* (A) and *Sphyrapicus ruber* (B). Panel B also indicates the two parameters that are used to describe incidence functions, the critical island area A^* and x, the slope of the incidence function at A^*. The following panels show the estimated values of x and A^* in a plant community (C; data from Moran 1983) and in a bird community (D; data from Thibault et al. 1990). The final panels show the species-area curves and their 95% confidence intervals for the plant (E) and the bird community (F) calculated on the basis of the single-species incidence functions as explained in the text (from Ovaskainen and Hanski 2003).

of the A_i^* and x_i values. There is no general theory from which these distributions could be inferred; hence we examined two large data sets for plants and birds (sources described in figure 7.2). In both cases, the exponential distribution fitted the $-\log A^*$ values reasonably well, while the $\log x$ values were normally distributed (Ovaskainen and Hanski 2003). Furthermore, in both cases there was a negative correlation between the $\log A_i^*$ and x_i values, which is perhaps expected, because species that are vulnerable to environmental stochasticity have small x_i and tend to require large areas to avoid extinction; hence they have large $\log A_i^*$. The negative correlation implies that species with large critical areas tend to respond more slowly to increasing island area than species with small A^* (the examples in figures 7.2a and 7.2b are thus representative).

The species-area relationship can be calculated either by estimating the parameters for each species separately and by summing up the predicted incidences, or by first generating a hypothetical community of species with parameters drawn from the estimated distributions of A_i^* and x_i and then summing up their incidence functions (Ovaskainen and Hanski 2003). Figures 7.2e and 7.2f show the latter result for the plant and the bird communities. The species-area relationships thus derived correspond closely to the power function species-area relationships fitted to the same data. Though similar regression lines were obtained, arguably the result based on the incidence functions for individual species is more fundamental, because it is based on properties of individual species, and it may bring new insight into the community-level pattern. For instance, the decomposition of the species-area relationship into the constituent incidence functions helps explain why it has been so hard to arrive at a meaningful biological interpretation of the slope parameter z (Connor and McCoy 1979 and many subsequent papers). Consider a situation where z is small. The present model indicates that z is small either because the rate at which new species reach their critical areas A^* with increasing island area is slow, which is a property characterizing the community of species, or because each species responds slowly to increasing island area (small x-value), which is a property characterizing individual species, or both.

Although the slope of the species-area relationship does not generally have a simple interpretation, in suitably circumscribed situations some progress can be made. As an example, assemblages of small-bodied birds and small mammals have a systematically smaller value of z than the corresponding assemblages of large-bodied species on the same set of islands (table 7.1). The explanation offered by Matter et al. (2002) relates to the greater impact of environmental stochasticity in the dynamics of small-bodied than large-bodied species, to which I referred above while discussing

TABLE 7.1
Estimated Slope Values (±SE) of the Power Function Species-Area Relationship for Five Assemblages of Birds and Small Mammals on Islands

Species assemblage	*Large-bodied z (±SE)*	*Small-bodied z (±SE)*	*P*
Great Basin birds	0.25±0.08	0.13±0.06	0.24
New Zealand birds	0.23±0.04	0.12±0.04	0.06
Torres Strait birds	0.23±0.04	0.14±0.03	0.08
Sea of Cortez birds	0.24±0.04	0.19±0.03	0.32
Lake Sysmä mammals	0.45±0.15	0.27±0.15	0.40

Source: Matter et al. (2002) which gives the sources of the data.
Notes: Each assemblage has been divided into the small-bodied and large-bodied species. The *P*-value is for a test of the difference in the slope values.

the scaling of extinction risk in shrews. Indeed, the respective explanations are the same: Matter et al. (2002) showed that the ranges of the critical island areas were about the same for both small-bodied and large-bodied species on the same set of islands, in which case the theory described above implies that the slope z of the species-area relationship directly reflects the average of the slopes ζ_{ext} of the species' incidence functions.

The Species-Area and Distribution-Abundance Relationships

The species-area relationship is one of the best-established generalizations in ecology. From the perspective of single-species incidence functions, the species-area relationship is obtained by summing up the rows of a matrix giving the occurrence (=1) of species (on columns) among a set of islands (on rows). The row sums give the numbers of species on islands; plotting these sums against the island areas gives the species-area relationship.

The island occurrences of species in the same matrix may be summed up along the columns to calculate on how many islands different species have occurred. The column sums then indicate the extent of species' distributions among the islands. Analogous to the plot of species number per island against island area, the distribution of a species may be plotted against its carrying capacity ("species size"), which in practice is measured by the average abundance on islands where the species occurs. This is called the distribution-abundance relationship, and it is also very widely reported and analyzed in the ecological literature (Hanski 1982, Brown 1984, Hanski et al. 1993, Gaston 2003). And what is it like? Just as bigger islands tend to have more species, species with bigger "size' (greater carrying capacity) tend to have more island occurrences (greater distribution) than species with smaller carrying capacity. There is no well-established functional form for the distribution-abundance relationship, but often the logistic function is used to model the fraction of islands or other sampling areas out of all islands or sampling areas that were occupied by a species as a function of its carrying capacity.

Given that the species-area relationship and the distribution-abundance relationship are obtained from the same matrix, by summing up the matrix elements either along the rows or along the columns, it is natural to ask how the two relationships might be related to each other in natural communities. Hanski and Gyllenberg (1997) derived both relationships from the spatially realistic metapopulation model given by equation (7.6). We assumed that the extinction rate is proportional to the inverse of the carrying capacity, and that different species have different population densities and hence different carrying capacities on the same set of islands. With these assumptions, small islands have fewer species than large islands because populations on small islands have smaller carrying capacities and hence greater risk of extinction. Likewise, species with lower density have narrower distributions than species with higher density, because the former have systematically smaller carrying capacities on the same set of islands and hence generally a greater risk of extinction. The model predicted several features of the observed species-area relationship, but the interesting point here is that species-area relationships with realistic slope values were predicted only when there were differences in the densities (abundances) among the species and when the more common species were more widely distributed than the less common ones, that is, when there was a realistic distribution-abundance relationship. In other words, the two relationships are so intimately related to each other that one does not occur without the other. I also note in passing that this model, with differences in species densities reflecting differences in their ecological requirements, effectively merges the two main hypotheses that have been proposed to explain the increasing number of species with increasing island area, namely, the extinction-colonization dynamics as in the MacArthur-Wilson model and habitat heterogeneity allowing more species with dissimilar ecological requirements to persist on larger islands (e.g., Rosenzweig 1995).

Discussion

Reading of the ecological literature suggests that the days of simple models, such as the MacArthur-Wilson island biogeographic model and the Levins metapopulation model, have been passed. Current modeling efforts concerning the spatial occurrence and dynamics of species at the landscape level tend to be quite specific, often including much information about species' life histories and information about the structure of the landscape. Two sorts of models of this type include statistical regression-type "habitat models" (e.g., Elith and Burgman 2003) and generic simulation-based models of population viability analysis (e.g., Akçakaya et al. 2004). Not surprisingly, in specific situations the predictive power of these models is much greater than that of simple general models. The cost is that the predictions are indeed specific, and hence these models are not that helpful in advancing our general understanding of the processes and phenomena at stake. Levins (1968) made the pertinent observation forty years ago: it is not possible to maximize simultaneously generality, realism, and precision, and therefore there is no single best-for-all-purposes model.

Concerning the more general theory in spatial ecology, one may discern a succession from the island model to metapopulation models to more general models of the spatial dynamics of species in any kind of environment, not just in patchy environments. Much of the general theory is concerned with the question of how spatiotemporal variation in population densities is generated and maintained by population processes (Durrett and Levin 1994, McGlade 1999, Dieckmann et al. 2000, Lande et al. 2003, Ovaskainen and Cornell 2006). This research has supported the early insight by Alan Turing (1952) that spatial dynamics may generate complex spatiotemporal patterns in species abundances in the absence of any environmental heterogeneity. Interest in such spatial pattern formation in ecology roughly parallels the previous excitement about nonlinear dynamics in single populations potentially generating complex temporal dynamics in the absence of any environmental stochasticity (May 1976a,b).

In this context, equation (7.6) and comparable deterministic models may appear overly simplistic, as these models make the mean-field approximation and thereby predict uniform density in a homogeneous environment. However, it should be remembered that, while most of the general theory about spatial pattern formation has been developed for homogeneous environments, real landscapes are always heterogeneous and include spatially fixed variation in habitat quality. In a patch network such as shown in figure 7.1d, spatial variation in patch areas, qualities, and connectivities greatly constrains population dynamics. In other words, the probability that a species is present in a particular part of the landscape may be influenced by its own dynamics and by interactions with other species, but it is also strongly influenced by the spatial structure of the landscape, which makes it less likely that spatial patterns due to population processes would dominate over patterns due to heterogeneous environment, especially in single-species models. This is the reason why the deterministic mean-field approximation often predicts surprisingly well the occurrence of metapopulations in fragmented landscapes.

The single-species metapopulation model introduced by Levins (1969) had been extended to competing species by the early 1970s (Levins and Culver 1971, Horn and MacArthur 1972, Slatkin 1974). This research soon demonstrated that the mean-field approximation led to a qualitatively wrong conclusion, namely, competitive exclusion of all but one of the competitors (Slatkin 1974). In contrast, in a model that properly accounts for the spatial correlation in the occurrences of competitors in a homogeneous patch network, two or more species may coexist in spite of strong competition, because strong competition effectively reduces the numbers of habitat patches in which the two species occur simultaneously (spatial pattern formation). Such spatial segregation enhances intraspecific competition in relation to interspecific competition and thereby facilitates regional coexistence. Indeed, the mean-field approximation fails badly for a homogeneous patch network and equal competitors, but if one or both of these assumptions are relaxed, and we examine the dynamics of at least somewhat dissimilar species in heterogeneous networks, the mean-field model predicts well the equilibrium distributions, including complementary distributions in the case of local migration (Hanski 2008). The message is that we should not be led astray by complex models examining interesting phenomena but in a context that is not relevant for populations in natural environments.

Turning from theory to one very practical issue, one legacy of the island biogeographic and metapopulation models is what is commonly called the habitat area and isolation paradigm in conservation biology. What is meant by this is that the spatial distribution of species is largely determined by the areas and isolations (more properly connectivities; see box 7.1) of habitat patches in a fragmented landscape. This prediction is often contrasted with the view that what really matters for the occurrence of species is not habitat area and isolation but habitat quality, and spatial variation in habitat quality from one patch to another. An extensive literature has grown around this issue (reviewed by Fahrig 1997, 2003, Hanski 2005). However, important as it is to know what really determines the occurrence of species in particular cases, not least for conservation and management, one should realize that there is no general answer beyond the observation that, of course, both the quality and the amount of habitat matter. How much they matter in particular situations must depend on the specific circumstances. Each empirical study is necessarily based on a limited number of habitat patches and variables that are measured, and exactly which patches are included makes a difference. Including more patches of very low quality will most likely increase the "significance" of habitat quality in explaining habitat occupancy; adding tiny patches (which an ecologist might be tempted to exclude because they do not often support a local population) would increase the "significance" of patch area; and including some very isolated patches might do the same for the "significance" of connectivity. The point is that there is no general answer, and one should not be misled into assuming that ten studies demonstrating the importance of habitat quality have somehow demonstrated the general unimportance of the spatial configuration of habitat for the dynamics of species living in fragmented landscapes. Incidentally, the literature on the species-area relationship contains a parallel debate about the importance of island area versus habitat heterogeneity on islands in explaining the increasing number of species on islands with increasing area (Williamson 1981, Rosenzweig 1995, Whittaker 1998).

I conclude by commenting on one striking difference between MacArthur and Wilson's island biogeographic model and Levins's metapopulation model—how were they received by the scientific community? The MacArthur-Wilson model quickly became very well known, it started to have great impact on basic research, and it was one of the building blocks upon which modern conservation biology was established in the mid-1970s (Simberloff 1988, Hanski and Simberloff 1997). In contrast, the Levins model remained little known and had very little impact on anything for the next ten years. Levins's 1969 paper received fewer than ten citations per year until 1991 (ISI Web of Knowledge), by which time the MacArthur-Wilson volume had accumulated more than 2,200 citations, an incredible number for those years (any ecologist would be glad to have papers with the same citation record as the common misspellings of the MacArthur-Wilson classic). In the past fifteen years, the difference has become much smaller, and while 34% of the pooled citations to MacArthur and Wilson (1967) are in papers published since 2000, the corresponding figure for Levins (1969) is a whopping 65%. Amazingly, both publications have received their highest annual number of citations to date in . . . 2007. This is amazing for a paper and a book published in the 1960s, even allowing for the ever-expanding literature and hence increasing annual number of total citations.

So why were MacArthur and Wilson so successful early on, and why was Levins not? MacArthur-Wilson (1967) was published as the inaugural volume in a monograph series that was bound to succeed, whereas

Levins (1969) was published as a short paper in a rather obscure journal (that is, obscure from the perspective of most ecologists). This difference surely mattered, but I suggest that another difference was even more important. From the very beginning, in fact from the introduction to the original description of the island biogeographic model in MacArthur and Wilson's paper published in *Evolution* in 1963, the theory became associated with the species-area relationship. This is important, because the species-area relationship was something that scores of biologists had been working on previously, and something for which more data could be easily gathered. The MacArthur-Wilson model appeared to provide a ready recipe for empirical studies, and for studies that would be highly doable and seemingly highly relevant for a current high-profile theory in ecology. No wonder that ecologists seized the opportunity. And not only that, soon that theory appeared to make a major contribution to conservation as well! In contrast, the Levins model must have appeared a rather abstract exercise to the few ecologists who noticed it. The Levins model did not lead to instructions as to what ecologists should, or could, do in practice.

We now know that the expectations concerning the MacArthur-Wilson model were too high, that demonstrating the species-area relationship does not critically validate, or refute, the island model, and that the conservation applications were simplistic. One could even argue that the excessive emphasis on the species-area relationship may have distracted attention from single-species incidence functions, which would have provided a much richer material for research, and for no extra cost at all, because with exactly the same data that were collected to study species-area relationships one could have calculated the incidence functions for individual species. But this is all wisdom based on hindsight. The Levins model has experienced a renaissance partly because it deals with situations that are now prevalent in the terrestrial world everywhere, highly fragmented habitats without a mainland, but also because it is the basis for the spatially realistic models described in this chapter, which have provided the blueprint for empirical studies of metapopulation dynamics. The works of MacArthur and Wilson and of Levins have had lasting impact in ecology and conservation because they succeeded so beautifully in capturing great ideas in simple mathematical models.

Acknowledgments

I thank Elizabeth Crone, Otso Ovaskainen, Robert Ricklefs, and two anonymous reviewers for comments, and Sami Ojanen for technical help.

Literature Cited

Akçakaya, H. R., M. A. Burgman, O. Kindvall, C. C. Wood, P. Sjögren-Gulve, J. S. Hatfield, and M. A. McCarthy, editors. 2004. *Species Conservation and Management. Case Studies.* Oxford: Oxford University Press.

Andrewartha, H. G., and L. C. Birch. 1954. *The Distribution and Abundance of Animals.* Chicago: University of Chicago Press.

Arrhenius, O. 1921. Species and area. *Journal of Ecology* 9:95–99.

Brown, J. H. 1984. On the relationship between abundance and distribution of species. American Naturalist 124:255–79.

Connor, E. F., and E. D. McCoy. 1979. The statistics and biology of the species-area relationship. *American Naturalist* 113:791–833.

Crowell, K. L. 1986. A comparison of relict versus equilibrium models for insular mammals of the Gulf of Maine. In *Island Biogeography of Mammals,* ed. L. R. Heaney and B. D. Patterson, 37–64. London: Academic Press.

Curtis, J. T. 1956. The modification of mid-latitude grasslands and forests by man. In *Man's Role in Changing the Face of the Earth*, ed. W. L. Thomas, 721–36. Chicago: University of Chicago Press.

Diamond, J. M. 1975. Assembly of species communities. In *Ecology and Evolution of Communities,* ed. M. L. Cody and J. M. Diamond, 342–444. Cambridge, MA: Harvard University Press.

Dieckmann, U., R. Law, and J. A. J. Metz. 2000. *The Geometry of Ecological Interaction: Simplifying Spatial Complexity*. Cambridge: Cambridge University Press.

Durrett, R., and S. Levin. 1994. The importance of being discrete (and spatial). *Theoretical Population Biology* 46:363–94.

Elith, J., and M. A. Burgman. 2003. Habitat models for population viability analysis.In *Population Viability In Plants,* ed. C. A. Brigham and M. W. Schwartz, 203–35. Ecological Studies vol. 165. Berlin: Springer-Verlag.

Etienne, R. S., C.J.F. van ter Braak, and C. C. Vos. 2004. Application of stochastic patch occupancy models to real metapopulations. In *Ecology, Genetics, and Evolution of Metapopulations*, ed. I. Hanski and O. E. Gaggiotti, 105–32. Amsterdam: Elsevier Academic.

Fahrig, L. 1997. Relative effects of habitat loss and fragmentation on population extinction. *Journal of Wildlife Management* 61:603–10.

———. 2003. Effects of habitat fragmentation on biodiversity. *Annual Review of Ecology, Evolution, and Systematics* 34:487–515.

Gaston, K. J. 2003. *The Structure and Dynamics of Geographical Ranges.* Oxford: Oxford University Press.

Hanski, I. 1982. Dynamics of regional distribution: The core and satellite species hypothesis. *Oikos* 38:210–21.

———. 1992. Inferences from ecological incidence functions. *American Naturalist* 139:657–62.

———. 1993. Dynamics of small mammals on islands. *Ecography* 16:372–75.

———. 1994. A practical model of metapopulation dynamics. *Journal of Animal Ecology* 63:151–62.

———. 1998a. Connecting the parameters of local extinction and metapopulation dynamics. *Oikos* 83:390–96.

———. 1998b. Metapopulation dynamics. *Nature* 396:41–49.

———. 1999. *Metapopulation Ecology*. New York: Oxford University Press.

———. 2001. Spatially realistic models of metapopulation dynamics and their implications for ecological, genetic and evolutionary processes. In *Integrating Ecology and Evolution in a Spatial Context*, ed. J. Silvertown and J. Antonovics, 139–56. Oxford: Blackwell Science.

———. 2005. *The Shrinking World: Ecological Consequences of Habitat Loss.* Oldendorf/Luhe: International Ecology Institute.

———. 2008. Spatial patterns of coexistence of competing species in patchy habitats. *Theoretical Ecology*. DOI:10.1007/s12080-007-0004-y.

Hanski, I., and M. Gyllenberg. 1997. Uniting two general patterns in the distribution of species. *Science* 275:397–400.

Hanski, I., J. Kouki, and A. Halkka. 1993. Three explanations of the positive relationship between distribution and abundance of species. In *Community Diversity: Historical and Geographical Perspectives,* ed. R. E. Ricklefs and D. Schluter, 108–16. Chicago: University of Chicago Press.

Hanski, I., and O. Ovaskainen. 2000. The metapopulation capacity of a fragmented landscape. *Nature* 404:755–58.

———. 2003. Metapopulation theory for fragmented landscapes. *Theoretical Population Biology* 64:119–27.

Hanski, I., and D. Simberloff. 1997. The metapopulation approach, its history, conceptual domain and application to conservation. In *Metapopulation Biology: Ecology, Genetics, and Evolution,* ed. I. Hanski and M. E. Gilpin, 5–26. San Diego: Academic Press.

Holyoak, M., M. A. Leibold, and R. D. Holt, eds. 2005. *Metacommunities: Spatial Dynamics and Ecological Communities.* Chicago: University of Chicago Press.

Horn, H. S., and R. H. MacArthur. 1972. Competition among fugitive species in a harlequin environment. *Ecology* 53:749–52.

Huffaker, C. B. 1958. Experimental studies on predation: dispersion factors and predator-prey oscillations. *Hilgardia* 27:343–83.

Lande, R., S. Engen, and B.-E. Saether, editors. 2003. *Stochastic Population Dynamics in Ecology and Conservation*. Oxford: Oxford University Press.

Levins, R. 1968. *Evolution in Changing Environments.* Princeton, NJ: Princeton University Press.

———. 1969. Some demographic and genetic consequences of environmental heterogeneity for biological control. *Bulletin of the Entomological Society of America* 15:237–40.

———. 1970. Extinction. *Lecture Notes in Mathematics* 2:75–107.

Levins, R., and D. Culver. 1971. Regional coexistence of species and competition between rare species. *Proceedings of the National Academy of Sciences U.S.A.* 68:1246–48.

Levins, R., and H. Heatwole. 1963. On the distribution of organisms on islands. *Caribbean Journal of Science* 3:173–77.

Levins, R., and R. MacArthur. 1966. Maintenance of genetic polymorphism in a heterogeneous environment: Variations on a theme by Howard Levene. *American Naturalist* 100:585–90.

Lomolino, M. V. 1993. Winter filtering, immigrant selection and species composition of insular mammals of Lake Huron. *Ecography* 16:24–30.

MacArthur, R., and R. Levins. 1964. Competition, habitat selection and character displacement in a patchy environment. *Proceedings of the National Academy of Sciences U.S.A.* 51:1207–10.

MacArthur, R. H., and E. O. Wilson. 1963. An equilibrium theory of insular zoogeography. *Evolution* 17:373–87.

———. 1967. *The Theory of Island Biogeography*. Princeton, NJ: Princeton University Press.

Matter, S. F., I. Hanski, and M. Gyllenberg. 2002. A test of the metapopulation model of the species-area relationship. *Journal of Biogeography* 9:977–83.

May, R. M. 1976a. Models for single populations. In *Theoretical Ecology: Principles and Applications,* ed. R. M. May, 5–29. Oxford: Blackwell Scientific.

May, R.M. 1976b. Simple mathematical models with very complicated dynamics. *Nature* 261:459–67.

McGlade, J., editor. 1999. *Advanced Ecological Theory.* Oxford: Blackwell Scientific.

Moran, R. 1983. Vascular plants of the Gulf Islands. In *Island Biogeography of the Sea of Cortez,* ed. T. J. Case and M. L. Cody, 348–81 Berkeley: University of California Press.

Nicholson, A. J. 1933. The balance of animal populations. *Journal of Animal Ecology* 2:132–78.

Nieminen, M., M. Siljander, and I. Hanski. 2004. Structure and dynamics of *Melitaea cinxia* metapopulations. In *On the Wings of Checkerspots: A Model System for Population Biology,* ed. P. R. Ehrlich and I. Hanski, 63–91. New York: Oxford University Press.

Ovaskainen, O. 2001. The quasi-stationary distribution of the stochastic logistic model. *Journal of Applied Probability* 38:898–907.

Ovaskainen, O., and S. J. Cornell. 2006. Space and stochasticity in population dynamics. *Proceedings of the National Academy of Sciences U.S.A.* 103:12781–86.

Ovaskainen, O., and E. Crone. 2009. Modeling movement behavior with diffusion. In *Spatial Ecology,* ed. S. Cantrell, C. Cosner, and S. Ruan. Boca Raton, FL: CRC Press.

Ovaskainen, O., and I. Hanski. 2001. Spatially structured metapopulation models: global and local assessment of metapopulation capacity. *Theoretical Population Biology* 60:281–304.

———. 2002. Transient dynamics in metapopulation response to perturbation. *Theoretical Population Biology* 61:285–95.

———. 2003. The species-area relation derived from species-specific incidence functions. *Ecology Letters* 6:903–9.

———. 2004. Metapopulation dynamics in highly fragmented landscapes. In *Ecology, Genetics, and Evolution in Metapopulations,* ed. I. Hanski and O. E. Gaggiotti, 73–104. Amsterdam: Elsevier Academic.

Pimm, S. L. 1991. *The Balance of Nature? Ecological Issues in the Conservation of Species and Communities*. Chicago: University of Chicago Press.

Pulliam, H. R. 1988. Sources, sinks, and population regulation. *American Naturalist* 132:652–61.

Rosenzweig, M. L. 1995. *Species Diversity in Space and Time*. Cambridge: Cambridge University Press.

Simberloff, D. 1988. The contribution of population and community biology to conservation science. *Annual Review of Ecology and Systematics* 19:473–512.

Sinclair, A.R.E. 1989. Population regulation in animals. In *Ecological Concepts*, ed. J. M. Cherrett, 197–242. Oxford: Blackwell.

Slatkin, M. 1974. Competition and regional coexistence. *Ecology* 55:128–34.

Strong, D. R. J., D. Simberloff, L. G. Abele, and A. B. Thistle, eds. 1984. *Ecological Communities: Conceptual Issues and the Evidence*. Princeton, NJ: Princeton University Press.

Tarantola, A. 2005. *Inverse Problem Theory and Methods for Model Parameter Estimation*. Philadelphia: SIAM.

Thibault, J.-C., J.-L. Martin, and I. Guyot. 1990. Les oiseaux terrestres nicheurs des iles mineures des Bouches-de-Bonifacio: Analyse du peuplement. *Alauda* 58:173–85.

Thomas, C. D., and T. M. Jones. 1993. Partial recovery of a skipper butterfly (Hesperia comma) from population refuges: Lessons for conservation in a fragmented landscape. *Journal of Animal Ecology* 62:472–81.

Turing, A. M. 1952. The chemical basis of morphogenesis. *Philosophical Transactions of the Royal Society of London, Series B* 237:37–72.

Whittaker, R. J. 1998. *Island Biogeography: Ecology, Evolution, and Conservation*. Oxford: Oxford University Press.

Wiegand, T., F. Jeltsch, I. Hanski, and V. Grimm. 2003. Using pattern-oriented modeling for revealing hidden information: A key for reconciling ecological theory and application. *Oikos* 100:209–22.

Williamson, M. 1981. *Island Populations*. Oxford: Oxford University Press.

Wright, S. 1940. Breeding structure of populations in relation to speciation. *American Naturalist* 74:232–48.

Beyond Island Biogeography Theory

UNDERSTANDING HABITAT FRAGMENTATION IN THE REAL WORLD

William F. Laurance

Island Biogeography Theory (IBT; MacArthur and Wilson 1963, 1967) has profoundly influenced the study of biogeography, ecology, and even evolution (Janzen 1968, Losos 1996, Heaney 2000), and has also had an enormous impact on conservation biology. The theory has inspired much thinking about the importance of reserve size and connectivity in the maintenance of species diversity, and stimulated an avalanche of research on fragmented ecosystems. But, like all general models, IBT is a caricature of reality, capturing just a few important elements of a system while ignoring many others. Does it provide a useful model for understanding contemporary habitat fragmentation?

Here I critically evaluate the conceptual utility and limitations of IBT for the study of fragmented ecosystems. I briefly encapsulate the historical background, considering how IBT has helped to shape our thinking about habitat fragmentation over the past forty years. I then describe how fragmentation research has transcended the theory, using findings from tropical and other ecosystems.

The Impact of IBT

Prior to MacArthur and Wilson's (1967) seminal book, habitat fragmentation was not high on the radar screen of most ecologists, land managers, and politicians. That all changed with IBT (Powledge 2003). The theory has helped to revolutionize the thinking of mainstream ecologists about habitat fragmentation and stimulated literally thousands of studies of fragmented and insular ecosystems (figure 8.1).

Before summarizing some key conceptual advances linked to IBT, I have two caveats. First, in discussing the impact of IBT on fragmentation research, it can be difficult to distinguish between the contributions of

Figure 8.1. Experimentally isolated forest fragments in central Amazonia, part of the Biological Dynamics of Forest Fragments Project (photo by R. O. Bierregaard). This long-term experiment was inspired by a heated debate over the relevance of Island Biogeography Theory to nature conservation.

the original theory itself—*sensu stricto*—versus the ancillary contributions of the many investigations it has helped to spawn. Rather than worrying overly about this, I have listed as many conceptual advances as occurred to me, and tried (no doubt inadequately) to give credit where credit is due. Second, an inherent problem with the burgeoning IBT literature is that it is a little like the Bible: so large, diverse, and eclectic that one can seemingly draw any lesson one wants. Casting such concerns aside I stride incautiously ahead.

Perhaps more than anything, IBT opened people's eyes to the importance of bigness for nature conservation (see also Preston 1960). Big reserves contain more species, lose species more slowly (MacArthur and Wilson 1967, Burkey 1995), and suffer fewer of the deleterious effects of habitat isolation than do smaller reserves (Terborgh 1974, Diamond 1975, May 1975, Diamond and May 1976). The main advantage of bigness, according to IBT, is that individual species can maintain bigger populations than in small areas, and that big populations go locally extinct less often than do small populations (Shafer 1981). Big reserves should also be better at preserving the full range of successional communities and patch dynamics within ecosystems (Pickett and Thompson 1978). The presumed importance of area-dependent extinctions has given rise to evocative terms such as "supersaturation," "species relaxation," "faunal collapse," and "ecosystem decay" that have collectively helped to cement the importance of bigness in the scientific and popular imaginations (e.g., Diamond 1972, Lovejoy et al. 1984, Quammen 1997). Indeed, the pendulum of thought has swung so far in favor of bigness that some authors have found it necessary to remind us that small reserves can be important too (Shafer 1995, Turner and Corlett 1996).

Of course, IBT helped to refine people's thinking about habitat isolation as well. Isolation is bad, connectivity is good. If a little isolation is a bad thing, then a lot of isolation is even worse. Hence, reserves that are isolated from other areas of habitat by large expanses of degraded, hostile landscape will sustain fewer species of conservation concern than those nearer to intact habitat (Lomolino 1986, Watling and Donnelly 2006). This occurs for two reasons: weakly isolated reserves are easily colonized by new species, and they receive immigrants whose genetic and demographic contributions can reduce local extinction rates within the reserve (Brown and Kodric-Brown 1977).

IBT has also spawned a highly dynamic view of fragmented ecosystems. A key prediction of IBT is that insular biota should be inherently dynamic, with species disappearing (from local extinction) and appearing (from colonization) relatively often. If extinction and colonization are largely governed by fragment size and isolation, respectively, then big, isolated fragments should have slower species turnover than do small, weakly isolated fragments. Demonstration of such relationships is a litmus test for IBT (Gilbert 1980, Abbott 1983) because other biogeographic phenomena, such as the species-area relationship, can arise for reasons aside from those hypothesized by IBT (for example, higher habitat diversity, rather than lower extinction rates, can cause species richness to increase on larger islands; Boecklen and Gotelli 1984, Ricklefs and Lovette 1999). Given its central importance, it is perhaps surprising that only a modest subset of all IBT studies has demonstrated elevated turnover (e.g., Diamond 1969, Wright 1985, Honer and Greuter 1988, Schmigelow et al. 1997—and even these have often been controversial (Simberloff 1976, Diamond and May 1977, Morrison 2003; reviewed in Schoener, this volume). As discussed below, population and community dynamics are often greatly amplified in habitat fragments relative to natural conditions (Laurance 2002), but a number of factors aside from those hypothesized by IBT can be responsible.

Habitat fragmentation affects different species in different ways. Some species decline sharply or disappear in fragments (figure 8.2), others remain roughly stable, and yet others increase, sometimes dramatically. Although IBT *sensu stricto* provides little understanding of the biological reasons for such differences, some insights have come from interpreting the slope (z) of species-area relationships in insular communities (Connor and McCoy 1979, Ricklefs and Lovette 1999). For instance, species at higher trophic levels (Holt et al. 1999), with lower volancy (Wright 1981), with greater ecological specialization (Krauss et al. 2003; Holt, this volume), and with greater taxonomic age (Rickefs and Cox 1972, Rickefs and Bermingham 2004) generally have steeper slopes, and thus respond more negatively to insularization than do those with opposite characteristics. Characteristics of fragmented landscapes can also affect species-area slopes (Wright 1981). For example, slopes are on average steeper for fauna on true islands than terrestrial fragments, presumably because agricultural or urban lands are less hostile to faunal movements than are oceans and lakes (Watling and Donnelly 2006).

Early proponents of IBT were keen to apply its principles to the design of protected areas, and used the theory (among other things) to advance the notion that a single large reserve was better for ensuring long-term species persistence than were several small reserves of comparable area (Terborgh 1974, Diamond 1975, May 1975, Wilson and Willis 1975). This idea, encapsulated in the famous acronym SLOSS (single large or several small reserves), became a remarkable flashpoint of controversy, following a pointed attack by Simberloff and Abele (1976a). Although of theoretical interest, the ensuing debate (e.g., Diamond 1976, Simberloff and Abele 1976b, Terborgh 1976, Whitcomb et al. 1976, Abele and

Figure 8.2. Forest specialists such as the lemuroid ringtail possum (*Hemibelideus lemuroides*), a restricted endemic in tropical Queensland, are often highly vulnerable to habitat fragmentation (photo by M. Trenerry).

Connor 1979, Higgs and Usher 1980) provided only a limited list of practical lessons for reserve managers (Soulé and Simberloff 1986, Zimmerman and Bierregaard 1986, Saunders et al. 1991). Perhaps the most important conclusion was that SLOSS depended on the degree of nestedness exhibited by an ecosystem (the extent to which the biota of small reserves was a proper subset of those in larger reserves; Patterson and Atmar 1986, Patterson 1987). The most extinction-prone species are often found only in large reserves, favoring the single-large-reserve strategy, although small reserves scattered across a region can sustain certain locally endemic species that would otherwise remain unprotected (see Ovaskainen 2002 and references therein). Thus, the answer to SLOSS is, "it depends."

Habitat Fragmentation in the Real World

By stimulating an avalanche of research on insular ecosystems, IBT has helped to teach us a great deal about habitat fragmentation. In a strict sense, however, IBT itself has only limited relevance to fragmentation because it fails to consider some of the most important phenomena in fragmented landscapes. Here I summarize some of the key lacunae.

Nonrandom Habitat Conversion

Habitat conversion is a highly nonrandom process. Farmers preferentially clear land in flatter lowland areas (Winter et al. 1987, Dirzo and Garcia 1992) and in areas with productive, well-drained soils (Chatelain et al. 1996, Smith 1997). Habitat loss also tends to spread contagiously, such that areas near highways, roads, and towns are cleared sooner than those located further from human settlements. In the Brazilian Amazon, for example, over 90% of all deforestation occurs within 50 km of roads or highways (Laurance et al. 2001a, Brandão et al. 2007).

As a consequence of nonrandom clearing, habitat remnants are often a highly biased subset of the original landscape. Remnants frequently persist in steep and dissected areas, on poorer soils, at higher elevations, and on partially inundated lands. In addition, habitat fragments near roads and townships are often older, more isolated, and smaller than those located further afield, where habitat destruction is more recent (Laurance 1997). The influence of nonrandom habitat loss on fragmented communities has been little studied, although Seabloom et al. (2002) concluded that species-area curves underestimate the magnitude of species extinctions when habitat destruction is contagious, as is typically the case. Regardless, it is important to recognize that the biota of habitat fragments are likely to have been influenced by nonrandom habitat loss long before the effects of fragmentation per se are manifested.

Distinguishing Habitat Loss and Fragmentation Effects

The process of habitat fragmentation involves two distinct but interrelated processes. First, the total amount of original habitat in the landscape is reduced. Second, the remaining habitat is chopped up into fragments of various sizes and degrees of isolation. Distinguishing the impacts of these two processes on biodiversity is challenging because they generally covary. For example, in forested landscapes in which most of the original habitat has been destroyed, the surviving fragments tend to be small and isolated from other forest areas, and the opposite is true in landscapes with little forest loss. Hence, strong declines of biodiversity reported for many fragmented landscapes might actually be mostly a consequence of habitat loss, rather than habitat fragmentation per se (Fahrig 2003).

IBT emphasizes analyses at the individual-fragment scale, but the best way to quantify the relative importance of habitat loss versus fragmentation is to conduct comparative analyses at the landscape scale. In a meta-analysis, Fahrig (2003) concluded that habitat loss generally had much stronger effects on biodiversity than did fragmentation per se, although she emphasized that much is uncertain, especially for tropical forests. Others have tried to distinguish effects of habitat loss and fragmentation, either by experimentally controlling for habitat amount while varying fragmentation (e.g., Collins and Barrett 1997, Caley et al. 2001) or by comparing many different landscapes and extracting indices of fragmentation that are not correlated with the amount of habitat in each landscape (e.g., McGarigal and McComb 1995, Villard et al. 1999). Results have varied, and disentangling the often confounded effects of habitat loss and fragmentation remains a challenge for those attempting to discern the mechanisms of biodiversity loss in fragmented landscapes.

Edge Effects

Edge effects are diverse physical and biological phenomena associated with the abrupt, artificial boundaries of habitat fragments (figure 8.3). They include the proliferation of shade-intolerant vegetation along fragment margins (Ranney et al. 1981, Lovejoy et al. 1986) as well as changes in microclimate and light regimes that affect seedling germination and survival (Ng 1983, Bruna 1999). Forest interiors often are bombarded by a "seed rain" of weedy propagules (Janzen 1983, Nascimento et al. 2006) and by animals originating from outside habitats (Buechner 1987).

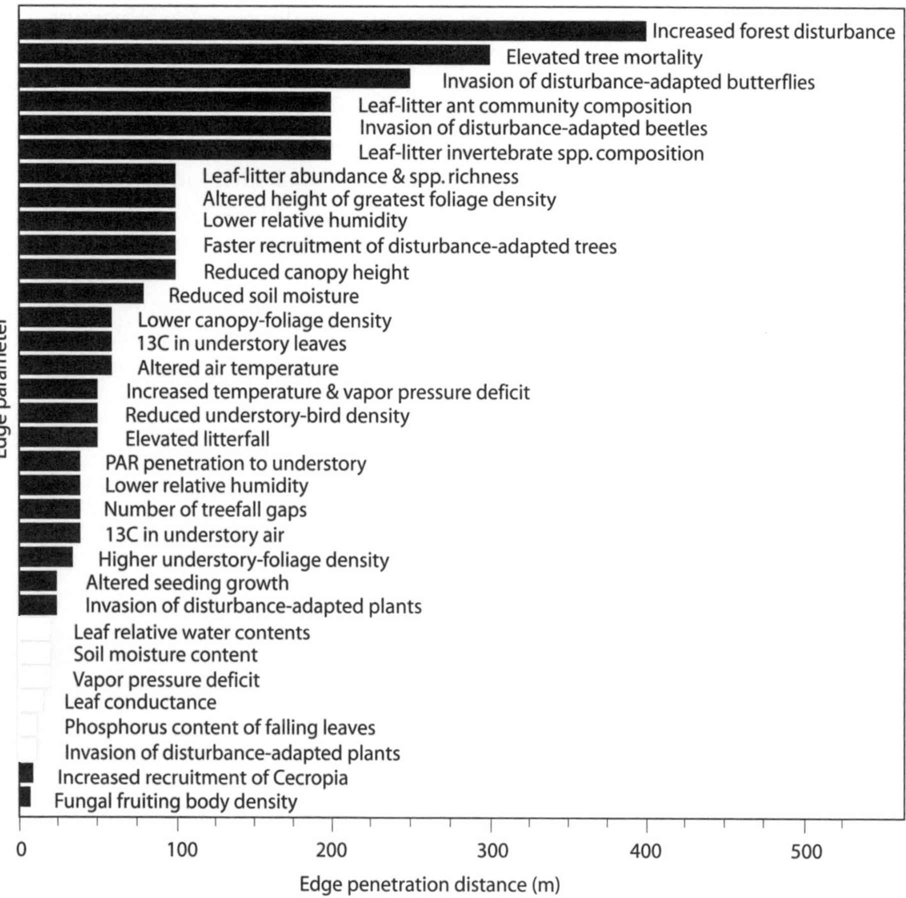

Figure 8.3. Edge effects documented in Amazonian forest fragments, showing the great diversity of edge phenomena and the varying distances they penetrate into forest interiors (after Laurance et al. 2002).

Increased windshear forces near edges can cause elevated rates of tree mortality that alter forest structure and composition (Chen et al. 1992, Laurance et al. 1997, 2000). Abundant generalist predators, competitors, or brood parasites in the vicinity of edges often impact forest-interior birds (Gates and Gysel 1978, Wilcove 1985) and mammals (Sievert and Keith 1985).

Edge effects can alter many aspects of the structure, microclimate, dynamics, and species composition of fragmented ecosystems (Lovejoy et al. 1986, Laurance et al. 2002, Lehtinen et al. 2003, Ries et al. 2004). Crucially, they are not addressed by IBT, which assumes that biota in fragments are influenced solely by the opposing forces of colonization and extinction. Edge effects may be especially important in fragmented rainforests, where the dense forest with its stable temperatures and dark, humid, nearly windless conditions contrasts starkly with the dry, harsh, windy conditions of surrounding pastures or croplands.

It can be challenging to discriminate edge and area effects in fragmentation studies. Edge phenomena tend to increase in intensity as fragment size diminishes, and this creates a confounding intercorrelation between edge and area effects in fragmented landscapes (Laurance and Yensen 1991). In fact, many putatively "area-related" species losses in habitat fragments probably have been caused by edge effects (Schonewald-Cox and Bayless 1986, Temple 1986) or a synergism between edge and area effects (Ewers et al. 2007).

Understanding the role of edge effects is important, because edge models yield different predictions than does IBT about the effects of fragmentation on ecosystems and biota. For example, unlike IBT, edge-effect models predict large ecological changes (1) in irregularly shaped as well as in small fragments, (2) along the margins of even very large fragments, and (3) especially in areas affected by two or more nearby edges (Laurance and Yensen 1991, Malcolm 1994, Laurance et al. 2006a). Edge models also provide useful predictions about species responses to fragmentation. For instance, (1) the abundances of individual forest-interior species should be positively correlated with the unaltered core areas of fragments (Temple 1986, Ewers and Didham 2007), (2) edge specialists should be correlated with the total length of fragment edges, and (3) edge-insensitive species that depend on primary habitat should be correlated with the total areas of fragments (Laurance and Yensen 1991). IBT yields none of these insights.

Matrix Effects

For all its conceptual utility, IBT has had a huge downside for understanding forest fragmentation: it ignores the matrix of modified lands surrounding fragments. Whether surrounded by corn fields, strip malls, water, or secondary forest, all fragments (including isolated nature reserves) are treated equally by IBT. Such fragments are not equivalent, of course—the matrix matters.

The matrix has a big influence on fragment connectivity (Ricketts 2001). Matrices that differ dramatically in structure and microclimate from the primary habitat tend to be most hostile to native species (Laurance and Bierregaard 1997). In the Amazon, forest fragments surrounded by cattle pastures suffer considerably greater species losses than do those surrounded by regrowth forest, and a variety of species—including certain primates, antbirds, obligate flocking birds, and euglossine bees—have been shown to recolonize fragments as young secondary forest regenerates around them (Becker et al. 1991, Stouffer and Bierregaard 1995, Gilbert and Setz 2001). Where hunting is pervasive, the matrix can become a population sink for exploited species (Woodroffe and Ginsberg 1998). By acting as a selective filter for animal and propagule movements, the matrix has pervasive effects on species composition in fragments.

The matrix can also influence the nature and magnitude of edge effects in fragments. In the Amazon, forest fragments surrounded by young regrowth forest experience less intensive changes in microclimate (Didham and Lawton 1999) and have lower edge-related tree mortality (Mesquita et al. 1999) than do similar fragments adjoined by cattle pastures. Edge avoidance by forest-interior birds is also reduced when fragments are adjoined by regrowth forest (Stouffer and Bierregaard 1995, S. G. Laurance 2004). Because fragments can receive a heavy seed rain from the nearby matrix, patterns of plant regeneration in forest fragments can be strongly influenced by the species composition of the matrix (Janzen 1983, Nascimento et al. 2006).

Correlates of Extinction Proneness

Whether on islands or habitat fragments, species vary enormously in their vulnerability to local extinction: some vanish rapidly, others more slowly, and yet others persist almost indefinitely. Why? Much effort has been expended in attempting to predict why certain species are especially extinction prone in insular habitats (e.g., Terborgh 1974, Pimm et al. 1989, Laurance 1991).

The traits associated with vulnerability may well differ between islands and habitat fragments. Studies of fauna on islands have often emphasized the importance of local rarity or its correlates, such as body size and trophic status, in determining species vulnerability (e.g., Terborgh 1974, Willis 1974, Wilcox 1980, Diamond 1984, Holt, this volume). Unlike islands, however, habitat fragments are surrounded by a matrix of modified habitats that permit dispersal or survival for species that can use the matrix, and matrix tolerance is often identified as a key predictor of vulnerability (Laurance 1990, 1991, Gascon et al. 1999, Nupp and Swihard 2000, Pires et al. 2002). On islands, or on other isolates surrounded by completely inhospitable habitat, matrix tolerance is necessarily a nonexistent predictor of extinction proneness, and effects of other predictors, such as rarity and its correlates, are likely to become more apparent.

Thus, as a model for predicting faunal extinctions in habitat fragments, studies of oceanic or land-bridge islands may (1) underestimate the importance of overland vagility and tolerance of modified habitats, and (2) overestimate the significance of factors such as rarity, body size, and trophic status. Insofar as IBT emphasizes true islands, its lessons for understanding species vulnerability in habitat fragments might be weak and even misleading.

Altered Ecosystem Processes

As a prism for understanding habitat fragmentation, IBT is woefully limited in scope: it concerns only the factors that affect species diversity. But habitat fragmentation has far broader effects on ecosystems, altering such diverse processes as forest dynamics, nutrient cycling, carbon storage, and forest-climate interactions.

In many forested landscapes, for example, habitat fragmentation leads to sharply elevated tree mortality, because trees near forest edges are particularly vulnerable to wind turbulence and increased desiccation (Chen et al. 1992, Laurance et al. 1997, 1998a). This fundamentally alters canopy-gap dynamics, forest structure, microclimate (Kapos 1989, Malcolm 1998), and the relative abundance of different plant functional groups (Laurance et al. 2001b, 2006a,b, Nascimento et al. 2006). Forest carbon storage is also reduced (figure 8.4) because large canopy and emergent trees, which contain a high proportion of forest biomass, are particularly vulnerable to fragmentation (Laurance et al. 2000). As the biomass from the dead trees decomposes, it is converted into greenhouse gases such as carbon dioxide and methane. In fragmented forests worldwide, many millions of tons of atmospheric carbon emissions are released each year by this process (Laurance et al. 1998b).

Fragmentation alters many aspects of the physical environment. Large-scale clearing of native vegetation can cause major changes in water and nutrient cycles, radiation balance, and wind regimes, which in turn affect communities in habitat remnants (Saunders et al. 1991, Laurance 2004). In Western Australia, the removal of most native vegetation for wheat production has reduced evapotranspiration and altered soil water flows. This has increased local flooding, brought the water table with its dissolved salts closer to the soil surface, and caused chronic waterlogging and salinization of the remaining vegetation (Hobbs 1993). Wind- or waterborne fluxes of agricultural chemicals (fertilizers, herbicides, pesticides) and other pollutants into habitat remnants (Cadenasso et al. 2000, Weathers et al. 2001) can also have long-term effects on ecosystems.

Fragmentation often drastically alters natural fire regimes. In some cases, burning declines sharply because fires are suppressed in the sur-

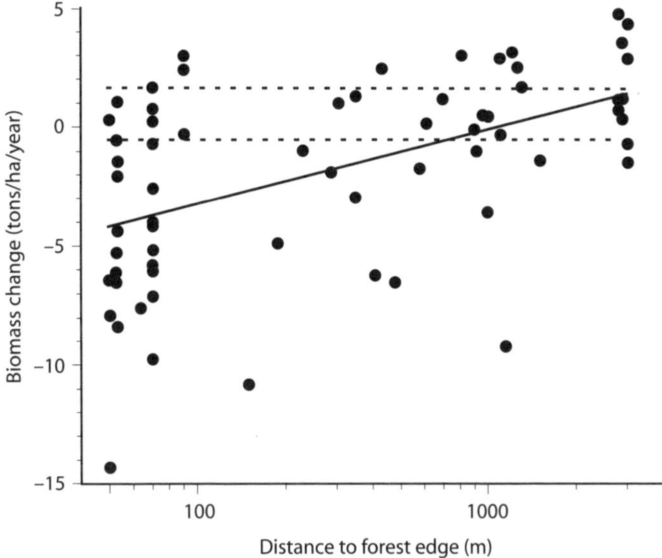

Figure 8.4. Collapse of aboveground biomass in Amazonian forest fragments. Shown is the annual loss of live tree biomass in 1 ha plots as a function of distance from forest edge. The dotted lines show the 95% confidence intervals for forest-interior plots (>500 m from edge) (after Laurance et al. 1997).

rounding matrix, leading to long-term changes in the composition and structure of remnant vegetation (Baker 1994). In other cases, fragmentation promotes burning in ecosystems that are highly vulnerable to fire, such as tropical rainforests (Cochrane et al. 1999, Gascon et al. 2000). In the Amazon, for example, fire frequency rises drastically in fragmented landscapes (figure 8.5) because forest remnants are juxtaposed with frequently burned pastures. These recurring burns have severe effects because the rainforest vegetation is poorly adapted for fire, and forest fragments can literally implode over time from recurring fires (Cochrane and Laurance 2002, 2008).

Environmental Synergisms

In the real world, habitat fragments are not merely reduced and isolated; they are also frequently affected by other perturbations that may interact additively or synergistically with fragmentation (Laurance and Cochrane 2001). Forest fragments in the tropics, for example, are often selectively logged, degraded by ground fires, and overhunted—changes that can dramatically alter fragment ecology (Peres 2001, Cochrane and Laurance

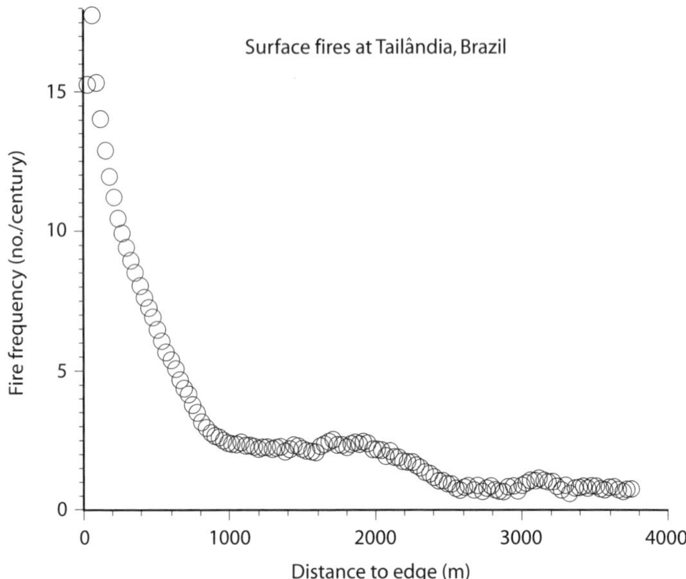

Figure 8.5. Fires can increase dramatically in fragmented forests. Shown here is the mean fire frequency (number per century) as a function of distance to forest edge for several hundred forest fragments in eastern Amazonia. Analyses were based on 14 years of satellite observations (adapted from Cochrane and Laurance 2002).

2002, Peres and Michalski 2006). In agricultural and urban areas, acid rain, pesticides and herbicides, hydrological changes, livestock grazing, and pressure from invading species can severely degrade fragments (Myers 1988, Apensperg-Traun et al. 1996, Hobbs and Huenneke 1992). In coming decades, anthropogenic climate change may emerge as an increasingly important threat to fragmented ecosystems, especially if droughts, storms, and other rare weather events increase in frequency or severity (Timmerman et al. 1999, Laurance and Curran 2008).

Thus, forest fragments and their biota are sometimes subjected to a withering array of environmental pressures that may be episodic or chronic in nature. A paradigm like IBT that considers only changes in fragment size and isolation while ignoring other anthropogenic effects (e.g., Curran et al. 1999, Laurance 2000) is dangerously inadequate for conservation purposes. It is also inadequate from a scientific perspective. A more realistic view of fragmented landscapes is one that explicitly recognizes the potential for interacting environmental changes to amplify and alter the ecological impacts of habitat fragmentation.

Elevated Dynamics

Finally, IBT postulates that fragmented ecosystems will be more dynamic than intact habitat, but only because of species relaxation and increased species turnover. In fact, a far wider range of phenomena promotes dynamism in fragmented landscapes, even to the extent that many fragments can be described as "hyperdynamic" (Laurance 2002).

Being a small resource base, a habitat fragment is inherently vulnerable to stochastic effects. Species abundances can fluctuate wildly in small communities, especially when immigration is low and disturbances are frequent (Hubbell 2001). The dynamics of plant and animal populations can be dramatically altered in fragmented habitats in response to edge effects, reduced dispersal, altered disturbance regimes, and changing herbivore or predation pressure (Lidicker 1973, Karieva 1987, Quintana-Ascencio and Menges 1996). Fragmented animal communities often pass through unstable transitional states that do not otherwise occur in nature (Terborgh et al. 2001). These can cause serious ecological distortions, such as a collapse of predator and parasite populations and a hyperabundance of herbivores (Mikkelson 1993, Terborgh et al. 2001, Holt, this volume, Terborgh, this volume). These and other instabilities plague small, dwindling populations in fragments.

As discussed above, habitat fragments are often strongly affected by external vicissitudes and disturbances in the human-dominated lands that surround it. For example, forest species that exploit edge or disturbed habitats often increase dramatically in fragmented landscapes (Margules and Milkovits 1994, Laurance et al. 2002). As habitat loss proceeds, displaced animals from surrounding degraded lands can flood into remaining habitat fragments, leading to sudden increases in local population densities (Lovejoy et al. 1986, Hagan et al. 1996, Curran et al. 1999, Holt, this volume). Modified landscapes can be a major source of recurring disturbances, with hunters, livestock, fires, smoke, and large abiotic fluxes penetrating into and destabilizing fragments.

Conclusions

If ideas were mountains, IBT would be a Mount Everest, towering above thousands of lesser ideas and concepts. The theory has provided a conceptual framework for understanding habitat fragmentation that continues to inform researchers today. The avalanche of research stimulated by IBT has dramatically advanced the study of fragmented and insular habitats.

That having been said, the study of fragmented ecosystems has now far transcended IBT. With perfect hindsight, the theory seems simplistic to the point of being cartoonish, and fails to address some of the most important phenomena affecting fragmented landscapes. Yet it would be churlish not to herald a theory of this importance, and unfair to expect it to do everything. Fragmentation research today has diversified enormously, touching on subdisciplines ranging from landscape ecology to metapopulation dynamics, and from conservation genetics to population viability analysis. Everyone working in these fields owes some allegiance to the original inspiration provided by IBT.

Acknowledgments

I thank Jonathan Losos, Robert Ricklefs, Robert Ewers, Susan Laurance, and an anonymous referee for many helpful comments on the manuscript. This is publication number 508 in the technical series of the Biological Dynamics of Forest Fragments Project.

Literature Cited

Abbott, I. 1983. The meaning of *z* in species/area regressions and the study of species turnover in island biogeography. *Oikos* 41:385–90.

Abele, L. G., and E. F. Connor. 1979. Application of island biogeography theory to refuge design: Making the right decision for the wrong reasons. In *Proceedings of the First Conference on Scientific Research in the National Parks*, ed. R. M. Linn, 89–94. Washington, DC: U.S. Department of the Interior.

Abensperg-Traun, M., G. T. Smith, G. Arnold, and D. Steven. 1996. The effects of habitat fragmentation and livestock-grazing on animal communities in remnants of gimlet *Eucalyptus salubris* woodland in the Western Australian wheatbelt *I. Arthropods*. *Journal of Applied Ecology* 33:1281–301.

Baker, W. L. 1994. Restoration of landscape structure altered by fire suppression. *Conservation Biology* 8:763–69.

Becker, P., J. B. Moure, and F. J. A. Peralta. 1991. More about euglossine bees in Amazonian forest fragments. *Biotropica* 23:586–91.

Boecklen, W. J., and N. Gotelli. 1984. Island biogeographic theory and conservation practice: species-area or specious-area relationships? *Biological Conservation* 29:63–80.

Brandão, A. O., C. M. Souza, J. G. Ribeiro, and M. H. Sales. 2007. Desmatamento e estradas não-officias da Amazônia. *Anais XIII Simpósio Brasiliero de Sensoriamento Remoto* April:2357–64.

Brown, J. H., and A. Kodric-Brown. 1977. Turnover rates in insular biogeography: effect of immigration on extinction. *Ecology* 58:445–49.

Buechner, M. 1987. Conservation in insular parks: Simulation models of factors affecting the movement of animals across park boundaries. *Biological Conservation* 41:57–76.

Bruna, E. M. 1999. Seed germination in rainforest fragments. *Nature* 402:139.

Burkey, T. V. 1995. Extinction rates in archipelagoes: Implications for populations in fragmented habitats. *Conservation Biology* 9:527–41.

Cadenasso, M., S. Pickett, and K. Weathers. 2000. Effects of edges and boundaries on the flux of nutrients, detritus, and organisms. In *Food Webs at the Landscape Level*, ed. G. Polis and M. Power, 128–35. Chicago: University of Chicago Press.

Caley, M. J., K. A. Buckley, and G. P. Jones. 2001. Separating ecological effects of habitat fragmentation, degradation, and loss on coral commensals. *Ecology* 82:3435–48.

Chatelain, C., L. Gautier, and R. Spichiger. 1996. A recent history of forest fragmentation in southwestern Ivory Coast. *Biodiversity and Conservation* 5:37–53.

Chen, J., J. F. Franklin, and T. A. Spies. 1992. Vegetation responses to edge environments in old-growth Douglas-fir forests. *Ecological Applications* 2:387–96.

Cochrane, M. A., A. Alencar, M. Schulze, C. Souza, Jr., D. Nepstad, P. Lefebvre, and E. Davidson. 1999. Positive feedbacks in the fire dynamic of closed canopy tropical forests. *Science* 284:1832–35.

Cochrane, M. A., and W. F. Laurance. 2002. Fire as a large-scale edge effect in Amazonian forests. *Journal of Tropical Ecology* 18:311–25.

———. 2008. Synergisms among fire, land use, and climate change in the Amazon. *Ambio* 37:522–27

Collins, R. J., and G. W. Barrett. 1997. Effects of habitat fragmentation on meadow vole (*Microtus pennsylvanicus*) population dynamics in experiment landscape patches. *Landscape Ecology* 12:63–76.

Connor, E. R., and E. D. McCoy. 1979. The statistics and biology of the species-area relationship. *American Naturalist* 113:791–833.

Curran, L. M., I. Caniago, G. Paoli, D. Astianti, M. Kusneti, M. Leighton, C. Nirarita, and H. Haeruman. 1999. Impact of El Niño and logging on canopy tree recruitment in Borneo. *Science* 286:2184–88.

Diamond, J. M. 1969. Avifaunal equilibria and species turnover rates on the Channel Islands of California. *Proceedings of the National Academy of Sciences U.S.A.* 64:57–63.

———. 1972. Biogeographic kinetics: estimation of relaxation times for avifaunas of southwest Pacific islands. *Proceedings of the National Academy of Sciences U.S.A.* 69:3199–203.

———. 1975. The island dilemma: lessons of modern biogeographic studies for the design of natural reserves. *Biological Conservation* 7:129–46.

———. 1976. Island biogeography and conservation: Strategy and limitations. *Science* 193:1027–29.

———. 1984. "Normal" extinctions of isolated populations. In *Extinctions,* ed. M. H. Nitecki, 191–246. Chicago: University of Chicago Press.

Diamond, J. M., and May, R. M. 1976. Island biogeography and the design of natural reserves. In *Theoretical Ecology: Principles and Applications*, ed. R. M. May, 163–186. Oxford: Blackwell Scientific.

———. 1977. Species turnover rates on islands: Dependence on census interval. *Science* 197:266–70.

Didham, R. K., and J. H. Lawton. 1999. Edge structure determines the magnitude of changes in microclimate and vegetation structure in tropical forest fragments. *Biotropica* 31:17–30.

Dirzo, R., and M. C. Garcia. 1992. Rates of deforestation in Los Tuxtlas, a neotropical area in southern Mexico. *Conservation Biology* 6:84–90.

Ewers, R. M., and R. K. Didham. 2007. The effect of fragment shape and species' sensitivity to habitat edges on animal population size. *Conservation Biology* 21:926–36.

Ewers, R. M., S. Thorpe, and R. K. Didham. 2007. Synergistic interactions between edge and area effects in a heavily fragmented landscape. *Ecology* 88:96–106.

Fahrig, L. 2003. Effects of habitat fragmentation on biodiversity. *Annual Review of Ecology and Systematics* 34:487–515.

Gascon, C., T. E. Lovejoy, R. O. Bierregaard, J. R. Malcolm, P. C. Stouffer, H. Vasconcelos, W. F. Laurance, B. Zimmerman, M. Tocher, and S. Borges. 1999. Matrix habitat and species persistence in tropical forest remnants. *Biological Conservation* 91:223–9.

Gascon, C., G. B. Williamson, and G. Fonseca. 2000. Receding edges and vanishing reserves. *Science* 288:1356–8.

Gates, J. E., and L. W. Gysel. 1978. Avian nest dispersion and fledgling outcome in field-forest edges. *Ecology* 59:871–83.

Gilbert, F. S. 1980. The equilibrium theory of island biogeography, fact or fiction? *Journal of Biogeography* 7:209–35.

Gilbert, K. A., and E. Z. F. Setz. 2001. Primates in a fragmented landscape: Six species in central Amazonia. In *Lessons from Amazonia: Ecology and Conservation of a Fragmented Forest,* ed. R. O. Bierregaard, Jr., C. Gascon, T. E. Lovejoy, and R. Mesquita, 262–70. New Haven, CT: Yale University Press.

Hagan, J. M., H. Vander, W. Matthew, and P. McKinley. 1996. The early development of forest fragmentation effects on birds. *Conservation Biology* 10:188–202.

Heaney, L. R. 2000. Dynamic disequilibrium: A long-term, large-scale perspective on the equilibrium model of island biogeography. *Global Ecology and Biogeography* 9:59–74.

Higgs, A. J., and M. B. Usher. 1980. Should nature reserves be large or small? *Nature* 285:568–69.

Hobbs, R. J. 1993. Effects of landscape fragmentation on ecosystem processes in the Western Australian wheatbelt. *Biological Conservation* 64:193–201.

Hobbs, R. J., and L. F. Huenneke. 1992. Disturbance, diversity, and invasion: Implications for conservation. *Conservation Biology* 6:324–37.

Holt, R. D., J. H. Lawton, G. A. Polis, and N. D. Martinez. 1999. Trophic rank and the species-area relationship. *Ecology* 80:1495–504.

Honer, D., and W. Greuter. 1988. Plant population dynamics and species turnover on small islands near Karpathos (South Aegean, Greece). *Vegetatio* 77:129–37.

Hubbell, S. P. 2001. *The Unified Neutral Theory of Biodiversity and Biogeography*. Princeton, NJ: Princeton University Press.

Janzen, D. H. 1968. Host plants as islands in evolutionary and contemporary time. *American Naturalist* 102:592–95.

———. 1983. No park is an island: Increase in interference from outside as park size decreases. *Oikos* 41:402–10.

Kapos, V. 1989. Effects of isolation on the water status of forest patches in the Brazilian Amazon. *Journal of Tropical Ecology* 5:173–85.

Kareiva, P. 1987. Habitat fragmentation and the stability of predator-prey interactions. *Nature* 26:388–90.

Krauss, J., I. Stefan-Dewenter, and T. Tscharntke. 2003. Local species immigration, extinction, and turnover of butterflies in relation to habitat area and habitat isolation. *Oecologia* 137:591–602.

Laurance, S. G. 2004. Responses of understory rainforest birds to road edges in central Amazonia. *Ecological Applications* 14:1344–57.

Laurance, W. F. 1990. Comparative responses of five arboreal marsupials to tropical forest fragmentation. *Journal of Mammalogy* 71:641–53.

———. 1991. Ecological correlates of extinction processes in Australian tropical rainforest mammals. *Conservation Biology* 5:79–89.

———. 1997. Responses of mammals to rainforest fragmentation in tropical Queensland: A review and synthesis. *Wildlife Research* 24:603–12.

———. 2000. Do edge effects occur over large spatial scales? *Trends in Ecology and Evolution* 15:134–35.

———. 2002. Hyperdynamism in fragmented habitats. *Journal of Vegetation Science* 13:595–602.

———. 2004. Forest-climate interactions in fragmented tropical landscapes. *Philosophical Transactions of the Royal Society of London, Series B* 359:345–52.

Laurance, W. F., and R. O. Bierregaard, eds. 1997. *Tropical Forest Remnants: Ecology, Management, and Conservation of Fragmented Communities*. Chicago: University of Chicago Press.

Laurance, W. F., and M. A. Cochrane. 2001. Synergistic effects in fragmented landscapes. *Conservation Biology* 15:1488–89.

Laurance, W. F., M. A. Cochrane, S. Bergen, P. M. Fearnside, P. Delamonica, C. Barber, S. D'Angelo, and T. Fernandes. 2001a. The future of the Brazilian Amazon. *Science* 291:438–39.

Laurance, W. F., and T. J. Curran. 2008. Impacts of wind disturbance on fragmented tropical forests: a review and synthesis. *Austral Ecology* 33:399–408.

Laurance, W. F., P. Delamonica, S. G. Laurance, H. L. Vasconcelos, and T. E. Lovejoy. 2000. Rainforest fragmentation kills big trees. *Nature* 404:836.

Laurance, W. F., L. V. Ferreira, J. M. Rankin-de Merona, and S. G. Laurance. 1998a. Rain forest fragmentation and the dynamics of Amazonian tree communities. *Ecology* 79:2032–40.

Laurance, W. F., S. G. Laurance, and P. Delamonica. 1998b. Tropical forest fragmentation and greenhouse gas emissions. *Forest Ecology and Management* 110:173–80.

Laurance, W. F., S. G. Laurance, L. V. Ferreira, J. Rankin-de Merona, C. Gascon, and T. E. Lovejoy. 1997. Biomass collapse in Amazonian forest fragments. *Science* 278:1117–18.

Laurance, W. F., T. E. Lovejoy, H. L. Vasconcelos, E. M. Bruna, R. K. Didham, P. C. Stouffer, C. Gascon, R. O. Bierregaard, S. G. Laurance, and E. Sampaio. 2002. Ecosystem decay of Amazonian forest fragments: A 22-year investigation. *Conservation Biology* 16:605–18.

Laurance, W. F., H. Nascimento, S. G. Laurance, A. Andrade, P. M. Fearnside, and J. Ribeiro. 2006a. Rain forest fragmentation and the proliferation of successional trees. *Ecology* 87:469–82.

Laurance, W. F., H. E. M. Nascimento, S. G. Laurance, A. Andrade, J.E.L.S. Ribeiro, J. P. Giraldo, T. E. Lovejoy, R. Condit, J. Chave, and S. D'Angelo. 2006b. Rapid decay of tree-community composition in Amazonian forest fragments. *Proceedings of the National Academy of Sciences U.S.A.* 103:19010–14.

Laurance, W. F., D. Perez-Salicrup, P. Delamonica, P. M. Fearnside, S. D'Angelo, A. Jerozolinski, L. Pohl, and T. E. Lovejoy. 2001b. Rain forest fragmentation and the structure of Amazonian liana communities. *Ecology* 82:105–16.

Laurance, W. F., and E. Yensen. 1991. Predicting the impacts of edge effects in fragmented habitats. *Biological Conservation* 55:77–92.

Lehtinen, R. M., J. Ramanamanjato, and J. G. Raveloarison. 2003. Edge effects and extinction proneness in a herpetofauna from Madagascar. *Biodiversity and Conservation* 12:1357–70.

Lidicker, W. Z., Jr. 1973. Regulation of numbers in an island population of the California vole, a problem in community dynamics. *Ecological Monographs* 43:271–302.

Lomolino, M. V. 1986. Mammalian community structure on islands: immigration, extinction and interactive effects. *Biological Journal of Linnean Society* 28:1–21.

Losos, J. B. 1996. Ecological and evolutionary determinants of the species–area relation in Caribbean anoline lizards. *Philosophical Transactions of the Royal Society of London, Series B* 351:847–54.

Lovejoy, T. E., R. O. Bierregaard Jr., A. B. Rylands, J. R. Malcolm, C. E. Quintela, L. H., Harper, K. S. Brown, Jr., A. H. Powell, G. V. N. Powell, H. O. Schubart, and M. B. Hays. 1986. Edge and other effects of isolation on Amazon forest fragments. In *Conservation Biology: The Science of Scarcity and Diversity*, ed. M. E. Soulé, 257–85. Sunderland, MA: Sinauer Associates.

Lovejoy, T. E., J. M. Rankin, R. O. Bierregaard, Jr., K. S. Brown Jr., L. H. Emmons, and M. E. Van der Voort. 1984. Ecosystem decay of Amazon forest fragments. In *Extinctions*, ed. M. H. Nitecki, 295–25. Chicago: University of Chicago Press.

MacArthur, R. H., and E. O. Wilson. 1963. An equilibrium theory of insular zoogeography. *Evolution* 17:373–87.

———. 1967. *The Theory of Island Biogeography*. Princeton, NJ: Princeton University Press.

Malcolm, J. R. 1994. Edge effects in central Amazonian forest fragments. *Ecology* 75:2438–45.

———. 1998. A model of conductive heat flow in forest edges and fragmented landscapes. *Climatic Change* 39:487–502.

May, R. M. 1975. Island biogeography and the design of wildlife preserves. *Nature* 254:177–78.

Margules, C. R., and G. A. Milkovits. 1994. Contrasting effects of habitat fragmentation on the scorpion *Cercophonius squama* and an amphipod. *Ecology* 75:2033–42.

McGarigal, K., and W. C. McComb. 1995. Relationships between landscape structure and breeding birds in the Oregon Coast Range. *Ecological Monographs* 65:235–60.

Mesquita, R., P. Delamônica, and W. F. Laurance. 1999. Effects of surrounding vegetation on edge-related tree mortality in Amazonian forest fragments. *Biological Conservation* 91:129–34.

Mikkelson, G. M. 1993. How do food webs fall apart? A study of changes in trophic structure during relaxation on habitat fragments. *Oikos* 67:539–47.

Morrison, L. W. 2003. Plant species persistence and turnover on small Bahamian islands. *Oecologica* 136:51–62.

Myers, N. 1988. Synergistic interactions and environment. *BioScience* 39:506.

Nascimento, H., A. Andrade, J. Camargo, W. F. Laurance, S. G. Laurance, and J. Ribeiro. 2006. Effects of the surrounding matrix on tree recruitment in Amazonian forest fragments. *Conservation Biology* 20:853–60.

Ng, F.S.P. 1983. Ecological principles of tropical lowland rain forest conservation. In *Tropical Rain Forest: Ecology and Management*, ed. S. L. Sutton, T. C. Whitmore, and A. C. Chadwick, 359–76. Oxford: Blackwell.

Nupp, T. E., and R. K. Swihart. 2000. Landscape-level correlates of small mammal assemblages in forest fragments of farmland. *Journal of Mammalogy* 81:512–26.

Ovaskainen, O. 2002. Long-term persistence of species and the SLOSS problem. *Journal of Theoretical Biology* 218:419–33.

Patterson, B. D. 1987. The principle of nested subsets and its implications for biological conservation. *Conservation Biology* 1:323–34.

Patterson, B. D., and U. Atmar. 1986. Nested subsets and the structure of insular mammalian faunas and archipelagoes. *Biological Journal of the Linnean Society* 28:6–82.

Peres, C. A. 2001. Synergistic effects of subsistence hunting and forest fragmentation on Amazonian forest vertebrates. *Conservation Biology* 15:1490–505.

Peres, C. A., and F. Michalski. 2006. Synergistic effects of habitat disturbance and hunting in Amazonian forest fragments. In *Emerging Threats to Tropical Forests,* ed. W. F. Laurance and C. A. Peres, 105–27. Chicago: University of Chicago Press.

Pickett, S.T.A., and J. N. Thompson. 1978. Patch dynamics and the design of nature reserves. *Biological Conservation* 13:27–37.

Pimm, S. L., H. L. Jones, and J. M. Diamond. 1989. On the risk of extinction. *American Naturalist* 132:757–85.

Pires, A. S., P. Lira, F. Fernandez, G. Schittini, and L. Oliveira. 2002. Frequency of movements of small mammals among Atlantic Coastal Forest fragments in Brazil. *Biological Conservation* 108:229–37.

Powledge, F. 2003. Island biogeography's lasting impact. *BioScience* 53:1032–38.

Preston, F. W. 1960. Time and space and the variation of species. *Ecology* 41:611–27

Quammen, D. 1997. *The Song of the Dodo: Island Biogeography in an Age of Extinctions.* New York: Scribner.

Quintana-Ascencio, P. F., and E. S. Menges. 1996. Inferring metapopulation dynamics from patch-level incidence of Florida scrub plants. *Conservation Biology* 10:1210–19.

Ranney, J. W., M. Bruner, and J. Levenson. 1981. The importance of edge in the structure and dynamics of forest islands. In *Forest Island Dynamics in Man-Dominated Landscapes,* ed. R. L. Burgess and D. M. Sharpe, 67–95. New York: Springer.

Ricketts, T. H. 2001. The matrix matters: Effective isolation in fragmented landscapes. *American Naturalist* 158:87–99.

Ricklefs, R. E., and E. Bermingham, E. 2001. Nonequilibrium diversity dynamics of the Lesser Antillean avifauna. *Science* 294:1522–24.

Ricklefs, R. E., and G. W. Cox. 1972. Taxon cycles in the West Indian avifauna. *American Naturalist* 106:195–219.

Ricklefs, R. E., and I. J. Lovette. 1999. The roles of island area per se and habitat diversity in the species-area relationships of four Lesser Antillean faunal groups. *Journal of Animal Ecology* 68:1142–60.

Saunders, D.A., R. J. Hobbs, and C. R. Margules. 1991. Biological consequences of ecosystem fragmentation: a review. *Conservation Biology* 5:18–32.

Schmigelow, F. K. A., C. S. Machtans, and S. J. Hannon. 1997. Are boreal birds resilient to forest fragmentation? An experimental study of short-term community responses. *Ecology* 78:1914–32.

Schonewald-Cox, C. M., and J. W. Bayless. 1986. The boundary model: A geographical analysis of design and conservation of nature reserves. *Biological Conservation* 38:305–22.

Seabloom, E. W., A. P. Dobson, and D. M. Stoms. 2002. Extinction rates under nonrandom patterns of habitat loss. *Proceedings of the National Academy of Sciences U.S.A.* 99:1129–34.

Shafer, C. L. 1995. Values and shortcomings of small reserves. *BioScience* 45:80–88.

Shafer, M. L. 1981. Minimum population sizes for species conservation. *BioScience* 31:131–34.

Sievert, P. R., and L. B. Keith. 1985. Survival of snowshoe hares at a geographic range boundary. *Journal of Wildlife Management* 49:854–66.

Simberloff, D. S. 1976. Species turnover and equilibrium island biogeography. *Science* 194:572–78.

Simberloff, D. S., and L. G. Abele. 1976a. Island biogeography theory and conservation practice. *Science* 191:285–86.

———. 1976b. Island biogeography and conservation: Strategy and limitations. *Science* 193:1032.

Smith, A. P. 1997. Deforestation, fragmentation, and reserve design in western Madagascar. In *Tropical Forest Remnants: Ecology, Management, and Conservation of Fragmented Communities,* ed. W. F. Laurance and R. O. Bierregaard, Jr., 415–41. Chicago: University of Chicago Press.

Soulé, M. E., and D. S. Simberloff. 1986. What do genetics and ecology tell us about the design of nature reserves? *Biological Conservation* 35:19–40.

Stouffer, P. C., and R. O. Bierregaard. 1995. Use of Amazonian forest fragments by understory insectivorous birds. *Ecology* 76:2429–45.

Temple, S. A. 1986. Predicting impacts of habitat fragmentation on forest birds: A comparison of two models. In *Wildlife 2000: Modeling Habitat Relationships of Terrestrial Vertebrates,* ed. J. Verner, M. L. Morrison, and C. J. Ralph, 301–4. Madison: University of Wisconsin Press.

Terborgh, J. 1974. Faunal equilibria and the design of wildlife preserves. In *Tropical Ecological Systems,* ed. F. B. Golley and E. Medina, 369–80. New York: Springer.

———. 1976. Island biogeography and conservation: Strategy and limitations. *Science* 193:1029–30.

Terborgh, J., L. Lopez, P. V. Nuñez, M. Rao, G. Shahabuddin, G. Orihuela, M. Riveros, R. Ascanio, G. Adler, T. Lambert, and L. Balbas. 2001. Ecological meltdown in predator-free forest fragments. *Science* 294:1923–26.

Timmerman, A., J. Oberhuber, A. Bacher, M. Esch, M. Latif, and E. Roeckner. 1999. Increased El Niño frequency in a climate model forced by future greenhouse warming. *Nature* 398:694–97.

Turner, I. M., and R. T. Corlett. 1996. The conservation value of small, isolated fragments of lowland tropical rain forest. *Trends in Ecology and Evolution* 11:330–33.

Villard, M.-A., M. K. Trzcinski, and G. Merriam. 1999. Fragmentation effects on forest birds: relative influence of woodland cover and configuration on landscape occupancy. *Conservation Biology* 13:774–83.

Watling, J. I., and M. A. Donnelly. 2006. Fragments as islands: a synthesis of faunal responses to habitat patchiness. *Conservation Biology* 20:1016–25.

Weathers, K., M. Cadenasso, and S. Pickett. 2001. Forest edges as nutrient and pollutant concentrators: potential synergisms between fragmentation, forest canopies, and the atmosphere. *Conservation Biology* 15:1506–14.

Whitcomb, R. F., J. F. Lynch, P. A. Opler, and C. S. Robbins. 1976. Island biogeography and conservation: strategy and limitations. *Science* 193:1030–32.

Wilcove, D. S. 1985. Nest predation in forest tracts and the decline of migratory songbirds. *Ecology* 66:1212–14.

Wilcox, B. A. 1980. Insular ecology and conservation. In *Conservation Biology: An Evolutionary-Ecological Perspective,* ed. M. E. Soule and B. A. Wilcox, 95–117. Sunderland, MA: Sinauer Associates.

Willis, E. 0. 1974. Populations and local extinctions of birds on Barro Colorado Island, Panama. *Ecological Monographs* 441:53–169.

Wilson, E. O., and E. O. Willis. 1975. Applied biogeography. In *Ecology and Evolution of Communities,* ed. M. L. Cody and J. M. Diamond, 522–34. Cambridge, MA: Harvard University Press.

Winter, J. W., F. C. Bell, L. I. Pahl, and R. G. Atherton. 1987. Rainforest clearing in northeastern Australia. *Proceedings of the Royal Society of Queensland* 98:41–57.

Woodroffe, R., and J. R. Ginsberg. 1998. Edge effects and the extinction of populations inside protected areas. *Science* 280:2126–28.

Wright, S. J. 1981. Intra-archipelago vertebrate distributions: the slope of the species-area relation. *American Naturalist* 118:726–48.

———. 1985. How isolation affects rates of turnover of species on islands. *Oikos* 44:331–40.

Zimmerman, B. L., and R. O. Bierregaard. 1986. Relevance of the equilibrium theory of island biogeography and species-area relations to conservation with a case from Amazonia. *Journal of Biogeography* 13:133–43.

Birds of the Solomon Islands

THE DOMAIN OF THE DYNAMIC EQUILIBRIUM THEORY AND ASSEMBLY RULES, WITH COMMENTS ON THE TAXON CYCLE

Daniel Simberloff and Michael D. Collins

BIRDS OF THE SOLOMON ISLANDS have played a prominent role in two of the most influential ecological theories of the last forty years. Robert MacArthur and Edward O. Wilson cited these birds in both their 1963 paper introducing the dynamic equilibrium theory of island biogeography and their 1967 monograph on the theory (MacArthur and Wilson 1963, 1967). In 1976, Jared Diamond, Ernst Mayr, and Michael Gilpin published three papers on Solomon Islands avifaunas, interpreting them in terms of dynamic equilibrium turnover, relating the area and isolation of islands to hypothesized immigration and extinction curves (Diamond and Mayr 1976, Diamond et al. 1976, Gilpin and Diamond 1976). At about the same time, Diamond (1975) elaborated his theory that assembly rules govern island species composition and are largely determined by resource competition but influenced by other factors (e.g., dispersal ability), based primarily on birds of the Bismarck Archipelago but with many examples from and references to birds of the Solomons. Remarkably, Philip J. M. Greenslade (1968) first applied the taxon cycle model (Wilson 1959, 1961) to birds, using the Solomon Islands avifauna.

For the equilibrium theory, four decades of research have cast doubt on its applicability to many natural systems (references in Whittaker and Fernández-Palacios [2007]; cf. Schoener, this volume). The range of systems described well by the assembly rules remains highly controversial. In a meta-analysis, Gotelli and McCabe (2002) find that certain distributional patterns predicted by the rules are more common in nature than a noncompetitive null model would predict, but for very few systems is there direct evidence on the reasons for these patterns. The notion of a taxon cycle has also been quite controversial, particularly as regards its applicability to birds (Ricklefs and Bermingham 2002; Ricklefs, this volume). Strikingly, distributions of Solomon Islands birds, though prominent in the development of all three theories, have barely been scrutinized after the original papers. This neglect is because the distributions—which species are on which islands—were unavailable until they were published by Mayr and Diamond (2001). Here we use these data to reassess whether these three theories apply to this biota and to address the implications of our results for the status of the theories and, more generally, for the nature of the evidence required to test them.

The iconic "crossed-curves" equilibrium model of MacArthur and Wilson (1963, 1967) focuses on demography of individual species, leading to stochastic extinction, and not on interactions among species. It does not account for species' identities, looking only at numbers of species. However, MacArthur and Wilson (1967) also stressed the possible role of diffuse competition in generating turnover and recognized that deterministic forces related to species composition and interactions may partly determine how many and which species are found on islands: "A closer examination of the composition and behavior of resident species should often reveal the causes of exclusion, so that random processes in colonization need not be invoked" (p. 121). Diamond's theory that assembly rules govern species composition is based on exactly that sort of examination of the identities and behavior of resident species. The two theories need not conflict so long as substantial turnover occurs and interactions are a major contributor to it. In fact, in an archipelago of islands in which all are conceived as potential sources for one another of multiple potentially interacting species, as in the birds of the Solomon Islands, the equilibrium theory describes what is now recognized as a metacommunity (Leibold et al. 2004). Several authors, beginning with Wilson (1969), have suggested extending the equilibrium theory to an evolutionary scale by adding adaptation and speciation, while the assembly rules were seen as acting in ecological time. As do the assembly rules, the taxon cycle model treats species identities and assigns a key role to competitive interactions: these drive the range and habitat contraction phase of the cycle (Ricklefs, this volume). However, unlike in the assembly rules and most interpretations of the equilibrium theory, evolution is prominent in the taxon cycle, with morphological differentiation aiding assignment of species to particular cycle phases and hypothesized behavioral and physiological changes driving species' trajectories through the phases.

The Equilibrium Theory

To calculate the immigration and extinction curves of the equilibrium theory, Gilpin and Diamond (1976) examined the 106 lowland breeding land and freshwater birds on 52 of the Solomon Islands,[1] including all

[1]We designate by "Solomon Islands" the geographic archipelago, not the nation of the Solomon Islands. We include Bougainville and Buka (part of Papua New Guinea) but not the Santa Cruz Islands, far to the east of the archipelago, just north of Vanuatu, but part of the nation of the Solomon Islands.

major islands. Some species that reach sea level on one island may be restricted to higher elevations on another (a pattern Mayr and Diamond [1976] ascribe to competition); the species pool for this exercise was all species reaching sea level on any island. Assuming all islands to be in equilibrium, they constructed immigration (I) and extinction (E) functions in terms of the area (A), distance (D), and number of species (S) for each island, set these functions equal, and sought functional forms such that variation in area and distance explained as large a fraction as possible of the variation in number of species. For islands with more than 50 species total, or for islands within 6 miles of such an island, distance was taken as 0. For other islands, the distance was the distance to the nearest island with more than 50 species. The upshot is that 37 islands had $D=0$.

As a benchmark, Gilpin and Diamond (1976) found a phenomenological model with five fitted parameters (a, b, c, d, and e) that explained 98% of the variance in S:

$$S=(a+b \log A) \exp(-D^c/dA^e). \tag{9.1}$$

However, the parameters have no straightforward biological interpretation. The goal was to equal this explanatory power with biologically reasonable immigration and extinction functions.

Thus, extinction (E) was assumed to be a function of A and S, and immigration (I) a function of A, D, and S. In addition, Gilpin and Diamond (1976) assumed that any valid extinction function should have at least three parameters:

R: a fitted constant

n: so that E is a concave upward function of S, proportional to S^n ($n>1$)

x: so that, with decreasing A, and extinctions solely the result of demographic fluctuations, E is a function of A^{-x}, with $x>1$

and any valid immigration function should have at least four parameters:

m: so that I is concave upward ($m>1$)

D_0: in accord with a model with a constant direction and risk of death per unit distance traversed (the exponential model of MacArthur and Wilson [1967])

y: accounting for differences among species in overwater flight distances ($y<1$)

v: because a bigger island will present a larger target to a disperser at sea level, and increasing island elevation may make the target more visible ($v\geq 0.5$).

Gilpin and Diamond (1976) found a best-fit model matching the phenomenological model in explaining 98% of the variation in S, even without one parameter (x):

$$E = RS^n/A, \quad I = (1 - S/P_0)^m \exp(-D^y/D_0 A^v). \tag{9.2}$$

Here P_0 is the size of the species pool, 106. S is then an implicit function when I is set equal to E.

Noteworthy in this exercise are four features:

1. No unequivocal bird extinctions in the Solomon Islands have been observed in historic times. However, this fact does not conflict with the theory because
2. Time is not a factor in any parameters and variables of the equations for I and E. That is, the immigration and extinction curves, plotted against S, are in arbitrary time units.
3. The island avifaunas are assumed to be at equilibrium.
4. The same data were used to produce the equations as to test them.

With respect to point 1 and the fact that the equations do not predict what the extinction and immigration rates are, only that they are equal, it is interesting to consider possible extinctions in the Solomons. Mayr and Diamond (2001) list four species (*Gallicolumba jobiensis*, *G. salamonis*, *Microgoura meeki*, and *Zoothera dauma*) not recorded in the archipelago since 1927 and a fifth (*Anas gibberifrons*) not seen since 1959. These may be extinct (some globally, others just in the Solomons). They also observe that all five are ground-nesters, "suggesting that introduced cats may have been the culprits" (p. 38).

Other introduced species may also have been involved. For example, the teal, *A. gibberifrons*, disappeared from the one island it occupied (Rennell) right after *Oreochromis* (*Tilapia*) *mossambica* was introduced (Mayr and Diamond 2001). Diamond (1984) surmised that the fish somehow eliminated the teal, and he may have been prescient. This species is the most ecologically damaging introduced tilapia (Pullin et al. 1997) and is believed to be one of several threats to the Eurasian white-headed duck, *Oxyura leucocephala*, by virtue of competition (Hughes et al. 2004). Rats are also present in the Solomon Islands and prey on birds. The Pacific rat, *Rattus exulans*, was introduced prehistorically by humans, probably to all inhabited islands. The black rat, *R. rattus*, present on many of the islands (Yom-Tov et al. 1999), was introduced at unknown times after European arrival in the sixteenth century. Other species than the above five may have been extirpated from particular islands during this period but remain on others (cf. BirdLife International 2000); there is no published record of such extirpations.

If these five species are extinct in the Solomons, then they are not examples of equilibrium turnover driven by the demography of small populations or diffuse competition. Rather, these would probably be deterministic extinctions caused by human activities. This is the same distinction Caughley (1994) drew in conservation biology between the small-population paradigm (focusing on inherent extinction risk for all small populations, by virtue of smallness) and the declining-population paradigm, which seeks for each dwindling species the specific, deterministic reasons for its decline. In any event, and returning to point 2 above, because the Gilpin-Diamond model lacks a time scale, it cannot conflict with any extinction rate data, including data that show few or no extinctions over a century.

With respect to point 3 above, the proposition that these avifaunas have been in *any* sort of equilibrium for tens of thousands of years is unconvincing because of enormous anthropogenic change. Although Pleistocene archeology is poorly known in the Solomons except for Buka, humans have occupied most or all of the main islands for at least 30,000 years; Kilu Cave on Buka has been well studied and anthropogenic deposits date to ca. 29,000 B.P. (Steadman 2006). On mid-sized Buka, the only island in the Solomons for which avian fossil evidence is not sorely lacking, 61% of the prehistoric avifauna is no longer present (Steadman 2006). This is a staggering figure, high even among massive post-human colonization extinctions widely documented among Pacific island birds. Steadman (2006) argues that most if not all absences today from the large islands, including Buka, are anthropogenic. An alternative in the spirit of the equilibrium theory is "faunal relaxation," in which the decrease in area (and, for Buka, separation from Bougainville) owing to higher sea levels since the end of the last Ice Age would, simply by the demography of smaller populations, have led ultimately to fewer species. Of the four species extinct on Buka but persisting elsewhere in the Solomons (Steadman 2006), two (*Nesasio solomonensis* and *Nesoclopeus woodfordi*) are present only on islands larger than Buka, while the other two (*Gallicolumba rufigula* and *Caloenas nicobarica*) are on many islands both smaller and larger than Buka (data in Mayr and Diamond [2001]), providing at most weak support for the relaxation hypothesis.

Arrival of the Lapita people to Pacific islands was particularly catastrophic to birds (Steadman 2006), and their colonization of the Solomons, ca. 3000 B.P., was probably devastating. There is almost no evidence for bird extinctions before human arrival throughout Oceania, including the Solomons (Steadman 2006). However, human population growth as well as animals and plants introduced by humans are believed to have massively affected island bird communities. In addition to cats and rats, humans deliberately introduced dogs and pigs to many islands. All prey on birds and/or their eggs. Also, pigs, introduced to many of the Solomon Islands (Long 2003), have greatly modified habitat in many places (Long 2003). Prehistoric humans also carried many alien plants to

Pacific islands, and there was rampant deforestation (often by burning) to cultivate these plants, most of which were of little use to native birds (Steadman 2006). Today there is tremendous habitat destruction by logging (BirdLife International 2000).

Native rodents on some larger islands in the Solomons may have rendered their avifaunas less vulnerable to introduced predators than were birds on remote Pacific islands (Steadman 2006). Nevertheless, the Buka data suggest that massive extinction did occur with human colonization. Not only was this extinction not a form of equilibrium turnover, but it left an avifauna that one could hardly expect to be in equilibrium. All the numbers of lowland bird species cited in the exercise of Gilpin and Diamond (1976) are lower, probably far lower, than those that obtained before humans arrived. And they are still falling rapidly. For land birds of the Solomon Islands (minus Bougainville and Buka), BirdLife International (2000) lists eighteen species as threatened and sixteen as near-threatened (a total of ca. one-fourth of the avifauna). The suspected threats listed in the individual species accounts in the same reference are overwhelmingly anthropogenic, with many citing logging; for only two species are "natural" causes even mentioned as a possibility.

Just as few (if any) nonanthropogenic extinctions are documented in the Solomons, neither is immigration of new species recorded. Given the difficulty of working in these islands, it would be difficult to attribute a new record to immigration rather than to better sampling. For instance, Kratter et al. (2001) recorded three new land bird species on Isabel in three weeks in a dry forest; they do not regard these as new immigrants. Notably, no instance is known in the Solomons of a species lost, then recolonizing on its own (Steadman 2006). Although it would not constitute equilibrium immigration, the Solomons, lacking the acclimatization societies that introduced entire avifaunas to such islands as New Zealand, the Hawaiian Islands, and the Mascarenes (cf. Lever 1992), do not even have many introduced bird species. At most three are established, and these are on very few islands (Long 1981). Thus, given the many documented extinctions (Steadman 2006), the Solomon Islands contradict the pattern noted by Sax et al. (2002), of an approximate equality of immigrations and extinctions for birds on oceanic islands.

Finally, the equations in (9.2) were derived from the data set that was then used to test them, with no attempt at cross-validation. It is not clear that any other biota could be used to test this model. Gilpin and Diamond (1976, p. 4134) observe that "a fauna or flora other than Solomon birds will certainly require parameter values, and maybe require functional forms, different from those of Eqs 7b and 7a [equations in (9.2)], respectively."

Assembly Rules

Just as Gilpin and Diamond (1976) attempted to demonstrate a process (turnover) from a static pattern, so the assembly rules (Diamond 1975) constituted an effort to use a more detailed static pattern (the species composition of each island) to implicate a process (competition) as far more important in generating the pattern than other alternatives (habitat requirements and dispersal limitation). Diamond (1975) assumed that the current island avifaunas are for the most part in a species-number equilibrium and that the processes yielding the assembly-rule patterns operated much more quickly than those yielding a species-number equilibrium.

Here we explore Diamond's basic assembly rule, number 5: "Some pairs of species never coexist, either by themselves or as part of a larger combination" (Diamond 1975, p. 423). Such "checkerboard" distributions have often been taken as evidence for interspecific competition (Gotelli and Graves 1996). Controversy has largely revolved around two issues. First, depending on the numbers of islands and species, some checkerboard distributions might have been expected even if species colonized islands independently of one another (Connor and Simberloff 1979). Second, even if some checkerboards are statistically unlikely to have resulted from independent colonization, other explanations than interspecific competition are possible (Connor and Simberloff 1979, Simberloff and Connor 1981). Two species might have distinct habitat requirements, for example, or might be sister species that have recently speciated allopatrically, or might have arrived in an archipelago by different routes and/or at different times.

We examined the Solomon Islands avifauna (45 islands, 142 species) as described by Mayr and Diamond (2001) for checkerboard distributions. To avoid the "dilution effect" (Diamond and Gilpin 1982; cf. Colwell and Winkler 1984), we looked only at the subset of species pairs in which competition would be expected. First we examined just congeneric pairs of species. Taxonomic groups are not always congruent with guilds (Diamond and Gilpin 1982, Simberloff and Dayan 1991), but many authors have argued that congeners are on average ecologically more similar to one another than are heterogeneric species, and many studies have partitioned biotas into guilds by taxonomy (e.g., MacArthur 1958). Also, all mapped checkerboards in Diamond (1975) consisted of congeners, so we feel this convention suffices for our purposes. We then examined checkerboards in four multigenus guilds (table 9.1) specified by Diamond (1975).

Table 9.1
Guild Memberships in the Solomon Islands for Multigenus Guilds Specifically Designated by Diamond (1975)

Guild	*Genera*	*No. of species*
Cuckoo dove	*Macropygia*	2
	Reinwardtoena	
Gleaning flycatcher	*Monarcha*	7
	Myiagra	
	Pachycephala	
Myzomela-sunbird	*Myzomela*	3
	Nectarinia	
Fruit pigeon	*Ducula*	8
	Ptilinopus	

Finally, Diamond (1975; cf. Mayr and Diamond 2001) defined as "supertramps" species found only on islands (generally small ones) with few species, a pattern he also attributed primarily to competition. However, a species could be a supertramp for other reasons (Simberloff and Martin 1991), for example, a preference for habitats especially common on small islands, or exclusion from larger islands by predators. Supertramps would dominate a search for checkerboards, even if the reasons for their status had nothing to do with the competitive interactions that are posited as causal. Because they are on islands with only a few species, they are likely automatically to comprise many checkerboards. We therefore conducted our entire analysis both with and without supertramps. Diamond (1975) did not provide quantitative criteria for qualification as a supertramp. We defined them statistically (Collins et al. in preparation). By our method, the three supertramps in the Solomons are *Ducula pacifica*, *Monarcha cinerascens*, and *Aplonis [feadensis]*.[2] To these, Mayr and Diamond (2001) add *Ptilinopus [purpuratus]*, *Caloenas nicobarica*, and *Pachycephala melanura*.

To evaluate the assembly rules, it is necessary to consider historical geography. According to Mayr and Diamond (2001), five island groups occur in the Solomons: (1) the Bukida group, or Main Chain—Greater

[2]We follow the convention of Mayr and Diamond (2001) in designating superspecies by square brackets. Taxa within superspecies in the Solomons have been assigned different ranks by different authors.

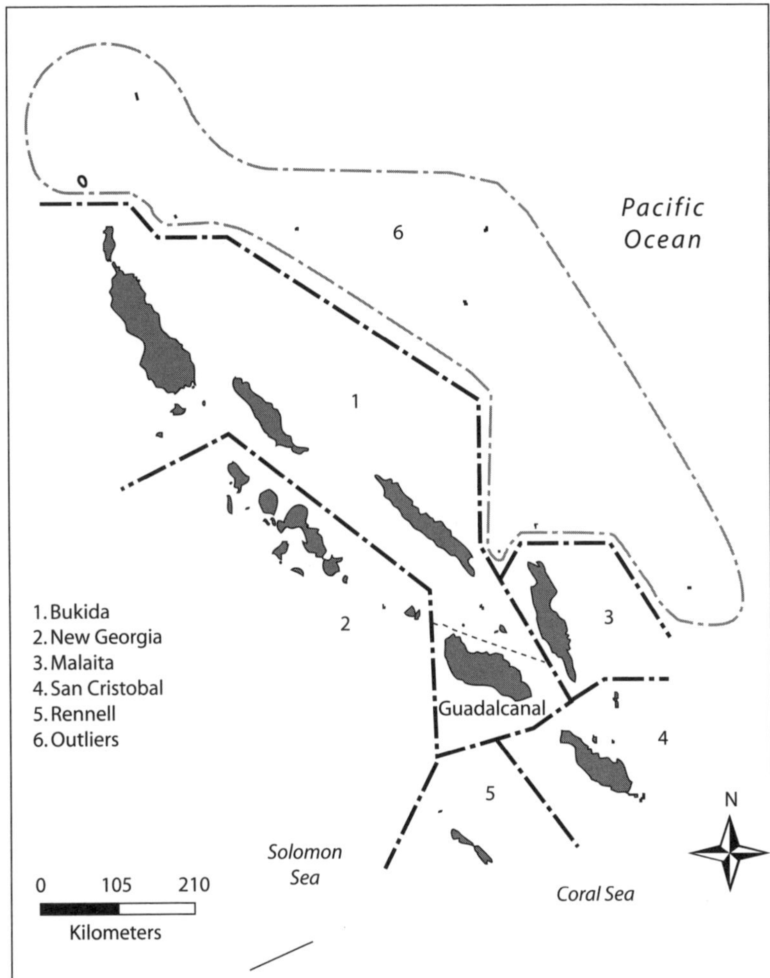

Figure 9.1. Island groups as currently configured in Solomons separated by hypothesized dispersal barriers (cf. Mayr and Diamond 2001).

Bukida, a Pleistocene land-bridge island running from Buka to Florida, and Guadalcanal, which was separated from Greater Bukida by a narrow channel (cf. Steadman 2006), (2) the New Georgia group—three Pleistocene land-bridge islands with current islands from Vella Lavella to Gatukai, and two unconnected islands (Gizo and Simbo), (3) Malaita, (4) the San Cristobal group—San Cristobal (Makira), Ulawa, Ugi, Three Sisters, Santa Anna, and Santa Catalina, and (5) the Rennell group—Rennell and Bellona (figure 9.1). Finally, a sixth group consists of outliers,

TABLE 9.2
Observed and Expected Numbers of Congeneric Checkerboards (CH) in the Solomon Islands (Including Supertramps)

Genus	*No. of taxa*	*ObservedCH*	*Expected CH*	*Probability*
Accipiter	5	5	1.52	<0.001
Aplonis	5	2	0.11	<0.001
Monarcha	3	2	<0.001	<0.001
Pachycephala	3	2	0.95	0.157
Rhipidura	6	3	3.06	0.659
Zosterops	5	8	4.23	0.006

Source: Matrix data extracted from Mayr and Diamond (2001).
Notes: Checkerboards derived by matrix randomization (see text). Depending on ranks of taxa within superspecies, observed and/or expected numbers of checkerboards may increase.

small, remote islands north and east of the archipelago (Fead, Kilimailau, Tauu, Nukumanu, Ontong Java, Ramos, Gower, Nissan, and Sikaina). Although the mega-islands of Greater Bukida, the expanded New Georgia, and the expanded San Cristobal would all have been within sight of each other during the late Pleistocene (Steadman 2006), Mayr and Diamond (2001) argue that, even during the Pleistocene when sea levels were much lower, these groups were separated by barriers to dispersal, differentially permeable to different species but sufficient to generate morphological differences among populations within species (or species groups) on islands in different island groups and compositional differences in bird communities on islands in different groups.

To assess the null probability of the observed numbers of checkerboards, we used the Miklós and Podani (2004) "trial-swap" method to randomize repeatedly the binary presence-absence matrix, maintaining column sums (species richness on each island) and row sums (number of islands occupied by each species). These conventions are explained by Gotelli and Graves (1996). We then sought tail probabilities for the observed numbers of congeneric checkerboards (and later for numbers of checkerboards in the multigenus guilds).

The Solomon Islands have 22 congeneric checkerboards in six genera (table 9.2); in four of these genera, these numbers appear improbably large if species were colonizing islands independently of each other. However, minus supertramps, which occur in two of these six genera, these two genera and four of the checkerboards disappear, and the numbers of

TABLE 9.3
Observed and Expected Numbers of Congeneric Checkerboards (CH) in the Solomon Islands with Supertramps Omitted

Genus	*No. of taxa*	*Observed CH*	*Experienced CH*	*Probability*
Accipiter	5	5	1.52	<0.001
Pachycephala	3	2	0.95	0.157
Rhipidura	6	3	3.06	0.659
Zosterops	5	8	4.23	0.006

Note: Depending on ranks of taxa within superspecies, observed and/or expected numbers of checkerboards may increase.

checkerboards are significantly large only in *Accipiter* and *Zosterops* (table 9.3).

At first blush then, it appears that at least some checkerboards are inconsistent with a hypothesis of independent colonization and in accord with the notion that they represent pairs mutually exclusive by virtue of competition. However, our close examination of all of these congeneric checkerboards, whether or not we include supertramps, yielded a surprise: the checkerboard metaphor, based on red and black squares filling an entire board, does not describe them. Usually there are very few representatives of one or both members of such a distribution, and rather than being spread throughout the Solomons, each representative is usually restricted to one or a few island groups. In other words, they are allopatric at a much broader scale than is implied by the metaphor (figure 9.2), and the boundaries of the allopatric regions coincide with the partitions that Mayr and Diamond (2001) describe as long-standing dispersal barriers. This fact plus the apparently relatively recent arrival of some members of checkerboards and the fact that many have never been seen flying over water suggest that history, in geological time, of the colonization of the archipelago may have led to many of these mutually exclusive distributions.

Of the five *Accipiter* species in the Solomons, *A. fasciatus* accounts for four of the five checkerboards and occurs only in the Rennell group; no other *Accipiter* is found there. Mayr and Diamond (2001) believe this population arrived in Rennell and Bellona from Australia via Vanuatu, bypassing the Bismarck Archipelago. *Accipiter fasciatus* may be excluded from other groups by competition with congeners, but it could also simply not have reached them, or reached them often enough to establish a population, because of the minimum 171 km it would have to fly to get there.

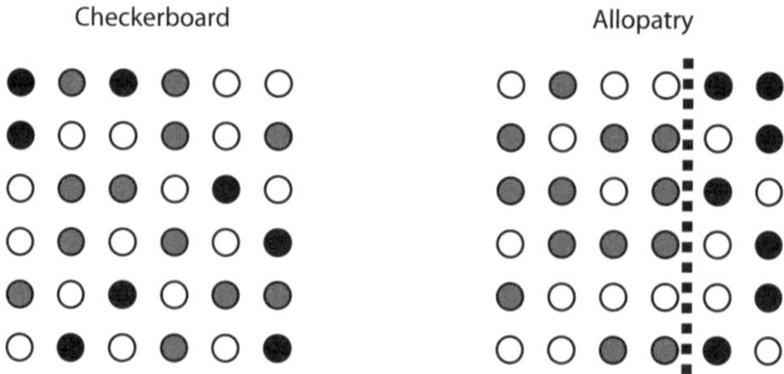

Figure 9.2. Contrast between checkerboard and allopatric conceptions of biogeographic patterns.

The fifth *Accipiter* checkerboard is between *A. imitator* and *A. meyerianus*, each occupying only three islands. *Accipiter imitator* is found only on Greater Bukida islands and has never been seen flying over water (Mayr and Diamond 2001). The three islands occupied by *A. meyerianus* include Guadalcanal of the Bukida group plus two islands in the New Georgia group. A goshawk, it is a strong flyer. It is quite possible that *A. imitator* is not on other islands for historical and behavioral reasons. Mayr and Diamond (2001) suggest it is not on Guadalcanal, though that island is in the Bukida group, because a small channel probably separated Guadalcanal from the rest of the chain. They also suggest that it probably was formerly on other islands that had been part of Greater Bukida but was subsequently extinguished. Competition with *A. meyerianus* would have been an unlikely cause for such extinctions, because (1) *A. meyerianus* is not found on any of these islands; (2) *A. meyerianus* is largely montane in the Solomons (Mayr and Diamond 2001) and *A. imitator* is not; (3) *A. meyerianus* is twice the size of *A. imitator*, suggesting a different diet and/or foraging mode.

Eight pairs among the five *Zosterops* taxa show checkerboard distributions in the Solomons. Except for the superspecies *Z.* [*griseotinctus*], all taxa are restricted to one or two island groups and each occupies six or fewer islands (table 9.4). Mayr and Diamond (2001) stress that, with only two exceptions (discussed below), none of the *Zosterops* taxa occupy the same island, and they see this as an assembly rule determined by competition. However, it is equally true that, with the same two exceptions, the *Zosterops* taxa do not occupy the same island groups, and they are highly restricted in the groups they occupy (table 9.4). Further, three of the species (*Z. stresemanni*, *Z. murphyi*, and *Z. metcalfii*) are believed to be sedentary and not to cross even narrow water gaps (Mayr and Diamond 2001). A plausible, parsimonious hypothesis is therefore that, his-

TABLE 9.4
Occupancy of Island Groups by Solomon Islands *Zosterops* Taxa

Species	*No. of islands*	*Island groups occupied*
Z. [*griseotinctus*]	14	New Georgia, Rennell, Nissan (outlier)
Z. murphyi	1	New Georgia
Z. metcalfii	6*	Bukida
Z. ugiensis	3	Bukida, San Cristobal
Z. stresemanni	1	Malaita

Source: Data from Mayr and Diamond (2001).
* +2 small islets in Bukida group.

torically, each species first reached the island group(s) it currently occupies and simply has not dispersed further.

In arguing for their competitive assembly-rule interpretation, Mayr and Diamond (2001) suggest that at least the three single-island-group species have occupied other, smaller islands (presumably in the same group, as they are not believed to cross water), went extinct, and failed to recolonize. However, no such extinctions have been documented. These hypothesized extinctions would have been facets of "equilibrium" turnover, the consequences of demographic variation in small populations (or perhaps "relaxation" with rising sea levels and decreasing area?). Above, we question the proposition of equilibrium turnover in this archipelago, especially the notion that extinction is "equilibrial." Here we can only add that white-eyes are often enormously abundant, and islands the size of Fauro (71 km^2) and Buena Vista (14 km^2) could have supported thousands of individuals, making extinction from demographic stochasticity unlikely. Of course, populations on smaller islands, such as these, might well be more susceptible to both anthropogenic pressures (cf. Steadman 2006) and the vagaries of environmental stochasticity and catastrophes. And equilibrial turnover might be more likely on islands still smaller than Fauro and Buena Vista (see below).

Two of the ten possible *Zosterops* pairs do not form checkerboards. *Zosterops murphyi* and *Z.* [*griseotinctus*] coexist on Kulambangra, while *Z. ugiensis* and *Z. metcalfii* coexist on Bougainville. Mayr and Diamond (2001) note that, in each pair, the first-named species is montane on the island of co-occurrence, while the other is found only in lowlands, a pattern they also ascribe to competition. This contention is buttressed by the fact that, on San Cristobal, where it is alone, *Z. ugiensis* is found in lowlands.

In any event, the elevational separation and the absence of species from certain islands *within*-island groups they occupy do not bear on the cause of the main pattern driving the number of checkerboards—the restriction of each species to a minority of island groups. This pattern is as compatible with an historical explanation as with one invoking present-day competition.

Three *Pachycephala* taxa occupy the Solomons (Mayr and Diamond 2001): the superspecies *P.* [*pectoralis*] occupies many islands in all five major groups, plus the isolated Russell Islands. *Pachycephala implicata* is a montane species on the Bukida islands of Bougainville and Guadalcanal, where it co-occurs with *P.* [*pectoralis*] but is segregated by elevation. The checkerboards are formed by each of these taxa with *P. melanura*, in the Solomons found only on the isolated island of Nissan plus several islets near Buka, Bougainville, and Shortland in the Bukida group (Mayr and Diamond 2001). *Pachycephala melanura* does not qualify as a supertramp by our statistical test, but Diamond (1975) and Mayr and Diamond (2001) designate it as a supertramp, and it would doubtless qualify statistically if avifaunas of many small islands it inhabits had been tabulated by Mayr and Diamond (2001). The montane habitat of *P. implicata* implies its checkerboard with *P. melaneura* is caused by habitat differences, not competition. However, the fact that islets occupied by *P. melanura* are close to large islands occupied by *P.* [*pectoralis*] suggested to Mayr and Diamond (2001) that competitive exclusion operated between these two species. Two considerations, both noted by Mayr and Diamond (2001), suggest that other factors may be at play.

First, even in allopatry, *P.* [*pectoralis*] does not use very small islands and *P. melanura* does not use large ones, a point also made by Lomolino (1999) for the Bismarck Archipelago. Mayr and Diamond (2001) suggest that this observation may imply the habitat preferences evolved in allopatry. If this were so, it would cast doubt on whether the Solomons checkerboard is competitively driven. Second, Mayr and Diamond (2001) believe *P. melanura* relatively recently invaded the Solomons and has not yet had time to spread beyond the Shortlands region. In that case, the checkerboard would at least partly reflect differing colonization histories. *Pachycephala melanura* has also never been seen flying over water (Mayr and Diamond 2001), again suggesting that, as a recent arrival in the Solomons, it may still be spreading. In Australia, Gotelli et al. (1997) found these species co-occurring less frequently than expected for individual colonization. However, their figure 6a shows the two taxa to be almost allopatric, with large ranges overlapping only in a small section of the northeast coast.

The two *Aplonis* checkerboards both include the supertramp *A.* [*feadensis*], which occupies small outlying islands plus Rennell. Neither of the two species exclusively distributed with it, *A. grandis* and *A. brunneicapilla*, is found on Rennell or any outlying island, so the checkerboard distributions also constitute regional allopatry. Why *A.* [*feadensis*] is a supertramp and is not found on other islands is uncertain; it is highly vagile. Mayr and Diamond (2001) suggest competition with *A. cantoroides* may exclude it from some islands, although these two species coexist on Rennell.

Rhipidura has six species in the Solomons, none supertramps. Of the fifteen possible two-species combinations, three form checkerboards. For all three checkerboards, the species occupy different island groups. *Rhipidura fuliginosa*, found only in the mountains of San Cristobal, forms checkerboards with *R. malaitae*, found only in the mountains of Malaita, and with *R. cockerelli*, found on Malaita and most of the big islands of Bukida and New Georgia. The third checkerboard is between *R. malaitae*, a montane endemic of Malaita, and *R.* [*spilodera*], found only on Bougainville and Guadalcanal in Bukida plus Rennell and San Cristobal. In sum, at least from the distributional data, history is as plausible as competition as an explanation for these checkerboards.

Last among genera with checkerboards, *Monarcha* in the Solomons consists of three taxa (*M. cinerascens*, *M.* [*melanopsis*], and *M.* [*manadensis*]). *Monarcha cinerascens*, a supertramp, coexists with neither of the other taxa. It occupies all nine outlier islands plus the small, isolated island of Borokua between the Bukida and New Georgia island groups, as well as small islets near major islands of the Bukida group, but not large islands. The other two taxa coexist on many large islands in all the other groups except Rennell. Mayr and Diamond (2001) point to competition with *M.* [*melanopsis*] as the likely reason *M. cinerascens* is a supertramp. Although it has not been seen flying over water (Mayr and Diamond 2001), surely *M. cinerascens* can reach at least the major Bukida islands, given its presence on nearby islets. Thus its colonization history cannot explain the checkerboards. However, *M. cinerascens* is a small-island specialist even where *M.* [*melanopsis*] is absent, as in the Bismarcks, so habitat preference may account for these checkerboards. The systematics of *M.* [*melanopsis*] and *M.* [*manadensis*] need revising, as the former is paraphyletic and the latter polyphyletic (Filardi and Smith 2005). Depending on the ranks of component taxa, the number of checkerboards with *M. cinerascens* may greatly exceed two. However, the habitat differences will remain.

Of the 22 congeneric checkerboards, then, 17 consist of pairs of taxa occupying different island groups, while for one (in *Accipiter*), historical dispersal limitation appears to account for the checkerboard even though the species are in the same group (table 9.5). For one checkerboard (in *Pachycephala*), a habitat difference seems to be the cause, while in the remaining three (one in *Pachycephala* and two in *Monarcha*), one taxon occupies very small islands and the other larger islands, and in each of

TABLE 9.5
Proposed Factors Explaining Congeneric Checkerboard Distributions of Solomon Islands Birds

Genus	*CH*	*DG*	*HI*	*HA*	*LS*
Accipiter	5	4	1		
Aplonis	2	2			
Monarcha	2				2
Pachycephala	2			1	1
Rhipidura	3	3			
Zosterops	8	8			
Totals	22	17	1	1	3

Notes: CH=number of checkerboards, DG=different island groups, HI=historical (other than different island groups), HA=habitat difference, LS=one species on small islands, the other on larger islands.

these instances the small-island specialist is still restricted to small islands in other regions where the other taxon is absent.

Among multigenus guilds defined by Diamond (1975), only one, the gleaning flycatchers, has checkerboard distributions in the Solomon Islands. Of the seven species in this guild, one (*Monarcha cinerascens*) is a supertramp by our statistical definition, while *Pachycephala melanura* is also classed as a supertramp by Mayr and Diamond (2001). If we exclude both of these species, there are no checkerboards. If we exclude only *M. cinerascens*, there are five. These all consist of *Pachycephala melanura* with another taxon: *P.* [*pectoralis*] and *P. implicata* as discussed above, plus *Monarcha* [*melanopsis*], *M.* [*manadensis*], and *Myiagra* [*rubecula*]. As observed above, *M.* [*melanopsis*] and *M.* [*manadensis*] are both found on many large islands in all groups except Rennell. *Myiagra* [*rubecula*] is also found on many large islands in those groups, and also on Rennell. We pointed out above that *P. melanura* inhabits small islands even outside the Solomons (including outside the range of *P.* [*pectoralis*], *Monarcha* [*melanopsis*], and *Myiagra* [*rubecula*]), it has also not been seen flying over water, and it is a recent arrival in the Solomons, possibly expanding its range there (Mayr and Diamond 2001). Therefore, both habitat preferences and the history of colonization may at least partly explain these checkerboards.

In sum, looking specifically at the subset of species pairs in which competition would be most expected, we found that no exclusively distributed pairs quite conformed to the checkerboard model and that the exclusive patterns might be explained by a combination of colonization history and timing, behavioral traits (especially propensity to fly over water), and habitat preferences. For three congeneric bird checkerboards in the Bismarck archipelago, Lomolino (1999) suggested a combination of interspecific interactions, habitat preferences, and propensity for over-water flight as causes, while Collins et al. (in preparation), examining all the congeneric and multigenus-guild checkerboards in the Bismarcks, found colonization history, habitat preferences, and propensity for over-water flight to be possible explanations for most of them. Gotelli et al. (1997), studying congeneric checkerboards of mainland Australian birds (including several genera found in the Solomons), saw a major role for habitat preferences and found competition to be unimportant.

Many Solomons checkerboards include one species found exclusively or almost exclusively on small islands, including supertramps. Some may be only on small islands because they are excluded elsewhere by competition. Other explanations are possible, however. They may prefer habitats disproportionately present on small islands (cf. Simberloff and Martin 1991). Holyoak and Thibault (1978) suggest that predation by *Accipiter* hawks may restrict one supertramp, *Ducula pacifica*, to small islands. That competition is unlikely to be the only factor restricting at least some of these supertramps to small islands is suggested by the fact that *Monarcha cinerascens*, *Aplonis* [*feadensis*], and *Pachycephala melanura* all occupy only small, remote, or recently volcanically disturbed islands throughout their ranges, including beyond the Solomons, even when possible competitors are absent.

Finally, the same caveat must be raised with respect to assembly rules in the Solomons as was raised with the respect to the equilibrium theory: anthropogenic extinction must have been staggering, but most of it cannot be specified. The overall picture with respect to checkerboard distributions might not have changed much, especially as regards restriction of species to particular island groups. However, it is also possible that some checkerboards have been created by undocumented anthropogenic extinction. Additionally, the possibility of incomplete censuses noted above should be borne in mind; some absences may be artifacts, and rectifying them would be more likely to obliterate checkerboards than to generate them.

Taxon Cycle

Classifying species by range, subspecific differentiation, and habitat use, Greenslade (1968) saw distributions of land and freshwater birds of the Solomons as reflecting a three-step process in accord with the taxon cycle of Wilson (1959, 1961) for Melanesian ants. First is expansion of a species to form a continuous range encompassing at least the major islands of groups 1–4 described above. This expansion is followed by range fragmentation, accompanied by extinction on small and/or isolated islands. As examples of second-stage species, Greenslade (1968) suggested *Pachycephala* [*pectoralis*] and *Rhipidura cockerelli*, both discussed above. The second stage also entails evolution of island endemics. The final stage consists of a highly fragmented, contracted distribution (often into mountains of the largest islands), presumed to have arisen by substantial extinction even on major islands. Noteworthy in this scenario are the assumption of much undocumented extinction in the second and third stages and the suggestion that restriction of many third-stage species to montane habitats may be due to competition at lower elevations. Greenslade (1968) did not elaborate on the causes of the hypothesized extinctions on small islands during the second stage but did refer to the ongoing extinction hypothesized by MacArthur and Wilson (1963).

Independently of Greenslade (1968), Mayr and Diamond (2001) also attempted to match bird distributions in the Solomon Islands, and Melanesia generally, to the taxon cycle of Wilson (1959, 1961), dividing the avifauna into temporal, evolutionary stages. However, the stages correspond only partially to those proposed by Greenslade (1968) (and by Wilson [1961]), and there is one major difference. The geographic distributions and their relationship to endemicity play a key role in assignment to stages, as for Greenslade (1968), but the habitat affiliations are generally not as strongly related to stage, in their view.

Unlike Greenslade (1968) and Wilson (1961), Mayr and Diamond (2001) see dispersal ability as characteristically differing among species in different stages and having many distributional consequences. Perhaps "dispersal propensity" describes the trait Mayr and Diamond (2001) stress more aptly than does "dispersal ability," as they focus on behavioral explanations rather than physiological and anatomical features. Mayr and Diamond (2001) also point to undocumented extinctions, especially on small islands, as key features of the later stages, but, at least with respect to the taxon cycle, they attribute these extinctions, and the resulting distributional patterns, to the loss of dispersal propensity, arguing that populations occasionally go extinct, but only vagile species "capable of reversing those extinctions" (p. 292) can persist on many islands or on small islands. Just as did Greenslade (1968), Mayr and Diamond (2001) suggest that some late-stage montane species are restricted to upper elevations by competition, an argument buttressed most forcefully by elevational distributions of species with some populations montane and others not, depending on co-occurring species (e.g., *Zosterops ugiensis*, discussed above).

For both Greenslade (1968) and Mayr and Diamond (2001), then, bird distributions in the Solomon Islands result from a cyclic process operating on an evolutionary time scale. The factors driving the process differ somewhat in the two conceptions, but in each, extinctions in the later stages of the cycle play a key role, including extinctions on both large and small islands. Neither proposal discusses evidence for such extinctions, though Mayr and Diamond (2001) call for an expanded search for fossil evidence to determine the extent and causes of past extinctions. Their preliminary assessment is that the hecatomb afflicting other Pacific islands with the arrival of humans may not have been as severe in northern Melanesia because of the presence of native predatory mammals and reptiles. Steadman (2006), by contrast, emphasizes the wave of anthropogenic extinctions and absence of evidence for nonanthropogenic ones.

Discussion

Birds of the Solomons

Our examination of the distributions of these birds, and of evidence and speculation regarding distributional changes, suggests that the processes regulating community composition on large islands may differ greatly from those operating on small ones. With respect to the equilibrium theory in the Solomons, Gilpin and Diamond (1976) probably erred in considering large and small islands together. For large islands in the Solomons, there is virtually no evidence for nonanthropogenic extinction over a time frame of millennia (Steadman 2006). This is not to say that extinctions never occur, or even that no equilibrium richness obtains, but if we are dealing with rare events over time scales of millions of years, it is unlikely that the stochastic demography originally envisioned as mainly driving the dynamism would be important, or that the original assumption of unchanging physical characteristics would be valid. For birds on these large islands, the dynamic equilibrium model may not be appropriate.

By contrast, birds of the small islets near the major islands of each group might operate as envisioned by the original equilibrium theory, though there are insufficient data on turnover to know. One potential disqualifier would be if populations on such islands are insufficiently isolated for persistence to result mainly from in situ reproduction rather than continuing recruitment from the mainland (the "rescue effect" of Brown and Kodric-Brown [1977]). One of the earliest sources of criticism of the applicability of the equilibrium theory was concern about this very point—do individuals in the various island populations constitute separate populations or are they just parts of one widely ranging population, what might now be termed a metapopulation (references in Hanski and Simberloff 1997)?

In the original model, for the equilibrium to be dynamic, another requirement is that extinction must occur, and it must be a consequence of equilibrium demographic processes and perhaps interactions of members of the species pool rather than change in the island environment. Because many small islands in the Solomons are uninhabited, the massive anthropogenic changes found on large islands might not be as severe, and introduced species may not be as numerous. Steadman (2006) describes a 7 km^2 forested island in the Marianas that appears unscathed by humans aside from the presence of Pacific rats, which still contains all bird species recorded from prehistoric sites except for two rails, and which might be able to support populations of other birds. Perhaps islets in the Solomons exist that are also relatively unaffected by humans, are small enough that extinction occasionally occurs, and are sufficiently remote that propagules rarely arrive.

If there were turnover on such small islands, this would clearly be in the spirit of MacArthur and Wilson's conception of turnover, even if competition as envisioned by the assembly rules accounted for at least some of it, as noted above. One would also want a substantial proportion of the species to engage in the turnover. A common knock against the wide applicability of the dynamic equilibrium model is captured by Schoener and Spiller (1987): "in general turnover involves only a subset of fugitive populations, with many others, mostly much larger, being permanent" (p. 477; cf. Simberloff 1976, Whittaker and Fernández-Palacios 2007, Schoener, this volume).

Such turnover could also be consistent with the assembly rules as originally posited by Diamond (1975). He was agnostic about how dynamic the competitive checkerboards are but often cited birds with sufficient dispersal ability to reach many islands from which they are absent, suggesting that such species must frequently arrive on islands occupied by their competitors, only to fail to establish or to suffer quick extinction. Small islands might be a far more likely locus than the large ones of the Solomons for competition to play a decisive role in presence and absence, as required by the assembly rules, and perhaps for a new arrival to persist and the resident to disappear rather than vice versa. The examples cited above from Mayr and Diamond (2001), of species they feel are competitively incompatible but can coexist on large islands by virtue of elevational separation, come immediately to mind: smaller islands would offer fewer opportunities than large ones for habitat partitioning not only in terms of elevational gradients but in other ways as well. Historical factors would also play less of a role on small islands near enough to large ones that immigration is not very rare.

The taxon cycle as envisioned by both Greenslade (1968) and Mayr and Diamond (2001) encompasses both large and small islands, but the evolution driving the cycle in both conceptions occurs on much larger islands than those we suggest may fit the equilibrium theory and the assembly rules. Avifaunas of small islands in the taxon cycle are epiphenomena of processes (evolution of morphology, habitat preference, and dispersal behavior) occurring on larger islands. Thus, should turnover and/or competitive exclusion be demonstrated on small islands in the Solomon archipelago (say, those smaller than $50\,km^2$), they would be consistent with the cycle but not strong evidence for it.

Both the equilibrium theory and the taxon cycle posit extinctions. The equilibrium theory envisions these as being relatively frequent, albeit less so the larger the island. In the taxon cycle, on small islands extinctions may be relatively rapid; Greenslade (1968) relates them to equilibrium turnover. On large islands, however, these take much longer, associated as they are with the evolution of island endemics and, for Mayr and Diamond (2001), behavioral evolution. Extinctions do not play such a major role in the assembly rules (except, perhaps, for rapid extinction of immigrants that form forbidden combinations), although Mayr and Diamond (2001) invoke extinctions in partial explanation for the *Zosterops* checkerboards and suggest that undocumented extinctions occurred among members of other checkerboards. However, as noted above, there is no direct evidence in the Solomons for any of these extinctions except on Buka. The geographic distributions among the islands themselves can be seen as indirect evidence of extinction, but it seems tautological to use the distributions to support theories that aim to explain the distributions.

Evidentiary Needs for Birds of the Solomons

What other sorts of evidence, in addition to many more fossils from many more sites, could one marshal to support claims of nonanthropogenic extinction? This same concern was voiced early in the most detailed attempt to apply the taxon cycle model to birds, by Ricklefs and Cox (1972) for land birds (exclusive of raptors) of the West Indies, especially the Lesser Antilles. The largest of these islands are much smaller than the largest of the Solomons, with areas in the range of that of Buka. Ricklefs and Cox (1972) hypothesized that extinctions occur on average every few million years on larger islands and much more frequently on smaller ones (cf. Ricklefs and Bermingham 1999; Ricklefs, this volume). They also worried about the confounding effects of anthropogenic extinction, arguing that at least a few documented recent extinctions in the Lesser Antilles cannot be attributed to humans. In response to a battery of criticisms by Pregill and Olson (1981), Ricklefs and Bermingham (1999) (cf. Ricklefs and Bermingham 2002) undertook molecular phylogenetic analyses of West Indian birds that supported many aspects of the hypothesized taxon cycle in the Lesser Antilles and adduced further evidence that anthropogenic impacts and late Pleistocene climatic events did not lead to so much extinction that evidence of a taxon cycle would be obliterated. They also showed that species restricted to few islands, interpreted as in the late (declining) phase of the taxon cycle, were in fact much older than other species. They observed that this fact and the fact that some assigned late-stage species have gaps between the few occupied islands are consistent with the hypothesis of extinction on some unoccupied islands. The argument that occupancy gaps represent extinction is identical to that of Mayr and Diamond (2001), but taxon ages constitute a different sort of evidence. The inference of higher extinction rates on small islands derives from the observation that older taxa also tend to be absent from small islands (Ricklefs and Bermingham 2004; Ricklefs, this volume).

The first item in the wish list of Mayr and Diamond (2001) for additional data to elucidate the distributional trajectories of northern Melanesian birds is molecular phylogenetic research, totally lacking as they published their book. Such research, combined with remedying the striking lack of avian fossil data for the Solomons, would go a long way toward testing claims that current bird distributions there have resulted from a taxon cycle. It would be striking to see if the pattern of older species having patchier distributions and being restricted to larger islands holds there as it does in the Lesser Antilles. Phylogenetic research could also aid in testing whether the timing of colonization (e.g., in *Pachycephala*) or of allopatric speciation (e.g., in *Zosterops*) can explain checkerboards. Molecular evidence might also determine whether populations on small islands are sufficiently isolated to fit the equilibrium model. Such research has just begun for Solomons birds (Filardi and Smith 2005, Smith and Filardi 2007).

Relevance of Solomons Birds to the Three Theories

That Solomon Islands bird distributions, at least on the islands for which data are available and at least since the late Pleistocene, appear not to be determined by the mechanisms envisioned by the dynamic equilibrium theory does not mean the theory does not accurately depict other systems. Similarly, that the checkerboard distributions of birds in the Solomons today do not seem to reflect the processes envisioned in the assembly rules does not mean the rules do not apply elsewhere.

Though the equilibrium theory seems not to apply to many systems (references in Whittaker and Fernández-Palacios 2007; cf Schoener, this volume), it has been enormously fruitful, forcing us to think in new ways about the determinants of extinction and diversity (Brown 1981, Haila and Järvinen 1982, Simberloff 1984, Haila 1986). Among other things, the theory led to (1) consideration of what sets minimum viable population sizes (Shaffer 1981, 1987) and the fate of small populations; (2) the concept of relaxation of insular biotas with changing conditions such as area reduction (Diamond 1972, Faeth and Connor 1979); (3) increased attention to the multiple possible contributors to the species-area relationship (Connor and McCoy 1979); and (4) development of metapopulation ecology, which partially superseded equilibrium theory in both ecology and conservation biology (Hanski and Simberloff 1997, Hanski, this volume). However, for large islands with mean time to extinction of species in the range of 10^6 years, we do not feel the equilibrium theory will be fruitful, as we suggest above for the Solomons. Aside from the likelihood of changing environments, forces that might operate on this time scale (e.g., evolution, plate tectonics, bolides; cf. Ricklefs, this volume) are unlikely to yield any sort of testable equilibrium number of species. The birds of the Solomons may be a particularly difficult system for testing the equilibrium theory because of the human footprint and paucity of fossils. However, the same problems surely arise for many other biotas (Steadman 2006).

As for the assembly rules, in addition to generating controversy, they have contributed to a proliferating literature on and increased understanding of binary matrices, even beyond biogeography (e.g., Snijders 1991, Rao et al. 1996). In instances where there are more checkerboards than expected by matrix randomization (cf. Gotelli and McCabe 2002), there is rarely detailed examination of the distributions or other research to elucidate the cause. This should be a fertile research area and will encompass a wide range of ecological and evolutionary approaches.

The number of systems explored from the standpoint of a taxon cycle pales compared to the many applications of the equilibrium theory and the assembly rules. However, the use of molecular techniques, opening a new avenue of inference about ages of taxa, may spur research on taxon cycles. There are other sorts of taxon cycles than that proposed by Wilson (1959, 1961). For instance, using phylogenetic reconstruction, Losos (1990) was able to refute a taxon cycle that predicted a particular direction of morphological change. Molecular research can also shed light on the possibility of endogenous forces leading to dynamism and extinction (e.g., parasite-host interactions) and singular events such as mass extinctions; Ricklefs (this volume) provides examples for Lesser Antillean birds.

Literature Cited

BirdLife International. 2000. *Threatened Birds of the World.* Barcelona and Cambridge: Lynx Edicions and BirdLife International.

Brown, J. H. 1981. Two decades of homage to Santa Rosalia: Toward a general theory of diversity. *American Zoologist* 21:877–88.

Brown, J. H., and A. Kodric-Brown. 1977. Turnover rates in insular biogeography: Effect of immigration on extinction. *Ecology* 58:445–49.

Caughley, G. 1994. Directions in conservation biology. *Journal of Animal Ecology* 63:215–44.

Colwell, R. K., and D. W. Winkler. 1984. A null model for null models in biogeography. In *Ecological Communities. Conceptual Issues and the Evidence,* ed. D. R. Strong Jr., D. Simberloff, L. G. Abele, and A. B. Thistle, 344–59. Princeton, NJ: Princeton University Press.

Connor, E. F., and E. D. McCoy. 1979. The statistics and biology of the species-area relationship. *American Naturalist* 113:791–833.

Connor, E. F., and D. Simberloff. 1979. The assembly of species communities: Chance or competition? *Ecology* 60:1132–40.

Diamond, J. M. 1972. Biogeographical kinetics: Estimation of relaxation times for avifaunas of southwest Pacific islands. *Proceedings of the National Academy of Sciences U.S.A.* 69:3199–203

———. 1975. Assembly of species communities. In *Ecology and Evolution of Species Communities,* ed. M. L. Cody and J. M. Diamond, 342–444. Cambridge, MA: Harvard University Press.

———. 1984. The avifaunas of Rennell and Bellona islands. *The Natural History of Rennell Island, British Solomon Islands* 8:127–68.

Diamond, J. M., and M. E. Gilpin. 1982. Examination of the "null" model of Connor and Simberloff for species co-occurrences on islands. *Oecologia* 52:64–74.

Diamond, J. M., M. E. Gilpin, and E. Mayr. 1976. Species-distance relation for birds of the Solomon Archipelago, and the paradox of the great speciators. *Proceedings of the National Academy of Sciences U.S.A.* 73:2160–64.

Diamond, J. M., and E. Mayr. 1976. Species-area relationship for birds of the Solomon Archipelago. *Proceedings of the National Academy of Sciences U.S.A.* 73:262–66.

Faeth, S. H., and E. F. Connor. 1979. Supersaturated and relaxing island faunas: A critique of the species-age relationship. *Journal of Biogeography* 6:311–16.

Filardi, C. E., and C. E. Smith. 2005. Molecular phylogenetics of monarch flycatchers (genus *Monarcha*) with emphasis on Solomon Island endemics. *Molecular Phylogenetics and Evolution* 37:776–88.

Gilpin, M. E., and J. M. Diamond. 1976. Calculation of immigration and extinction curves from the species-area-distance relation. *Proceedings of the National Academy of Sciences U.S.A.* 73:4130–34.

———. 1984. Are species co-occurrences on islands non-random, and are null hypotheses useful in community ecology? In *Ecological Communities: Conceptual Issues and the Evidence,* ed. D. R. Strong, Jr., D. Simberloff, L. G. Abele, and A. B. Thistle, 297–315. Princeton, NJ: Princeton University Press.

Gotelli, N. J., and G. R. Graves. 1996. *Null Models in Ecology*. Washington, DC: Smithsonian Institution Press.

Gotelli, N. J., and D. J. McCabe. 2002. Species co-occurrence: A meta-analysis of J. M. Diamond's assembly rules model. *Ecology* 83:2091–96.

Gotelli, N. J., N. J. Buckley, and J. A. Wiens. 1997. Co-occurrence of Australian land birds: Diamond's assembly rules revisited. *Oikos* 80:311–24.

Greensland, P.J.M. 1968. Island patterns in the Solomon Islands bird fauna. *Evolution* 22:751–61.

Haila, Y. 1986. On the semiotic dimension of ecological theory: The case of island biogeography. *Biology and Philosophy* 1:377–87.

Haila, Y., and O. Järvinen. 1982. The role of theoretical concepts in understanding the ecological theatre: A case study on island biogeography. In *Conceptual Issues in Ecology,* ed. E. Saarinen, 261–78. Dordrecht, Netherlands: D. Reidel.

Hanski, I., and D. Simberloff. 1997. The metapopulation approach, its history, conceptual domain, and application to conservation.In *Metapopulation Biology. Ecology, Genetics, and Evolution,* ed. I. A. Hanski and M. E. Gilpin, 5–26. San Diego: Academic Press.

Holyoak, D. T., and J. C. Thibault. 1978. Notes on the phylogeny, distribution, and ecology of frugivorous pigeons in Polynesia. *Emu* 78:201–6.

Hughes, B., J. A. Robinson, A. J. Green, Z.W.D. Li, and T. Mundkur. 2004. *International Single Species Action Plan for the White-headed Duck* Oxyura leucocephala. New York: United Nations Environment Programme.

Kratter, A. W., D. W. Steadman, C. E. Smith, C. E. Filardi, and H. P. Webb. 2001. Avifauna of a lowland forest site on Isabel, Solomon Islands. *Auk* 118: 472–83.

Leibold, M. A., M. Holyoak, N. Mouquet, P. Amarasekare, J. M. Chase, M. F. Hoopes, R. D. Holt, J. B. Shurin, R. Law, D. Tilman, M. Loreau, and A. Gonzales. 2004. The metacommunity concept: a framework for multi-scale community ecology. *Ecology Letters* 7:601–13.

Lever, C. 1992. *They Dined on Eland. The Story of Acclimatisation Societies.* London: Quiller.

Lomolino, M. V. 1999. A species-based, hierarchical model of island biogeography. In *Ecological Assembly Rules. Perspectives, Advances, Retreats,* ed. E. Weiher and P. Keddy, 272–310. Cambridge: Cambridge University Press.

Long, J. L. 1981. *Introduced Birds of the World.* New York: Universe Books.

———. 2003. *Introduced Mammals of the World.* Wallingford, UK: CABI International.

Losos, J. B. 1990. A phylogenetic analysis of character displacement in Caribbean *Anolis* lizards. *Evolution* 44:558–69.

MacArthur, R. H. 1958. Population ecology of some warblers of northeastern coniferous forests. *Ecology* 39:599–619.

MacArthur, R. H., and E. O. Wilson. 1963. An equilibrium theory of insular zoogeography. *Evolution* 17:373–87.

———. 1967. *The Theory of Island Biogeography.* Princeton: NJ: Princeton University Press.

Mayr, E., and J. M. Diamond. 1976. Birds on islands in the sky: Origin of the montane avifauna of northern Melanesia. *Proceedings of the National Academy of Sciences U.S.A.* 73:1765–69.

———. 2001. *The Birds of Northern Melanesia. Speciation, Ecology, and Biogeography.* Oxford: Oxford University Press.

Miklós, I., and J. Podani. 2004. Randomization of presence-absence matrices: Comments and new algorithms. *Ecology* 85:86–92.

Pregill, G. K., and S. L. Olson. 1981. Zoogeography of West Indian vertebrates in relation to Pleistocene climatic cycles. *Annual Review of Ecology and Systematics* 12:75–98.

Pullin, R., M. L. Palomares, C. Casal, M. Dey, and D. Pauly. 1997. Environmental impacts of tilapias. In *Tilapia Aquaculture. Proceedings from the Fourth International Symposium on Tilapia in Aquaculture,* ed. K. Fitzsimmons, vol. 2, 554–70. Ithaca, NY: Northeast Regional Agricultural Engineering Service Cooperative Extension.

Rao, A. J., R. Jana, and S. Bandyopadhyay. 1996. A Markov chain Monte Carlo method for generating random (0,1)-matrices with given marginals. *Sankhyā* 58:225–42.

Ricklefs, R. E., and E. Bermingham. 1999. Taxon cycles in the Lesser Antillean avifauna. *Ostrich* 70:49–59.

———. 2002. The concept of the taxon cycle in biogeography. *Global Ecology and Biogeography* 11:353–61.

———. 2004. History and the species-area relationship in Lesser Antillean birds. *American Naturalist* 163:227–39.

Ricklefs, R. E., and G. W. Cox. 1972. Taxon cycles in the West Indian avifauna. *American Naturalist* 106:195–219.

Sax, D. F., S. D. Gaines, and J. H. Brown. 2002. Species invasions exceed extinctions on islands worldwide: A comparative study of plants and birds. *American Naturalist* 160:766–83.

Schoener, T. W., and D. A. Spiller. 1987. High population persistence in a system with high turnover. *Nature* 330:474–77.

Shaffer, M. L. 1981. Minimum population sizes for species conservation. *BioScience* 31:131–34.

———. 1987. Minimum viable populations: Coping with uncertainty. In *Viable Populations for Conservation,* ed. M.E. Soulé, 69–86. Cambridge: Cambridge University Press.

Simberloff, D. 1976. Species turnover and equilibrium island biogeography. *Science* 194:572–78.

———. 1984. This week's citation classic. *Current Contents* 15:12.

Simberloff, D., and E. F. Connor. Missing species combinations. *American Naturalist* 118:215–39.

Simberloff, D., and T. Dayan. 1991. The guild concept and the structure of ecological communities. *Annual Review of Ecology and Systematics* 22:115–43.

Simberloff, D., and J. L. Martin. 1991. Nestedness of insular avifaunas: Simple summary statistics masking complex species patterns. *Ornis Fennica* 68:178–92.

Smith, C. E., and C. E. Filardi. 2007. Patterns of molecular and morphological variation in some Solomon Island land birds. *Auk* 124:497–93.

Snijders, T.A.B. 1991. Enumeration and simulation methods for 0-1 matrices with given marginals. *Psychometrika* 56:397–417.

Steadman, D. W. 2006. *Extinction and Biogeography of Tropical Pacific Birds.* Chicago: University of Chicago Press.

Whittaker, R. J., and J. M. Fernández-Palacios. 2007. *Island Biogeography: Ecology, Evolution, and Conservation,* 2nd ed. New York: Oxford University Press.

Wilson, E. O. 1959. Adaptive shift and dispersal in a tropical ant fauna. *Evolution* 13:122–44.

———. 1961. The nature of the taxon cycle in the Melanesian ant fauna. *American Naturalist* 95:169–93.

———. 1969. The species equilibrium. In *Diversity and Stability in Ecological Systems*, ed. G. M. Woodwell and H. H. Smith, 38–47. Brookhaven Symposia in Biology no. 22. Upton, NY: Brookhaven National Laboratory.

Yom-Tov, Y., S. Tom-Tov, and H. Moller. 1999. Competition, coexistence, and adaptation amongst rodent invaders to Pacific and New Zealand Islands. *Journal of Biogeography* 26:947–58.

Neutral Theory and the Theory of Island Biogeography

Stephen P. Hubbell

FORTY YEARS AGO the theory of island biogeography challenged the Huchinsonian niche assembly paradigm in community ecology by postulating that ecological communities on islands were nonequilibrium collections of species assembled and disassembled solely by immigration and local extinction. Although the implications of this postulate were not fully appreciated at that time, the theory's elegantly simple graphical representation of the immigration-extinction equilibrium implied that species were ecologically equivalent—symmetric—in their probabilities of immigrating to an island and going extinct once there. Recasting the symmetry assumption on a per capita basis and adding speciation, the extended theory predicts not only species richness but also relative species abundance. The symmetry assumption is equivalent to asking how many of the properties of ecological communities are captured by the mean, ignoring species differences. Clearly the mean can only give us a first approximation, but how good an approximation is it? This paper examines this question in a species-rich tropical tree community on Barro Colorado Island (BCI) in a plot whose dynamics my colleagues and I have followed for the past quarter century. Before examining the BCI results, however, I explain the underlying symmetry assumption of the theory of island biogeography, first, because there is some disagreement whether the theory makes this assumption, and second, because this assumption is the theoretical foundation for extending the theory to predict relative species abundance.

Although it is called an equilibrium theory, the theory of island biogeography can only be narrowly construed as such because it predicts continual species turnover, rather than a stable species composition in ecological communities. This is quite a radical idea that then—as now—flies squarely in the face of prevailing theory in community ecology. Contemporary theory is largely based on the Hutchinsonian niche paradigm, which states that each species has a unique niche or functional role that it performs better than any other species (Chase and Leibold 2003). According to this hypothesis, ecological communities are limited-membership, closed sets of species coexisting in competitive equipoise and that resist invasion of all other species. In contrast, the theory of island biogeography—in its famous graphical representation of crossing immigration and extinction curves as a function of island species richness—asserts that ecological communities are open assemblages of species that approach a steady state species richness that is dynamic, not a static species composition. The species are not labeled in the theory, which means the theory assumes that species are essentially interchangeable, i.e., equivalent in their likelihood of arriving on an island, or of going extinct after arrival.

MacArthur and Wilson did not discuss the fact that their theory assumes species symmetry. Indeed, much of the latter half of their monograph was devoted to discussing topics such as differences among species in the timing and order of immigration events or in probabilities of extinction once established on the island. In the original presentation, the immigration and extinction curves were drawn concave downward, which MacArthur and Wilson explained as follows: Immigration rate should slow with increasing numbers of species on the island because rapidly dispersing species should arrive sooner than slowly dispersing species, because competition from already established species reduces the colonization success of later arriving species, and because immigrants can no longer be counted if their species is already present on the island. Extinction rates, on the other hand, should accelerate with increasing numbers of species due to a larger number of potential competitive interactions among species and decreasing average population sizes as the island filled up (MacArthur and Wilson 1967, Schoener, this volume). Later, MacArthur and Wilson introduced a second version of the graphical representation of the equilibrium in which the immigration and extinction lines were linear (Schoener, this volume).

The graphical representation of island biogeography theory implies symmetry because, according to the theory, it does not matter which species contribute to balancing immigration and extinction rates on any given island. The single state variable in the model is the number of species on the island. All species in the original theory are treated as identical. Without this assumption, the model's reduction of island community dynamics to counting species does not logically work. This is true even of the version of the theory with downwardly concave immigration and extinction curves. This concavity makes late-arriving species experience lower successful immigration rates and higher extinction rates. However, this modification does not alter the basic fact that any species arriving late, regardless of whether it is a good colonizer or competitor, will exhibit the same rate changes (Hubbell 2001). Likewise, all species respond in

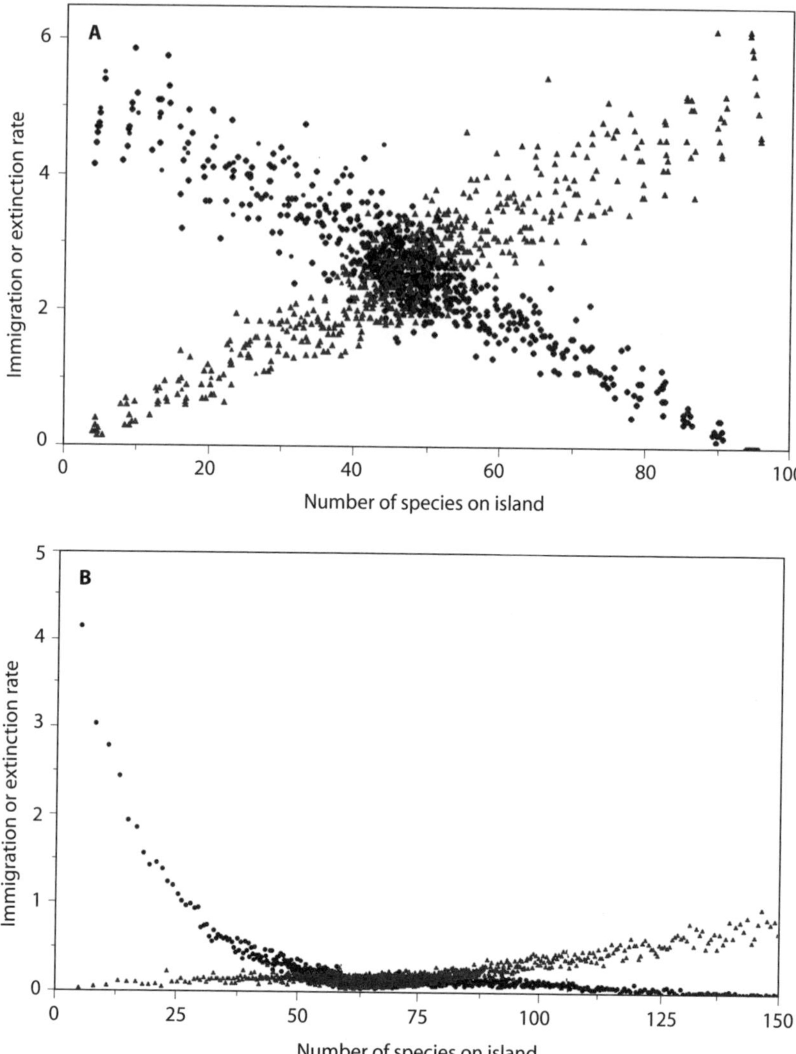

Figure 10.1. The classical immigration-extinction graph of the equilibrium species richness on an island generated by two versions of neutral theory. Immigration rates are circles and extinction rates are triangles. *Panel A*: The linear version is mathematically expected when the symmetry assumption is made at the species level, in which each source area species has an equal probability of immigrating to the island, and of going extinct once there. In this example there were 100 species in the source area To estimate rates of immigration and extinction, individual immigration and extinction events were binned into short (10 unit) time intervals and plotted against mean number of species on the island in that time interval. Points represent the scatter over an ensemble of 10 stochastic runs. *Panel B*: The curvilinear version (arises when the symmetry assumption is made at the individual

an identical manner to variation in the size of the island and its distance from the mainland source area.

It is easy to demonstrate how both the linear and curvilinear graphical versions of the immigration-extinction equilibrium in island biogeography arise from symmetric neutral theory (figure 10.1). The difference between the two versions is due to the level at which one makes the symmetry assumption, either at the species level—the level of the assumption in the theory of island biogeography—or at the individual level, which is the level of the assumption in neutral theory. If one makes the symmetry assumption at the species level, then species per se are equally likely to immigrate or go extinct, and in this case, one obtains the linear immigration-extinction graph. Figure 10.1a presents the results of an ensemble of ten stochastic simulations of the colonization of an island assuming equivalence at the species level. However, one can change the symmetry assumption to apply at the individual level, not at the species level, a change which means that each source-area *individual*—not each species—has an equal probability of immigrating, irrespective of the species to which it belongs. With this change in the level of the symmetry assumption, neutral theory is able to extend the theory of island biogeography to encompass relative species abundance both in the source area and on the island. One can prove that the expected distribution of relative species abundance in a continuous source area (the "metacommunity") is Fisher's logseries (Hubbell 2001, Volkov et al 2003). There is recent empirical evidence that the logseries distribution applies at large landscape scales in Amazonia (Hubbell et al. 2008).

When species have different relative abundances in the source area, then species no longer have equal probabilities of immigrating to the island. Common species are more likely to arrive before rare species. The individual-level symmetry assumption gives rise to the concave-curvilinear immigration curve in which the probability of species immigrations are

---

level, in which case the relative abundances of the species in the source area affect the probability of immigration (common species are more likely to immigrate than rare species). Extinction rates are a function of local species abundance on the island and accelerate as species become rarer with increasing numbers of species on the island. In this example there were 150 species in the source area ("metacommunity") whose abundances were determined by a value of 20 for biodiversity number θ of neutral theory (Fisher's α) and a metacommunity (source area) size of 10,000. The immigration rate m was 0.5. The degree of asymmetry of the immigration and extinction curves varies and is a function of the immigration rate, the size of the island (measured by the sum of the population sizes on the island), and the value of θ.

ordered stochastically by ranked species abundances in the source area (figure 10.1b). Neutral theory requires no assumptions about differing dispersal abilities of species or interactions among species to generate curvilinear immigration and extinction curves, so it is a very parsimonious theory—more so even than the theory of island biogeography, because one no longer needs to specify the extinction rate, which is a prediction of the theory and arises through the demographic stochasticity of island populations. In the example shown, the immigration curve is much higher than the extinction curve; the degree of asymmetry is a function of the immigration rate and is less for slower immigration rates. However, the approach to equilibrium from an empty island is generally much faster than the loss of species through extinction from an "oversaturated" island. Asymmetric curves tend to occur because a colonization event requires the arrival of only *one* individual of a given species, whereas extinction requires the death of *all* individuals of a species on the island.

As MacArthur and Wilson point out, many species do differ in their colonizing ability and in their susceptibility to extinction. However, neutral theory says that simply observing curvilinear immigration and extinction rates is not sufficient evidence because species differences in immigration and extinction could be due primarily to differences in species abundance. It is possible that differences in source-area abundance of species may be many orders of magnitude greater than differences in dispersal ability and therefore could dominate the immigration process; this is an important open question for future research. In fact, neutral theory is a rich source of many detailed predictions about how the actual shapes of the immigration rate and extinction rate curves should change as a result of island size and immigration rate and the distribution of relative species abundance in the source area. For example, under low rates of immigration, the theory predicts that one may observe a bimodal extinction cure as a function of number of species on the island. This can happen because, under low immigration rates, some island species that colonized the island early have a chance to build to large population sizes, causing a bump in early extinctions in rare species before equilibrium species diversity is reached. To my knowledge, this result was not anticipated by island biogeography theory; it is a prediction that has never been made before, and has yet to be tested empirically.

I turn now to discussing the BCI results and evaluating their consistency with the theory of island biogeography and its extension the symmetric neutral theory. This paper discusses the following findings from empirical and theoretical studies of the BCI plot. (1) Although tree species in the BCI forest exhibit many differences, nevertheless island biogeography theory—and its neutral theory extensions—does quite a good job fitting both the aggregate community static and dynamic data. (2) Density dependence—the supposed signature of a diversity-regulated, niche-differentiated community—although strong and pervasive in the BCI tree community, especially in the early life history stages, nevertheless is not strong enough to regulate tree populations at the scale of the entire 50 ha plot. (3) A key ingredient in island biogeography is dispersal limitation, and all BCI species are strongly dispersal and recruitment limited. (4) Virtually all BCI tree species are ecological equivalents or near equivalents in their nutrient niches, so R^* competition theory, the iconic niche-assembly theory in plat ecology, does not work for BCI trees. (5) Contrary to popular belief, simple evolutionary models show that ecological equivalence, the key concept of neutral theory, can evolve easily and often in communities of competing, dispersal-limited species. I discuss each of these findings, but in reverse order.

Evolution of Ecological Equivalence

The core idea of neutral theory is ecological equivalence or near equivalence. A legitimate question is whether ecological equivalence among competing species can evolve, and if it can, how likely is it to do so. I have argued that ecological equivalence can and will arise easily and often under selective regimes that should be commonplace (Hubbell 2006). To study this problem, I adapted a model from Hurtt and Pacala (1995), who studied a model community of dispersal-limited competing species, each of which was the best competitor for some set of microsites. When dispersal was not limiting, such that offspring of each species reached every site, then each species won those sites for which it was the best competitor. However, under dispersal limitation, many species won by default sites for which they were not the best competitor—because the best competitor did not reach the site. Hurtt and Pacala showed that dispersal limitation can delay competitive exclusion nearly indefinitely, even of species that were inferior competitors to some other species in every microsite. Dispersal limitation delayed competitive exclusion longer the more species-rich the community became.

Hurtt and Pacala (1995) did not study the evolution of niches, however, so I added genetics, modeling the evolution of a quantitative trait of many genes of small, additive effect that adapted species to particular microsites (Hubbell 2006). I considered three selective scenarios under chronic dispersal limitation (figure 10.2). In scenario 1, environmental (microsite) variation was fine-grained, and each species experienced the full range of microsite variation over the range of the species. Under this scenario, species exhibited convergent evolution, converging on nearly identical

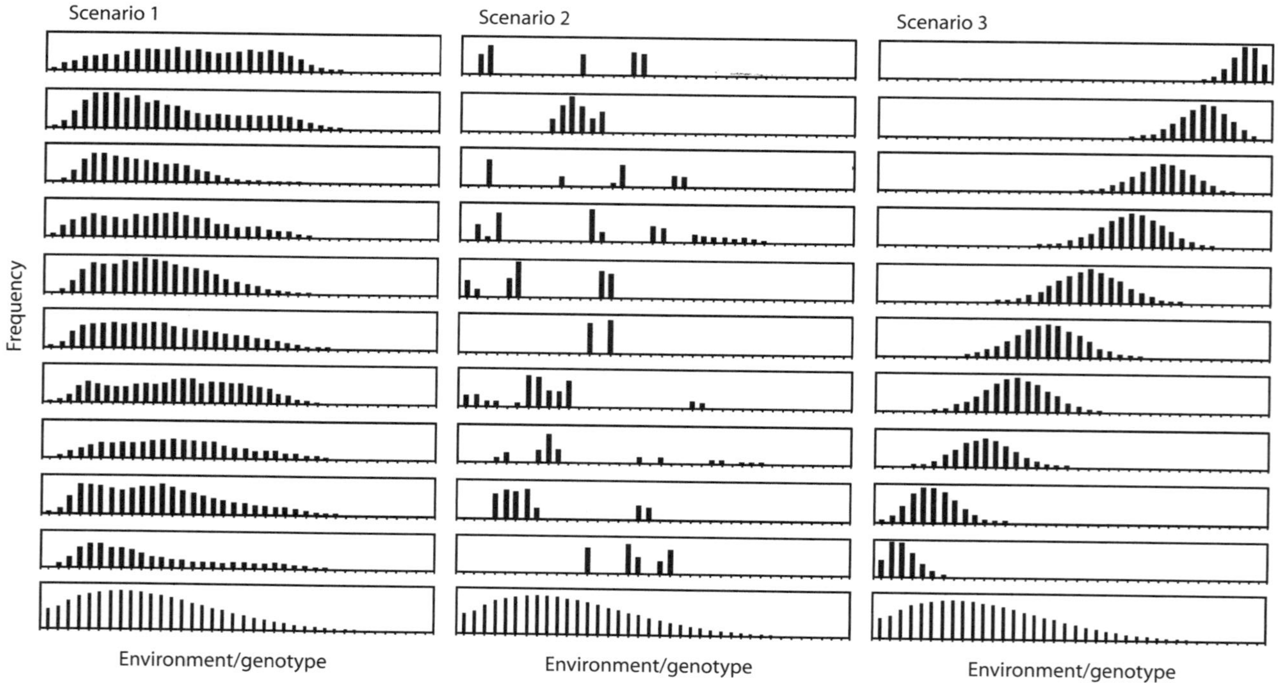

Scenario 1

Scenario 2

Scenario 3

Frequency

Environment/genotype

Environment/genotype

Environment/genotype

Figure 10.2. Three scenarios for the evolution of ecological niches in a dispersal-limited community of 10 species. Top 10 panels under each scenario are the distributions of genotype frequencies (percentages) for a metric trait with values ranging from 0-40 in each of the 10 species after 10,000 generations. Bottom panel under each scenario is the frequency distribution of environmental states, which ranged in value from 0 to 40. Selection favored juveniles having the genotype value (number of alleles) most closely matching the environmental state value. *Scenario 1*: environment is fine-grained, each species is exposed to the full range of environmental variation; result: convergent evolution, broadly overlapping niches with genotype frequencies similar to the frequencies of environmental states encountered. *Scenario 2:* environment is coarse-grained and patchy, such that local populations of a species are not fully exposed to all environmental variation, but the range of the species span the full range of environmental variation; result: species evolve into polymorphic generalists with local ecotypes, but no limiting niche similarity between species. *Scenario 3:* environment is coarse-grained and patchy and species are not exposed to the full range of environmental variation over their evolutionary history; result: classical niche differentiation. Under this scenario, the species were ordered for illustration to better reveal the staggered niche distributions.

distributions of genotypes matching the frequency of the different microsites they encountered, irrespective of the number of other species doing the same thing, and regardless of starting conditions. Under scenario 2, environmental (microsite) variation was coarse-grained and spatially autocorrelated, but nevertheless all species still experienced the full range of microsite variation over their geographic range. Under this scenario, species evolved into polymorphic generalists, consisting of locally adapted ecotypes. This case might seem like niche differentiation, but it is fundamentally different because the niches of all species overlapped broadly across their polymorphisms, and there was no limiting similarity (figure 10.2). Finally, in scenario 3, environmental (microsite) variation was again coarse-grained and spatially autocorrelated, but in this case, the species did not encounter the full range of environments (microsites). Only under this scenario did species evolve classical niche differentiation with limiting similarity. I expect all three selective regimes to be commonplace, and ecological equivalence or near-equivalence evolved under two out of three selective regimes. These ecologically equivalent or near-equivalent species persisted without extinction for at least 10,000 generations, the duration of the model simulations (Hubbell 2006).

A question might arise as to whether these results are obtained only on local spatial scales. Subsequent to the analyses in Hubbell (2006), Jeff Lake, Luís Borda-de-Agua, and I (unpublished) have explored these models on much larger spatial scales and with more explicit functional traits. As long as strong dispersal limitation applies (which becomes stronger on larger spatial scales), and the selective regimes are the same, then we obtain the same qualitative results on large scales. Of course, real environments are spatially autocorrelated, and they are more likely to differ the farther apart they are separated. Therefore it is not surprising that niche differentiation should generally be greater among species separated by larger distances.

Nutrient Niches: Empirical Evidence of Equivalence or Near-Equivalence

R^* competition theory (Tilman 1982, 1988) postulates that plant species coexist by virtue of partitioning limiting nutrients through an interaction of spatially variable nutrient supply rates and species-specific uptake rates for these nutrients. R^* theory is also called resource-ratio theory because plants use nutrients in relatively fixed tissue ratios, and in R^* theory the outcome of competition for multiple nutrients depends on ratios of supply rates of limiting nutrients in relation to ratios of consumption rates by competing species. R^* theory is very parameter-rich, and the species-specific values of these parameters are unknown for BCI tree species. Nevertheless, there are strong qualitative predictions that we can test, and we summarize our findings for three of these predictions here.

We mapped all soil macronutrients except S and most micronutrients across the BCI plot (John et al. 2007). We analyzed species richness across three primary gradients of macronutrients in the BCI plot, ratios of N/P, Ca/K, and Mn/Mg, chosen because they were statistically independent from each other, and because these six macronutrients are generally thought to include the nutrients that are most often limiting. Spatial variogram analysis revealed that virtually all of the spatial autocorrelation in nutrients occurs on spatial scales of 200 m or less, so the appropriate scale for testing the effects of variation in nutrients on species richness is on spatial scales of less than 4 hectares. There is one to two orders of magnitude variation in these nutrient ratios across the plot. Here we report only the results for the N/P gradients, but the conclusions are identical to those reached from considering the nutrient ratios of Ca/K and Mn/Mg. We will publish the full results elsewhere (Hubbell et al. unpublished).

The first prediction of R^* theory is that species richness should increase with the spatial variance in nutrient ratios. There is considerable variation in local species richness to explain. For example, on a scale of 400 m^2, species richness varies from 26 to 81 species. Does local variation in nutrient ratios explain this variation in species diversity and composition? The answer appears to be *no*. We found no relationship between species richness and spatial variance in nutrient ratios (Hubbell et al. unpublished). Figure 10.3 shows the results on the N/P gradient at a spatial scale of 400 m^2, and we obtained similar results at all spatial scales and for other nutrient ratios. We did find by principal-component analysis that a linear combination of Ca, P, and Zn explained over 40% of the variation in species richness (John et al. 2007). However, this nutrient interaction is not predicted by R^* theory, but probably reflects an underlying interaction between these nutrients that is not captured by R^* theory, as discussed below.

The second qualitative prediction is that if one moves across gradients of limiting nutrients or their ratios, there should be a sequence of species replacements (figure 10.4). We tested this prediction on the ten most abundant species, which constitute 52% of all individuals. One would expect competition to be the most intense, and nutrient partitioning to be the most evident, among these very abundant species. However, these species remain relatively invariant, with some fluctuations in abundance across the three primary gradients of macronutrients (figure 10.4) (Hubbell et al. unpublished).

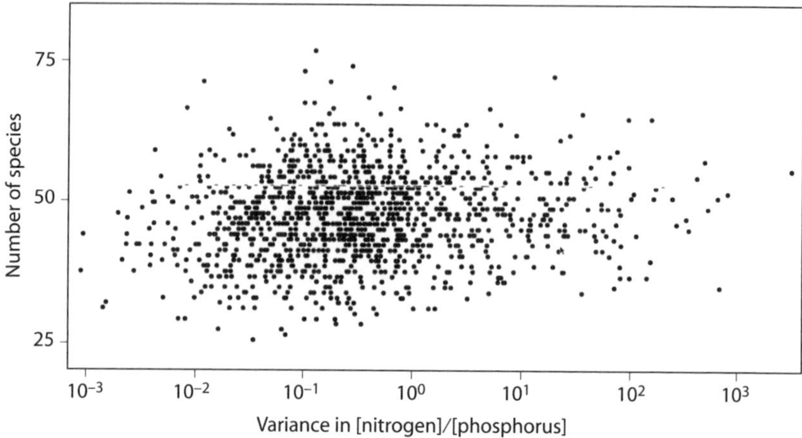

Figure 10.3. Lack of relationship between species richness per 400 m² in the BCI plot and position on the N/P gradient across the plot. Species richness varies from 26 to 81 species on this spatial scale. similar qualitative results were obtained on Ca/K and Mn/Mg gradients, and on different spatial scales, ranging um to 4 ha.

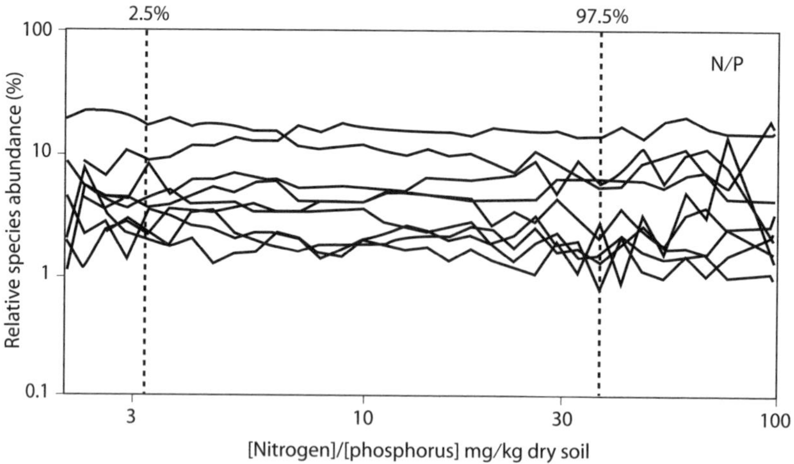

Figure 10.4. Lack of species replacements over the N/P gradient among the 10 most abundant species in the BCI plot. These 10 species represent more than half of all stems in the forest, and should show niche differentiation for limiting nutrients if it exists. Similar results were obtained on the Ca./K and Mn/Mg gradients.

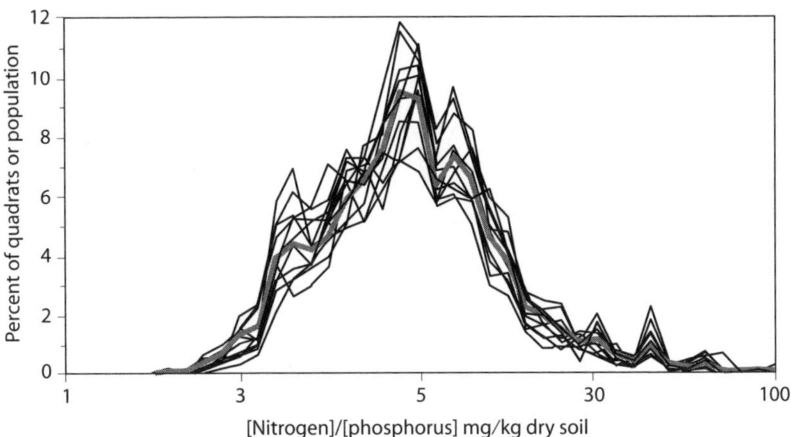

Figure 10.5. Evidence of nutrient niche generalization over the N/P gradient among the 10 most abundant species in the BCI plot. The heavy gray line is the distribution of the proportion of quadrats having a given value of the N/P ratio in the plot, which is the null distribution of the proportion of species abundance that is expected if they are indifferent to the nutrient ratio variation. The thin lines are for each of the 10 species. The species lines do not differ significantly from the null distribution. Similar results were obtained on the Ca/K and Mn/Mg gradients.

A third prediction is that the nutrient niches of BCI species should minimally overlap on nutrient gradients. The null expectation is that the proportion of the individuals of a given species occurring at a given nutrient ratio should match the proportion of plot area exhibiting that nutrient ratio. The most common species should exhibit strong nutrient niche differentiation. However, this is not what we observe. Virtually all species show very broad niche overlap in their distributions, many species conforming very closely to the null expectation. For example, the ten most abundant species are all nutrient generalists on the three gradients; we illustrate these results for the N/P gradient in figure 10.5. The distributions conform to the null expectation, i.e., they are indifferent to position on the nutrient gradient. This said, about 70% of BCI species distributions deviate significantly from the null distribution, consistent with our previous findings (John et al. 2007). However, our very large sample sizes allow us to detect significance in quantitatively small deviations from the null. Moreover, many of the species that deviate from the null expectation do not differ from each other (e.g., figure 10.6). In fact, all BCI species overlap to a very large extent in niche breadth on all three nutrient gradients. Of the 187 species abundant enough to test, in 155 species the intersection of their niche breadths was >95% of the union of their niche breaths on these nutrient gradients and in 139 species it was

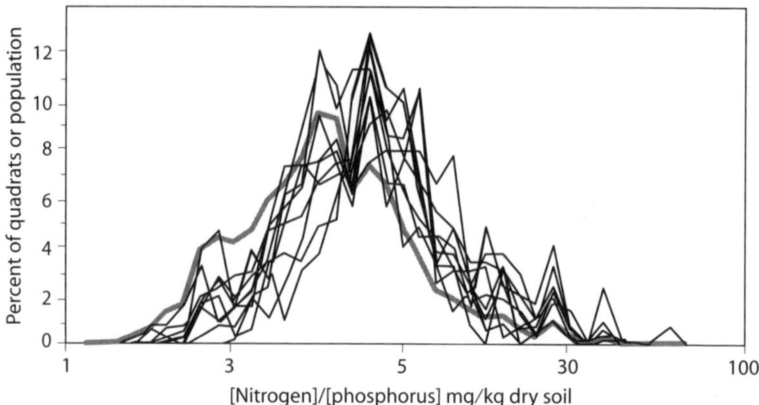

Figure 10.6. About 70% of BCI species deviate from the null distribution of one or more nutrient ratio gradients. However, many of these species, although they differ from the null distribution, are not distributed differently from each other. For example, here are the distributions of 10 species that show a slight skewing of abundance toward the high end of the N/P gradient in the plot, but do not differ from each other. The heavy line is the null distribution.

>99% (Hubbell et al. unpublished). We conclude that BCI species are nearly ecologically equivalent for the major macronutrients likely to be limiting to them, and that the primary explanation for the coexistence of so many BCI tree species is not likely to lie in niche partitioning of nutrient gradients.

The mathematics of R^* theory is internally consistent, so what is going on? One possibility is that BCI tree species do not actually compete for these nutrients, but this seems very unlikely. A second possibility is that the niche differentiation is in regard to other macro- and micronutrients not yet examined, which remains to be tested. A third possibility is that our measurements of soil nutrient concentrations do not accurately reflect the supply rates of these nutrients; but we have tested this possibility, and there is a very high positive correlation (>0.9) between soil concentrations and levels of nutrient availability to plants (Dalling, personal communication).

A fourth possibility is that BCI tree species do not conform to one or more assumptions of the mathematics of R^* theory. One assumption is that species are nutrient specialists, but this is not true of the vast majority of BCI tree species. Another false assumption is that the essential macro- and micronutrients are taken up independently. Over the past quarter century since R^* theory was developed, there have been major advances in understanding of the mineral nutrition of plants (Epstein and Bloom 2005) that have not yet been incorporated into the theory of resource competition. One of the main research findings is that many nutrients are not taken up independently. For example, Ca facilitates the uptake of many cations and anions. Another false assumption of R^* theory is that nutrient uptake and growth parameters are invariant over time and the same among all individuals of a given species. Nutrient uptake parameters vary among individuals and even in the same individual over time. Plants regulate their internal tissue stoichiometry of macro- and micronutrients against concentration gradients in the environment, and they do this by changing enzymatic pathways and affinities in nutrient uptake depending on the concentrations to which they are exposed. Plants can also adaptively change their mycorrhizal associates as nutrient environments change, favoring associates that are better at facilitating uptake of nutrients such as P over different concentration ranges.

These and other findings suggest that we need a new resource-based theory for testing the importance of nutrients to coexistence of species in plant communities, including tropical tree communities. Regardless of the development of new theory, there is little doubt that most BCI species are nutrient generalists with broadly overlapping niches. In terms of the model of the evolution of ecological equivalence summarized above (Hubbell 2006), the origin of this near-ecological equivalence is presumably response to selection from similarly variable nutrient regimes over the evolutionary history of these species.

What about niche differentiation along other niche dimensions, such as light and water availability gradients? There is a strong axis of niche differentiation at the guild level with regard to light. However, there are many nearly equivalent shade-tolerant species, many more than the number of shade-intolerant species (figure 10.7). The large number of shade-tolerant species could be a problem for niche theory because one would expect light to be more finely partitioned when it is abundant than when it is scarce (Hubbell 2005). Although competition for light is intense in the closed-canopy BCI forest, shade is not species-specific nor a resource to be partitioned. The most parsimonious hypothesis to explain these results is simply that most BCI tree species have experienced shady environments over their evolutionary history, each converging on adaptations for tolerating shade stress, irrespective of the number of species following the same adaptive trajectory. We therefore do not believe that light partitioning is a strong candidate to explain the high tree species richness of the BCI forest.

What about hydrological niches, as in the hypothesis made by Silvertown et al. (1999) that different species have different drought tolerances? We do find a considerable range in seedling drought sensitivity among Panamanian tree species (Englebrecht et al. 2007). We have tested

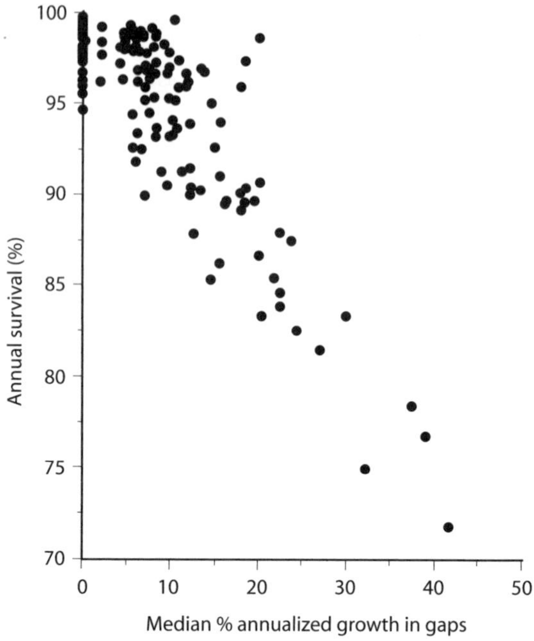

Figure 10.7. Axis of niche differentiation with respect to light availability. Each point represents the mean phenotype of a single species. The species at the upper left are shade tolerant (high survival in shade, low maximal growth rate in high light), whereas species' at the lower right are shade intolerant (low survival in shade, high maximum growth rate in full sun). There are many more shade-tolerant species than shade-intolerant species, posing a potential difficulty for niche theory in explaining why low light environments would be more finely partitioned than high light environments. A simple hypothesis is that species have niche-converged on shade tolerance because more species have experienced shady environments more persistently over evolutionary time than sunny environments, irrespective of the number of species following the same evolutionary trajectory.

drought tolerance in about 70 species across the isthmus of Panama, from the wet Caribbean side to the more seasonal and drier Pacific side, and the ratio of population density of species in dry versus wet sites across the isthmus is significantly correlated, although weakly, with drought sensitivity (R^2<0.2) (Englebrecht et al. 2007). On small spatial scales (the 50 ha BCI plot), seasonal water availability appears to act as an environmental filter determining which species can persist in the seasonally drier parts of the plot (the plateau). However, virtually all of the more drought-resistant species also are present in (i.e., not excluded from) the wetter areas (slopes) of the plot and grow right alongside the less drought-tolerant species.

In summary, if one examines the nutrient, light, and hydrological gradients in the BCI plot, there are many nearly equivalent species at each point along each gradient, and I am unaware of any niche-based theory that predicts how many species will be found at any given position along these gradients (Hubbell 2005). This is not to say that new dimensions of niche differentiation will not be discovered in the future to explain all of these locally co-occurring species; but at the moment, a simpler hypothesis suffices, namely, that species in each guild have been subject to similar environments and selection pressures over their evolutionary history and have converged on a similar suite of traits that adapt them to these shared environments, irrespective of the number of other species evolving the same, or a very similar, suite of traits. If Hurtt and Pacala (1995) are correct, dispersal limitation prevents competitive exclusion among these niche-convergent species. According to this view, the number of tree species in the BCI is more a reflection of larger-scale evolutionary-biogeographic processes that dictate the number of species in the regional species pool.

Dispersal and Recruitment Limitation: Empirical Evidence

We have already discussed the theoretical evidence that dispersal limitation can promote long-term species coexistence in communities (Tilman 1994, Hurtt and Pacala 1995). Dispersal limitation is the failure of seeds to arrive at all sites favorable for the growth and survival of a given species, and recruitment limitation is the failure to recruit germinated seedlings in a site similarly favorable for growth and survival. I will lump both processes under the rubric of dispersal limitation for purposes of the present discussion. We have been studying seed dispersal in BCI trees in the 50 ha plot for the past 21 years, sampling seed rain biweekly in a network of 200 seed traps, and following seedling germination in three 1 m^2 quadrats next to each of the traps (Hubbell et al. 1999, Muller-Landau et al. 2002, Dalling et al. 2002, Wright et al. 2002). The results show that only a small number of species managed to deposit seeds in a substantial fraction of the traps. In the first decade, only 5 species deposited at least one seed in over half of the traps, whereas 50% of the >200 species whose seeds were collected somewhere at least once, managed to deliver at least one seed to only 5 or fewer traps over a decade (figure 10.8) (Hubbell et al. 1999).

Jacaranda copia (Bignoniaceae) is the best disperser of any species whose seeds were collected in the seed traps. At least one seed of this species arrived in every trap during the first decade, and no other species came close to this record. Despite this, even *J. copaia* is recruitment limited

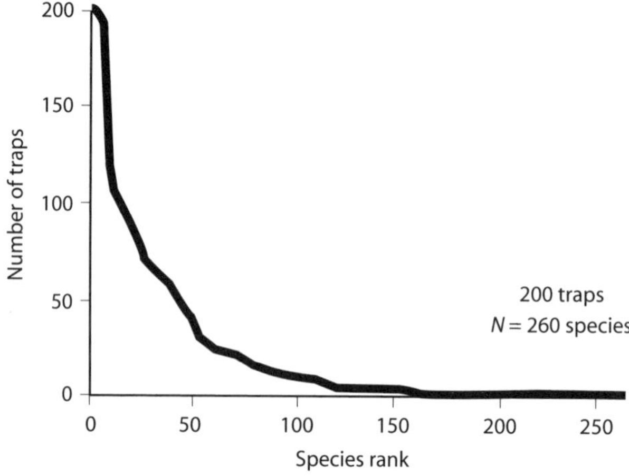

Figure 10.8. Evidence for community-wide dispersal limitation among BCI trees. Seeds were collected weekly in a network of 200 traps throughout the BCI plot. Of the 260 species collected over a decade, only a dozen species deposited seeds in more than half of the traps, whereas half of all species dispersed seeds to 5 or fewer traps in a decade. After Hubbell et al. (1999).

because it requires very large light gaps to survive, and gaps of sufficient size for successful regeneration of this species average more than 100 m from adults of this species in the BCI forest. We studied dispersal in this species using microsatellite markers (Jones et al. 2006). We genotyped potential parents and maternal tissue from seeds collected after dispersal. This is a light-demanding canopy emergent that is under strong selection for dispersal because the large gaps it requires to regenerate are few and far apart. The genetic data indicated that, although more than 91% of the seeds landed within 100 m of the mother, 57% of sapling recruits (reaching the census size of 1 cm DBH) were from the tail of the dispersal kernel, more than 100 m from the mother.

In summary, the trap data and the genetic results indicate that all BCI tree species are dispersal and recruitment limited. This is a key assumption of the theory of island biogeography and of neutral theory.

Density Dependence: Theoretical and Empirical Evidence

A great deal of attention has been paid to the question of density dependence in tropical forests, particularly to the hypotheses of Janzen (1970) and Connell (1971) about the role of enemies in maintaining high species diversity in tropical forests. Janzen and Connell independently proposed that an interaction of dispersal and seed predation would prevent monodominance by any single species by lowering the probability of self-replacement of a given species at the same location. We have been testing a generalization of this hypothesis, measuring not only losses in the seed-to-seedling transition, but also density dependence in subsequent growth and survival of juvenile individuals, as a function of local conspecific population density. Using data from the seed rain/seedling germination study, Harms et al. (2000) demonstrated that there was pervasive density dependence throughout the BCI tree community in the seed-to-seedling transition. If a species deposited more seeds in a given trap, it had lower per capita seedling germination in the adjacent seedling plots than when a species deposited fewer seeds in a given trap. This effect was species-specific: traps with more seeds of other species did not increase the mortality of seeds of a given focal species.

In 2001, to study density dependence in a spatially stratified sampling design covering the entire 50 ha plot, we began a study of seedling recruitment, growth, and survival in 20,000 1 m^2 seedling plots in a 5 m grid over the entire plot. This grid puts 2 to 5 traps under the crown of every single canopy tree in the plot. We have analyzed seedling survival during the first three years of this study in 48,956 established seedlings and small saplings of 235 species (Comita and Hubbell, 2009). When we tested for density dependence across all species, there was a significant negative effect of conspecific seedling and adult densities on conspecific growth and survival. In contrast, heterospecific neighbors had no effect on seedling growth and a positive effect on survival. At the species level, the density of conspecific neighbor seedlings had a significant negative effect on survival for 45 of the 59 species (76%) that were sufficiently abundant to test. We expect the percentage of species showing negative density dependence to increase as the length of the study increases. The expectation is based on the fact that we know that density-dependent effects on growth and survival persist into the sapling and subadult stages of BCI tree species as well (Hubbell et al. 2001, Ahumada et al. 2004). Smaller saplings show a greater depression of relative growth rate than do larger subadult trees from conspecific neighbors. These juvenile life stages last for decades in many species, so even small effects can accumulate over the lifespan of individual trees. Pervasive interspecific frequency dependence, although weak in comparison with intraspecific density dependence, has also been detected at the community level (Wills et al. 1997, 2006).

However, the primary question we are posing here is, do Jansen-Connell density-dependent effects regulate BCI tree populations? Given the strength, pervasiveness, and persistence of the negative conspecific density effects in the BCI community, there is no doubt any longer that

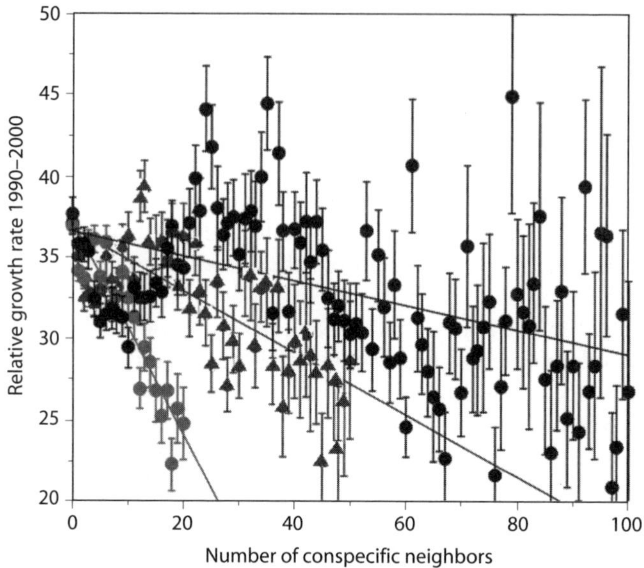

Figure 10.9. Effect of number of conspecific neighbors on relative growth rate (percentage growth) over the decade, 1990–2000, as a function of distance from a focal plant, for focal plants 1–4 cm DBH. Light gray circles: Effect of close conspecific neighbors, within 5 m of the focal plant. Dark gray triangles: Effects of conspecific neighbors from 5 to 10 m from the focal plant. Black circles: Effects of conspecific neighbors from 15 to 20 m from the focal plant. The data for 10 to 15 m are not shown for graph clarity. The negative effect of a conspecific neighbor on the growth rate of a focal plant is about an order of magnitude weaker at a distance of 15–20 m than it is at a distance of 0–5 m.

these effects promote local diversity in the BCI forest by reducing the probability of conspecific self replacement. However, this is different from the question of whether these Janzen-Connell effects regulate the adult population sizes of BCI tree species. Several empirical observations and theoretical considerations cast serious doubt on this possibility.

The most important of these observations is that the strength of the negative density dependence on conspecific recruitment, growth, and survival decays to background levels over very short distances, measured in a few tens of meters, usually less than 20 m (figure 10.9) (Ahumada et al. 2004, Hubbell et al. 2001). Therefore, there is little or no force of density dependence acting at the scale of the entire plot on adult tree population densities—or even on spatial scales of a few hectares. Janzen-Connell effects do reduce the probability that a given tree will replace itself at the same location in the forest, so they increase the mixing of species and species richness on a local spatial scale. However, they are not sufficiently strong and spatially extensive to regulate adult population abundances on landscape scales. This conclusion is consistent with the observation that, despite locally negative effects on survival of conspecific neighbors, seedling survival is positively correlated with species abundance in the BCI tree community at the whole plot level (Comita and Hubbell, 2009).

One can reach the same conclusion on theoretical grounds (Zillio et al. 2005, Hubbell 2008). Without delving into the mathematics, the logic is clear from a simple verbal argument. Consider a perfect Janzen-Connell effect, such that no species can replace itself in the same location. However, suppose that species i can replace any of the other S–1 species in the forest. Turning this around, any of the S–1 species in the forest can replace the ith species at a given location. Unless and until a species approaches monodominance, this constraint on the population growth of the ith species is very weak in a species-rich forest such as BCI. It is weak even in a forest consisting of only a few dozen species, such as a typical mid-latitude temperate forest. Janzen-Connell effects are also prevalent in relatively species-poor temperate forests, so one must also conclude that these effects are not responsible for the latitudinal gradient in tree species richness either (HilleRisLambers et al. 2000). These findings mean that Janzen-Connell effects are not the "cause" of tree species richness in tropical forests. What these effects do is mix species more thoroughly in a small area and maintain whatever species are present, but they do not dictate how many species participate overall in this mixing.

The relevance of these findings regarding the application of neutral theory to plant communities—and also probably to many animal communities—is that density dependence is a very local-scale phenomenon that becomes an unimportant force in population dynamics at larger spatial scales. Zillio et al. (2005) showed that patterns of beta diversity in tropical forests on local to biogeographic spatial scales are consistent with a loss of density dependence on scales of a few tens of meters. Patterns of relative species abundance in the BCI plot are also consistent with a loss of density dependence at densities above a few tens of trees (Volkov et al. 2005). These conclusions on density dependence have profound implications for ecology, biogeography, and conservation biology, namely. that our familiar notions of population regulation do not apply in macroecology on landscape spatial scales, scales on which population growth becomes very close to, and indistinguishable from, density independence (i.e., neutrality).

Testing the Theory on the Dynamical Data

There are currently two mechanistic versions of neutral theory. The original version (Hubbell 2001, Volkov et al. 2003, Vallade and Houchmandzadeh 2003, McKane et al. 2004, Etienne 2005) embodies the mechanism in the theory of island biogeography, namely, dispersal limitation. According to this mechanism, relative species abundances are dictated by the steady state between the arrival of immigrants to a particular community and their local extinction. The loss of all diversity is prevented by adding a slow trickle of new species into the source area or metacommunity, from which the immigrants to the local community are drawn. Under this version of neutral theory, rare species are less frequent than species of intermediate abundance in the local community because they are more prone to local extinction and, once they go locally extinct, they take longer to reimmigrate than do common species.

The other version of neutral theory embodies a mechanism of symmetric density and frequency dependence (Volkov et al. 2005). In this version, there are fewer rare species in the community because they have a higher per capita growth rate than do common species. Thus populations of rare species tend to grow in abundance relative to common species and thereby graduate out of the rare abundance categories, depleting the steady-state frequency of rare species in the community. This rare species advantage is captured in the ratio of the average per capita birth rate to the death rate, b/d (Volkov et al. 2005). At low population sizes, the birth rate exceeds the death rate ($b/d>1$), but at higher population rates, b/d is very close to, but slightly less than, unity. In the theory there is a parameter c which determines the strength of the density dependence. The larger the value of c, the higher the threshold abundance of species that enjoy a growth rate advantage (Volkov et al. 2005).

Dispersal limitation and density dependence are independent mechanisms, and both can operate simultaneously to varying degrees. Remarkably, both mechanisms under neutrality fit the static data on relative tree species abundance in the BCI plot equally well, and data from other 50 ha plots as well (Volkov et al. 2005) (figure 10.10). Although we cannot distinguish the quality of their fits to the static relative abundance data, we can do so in the fit to the dynamic data from the BCI plot. One of the surprising findings over the past quarter century is just how dynamic the BCI forest is (Hubbell 2008). More than half (55.8%, 179) of BCI species have changed by more than 25% in total abundance since 1982, and 36 species (11.2%) have changed by more than 100%. Large changes were not restricted to just uncommon or rare species, but also occurred in common to very common species (Hubbell 2008). The dynamism of

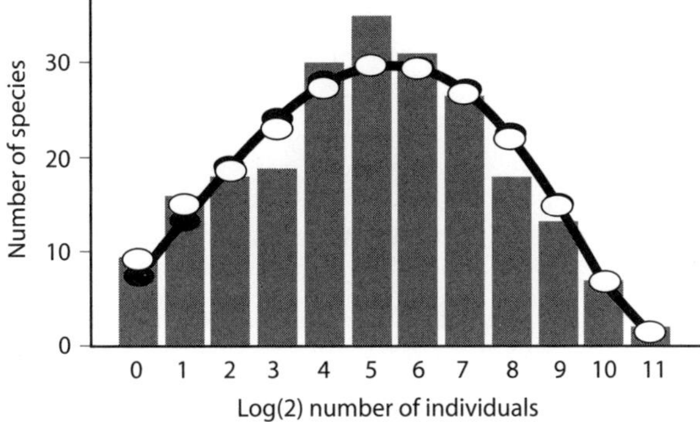

Figure 10.10. Fits of the two versions of neutral theory to the static BCI relative species abundance data. Observed relative abundance data are given by the bar histogram. Species are binned into doubling classes of abundance. The light gray line and ovals is the fit of the dispersal limitation version of the theory, which is the original version in Hubbell (2001) and the generalization of island biogeography theory. The unconnected dark gray ovals are the fit of the symmetric density dependence version of the theory. The quality of the fits is equally good and cannot be distinguished from the static data alone. After Volkov et al. (2005).

the BCI tree community gives us considerable power to test the two versions of neutral theory.

We can compare the predictions of a neutral model community in which species are stabilized by stochastic density dependence versus one in which species drift in abundance solely under the influence of immigration and extinction and demographic stochasticity. We compare the two model predictions for what should happen to the decay in community similarity over time. There are a number of possible ways to measure community similarity, but a simple way is to regress the logarithm of species abundance at time $t+\tau$ on the logarithm of the abundance of the same species at time t, where τ is the time lag separating the abundance snapshots of the tree community. We then can use the R^2 of this regression as a measure of community similarity, i.e., the proportion of variance in log abundance of species at time $t+\tau$ explained by the log abundance of the same species at time t (we add one individual to the abundances before log transforming them so we can include species that are not present at a particular census). Under both versions of neutral theory, the R^2 decays over time, reaching an asymptotic low R^2 value after some time period. Under the stochastic density dependence model, this asymptote is reached quite quickly, and theory predicts the R^2 decay curve to be obviously curvilinear and asymptoting even on short time scales such as a

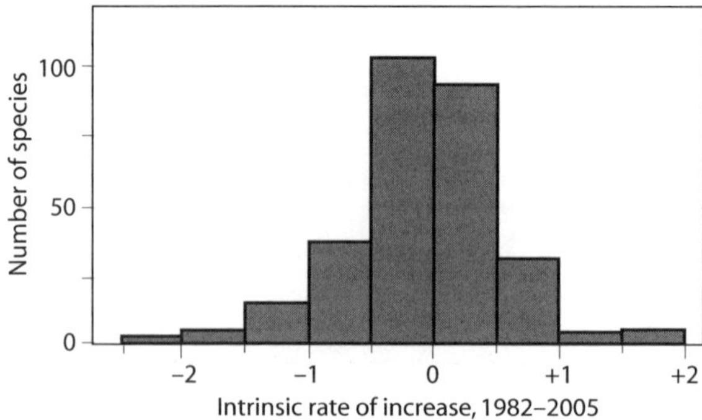

Figure 10.11. Observed near-normal distribution of the intrinsic rates of increase of BCI tree species, centered on $r=0$, over the 23 year time interval, 1982–2005.

quarter of a century. However, under the immigration-extinction model, the original theory of island biogeography, the R^2 decay curve is expected to take much longer to reach its asymptote, on the order of 3,000 years (Azaele et al. 2006), and the curve is predicted to decay essentially as a straight line for periods as short as 25 years (Hubbell 2008). Which curve do we observe?

We can compute the expected curve under density dependence by assuming that populations are fluctuating stochastically around fixed carrying capacities. The intrinsic rates of increase of BCI tree species over the past 25 years are nearly normally distributed around zero (figure 10.11). We can sample this distribution to produce expected changes in abundance of BCI tree species and project changes in their abundances from 1980 to 2005 in five-year intervals, matching the census intervals. I did this in an ensemble of 100 runs and calculated the mean decay curve in R^2 that resulted. To compute the expected decay curve under immigration-extinction, I simulated the changes expected in species abundances assuming the average per capita death rate observed in the BCI plot, and the fundamental biodiversity number θ and the dispersal parameter m of neutral theory, estimated from the static relative abundance data from the first census of the plot (Hubbell 2001, Volkov et al. 2003). The R^2 obtained for each lag interval was averaged with all lags of similar length, e.g., all five-year lags between censuses, all ten-year lags, and so on. I then compared the fit of the two model decay curves to the actual decay curve observed in the BCI tree community.

The conclusion from fitting the two versions of neutral theory is clear-cut and unambiguous: the immigration-extinction version fits the observed dynamic data on decay in community similarity with time, and the density-dependence version does not (figure 10.12). The observed

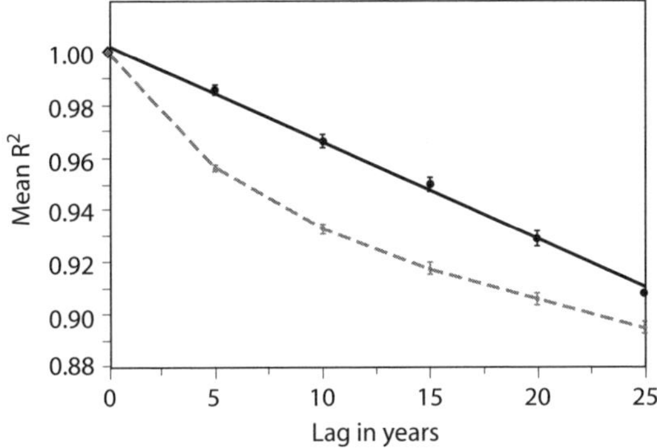

Figure 10.12. Predicted curves for the decay of community similarity under the dispersal limitation version of neutral theory (straight solid line), and under a model of symmetric density dependence, in which species are assumed to be stochastically fluctuating around fixed carrying capacities (curved dashed line). The two curves represent the expected decay in community similarity as measured by the decline in R^2 over time of the autoregression of log species abundances at time $t+\tau$ on the log of the abundances of the same species at prior time t for all possible combinations of 5-year inter-census time lags. The observed decay in R^2 is almost perfectly linear (top solid straight line) (coefficient of determination is 0.997); the error bars are 1 standard error of the mean across all inter-census time lags. The curve for density-dependence (bottom curved line) is the mean of an ensemble of 1000 runs. The error bars are one standard error of the mean. The line fit through the linear decay data is not a regression but is the prediction of the dispersal limitation (island biogeography) version of neutral theory. The values of the fundamental biodiversity number θ and dispersal parameter m were 40 and 0.09, respectively, and were obtained independently from fitting the static relative abundance data from the first census in 1982.

decay curve is nearly perfectly linear, not curvilinear, with an R^2 of 0.997. The fit of the immigration-extinction model is impressive, especially considering that the fit is not a regression, but the fitted line was derived completely independently by estimating the values of θ and m from the static relative abundance data of the first BCI census—completely independently from the dynamic data of changes in the BCI tree community over the subsequent quarter century.

Conclusions

These results do not "prove" that the BCI tree community is dynamically neutral. Indeed, we have presented evidence that the life histories of BCI tree species are not all ecologically equivalent. Moreover, when species names are attached to BCI trees, there are emerging signs of directional, non-neutral change in species composition of the BCI tree community (Feeley et al., unpublished). Species of higher wood density and slower growth rates are slowly and steadily increasing in abundance, possibly as a result of climate change, but the cause is not completely proven yet. Neutral theory assumes constant environments, and if environments change, then the competitive balance among species that had neutral or near-neutral dynamics under the old environmental regime may expose species differences that previously went unrecognized as important to determining which species persist and which ones do not under changing environments. Nevertheless, despite the slow, directional changes in the BCI forest, neutral theory still does a very good job of fitting the static and dynamic data on relative species abundance in the BCI plot. The precision of the fits of neutral theory to both the static and dynamic data must mean that neutral theory—as a first-moment approximation to be sure—captures much of the true behavior of the BCI tree community. Arguments that the theory of island biogeography and its neutral theory extensions, are "cartoonish" (Laurance 2008, this volume) are a mischaracterization of the theory's continuing utility. For an application of neutral theory to a question in conservation biology, namely, how many tree species there are in the Amazon, and how many of them are likely to go extinct, see Hubbell et al. (2008).

Indeed, I would argue that neutral theory provides a solid theoretical foundation on which to build a new non-neutral, niche-based theory of ecology from the perspective of statistical mechanics (Hubbell 1995, 1997, 2001, Bell 2001, Volkov et al. 2003, 2005, 2007, Vallade and Houchmandzadeh 2003, Alonzo and McKane 2004, McKane et al. 2004, Etienne 2005, He 2005, Azaele et al. 2006, Volkov et al., in press). These developments will add "higher-moment" processes as needed to achieve new levels of realism and precision. However, the guiding principle in theory development should always be to start simple and add complexity slowly, step by step, but only when absolutely necessary, kicking and screaming the whole time.

Acknowledgments

I thank Jonathan Losos, Bob Ricklefs, and Patty Gowaty for valuable comments on the first draft of this paper. The BCI forest dynamics research project was made possible by National Science Foundation grants to Stephen P. Hubbell: DEB-0640386, DEB-0425651, DEB-0346488, DEB-0129874, DEB-00753102, DEB-9909347, DEB-9615226, DEB-9615226, DEB-9405933, DEB-9221033, DEB-9100058, DEB-8906869,

DEB-8605042, DEB-8206992, and DEB-7922197, support from the Center for Tropical Forest Science, the Smithsonian Tropical Research Institute (STRI), the John D. and Catherine T. MacArthur Foundation, the Andrew Mellon Foundation, the Celera Foundation, and numerous private individuals, and through the hard work of over 100 students, postdocs, and assistants from ten countries over the past quarter century. The nutrient mapping of the BCI plot was made possible by an NSF grant to Jim Dalling and Kyle Harms, and the analysis of the soil chemistry was done by Joe Yavitt. The plot project is part the Center for Tropical Forest Science (CTFS), a pantropical network of large-scale forest dynamics plots modeled after the BCI project. I am especially grateful to Robin Foster, who began the project with me in 1980 and who met the botanical identification needs of the census through the early years, to Salomon Aguilar, Rick Condit, Jim Dalling, Kyle Harms, Suzanne Loo de Lau, Rolando Perez, and Joe Wright, for their long-term collaboration on the BCI project, and to Ira Rubinoff, Director of STRI, for his constant support of and belief in the project. I thank Liz Losos and Stuart Davies for their management of CTFS.

Literature Cited

Ahumada, J. A., S. P. Hubbell, R. Condit, and R. B. Foster. 2004. Long-term tree survival in a neotropical forest: The influence of local biotic neighborhood. In *Forest Diversity and Dynamism: Findings from a Network of Large-Scale Tropical Forest Plots,* ed. E. Losos and E. G. Leigh Jr., 408–32. Chicago: University of Chicago Press.

Alonzo, D., and A. J. McKane. 2004. Sampling Hubbell's neutral theory of biodiversity. *Ecology Letters* 7:901–10.

Azaele, S., S. Pigolotti, J. R. Banavar, and A. Maritan. 2006. Dynamical evolution of ecosystems. *Nature* 444: 926–28.

Bell, G. 2001. Neutral macroecology. *Science* 293:2413–16.

Chase, J. M., and M. A. Leibold. 2003. *Ecological Niches: Linking Classical and Contemporary Approaches.* Chicago: University of Chicago Press.

Comita, L., and S. P. Hubbell. 2009. Local neighborhood and species' shade tolerance influence survival in a diverse seedling bank. *Ecology* 90:328–34.

Connell, J. H. 1971. On the role of natural enemies in preventing competitive exclusion in some marine animals and in rain forest trees. In *Dynamics of Populations,* ed. P. J. den Boer and G. R. Gradwell, 298–312. Wageningen, The Netherlands: Centre for Agricultural Publishing and Documentation.

Dalling, J. W., H. C. Muller-Landau, S. J. Wright, and S. P. Hubbell. 2002. Role of dispersal in recruitment limitation in six pioneer species. *Journal of Ecology* 90: 714–27.

Donnelas, M., S. R. Connolly, and T. P. Hughes. 2006. Coral reef diversity refutes the neutral theory of biodiversity. *Nature* 440: 80–82.

Engelbrecht, B.M.J., L. Comita, R. Condit, T. A. Kursar, M. T. Tyree, and S. P. Hubbell 2007. Drought sensitivity shapes species distribution patterns in tropical forests. *Nature* 447:80–82.

Epstein, E., and A. J. Bloom. 2005. *Mineral Nutrition of Plants*. Sunderland, MA: Sinauer Associates.

Etienne, R. 2005. A new sampling formula for neutral biodiversity. *Ecology Letters* 8: 493–504.

Feeley, K. J., S. J. Davies, R. Peres, R. Condit, R. B. Foster, and S. P. Hubbell. Unpublished. Directional changes in the composition of a tropical forest due to climate change. *Proceedings of the National Academy of Sciences U.S.A.* (submitted).

Harms, K. E., J. S. Wright, O. Calderón, A. Hernández, and E. A. Herre. 2000. Pervasive density-dependent recruitment enhances seedling diversity in tropical forests. *Nature* 404: 493—95.

He, F. L., 2005. Deriving a neutral model of species abundance from fundamental mechanisms of population dynamics. *Functional Ecology* 19: 187–93.

HilleRisLambers, J., J. S. Clark, and B. Beckage. 2000. Density-dependent mortality and the latitudinal gradient in species diversity. *Nature* 417:732–35.

Hubbell, S. P. 1979. Tree dispersion, abundance and diversity in a tropical dry forest. *Science* 203: 1299–309.

———. 1995. Towards a theory of biodiversity and biogeography on continuous landscapes. In *Preparing for Global Change: A Midwestern Perspective,* ed. G. R. Carmichael, G. E. Folk, and J. L. Schnoor, 173–201. Amsterdam: Academic.

———. 1997. A unified theory of biogeography and relative species abundance and its application to tropical rain forests and coral reefs. *Coral Reefs* 16 (suppl.):S9–S21.

———. 2001. *The Unified Neutral Theory of Biodiversity and Biogeography*. Princeton Monographs in Population Biology, Princeton, NJ: Princeton University Press.

———. 2005. Neutral theory in ecology and the hypothesis of functional equivalence. *Functional Ecology* 19: 166–77.

———. 2006. Neutral theory in ecology and the evolution of ecological equivalence. *Ecology* 87: 1387–98.

———. 2008. Approaching tropical forest complexity, and ecological complexity in general, from the perspective of symmetric neutral theory. In *Tropical Forest Community Ecology*, ed. W. Carson and S. Schnitzer, 143–59. Oxford: Blackwell.

Hubbell, S. P., J. A. Ahumada, R. Condit, and R. B. Foster. 2001. Local neighbourhood effects on long-term survival of individual trees in a neotropical forest. *Ecological Research* 16: 859–75.

Hubbell, S. P., R. B. Foster, S. O'Brien, B. Wechsler, R. Condit, K. Harms, S. J. Wright, and S. Loo de Lau. 1999. Light gaps, recruitment limitation and tree diversity in a Neotropical forest. *Science* 283: 554–57.

Hubbell, S. P., K. E. Harms, J. W. Dalling, R. John, and J. B. Yavitt. Unpublished. Nutrient-based niches in a neotropical tree community: A test of nutrient-ratio (R^*) competition theory. *Science* (submitted).

Hubbell, S. P., F-L. He, R. Condit, L. Borda-de-Agua, J. Kellner, and H. ter Steege. 2008. How many tree species are there in the Amazon, and how many of then will go extinct? *Proceedings of the National Academy of Sciences U.S.A.* 105 (Suppl. 1): 11498–504.

Hurtt, G. C., and S. W. Pacala. 1995. The consequences of recruitment limitation: Reconciling chance, history, and competitive differences between plants. *Journal of Theoretical Biology* 176:1–12.

Janzen, D. 1970. Herbivores and the number of tree species in tropical forests. *American Naturalist* 104:501–28.

John, R., J. Dalling, K. E. Harms, J. B. Yavit, R. F. Stallard, M. Mirabello, S. P. Hubbell, R. Valencia, H. Navarrete, M. Vallejo, and R. B. Foster. 2007. Soil nutrients influence spatial distributions of tropical tree species. *Proceedings of the National Academy of Sciences U.S.A.* 104:864–69.

Jones, F. A., J. Chen, G.-J .Weng, and S. P. Hubbell. 2005. A genetic evaluation of long distance and directed dispersal in the Neotropical tree, *Jacaranda copaia* (Bignoniaceae). *American Naturalist* 166:543–55.

Laurance, W. H. 2008. Theory meets reality: How habitat fragmentation research has transcended island biogeographic theory. *Biological Conservation* 141:1731–44.

MacArthur, R. H., and E. O. Wilson. 1967. *The Theory of Island Biogeography*. Princeton, NJ: Princeton University Press.

McKane, A. J., D. Alonso, and R. V. Solé. 2004. Analytical solution of Hubbell's model of local community dynamics. *Theoretical Population Biology* 65:67–73.

Muller-Landau, H. C., S. J. Wright, O. Calderon, S. P. Hubbell, and R. B. Foster. 2002. Assessing recruitment limitation: Concepts, methods, and case-studies from a tropical forest. In *Seed Dispersal and Frugivory: Ecology, Evolution and Conservation,* ed. D. J. Levey, W. R. Silva, and M. Galetti, 35–53. New York: CAB International.

Silvertown, J. M., E. Dodd, J.J.G. Gowing, and J. O. Mountford. 1999. Hydrologically defined niches reveal a basis for species richness in plant communities. *Nature* 400:61–63.

Tilman, D. 1982. *Resource Competition and Community Structure*. Princeton Monographs in Population Biology. Princeton, NJ: Princeton University Press.

———. 1988. *Plant Strategies and the Dynamics and Structure of Plant Communities*. Princeton, NJ: Princeton University Press.

———. 1994. Competition and biodiversity in spatially structured habitats. *Ecology* 75: 2–16.

Vallade, M., and B. Houchmandzadeh. 2003. Analytical solution of a neutral model of biodiversity. *Physical Review E* 68:061902.

Volkov, I., J. R. Banavar, F.-L. He, S. P. Hubbell, and A. Maritan. 2005. Density dependence explains tree species abundance and diversity in tropical forests. *Nature* 438: 658–61.

Volkov I., J. R. Banavar, S. P. Hubbell, and A. Maritan 2003. Neutral theory and the relative abundance of species in ecology. *Nature* 424:1035–37.

———. 2007. Patterns of relative species abundance in rain forests and coral reefs. *Nature* 450: 45–49.

Volkov I., J. R. Banavar, S. P. Hubbell, and A. Maritan. 2009. Infering species interactions in tropical forests. *Procedings of the National Academy of Sciences* (in press).

Wills, C., R. Condit, R. Foster, and S. P. Hubbell, 1997. Strong density- and diversity-related effects help to maintain tree species diversity in a Neotropical forest. *Proceedings of the National Academy of Sciences U.S.A.* 94: 1252–57.

Wills, C., K. E. Harms, R. Condit, D. King, J. Thompson, F. L. He, H. Muller-Landau, P. Ashton, E. Losos, L. Comita, S. Hubbell, J. LaFrankie, S. Bunyavejchewin, H. S. Dattaraja, S. Davies, S. Esufali, R. Foster, R. John, S. Kiratiprayoon, S. Loo de Lau, M. Massa, C. Nath, Md. Nur Supardi Noor, A. R. Kassim, R. Sukumar, H. S. Suresh, I.-F. Sun, S. Tan, T. Yamakura, and J. Zimmerman. 2006. Non-random processes maintain diversity in tropical forests. *Science* 311: 527–31.

Wright, S. J., H. Muller-Landau, O. Calderón, and A. Hernandez. 2002. Annual and spatial variation in seedfall and seedling recruitment in a neotropical forest. *Ecology* 86: 848–50.

Zillio, T., I. Volkov, S. P. Hubbell, J. R. Banavar, and A. Maritan. 2005. Spatial scaling relationships in ecology. *Physical Review Letters* 95:098101.

Evolutionary Changes Following Island Colonization in Birds

EMPIRICAL INSIGHTS INTO THE ROLES OF MICROEVOLUTIONARY PROCESSES

Sonya Clegg

DIVERGENCE FOLLOWING ISLAND COLONIZATION stems from the action of microevolutionary processes, including drift, selection, gene flow, and mutation (Mayr 1954, Lande 1980, Barton 1998, Grant 1998). The suggestion that all of these processes can play a role in divergence, potentially acting separately or in concert, is uncontroversial. However the relative importance of each in natural systems is not generally agreed (Provine 1989, Barton 1998, Price 2008). Islands are regularly referred to as natural laboratories, and as such, studies of island forms have made major contributions to the development of general evolutionary theory (Grant 1998). Although the microevolutionary processes mentioned above are not unique to islands, the way that particular processes operate in insular versus continental situations may be fundamentally different due to consistent biotic and abiotic differences between the two geographic circumstances (MacArthur and Wilson 1967). Given an accumulating number of empirical studies, we can assess if particular microevolutionary processes are of more general importance than others in generating the diversity of island forms.

In their landmark book formalizing island biogeography as a field in its own right, MacArthur and Wilson (1967) devoted a chapter to evolutionary changes following colonization. This chapter is rich with ideas about how microevolution could proceed on islands, with reasoning largely based on the limited empirical data available at the time. Since then, empirical evidence for the importance of various microevolutionary processes has appreciated considerably, allowing a reassessment of MacArthur and Wilson's views. Here I discuss a number of mechanisms by which drift and selection can influence divergence of island-colonizing birds. I examine three concepts: (1) whether founder-mediated drift is more effective than long-term gradual drift in shaping levels of diversity and divergence as evidenced by neutral genetic markers, (2) whether morphological divergence is consistent with drift or selective mechanisms, and (3) how frequent shifts in competitive regimes on islands could affect common patterns of morphological divergence associated with insularity in passerine birds.

Founder Events and Gradual Drift

The establishment of a new population involves phases of founding and recovery leading to longer-term persistence. During each stage, the random sampling effect of drift has the potential to affect the degree of diversity and divergence exhibited by a population. The effects of drift are more pronounced when effective founding population sizes are smaller, recovery times are longer and long-term effective population sizes are limited (Wright 1931, Nei et al. 1975). Drift is particularly relevant to island populations as it has the potential to prevail over selective mechanisms due to the vulnerability of small isolated populations to stochastic events. The potential significance of founder-mediated drift was emphasized by Mayr (1942, 1954). MacArthur and Wilson (1967) considered how founder events could potentially impact the evolution of a newly established population, but in the absence of empirical data concluded that "the evolutionary effects of initially small population size can only be guessed at this time" (p. 154). However, their general skepticism of the relative importance of founder events is illustrated in the passage: "evolution due to genetic sampling error is an omnipresent possibility but one easily reduced to relative insignificance by small increases in propagule size, immigration rate or selection pressure" (p. 156). Despite this relatively unenthusiastic view, founder-effect ideas have had a prevailing influence on the development of divergence and speciation models on islands (reviews in Provine 1989, Grant 2001).

In the literature, the term "founder effect" has been applied very broadly, encompassing any change associated with population founding. These include changes in diversity measures or allele frequencies (e.g., Reiland et al. 2002, Abdelkrim et al. 2005, Hawley et al. 2006), the particular phenotypic attributes of the founders themselves (e.g., Grant and Grant 1995a, Berry 1998, Kliber and Eckert 2005, Baker et al. 2006) and more complex founder-induced speciation models that invoke a role of founder events in reorganizing quantitative genetic variation and catalyzing speciation (Mayr 1954, Carson and Templeton 1984). Debate has ensued over the theoretical grounding (Barton and Charlesworth 1984, Carson and Templeton 1984, Slatkin 1996) and empirical likelihood (Rice and Hostert 1993, Templeton 1996, Coyne and Orr 2004, Walsh et al. 2005, Templeton 2008) of specific founder-induced speciation models. However, when considering natural situations, it may be unfeasible to determine if all requirements of different founder-induced speciation models were met at the time of divergence (Barton and Charlesworth 1984). Many studies of founder effects have instead focused on the effects on neutral genetic variation as a tangible indicator of the strength of drift associated with founding. Two measures of diversity are usually considered, allelic diversity and heterozygosity, with the former being more sensitive to sampling effects due to the loss of rare alleles (Nei et al. 1975). Therefore, milder founder events are indicated by decreases in allelic diversity but not heterozyosity. Immediate and large-scale loss of both measures of diversity along with the appearance of instantaneous levels of differentiation would indicate a stronger perturbing effect of a founding event. While these measures do not address loci under selection, neutral marker heterozygosity can reflect fitness (Coltman and Slate 2003). The mechanisms of such a relationship are debated (Balloux et al. 2004), however in bottlenecked populations the association between neutral and selected loci may be largely due to increased linkage disequilibrium resulting in hitch-hiking effects for neutral loci (Hansson et al. 2004).

Studies of rapid population declines in a range of species have demonstrated that loss of diversity can be severe when declines are sizable and persist for an extended time (e.g., Pastor et al. 2004, Weber et al. 2004, Roques and Negro 2005, but see Hailer et al. 2006). Similar effects might be expected of colonizing populations that go through a bottleneck during founding. However, there are key differences between a colonization event and a population crash. In species that successfully colonize and establish a population in a new location, there may be greater opportunity for rapid recovery following founding and the possibility for continued immigration from the original source or multiple sources, limiting the genetic effects of a bottleneck. The establishment of a new population is therefore not necessarily accompanied by a strong genetic founder effect, a conclusion reached in studies that report similar levels of diversity in long separated mainland and island-dwelling taxa (Seutin et al. 1993, Illera et al. 2007). However, island populations generally do have lower genetic diversity than those on mainlands (Frankham 1997), a feature variously attributed to combinations of founder events (Pruett and Winker 2005), repeated population bottlenecks following establishment (Bollmer et al. 2007), and gradual drift in small populations over extended time periods (Mundy et al. 1997, Bollmer et al. 2005, 2007, Ohnishi et al. 2007). In populations that represent an ancient colonization, distinguishing between the genetic effects of a pulse of drift associated with a founder event and long-term persistent drift over time is difficult because both mechanisms can result in decreased diversity and increased differentiation. Situations where colonization dates are recent and recorded, such as historically documented natural colonization events or artificial introductions, are therefore required to determine if colonization and population establishment results in an immediate and substantial effect on neutral genetic diversity.

Empirical Examples of Founder Events

Population size changes can result in varying genetic signatures depending on the type of genetic markers utilized, and ideally information from multiple types of markers would be considered when assessing the genetic impacts of population founding (Hawley et al. 2008). However, in the absence of a full suite of genetic markers, microsatellites are a suitably sensitive marker for assessing variation associated with founder events and population bottlenecks (Hawley et al. 2008), and have frequently been applied to founder event scenarios (table 11.1). I first discuss microsatellite studies of rare natural situations where information on the timing and sequence of single and multiple colonization events is available for colonizing bird species. Further examples of artificially introduced bird populations are reviewed to assess current empirical evidence of founding events as a perturbing force in island-colonizing birds.

The historically documented sequential colonization by the Tasmanian silvereye (*Zosterops lateralis*) to New Zealand and outlying islands over the last 180 years is a classic of ornithological literature (Mayr 1942, Lack 1971; see figure 11.1). In addition to recently colonized populations, successively older populations are represented by *Z. l. chlorocephalus* on Heron Island which is at most 4,000 years old (based on the length of time the island has been vegetated and mitochondrial DNA divergence [Hopley 1982, Degnan and Moritz 1992]) and extant endemics on Norfolk Island (*Z. tenuirostris*) and Lord Howe Island (*Z. tephropleuris*). The latter two populations are in the order of millions and hundreds of thousands of years old, respectively, based on mitochondrial DNA divergence estimates (Phillimore 2006). The combination of documented colonizations and evolutionarily older populations provided an opportunity to contrast the role of founder events versus long-term gradual drift in shaping neutral genetic diversity (Clegg et al. 2002a).

The quantification of neutral genetic diversity and divergence using microsatellites in *Zosterops* populations revealed that single founder events did not result in significant reductions in genetic diversity as measured by allelic diversity or expected heterozygosity (figures 11.2a and 11.2b). Nor did significant levels of population differentiation arise as a consequence of single founding events (figure 11.2c) (from Clegg et al. 2002a). While no pairwise test showed a significant reduction in diversity,

Table 11.1
Comparisons of Microsatellite Genetic Variability Between Source and Naturally Colonized or Translocated Bird Populations

Species	*Source*	*New Pop.*	*Type (order)*	*No. of loci*	*%AD*	*%He*	F_{ST}	*Ref*
			(a) Natural colonizations					
Large ground finch[a] *Geospiza magnirostris*	Other Galápagos Is.	Daphne Major	S (source to 1st)	16	32 ns	ns	na	1
Silvereye[b] *Zosterops l. lateralis*	Tasmania	South Is	S (source to 1st)	6	0.2 ns	+0.4 ns	0.004 ns	2
	South Is.	Chatham	S (1st to 2nd)	6	17.0 ns	+1.8 ns	0.007 ns	2
	South Is.	P. North	S (1st to 2nd)	6	21.5 ns	4.9 ns	0.003 ns	2
	Palmerston North	Auckland	S (2nd to 3rd)	6	6.8 ns	3.1 ns	0.021 ns	2
	Auckland	Norfolk Is.	S (3rd to 4th)	6	18.6 ns	+2.1 ns	0.093 sig	2
	Tasmania	Chatham	D (source to 2nd)	6	17.2 ns	+2.2 ns	0.003 ns	2
	Tasmania	P. North	D (source to 2nd)	6	21.6 ns	4.5 ns	0.006 ns	2
	South Is.	Auckland	D (1st to 3rd)	6	26.9 ns	7.8 ns	0.027 sig	2
	Palmerston North	Norfolk Is.	D (2nd to 4th)	6	24.2 ns	1.1 ns	0.092 sig	2
	Tasmania	Auckland	T (source to 3rd)	6	27.0 ns	7.5 ns	0.027 sig	2
	South Is.	Norfolk Is.	T (1st to 4th)	6	40.5 ns	5.9 ns	0.079 sig	2
	Tasmania	Norfolk Is.	Q (source to 4th)	6	40.6 ns	5.6 ns	0.088 sig	2
Dark-eyed junco *Junco hyemalis*	Mountain pops.[c]	UC San Diego	S (source to 1st)	5	37.3 sig	12.5 sig	0.06–0.09 sig	3
			(b) Artificial introductions: Island examples					
Laysan finch *Telespiza cantans*	Laysan	Southeast	S (source to 1st)	9	7.1[d] ns	15.0 sig	0.055 sig	4
	Southeast	Grass	S (1st to 2nd)	9	27.0[d] ns	17.8 ns	0.266 sig	4
	Southeast	North	S (1st to 2nd)	9	27.0[d] ns	28.0 ns	0.150 sig	4

(*continued*)

Table 11.1 (continued)

Species	*Source*	*New Pop.*	*Type (order)*	*No. of loci*	*%AD*	*%He*	F_{ST}	*Ref*
	Laysan	Grass	D (source to 2nd)	9	32.2 sig	30.1 sig	0.147 sig	4
	Laysan	North	D (source to 2nd)	9	32.2 sig	38.8 sig	0.166 sig	4
North Is Saddleback[e]	Hen	Red Mercury	S (source to 1st)	6	na	12.3 ns	0.069 sig	5
Philisturnus c. rufaster		Cuvier	S (source to 1st)	6	na	4.2 ns	0.016 sig	5
		Whatupuke	S (source to 1st)	6	na	9.5 ns	0.065 ns	5
	Cuvier	Tiritiri	S (1st to 2nd)	6	na	+3.7 ns	0.018 sig	5
		LittleBarrier	S (1st to 2nd)	6	na	0.31 ns	0.012 ns	5
		Stanley	S (1st to 2nd)	6	na	+0.8 ns	0.048 ns	5
	Whatupuke	Lady Alice	S (1st to 2nd)	6	na	+4.8 ns	0.031 ns	5
		Coppermine	S (1st to 2nd)	6	na	6.7 ns	0.064 sig	5
	Hen	Tiritiri	D (source to 2nd)	6	na	0.6 ns	0.056 sig	5
	Hen	Little Barrier	D (source to 2nd)	6	na	4.4 ns	0.004 ns	5
	Hen	Stanley	D (source to 2nd)	6	na	3.4 ns	0.026 ns	5
	Hen	LadyAlice	D (source to 2nd)	6	na	5.2 ns	0.054 sig	5
	Hen	Coppermine	D (source to 2nd)	6	na	15.5 ns	0.060 sig	5
South Is Saddleback	Big South Cape	Big	S (source to 1st)	6	13.2 ns	8.0 ns	0.032 sig	6
P. c. carunculatus	Big South Cape	Kaimohu	S (source to 1st)	6	19.5 ns	26.6 ns	0.132 sig	6
	Big	Putauhinu	S (1st to 2nd)	6	5.3 ns	+4.8 ns	0.092 sig	6
	Big	Ulva	S (1st to 2nd)	6	1.9 ns	+3.0 ns	0.003 ns	6
	Big South Cape	Putauhinu	D (source to 2nd)	6	17.8 ns	3.5 ns	0.029 sig	6
	Big South Cape	Ulva	D (source to 2nd)	6	14.9 ns	5.2 ns	0.025 sig	6
	Big	Breaksea	D (1st to 3rd)	6	0.8 ns	3.0 ns	0.019 sig	6
	Kaimohu	Motuara	D (1st to 3rd)	5/6	+13.9 ns	+12.9 ns	0.205 sig	6
	Big South Cape	Motuara	T (source to 3rd)	6	8.3 ns	17.1 ns	0.110 sig	6
	Big South Cape	Breaksea	T (source to 3rd)	6	13.9 ns	10.7 ns	0.006 ns	6

Ruddy duck *Oxyura jamaicensis*	North America	Europe	S (source to 1st)	11	45.51 sig	26.0 sig	0.241–0.325 sig	7
South Is. robin[f] *Petroica a. australis*	Nukuwaiata Is	Motuara Is	S (source to 1st)	10	8.3 ns	+6.2 ns	0.117 sig	8
	Stewart Is[f]	Ulva Is	S (source to 1st)	10	+8.3 ns	+22.9 ns	0.221 sig	8
		(c) Artificial introductions: Continental examples						
Merriam's wild turkey *Meleagris g. merriami*	MSL, Arizona	MNK, Arizona	S (source to 1st)	9	5.8 sig	14.8 sig	na[g]	9
	MSL, Arizona	MMT, Arizona	S (source to 1st)	9	22.2 sig	18.7 sig	na[g]	9
	MSP, Colorado	MLC, Colorado	S (source to 1st)	9	23.7 sig	+1.3 ns	na[g]	9
Wild turkey	Indiana, Missouri, and Iowa	northern Indiana	S (source to 1st)	10	5.2 sig	2.6 ns	na[h]	10
M. g. silvestris	Indiana, Missouri, and Iowa	southern Indiana	S (source to 1st)	10	1.7 ns	1.3 ns	na[h]	10
House finch[i] *Carpodacus mexicanus*	west USA	east USA	S (source to 1st)	10	17.5 sig	4.9 sig	0.016–0.039 sig	11
Griffon vulture (*Gyps fulvus*)	Spain, France, and captive stock[j]	Causses, France	S (source to 1st)	10	+5.4 ns	1.4 ns	na[j]	12

Sources: 1=Grant et al. (2001), 2=Clegg et al. (2002a), 3=Rasner et al. (2004), 4=Tarr et al. (1998), 5=Lambert et al. (2005), 6=Taylor and Jamieson (2008), 7=Muñoz-Fuentes et al. (2006), 8=Boessenkool et al. (2007), 9=Mock et al. (2004), 10=Latch and Rhodes (2005), 11=Hawley et al. (2006), 12=Le Gouar et al. (2008). *Notes*: Type (order) indicates the number of founder events separating populations: single (S), double(D), triple (T) or quadruple (Q), and the order of the comparison (between combinations of original source and sequentially founded populations represented by first order, second order, third order, and fourth order). Percentage reductions in variation: %AD=% decrease in allelic diversity, %He=% decrease in expected heterozygosity (+sign indicates cases of increased variation), F_{ST}=pairwise F_{ST} between source and founded population. sig=significant, ns=nonsignificant, na=not assessed.

(*continued*)

TABLE 11.1 (continued)

[a] No direct comparison with source populations. Multiple source populations indicated by genotype assignments. Diversity assessed in the Daphne Major population over 18 years following founding. Allelic diversity increased and heterozygosity remained relatively constant. Initial reduction in allelic diversity followed by increasing trend. Heterozygosity remained constant.

[b] All pairwise comparisons of diversity nonsignificant after correcting for multiple comparisons, but significant decreasing trend in allelic variation as number of founder steps increased.

[c] Potential multiple source populations.

[d] Allelic diversity not corrected for sample size.

[e] All heterozygosity estimates for North Island saddleback calculated from Lambert et al. 2005, table 4.

[f] Stewart Island population extinct and not sampled. Comparisons made with Breaksea population.

[g] Allele frequency differences reported.

[h] Significant F_{ST} among sampling sites within source and each introduced population area. Multiple source populations.

[i] Multiple populations considered within east and west. Observed heterozygosity reported.

[j] Source values from Ossau, French Pyrenees. F_{ST} between captive founded populations and Ossau were not significant.

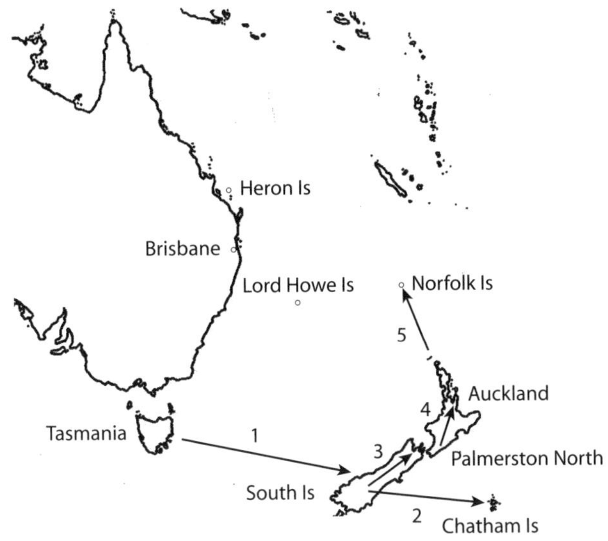

Figure 11.1. Map of the southwest Pacific showing the historically documented colonization of the Tasmanian silvereye, *Zosterops lateralis lateralis*, to New Zealand and outlying islands. Numbered arrows show colonization sequence. Years: 1=1830s, 2 and 3=1856, 4=1865, 5=1904. Other *Zosterops* species and subspecies included in the genetic analysis occur on Norfolk Island, Lord Howe Island, Heron Island, and mainland Australia, represented by Brisbane.

sequential founder events were associated with a significant decreasing trend in allelic diversity, corresponding to a 40% reduction overall. No significant trend in heterozygosity was observed. The level of differentiation was associated with the number of founder events separating any two populations (figure 11.2c). When comparisons were restricted to those occurring in sequence, significant F_{ST} values were recorded in one of five cases where populations were separated by a single founder event, two of four separated by double founder events, two of two separated by triple founder events, and between the two populations separated by four founder events (table 11.1) (Clegg et al. 2002a). Three to four sequential founder events were required for allelic diversity to approach that seen in the older populations (figure 11.2a). In contrast, the decreased levels of heterozygosity seen in the older forms (on Norfolk Island and Lord Howe Island) were not even approached (figure 11.2b), despite the potential for sequential founder events to affect this measure (Motro and Thomson 1982, LeCorre and Kremer 1998). Lower diversity and increased divergence of old populations when compared to the mainland population resulted from loss of alleles along with often dramatic shifts

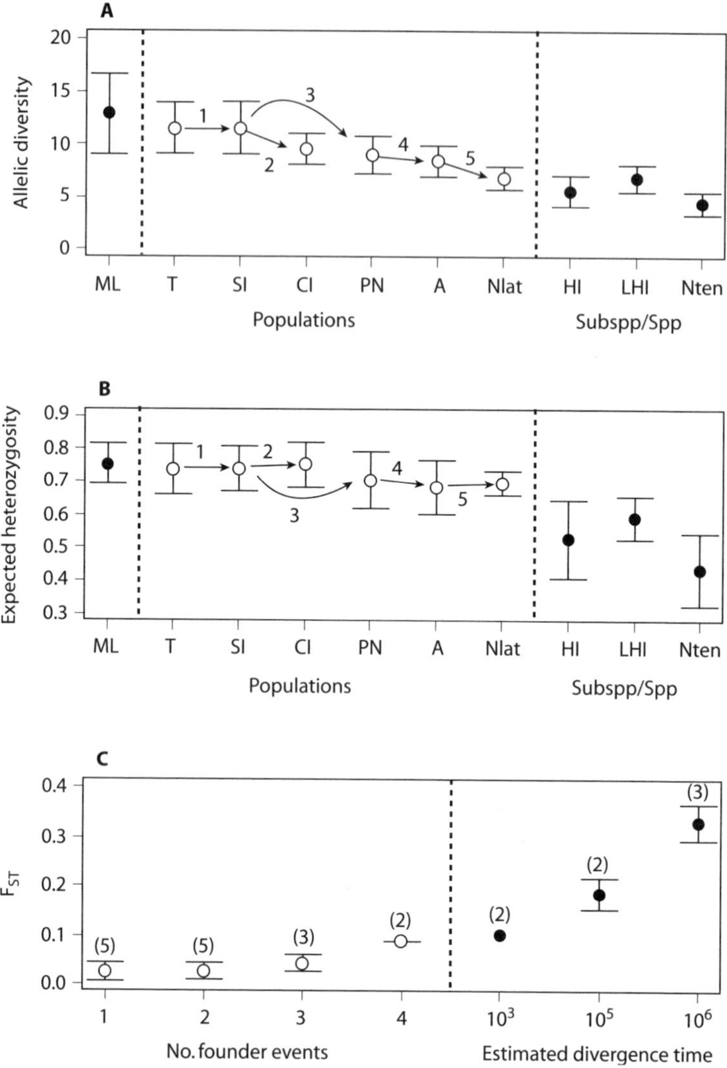

Figure 11.2. Genetic diversity and divergence, (±standard errors), of *Zosterops* forms as measured by (A) allelic diversity, (B) heterozygosity, and (C) pairwise F_{ST}. Number of founder events is the number of island colonizations separating two populations. Numbered arrows refer to colonization sequence in figure 11.1. Numbers in parentheses are the number of pairwise comparisons among populations or among subspecies/species. Locations are ML=mainland (Brisbane, Australia), T=Tasmania, SI=South Island, New Zealand, CI=Chatham Island, PN=Palmerston North, A=Auckland, NIlat=Norfolk Island *Z. lateralis*, HI=Heron Island, LHI=Lord Howe Island, NIten=Norfolk Island *Z. tenuirostris*. Modified from Clegg et al. (2002a).

in frequencies of the remaining alleles, resulting in fewer alleles with higher average frequencies in older populations. In half (9/18) of the locus/old population combinations, one or two alleles, not found in the mainland population, were detected. Some of these may represent replacement by mutation; however, mutation has not been sufficient to make up for allelic losses occurring in small, old populations. The level of diversity in old populations was not strictly related to island age, although the oldest population (Norfolk Island) had the lowest levels of diversity. A number of factors may account for this incongruity, including differences in long-term effective population sizes and the potential for rare immigration events to introduce alleles. The level of divergence between evolutionary old taxa was related to divergence time, with measures from the shortest divergence time being comparable to those recent populations that had experienced the most sequential founding events. Divergence among old forms separated for longer times far exceeded any level of divergence achieved via repeated founder events (figure 11.2c).

In the *Zosterops* system, the ineffectiveness of single founder events to perturb genetic diversity and divergence is repeatedly demonstrated. Differences accrued with sequential founder events, but in general a comparison of recent and old island forms pointed to a stronger influence of gradual drift over time on neutral genetic variation. Bayesian simulations of the founder events indicated that this result was likely due to a combination of substantial effective founder population size followed by rapid increases in population size (Estoup and Clegg 2003). Therefore, in island colonizations of *Zosterops*, founder events are neither long nor strong, and these features may be typical of bird species that colonize islands in small flocks.

Studies of the radiation of Darwin's finches in the Galápagos have generated important insight into the evolution of island forms. Within this dynamic system, the opportunity to study effects of colonization was provided by the large ground finch, *Geospiza magnirostris*, which established a population on Daphne Major in 1982, with founding individuals derived from a number of other Galápagos islands (Grant and Grant 1995a, Grant et al. 2001). Allelic diversity and heterozygosity were tracked across 18 generations following the founding event (Grant et al. 2001) (see table 11.1). Initially, allelic variation decreased by approximately 32%, but this trend was reversed with the continued arrival of breeding immigrants. In contrast, there was no observed initial effect on heterozyosity in the generations immediately following founding, or after input from new immigrants (Grant et al. 2001). This example highlights both the robustness of heterozygosity to population change and the importance of low but continued immigration to island populations.

Similar island-type situations can develop when disjunct populations establish outside of a species range. Hansson et al. (2000) characterized the level of genetic similarity in pairs of great reed warblers, *Acrocephalus arundinaceus*, which founded a new population in southern Sweden in the late 1970s. They found that, over a period of 8 years, the level of genetic similarity between breeding pairs declined, as measured by microsatellite variation and multilocus DNA fingerprinting. While no comparison with a source population could be made, the temporal increase in genetic variation among individuals suggested that continued immigration into the population lessened the impact of the founder event.

In a final natural example, a small disjunct population of the dark-eyed junco, *Junco hyemalis thurberi*, established from an estimated seven effective founders in the 1980s outside of its natural range in California (Rasner et al. 2004). This population had significantly lower allelic richness (37% decrease) and to a lesser degree, lower heterozygosity (12% decrease) compared to populations in the natural range (Rasner et al. 2004) (table 11.1). In contrast to the *Zosterops* and *Geospiza* examples, both types of diversity measures were significantly affected. The decreased diversity was attributed to the small effective size of the population (32 individuals) averaged over the eight generations since founding (Rasner et al. 2004).

There are only a small number of natural colonization events that have been genetically characterized in birds; however, artificially introduced populations are potentially informative about the genetic effects of population founding. Merilä et al. (1996) summarized isozyme studies of introduced bird species, and concluded that there was "little or no evidence for reduced levels of genetic variability in introduced populations." However, the inverse relationship between founder population size and genetic diversity was noted. Additional isozyme, minisatellite, and MHC studies of introduced bird populations have likewise reported maintenance of moderate levels of diversity (Ardern et al. 1997, Cabe 1998, Miller and Lambert 2004, Lambert et al. 2005).

Since Merilä et al.'s (1996) summary, studies of introduced bird populations have mostly used microsatellites as the genetic marker of choice (table 11.1). In general, the patterns seen are similar to natural colonizations, although each case also has its own idiosyncrasies. Allelic diversity was often affected, as seen in the ruddy duck introduction to Great Britain (Muñoz-Fuentes et al. 2006), one of the South Island robin introductions (Boessenkool et al. 2007), the house finch introduction to the eastern United States (Hawley et al. 2006), and one of the two wild turkey introductions in Indiana (Latch and Rhodes 2005). However, examples remain where allelic diversity was maintained (South Island saddleback, Taylor and Jamieson [2008]), or only eroded following multiple founder events (Laysan finch; Tarr et al. [1998]). Where heterozygosity was reduced, the extent was much less than for allelic diversity (e.g., house finch; Hawley et al. [2006]). Even multiple founder events often failed to perturb heterozygosity, as seen for the two saddleback subspecies (Lambert et al. 2005, Taylor and Jamieson 2008). Significant genetic divergence can appear quickly due to allele frequency differences, and the case of the North Island saddleback again demonstrates the amplifying effects of sequential bottlenecks in this regard. Three of eight single translocations resulted in significantly positive F_{ST} values, and a further three of five populations separated by two translocation events had significant and more pronounced F_{ST} values.

A common theme among the avian cases discussed here, whether natural or artificial colonizations, sourced from large outbred populations or small, threatened populations, is that single founder events rarely have a sizable impact on neutral genetic diversity. Loss of rare alleles can result in reduced allelic diversity, and is most evident after sequential founder events. Heterozygosity is not easily perturbed by single or multiple founder events. Shifts in allele frequency differences often result in significant divergence as measured by F_{ST}, but it is likely to be only a small fraction of what can accrue more gradually over time.

Multiple mechanisms could account for the generally mild effects on genetic variation noted in avian studies. One consideration is that species translocations often occur for conservation reasons, as exemplified by all but two of the artificial introduction examples (ruddy duck and house finch) in table 11.1. Such species are likely to have experienced reduced population size for some period of time to warrant conservation efforts. Therefore translocated populations, necessarily sourced from already depauperate populations, may not be expected to experience further significant losses of diversity (Taylor and Jamieson 2008). These situations may therefore provide more limited inference for understanding divergence of populations arising from natural colonization events.

In other cases, biological attributes of a species may buffer founded populations from loss of genetic diversity. In two of the documented natural colonizations mentioned above, continued immigration was identified as an important factor resulting in increased population variation (Grant et al. 2001, Hansson et al. 2000). Other studies of established populations note the positive effects of even limited gene flow in bolstering diversity in small populations (Keller et al. 2001, Ortego et al. 2007, Baker et al. 2008). The relatively high vagility of colonizing bird species may therefore limit genetic founder effects. In cases where continued immigration is less likely due to isolation, ample founder sizes may minimize founder effects, as suggested for the recent *Zosterops* colonizations (Estoup and Clegg 2003). Rapid recovery from small population size is theoretically one of the most important mechanisms to minimize loss of variation (Nei et al. 1975), and empirical results attest to its importance (Estoup and Clegg 2003, Miller and Lambert 2004, Brown et al. 2007). A comparison of MHC variation in two robin species in New Zealand, the Chatham Island black robin (*Petroica traversi*) and the South Island robin (*Petroica australis australis*), which both experienced population bottlenecks, found that the former species was monomorphic at MHC loci, whereas the latter species maintained moderate levels of MHC variation (Miller and Lambert 2004). This difference was attributed to the different types of bottlenecks experienced by the two species. The bottleneck in the Chatham Island black robin extended over 100 years of low population size before human-assisted recovery, whereas bottlenecks induced by translocation of South Island robins were short as the populations recovered quickly (Miller and Lambert 2004).

Mild neutral genetic effects of population founding have been reported in other fauna, including numerous mammals (e.g., rabbit [Zenger et al. 2003], brushtail possum [Taylor et al. 2004], ship rat [Abdelkrim et al. 2005], Rodrigues fruit bat [O'Brien et al. 2007], mouflon sheep [Kaeuffer et al. 2007], and Corsican red deer [Hajji et al. 2008]), and amphibians (natterjack toad [Rowe et al. 1998] and marsh frog [Zeisset and Beebee 2003]). As with the bird examples, allelic diversity in these studies was often impacted and heterozygosity less so. The minimal effects of founding were attributed to combinations of substantial numbers of founders, multiple introductions, and rapid recovery times (e.g., Rowe et al. 1998, Zeisset and Beebee 2003, Zenger et al. 2003, Taylor et al. 2004), introduction from an already depauperate source (Hajji et al. 2008), or selection at linked loci (Kaeuffer et al. 2007). In other studies, significant reductions in heterozygosity have indicated a relatively stronger impact of the founding event (e.g., Bennett's wallabies [Le Page et al. 2000] and Caribbean anoles [Eales et al. 2008]).

In species that are less vagile, tend to colonize in very small numbers, or are less capable of rapid recovery from small population sizes, narrower and longer bottlenecks can amplify the loss of genetic variation and result in severe founder effects. Colonization by a single gravid female represents an extreme case and is a situation that could feasibly occur. Indeed, examples of more sizable neutral genetic impacts of founding have been reported for animals and plants. Severe founder events in introduced *Drosophila pseudoobscura* population in New Zealand (Reiland et al. 2002), and in an aquatic plant (*Butomus umbellatus*) invasion to North America (Kliber and Eckert 2005) were explained in part by small numbers of successful founders. Reductions in mitochondrial DNA diversity in introduced bluegill sunfish (*Lepomis macrochirus*) populations in North America were likewise attributed to a small number of founders in combination with subsequent stochastic processes (Yonekura et al. 2007). The impact of serial founder events may be greater for some species, such as that reported for dice snakes (*Natrix tessellata*) in Europe (Gautschi et al. 2002).

MacArthur and Wilson did not consider founder effect mechanisms to be of crucial importance in driving divergence of insular forms in general. In birds, empirical assessments of variation at neutral genetic loci in colonized and translocated populations support the conjecture that losses of genetic diversity do not occur on a scale that would precipitate a "genetic revolution." While inferences from neutral genetic markers do not address loci under selection they can nevertheless be indicative of genome wide perturbations caused by founder effects. Bird colonizations and introductions are generally robust to losses in heterozygosity, suggesting that overall fitness is not compromised by founder events, at least when sourced from outbred populations. With respect to management of endangered, translocated populations, general increases in inbreeding can have important conservation implications (e.g., Jamieson et al. 2006, Hale and Briskie 2007). Losses in allelic diversity are often mild, although the effects of losing a few selectively advantageous alleles could have more serious effects. Allele frequency differences often translate into significant genetic divergence as measured by F_{ST}, but far more substantive divergence is likely to accrue over time. In the context of explaining divergence in naturally colonized and successfully established bird populations, an important or prevalent role for founder events as a divergence mechanism remains empirically unsupported.

Divergence via Gradual Drift and Selection

Given time and isolation, gradual drift can result in divergence without needing to invoke the action of other mechanisms. Despite this, the role that neutral mechanisms play in promoting evolutionary change is often overlooked in favor of adaptive explanations (see Barton 1998, Lynch 2007). Divergence at neutral loci that are not subject to selective pressures illustrates how effective drift can be in gradually increasing levels of divergence in island forms. In contrast, divergence at morphological characters is often assumed to have a selective basis. Few studies have examined whether or not patterns of variation can be explained solely by drift without recourse to selective explanations (but see Lynch 1990, Westerdahl et al. 2004, Renaud et al. 2007).

There are a number of types of data that can be used to assess whether drift is sufficient to explain divergence in island environments. First, the random nature of drift is not expected to produce recurring patterns of morphological change in species that repeatedly colonize islands. Selection has been invoked when repeated patterns are observed, for example, the production of similar ecomorphs in Anoles lizards on Caribbean islands (Losos et al. 1998, this volume) and a tendency for dwarfism in insular sloths (Anderson and Handley 2002). Second, a decoupling of phenotypic and neutral genetic measures of divergence can be interpreted as evidence of selection acting on phenotypic traits (Barrowclough 1983, Spitze 1993, Leinonen et al. 2007, Renaud et al. 2007). In birds, this logic has been applied to reject drift as the sole mechanism of morphological differentiation in a geographically restricted set of song sparrow (*Melospiza melodia*) subspecies in the San Francisco Bay region (Chan and Arcese 2003), and also, with mixed results, for Atlantic island populations of Berthelot's pipit (*Anthus berthelotii*) (Illera et al. 2007). Third, where time frames and effective population sizes are known or can be inferred, the rate at which a shift has occurred can be used to accept or reject drift as a sole mechanism of change (Lande 1976, Turelli et al. 1988, Lynch 1990). Small effective population sizes and low trait heritability can potentially result in large morphological shifts via drift alone (Turelli et al. 1988). If actual effective population sizes exceed the maximum effective population size that would explain the shift by drift alone, then additional microevolutionary mechanisms are required to explain the observed shift. This rationale has been used to reject drift as the sole mechanism of change in a sexually selected plumage trait (Yeh 2004) and morphometric traits (Rasner et al. 2004) in the recently founded *Junco* population in California discussed previously.

In insular *Zosterops*, repeated patterns and rates of morphological change imply a role for selection. *Zosterops* species show a tendency toward increased body size in island representatives (Lack 1971), a recurrent pattern also seen within the *Zosterops lateralis* species complex (figure 11.3a) (Mees 1969, Clegg et al. 2002b). Morphological shifts towards larger body size or bill size have occurred in most of the recent colonization events by *Z. l. lateralis* (figure 11.3b). Size increases are not universal however, with one population being significantly smaller in overall size and bill size. Additionally, morphological and genetic measures of differentiation in the recently colonized populations appear decoupled (Clegg et al. 2002b). The magnitude and rate of univariate shifts were often too large, whether toward increased or decreased size, to be accomplished by drift alone, with estimates of effective population sizes frequently too high for a chance mechanism of drift to completely account for the observed shifts (Clegg et al. 2002b). Selection is therefore required to explain morphological change in recently colonized populations. In contrast, rates of change in evolutionarily older *Zosterops* populations were consistent with a drift-alone mechanism when assum-

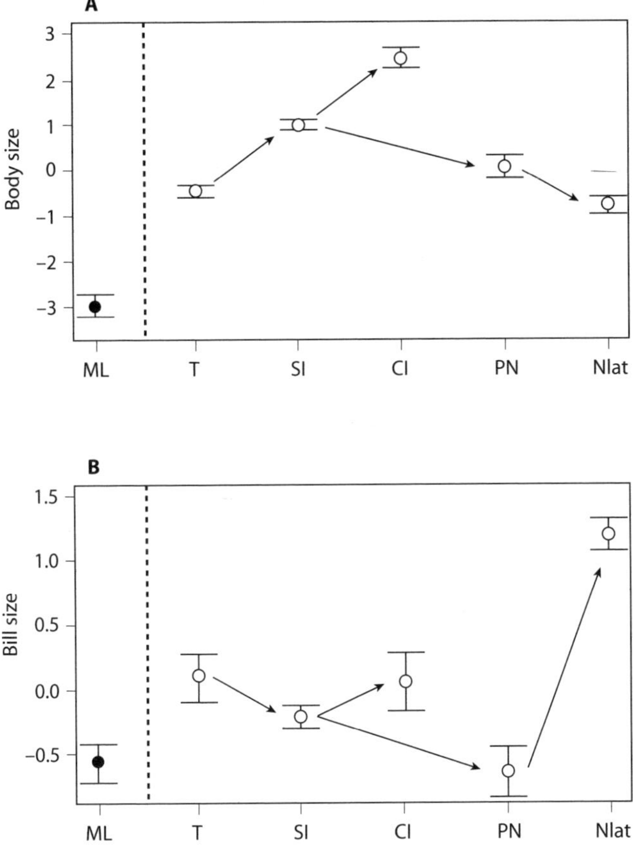

Figure 11.3. Multivariate representation (mean canonical variate (CV) scores summarized from 10 univariate traits) of shifts in morphology for the recently colonized *Z. l. lateralis* populations compared to the mainland subspecies (ML). A. Body size (CV1). B. Bill size (CV2). Arrows refer to colonization sequence. Location abbreviations as in figure 11.2. Modified from Clegg et al. (2002b).

ing a consistent rate of change since separation from the ancestor (Clegg et al. 2002b). This is unlikely to represent a difference in divergence mechanism between recently colonized and evolutionarily older forms. Rather, it becomes difficult to reject the null hypothesis of drift when considering divergence over long timescales because selection is unlikely to be consistent in strength or direction and effects are therefore averaged out over time (Kinnison and Hendry 2001). Indeed, divergent selection may be most effective early in the colonization history (Reznick et al. 1997). An alternative model applicable to island-colonizing species experiencing a novel environment is one of rapid displacement driven by directional selection followed by long periods of little change (Lande 1976, Estes and Arnold 2007). This type of model is consistent with divergence of the Capricorn silvereye on Heron Island when comparing patterns of morphological change over millennia, decades, and years (Clegg et al. 2008).

While drift does have the potential to contribute to morphological diversification, natural selection is often required to explain morphological shifts in birds (Price 2008, chapter 3). Studies of patterns and rates of change, in combination with direct measurement of natural selection currently acting in bird populations (e.g., Grant 1985, Grant and Grant 1995b, Merilä et al. 2001, Grant and Grant 2002, Frentiu et al. 2007) and translocation or common garden studies showing that morphological differences among populations are likely to have a genetic basis (Merilä and Sheldon 2001, Price 2008), point to the importance of natural selection in driving morphological divergence in island bird populations. Other phenotypic characters may have a plastic rather than heritable response to a new environment and it remains important to continue to consider whether adaptive explanations of divergence are necessary for different traits and different organisms.

Avian Body Size and Insular Shifts in Competitive Regimes

If we accept the contention that natural selection is a prominent microevolutionary process underlying divergence of island birds generally, a second line of questioning relates to how selection acts differently on islands compared to the mainland (MacArthur and Wilson 1967, p. 145). Specifically, do recurring abiotic and biotic features associated with island dwelling result in similar selection pressures across different islands?

There are numerous reasons why selective regimes on islands may systematically differ from the mainland. Island biota may be subject to reduced interspecific competition (Crowell 1962, Diamond 1970, Keast 1970), increased intraspecific competition (MacArthur et al. 1972, Blondel 1985), reduced predator pressure (Schoener and Toft 1983, Michaux et al. 2002, Blumstein 2002), changes in parasite prevalence and diversity, and disease susceptibility (Lindström et al. 2004, Fallon et al. 2005, Matson 2006), and various other shifts in biotic (e.g., resource availability and physical habitat structure; Abbott 1980, Martin 1992, Wu et al. 2006) and abiotic features (e.g., milder environments; Abbott 1980). These differences have been incorporated into adaptive explanations of diversification of island forms. Here I focus on how changes in inter- and intraspecific competition regimes have been used to explain the pattern of increased body size in island-dwelling passerines (Grant 1965, Clegg and Owens 2002) and whether empirical data are consistent with the proposed hypotheses.

One scenario linking competition shifts to body size changes is that reduced interspecific competition results in wider ecological niches and an increase in generalist behavior (Grant 1965, Van Valen 1965, Lack 1969, Carlquist 1974). Large body size, for example, may facilitate an increase in generalist behavior by increasing accessibility to a wider range of resources (Amadon 1953, Grant 1965, Keast 1970, Cody 1974, Grant 1979, Schlotfeldt and Kleindorfer 2006). Empirical support of an association between body size and generalist feeding was demonstrated in seed-eating medium ground finches (*Geospiza fortis*), where large-billed birds had access to a wider range of seed sizes than small-billed birds (Grant et al. 1976). Directional selection favoring larger forms might therefore be expected when there is an increase in generalist foraging behavior. Scott et al. (2003) outlined three expectations that need to be satisfied for an increase in generalist foraging behavior to provide a general explanation for increased body size in island populations of birds. First, it needs to be established that island populations are more generalist foragers; second, population-level generalist behavior needs to be achieved via individual-level generalist behavior rather than an amalgamation of different types of individual specialists; and finally there should be a positive association between degree of generalist behavior and body size.

The accumulation of studies that have quantified and compared aspects of niche width between island forms and their mainland relatives (e.g., Cox and Ricklefs 1977, Blondel et al. 1988, Carrascal et al. 1994, Scott et al. 2003, Föershler and Kalko 2006, Schlotfeldt and Kleindorfer 2006) support the view that increases in niche width and a shift toward more generalist foraging behavior in island birds is a common phenomenon (Diamond 1970, Keast 1970). The extent to which population-level generalist behavior can be explained by the presence of individual generalists or different types of individual specialists has long been recognized as an important ecological and evolutionary consideration (Van Valen 1965, Roughgarden 1974, Grant et al. 1976, Price 1987). However, few studies of island birds have established how population-level generalist behavior is achieved, most likely because it can be logistically difficult in natural situations to record ecological preferences of individually recognized birds.

Three examples where individual behavior has been quantified in island bird populations are the Capricorn silvereye (*Zosterops lateralis chlorocephalus*) on Heron Island, Australia (Scott et al. 2003), the Cocos

Island finch (*Pinoroloxias inornata*) on Cocos Island, Costa Rica (Werner and Sherry 1987), and the Darwin's medium ground finch (*Geospiza fortis*) on Daphne Major, Galápagos (Grant et al. 1976, Price 1987). Scott et al. (2003) showed that island *Zosterops* populations are more generalist with respect to foraging height and substrate than their mainland counterparts. However, detailed examination of the Capricorn silvereye on Heron Island revealed that the generalist population was composed of individuals that were more specialized foragers than expected by chance (Scott et al. 2003). The Cocos Island finch was found to be a highly generalist population with respect to foraging methods and this was achieved via individuals using an extremely limited range of resources compared to the population as a whole (Werner and Sherry 1987). In Darwin's medium ground finch, Price (1987) reported that the population was generalist with respect to use of three seed types, but individuals exhibited some degree of specialization, utilizing only a subset of the seed types available to the population as a whole. The degree to which this occurred was influenced by food availability with more specialist individuals present when food was short (Price 1987).

The degree to which there was a positive relationship between generalist behavior and morphological size varied across the three studies. Capricorn silvereyes showed no relationship between morphology and degree of foraging generalization (Scott et al. 2003). Likewise, individual Cocos Island finches showed no relationship between morphology (or sex or age) and foraging behavior (Werner and Sherry 1987). In contrast, a relationship between morphology and foraging in Darwin's medium ground finch was observed. Individuals with significantly larger bills utilized large and hard seeds that were unavailable to smaller-billed individuals, thereby displaying a positive association between a morphological character and one aspect of niche width (Grant et al. 1976, Price 1987). In this species, seeds are the predominant food source and are particularly relied upon when environmental conditions deteriorate (Price 1987). Grant et al. (1976) found no such relationship between bill size of medium ground finches and another, more easily accessed resource, *Bursera* berries. The relationship between morphology and foraging may be more likely to occur in cases where access to the food item is very tightly restricted by physical capabilities of the feeding apparatus. Such strong associations between bill size and resource have been reported in other seed-eaters, e.g., *Pyrenestes* finches in Africa (Smith 1987).

Of the limited examples available to examine individual niche width in island birds, each is a generalist population made up to some degree of individual specialists (with respect to all food types for the Capricorn silvereye and Cocos Island finch, or seed types for Darwin's medium ground finch). Further empirical results for island populations are required before generalizations are made; however, Werner and Sherry (1987) point out that the conditions under which individual specialization is likely to arise, including high food availability, variety, and predictability, high population density, low interspecific competition, and low territoriality, are often met on oceanic islands. More broadly, generalist populations made up of individual specialists may be more common than previously appreciated (Bolnick et al. 2007). In the cases presented here, there is variation in the degree of individual specialization, being more pronounced in the case of the Cocos Island finch than the other two examples, or when food availability decreases in the case of the medium ground finch. A link between foraging characteristics and morphology was found for the medium ground finch only. The idea that an increase in generalist behavior favors selection for a large generalist form is not consistent with the occurrence of individual specialists, and the lack of morphological association with generalist foraging behavior in two of the three cases. While changes in interspecific competition regimes may influence body size evolution of island birds in other ways, direct links between reduced interspecific competition, increased generalist behavior, and selection for a generalist (large) body type are not strongly supported by the limited empirical evidence available.

A second scenario linking competition shifts to body size changes centers on the effects of increased intraspecific competition. Population density increases within a species are often a feature of island populations (MacArthur et al. 1972). This phenomenon has been observed in a range of taxa, including birds (Crowell 1962, Kikkawa 1976, Thiollay 1993, George 1987, Blondel et al. 1988), mammals (Adler and Levins 1994, Goltsman et al. 2005), and herpetofauna (Rodda and Dean-Bradley 2002, Buckley and Jetz 2007, Wu et al. 2006). Population density increases plausibly lead to increased intraspecific competition. In birds an increase in agonistic encounters can often occur (Stamps and Buechner 1985) and, in such a situation, some have proposed that selection should favor traits that provide an advantage in agonistic interactions, the outcome of which may ultimately affect survival or fecundity (Kikkawa 1980, Robinson-Wolrath and Owens 2003). One such potentially favorable factor is increased body size. At the interspecific level, the relationship between body size and the order of dominance or aggressive superiority has been demonstrated (e.g., Piper and Catterall 2003, Rychlik and Zwolak 2006). Within species, the relationship between body size and aggressive behavior is less clear; for example, aggression in bluebirds is not related to body size (Duckworth 2006). However, in the Capricorn silvereye on Heron Island, a study of agonistic encounters within juveniles during a single over-winter period found a significant positive relationship between body size and proportion of aggressive encounters won (Robinson-Wolrath and Owens 2003). The addition of data taken across a three-year period on birds of all ages showed that, after taking into account the strong effects of age and sex, where males and adults win more often, body size remains a significant predictor of the outcome of aggressive interaction (Clegg and Owens, unpublished). Such individual variation in aggression and morphology could be an important target of selection in this population. Whether or not selection for large aggressive individuals is a general phenomenon in densely populated insular settings remains to be seen.

Concentrating on the role of either intra- or interspecific competition may help to identify the direct selective mechanism producing a morphological pattern. In the examples presented here, reduced interspecific competition is unlikely to be a direct cause of increased body size in small island birds via a feeding generalization mechanism, whereas increased intraspecific competition may have more direct selective effects on body size via behavioral mechanisms. However, it is the shift in balance between inter- and intraspecific competition, where reduced interspecific competition facilitates increased intraspecific competition, that may be at the base of a sequence of changes that occur on islands and ultimately result in morphological changes. Further, the relationships between body size, niche width, and aggressive tendencies discussed here are unlikely to operate in isolation from other insular features of changes in predation, parasites, and other abiotic and biotic differences. Additionally, changes in sexual rather than natural selection regimes offer an alternative explanation for large body size. If strong genetic correlations exist between the sexes, then sexual selection for large male body size may drive larger size overall (Price 1984, Merilä et al. 1998). The interplay among these factors awaits further empirical investigations.

Conclusions

Drift and natural selection are two of the microevolutionary processes that can cause divergence in island forms. Population genetic studies of naturally colonized and introduced island bird populations demonstrate that drift during the founding event often does not have severe consequences for diversity and divergence. Sequentially founded populations are more susceptible to cumulative effects of founder-mediated drift, but, even then, loss of diversity can be surprisingly mild. As development of molecular markers continues, future studies will have the opportunity to address loci under selection and to track the impact of founding events on selectively advantageous alleles. Drift, either during founding or over longer time frames, can conceivably contribute to morphological divergence. Situations of extreme isolation due to geographic distance or dispersal limitations will provide greater opportunity for drift to be an effective mechanism. However, evidence of patterns and magnitudes of morphological differentiation suggests that natural selection is a relatively more important microevolutionary process than neutral mechanisms, and may be particularly important in generating divergence in the early stages of colonization history. Common biotic and abiotic factors associated with insularity could produce congruent selection regimes on islands. The extent to which this produces general patterns of diversification and the particular selective pressure responsible requires more case studies. In particular, more studies at the individual level would be valuable for understanding the interplay among different selection pressures, and which may be of more direct influence in producing evolutionary change in island birds.

Acknowledgments

I thank Ian Owens, Jiro Kikkawa, Craig Moritz, Sandie Degnan, Susan Scott, and Sarah Robinson-Wolrath for discussions on topics presented in this chapter and Robert Ricklefs, Jonathan Losos, Peter Grant, Albert Phillimore, Jessica Worthington Wilmer, and an anonymous reviewer for helpful comments on the manuscript.

Literature Cited

Abbott, I. 1980. Theories dealing with the ecology of landbirds on islands. *Advances in Ecological Research* 11:329–71.

Abdelkrim, J., M. Pascal, and S. Samadi. 2005. Island colonization and founder effects: The invasion of the Guadeloupe islands by ship rats (*Rattus rattus*). *Molecular Ecology* 14:2923–31.

Adler, G. H., and R. Levins. 1994. The island syndrome in rodent populations. *Quarterly Review of Biology* 69:473–90.

Amadon, D. 1953. Avian systematics and the evolution in the Gulf of Guinea. *Bulletin of the American Museum of Natural History* 100:393–452.

Anderson, R. P., and C. O. Handley, Jr. 2002. Dwarfism in insular sloths: Biogeography, selection and evolutionary rate. *Evolution* 56:1045–58.

Ardern, S. L., D. M. Lambert, A. G. Rodrigo, and I. G McLean. 1997. The effects of population bottlenecks on multilocus DNA variation in robins. *Journal of Heredity* 88:179–86.

Baker, M. C., M.S.A. Baker, and L. M. Tilghman. 2006. Differing effects of isolation on evolution of birds songs: Examples from an island-mainland comparison of three species. *Biological Journal of the Linnean Society* 89: 331–42.

Baker, A. J., A. D. Greenslade, L. M. Darling, and J. C. Finlay. 2008. High genetic diversity in the blue-listed British Columbia population of the purple martin maintained by multiple sources of immigrants. *Conservation Genetics* 9:495–505.

Balloux, F., W. Amos, and T. Coulson. 2004. Does heterozygosity estimate inbreeding in real populations? *Molecular Ecology* 13:3021–31.

Barrowclough, G. F. 1983. Biochemical studies of microevolutionary processes. In *Perspectives in Ornithology,* ed. A. H. Brush and J.G.A. Clark, 223–61. New York: Cambridge University Press.

Barton, N. H., and B. Charlesworth. 1984. Genetic revolutions, founder effects and speciation. *Annual Review of Ecology and Systematics* 15:133–64.

Barton, N. H. 1998. Natural selection and random genetic drift as causes of evolution on islands. In *Evolution on Islands,* ed. P. R. Grant, 102–23. Oxford: Oxford University Press.

Berry, R. J. 1998. Evolution of small mammals. In *Evolution on Islands,* ed. P. R. Grant, 35–50. Oxford: Oxford University Press.

Blondel, J. 1985. Habitat selection in island versus mainland birds. In *Habitat Selection in Birds,* ed. M. Cody, 477–516. Orlando, FL: Academic Press.

Blondel, J., D. Chessel, and B. Frochot. 1988. Bird species impoverishment, niche expansion and density inflation in Mediterranean island habitats. *Ecology* 69:1899–917.

Blumstein, D. 2002. Moving to suburbia: Ontogenetic and evolutionary consequences of life on predator-free islands. *Journal of Biogeography* 29:685–92.

Boessenkool, S., S. S. Taylor, C. K. Tepolt, J. Komdeur, and I. G. Jamieson. 2007. Large mainland populations of South Island robins retain greater genetic diversity than offshore island refuges. *Conservation Genetics* 8:705–14.

Bollmer, J. L., N. K. Whiteman, M. D. Cannon, J. C. Bednarz, T. De Vries, and P. G. Parker. 2005. Population genetics of the Galapagos hawk (*Buteo galapagoensis*): genetic monomorphism within isolated populations. *Auk* 122:1210–24.

Bollmer, J. L., F. Hernán Vargas, and P. G. Parker. 2007. Low MHC variation in the endangered Galápagos penguin (*Spheniscus mendiculus*). *Immunogenetics* 59:593–602.

Bolnick, D. I. R. Svanbäck, M. S. Araujo, and L. Persson. 2007. Comparative support for the niche variation hypothesis that more generalized populations also are more heterogeneous. *Proceedings of the National Academy of Sciences U.S.A.* 104:10075–79.

Brown, J. W, P. J. Van Coeverden de Groot, T. P. Birt, G. Seiutin, P. T. Boag, and V. L. Friesen. 2007. Appraisal of the consequences of the DDT-induced bottleneck on the level and geographic distribution of neutral genetic variation in Canadian peregrine falcons, *Falco peregrinus. Molecular Ecology* 16:327–43.

Buckley, L. B., and W. Jetz. 2007. Insularity and the determinants of lizard population density. *Ecology Letters* 10:481–89.

Cabe, P. R. 1998. The effects of founding bottlenecks on genetic variation in the European starling (*Sturnus vulgaris*) in North America. *Heredity* 80:519–25.

Carlquist, S. 1974. *Island Biology*. New York: Columbia University Press.

Carrascal, L. M., E. Moreno, and A. Valido. 1994. Morphological evolution and changes in foraging behaviour of island and mainland populations of Blue Tit (*Parus caeruleus*)—a test of convergence and ecomorphological hypotheses. *Evolutionary Ecology* 8:25–35.

Carson, H. L., and A. R. Templeton. 1984. Genetic revolutions in relation to speciation phenomena: the founding of new populations. *Annual Review of Ecology and Systematics* 15:97–131.

Chan, Y., and P. Arcese. 2003. Morphological and microsatellite differentiation in *Melopsiza melodia* (Aves) at a microgeographic scale. *Journal of Evolutionary Biology* 16:939–47.

Clegg, S. M., S. M. Degnan, J. Kikkawa, C. Moritz, A. Estoup and I. P. F. Owens. 2002a. Genetic consequences of sequential founding events by and island-colonizing bird. *Proceedings of the National Academy of Sciences U.S.A.* 99:8127–32.

Clegg, S. M., S. M. Degnan, C. Moritz, A. Estoup, J. Kikkawa, and I. P. F. Owens. 2002b. Microevolution in island forms: the roles of drift and directional selection in morphological divergence of a passerine bird. *Evolution* 56:2090–99.

Clegg, S. M., D. F. Frentiu, J. Kikkawa, G. Tavecchia, and I. P. F. Owens. 2008. 4000 years of phenotypic change in an island bird: heterogeneity of selection over three microevolutionary timescales. *Evolution* 62:2393–410.

Clegg, S. M., and I.P.F. Owens. 2002. The 'island-rule' in birds: medium body size and its ecological explanation. *Proceedings of the Royal Society of London, Series B* 269:1359–65.

Cody, M. 1974. *Competition and the Structure of Bird Communities*. Princeton, NJ: Princeton University Press.

Coltman, D., and J. Slate. 2003. Microsatellite measures of inbreeding: A meta-analysis. *Evolution* 57:971–83.

Cox, G. W., and R. E. Ricklefs. 1977. Species diversity and ecological release in Caribbean land bird faunas. *Oikos* 28:113–22.

Coyne, J. A., and H. A. Orr. 2004. *Speciation*. Sunderland, MA: Sinauer Associates.

Crowell, K. 1962. Reduced interspecific competition among the birds of Bermuda. *Ecology* 43:75–88.

Degnan, S. M., and C. Moritz. 1992. Phylogeography of mitochondrial DNA in two species of white-eyes in Australia. *Auk* 109:800–811.

Diamond, J. M. 1970. Ecological consequences of island colonization by Southwest Pacific Birds, I: Types of niche shifts. *Proceedings of the National Academy of Sciences U.S.A.* 67:529–36.

Duckworth, R. 2006. Aggressive behaviour affects selection on morphology by influencing settlement patterns in a passerine bird. *Proceedings of the Royal Society of London, Series B* 273:1789–95.

Eales, J., R. S. Thorpe, and A. Malhotra. 2008. Weak founder effect signal in a recent introduction of Caribbean *Anolis*. *Molecular Ecology* 17:1416–26.

Estes, S., and S. J. Arnold. 2007. Resolving the paradox of stasis: Models with stabilizing selection explain evolutionary divergence on all timescales. *Evolution* 169:227–44.

Estoup, A., and S. M. Clegg. 2003. Bayesian inferences on the recent island colonization history by the bird *Zosterops lateralis lateralis*. *Molecular Ecology* 12:657–74.

Fallon, S. M., E. Bermingham, and R. E. Ricklefs. 2005. Host specialization and geographic localization of avian malaria parasites: a regional analysis in the Lesser Antilles. *American Naturalist* 165:466–80.

Förschler, M. I., and E.K.V. Kalko. 2006. Breeding ecology and nest site selection in allopatric mainland Citril finches *Carduelis [citrinella] citrinella* and insular Corsican finches *Carduelis [citrinella] corsicanus*. *Journal of Ornithology* 147:553–64.

Frankham, R. 1997. Do island populations have less genetic variation than mainland populations? *Heredity* 78:311–27.

Frentiu, F. D., S. M Clegg, M. W. Blows, and I. P. F. Owens. 2007. Large body size in an island-dwelling bird: a microevolutionary analysis. *Journal of Evolutionary Biology* 20:639–49.

Gautschi, B., A. Widmer, J. Joshi, and J. C. Koella. 2002. Increased frequency of scale anomalies and loss of genetic variation in serially bottlenecked populations of dice snake, *Natrix tessellata*. *Conservation Genetics* 3:235–45.

George, T. L. 1987. Greater land bird densities on island vs. mainland: relation to nest predation level. *Ecology* 68:1393–1400.

Goltsman, M., E. P. Kruchenkova, S. Sergeev, I. Velodin, and D. W. Macdonald. 2005. 'Island syndrome' in a population of Arctic foxes (*Alopex lagopus*) from Mednyi Island. *Journal of Zoology* 267:40–18.

Grant, B. R. 1985. Selection on bill characteristics in a population of Darwin's finches: *Geospiza conirostris* on Isla Genovesa, Galápagos. *Evolution* 39:523–32.

Grant, P. R. 1965. The adaptive significance of some size trends in island birds. *Evolution* 19:355–67.

———. 1979. Ecological and morphological variation of Canary Island blue tits, *Parus caeruleus* (Aves: Paridae). *Biological Journal of the Linnean Society* 11:103–29.

——— 1998. Speciation. In *Evolution on Islands,* ed. P. R. Grant, 83–101. Oxford: Oxford University Press.

——— 2001. Reconstructing the evolution of birds on islands: 100 years of research. *Oikos* 92:385–403.

Grant, P. R., and B. R. Grant. 1995a. The founding of a new population of Darwin's finches. *Evolution* 49:229–40.

———. 1995b. Predicting microevolutionary responses to directional selection on heritable variation. *Evolution* 49:241–51.

———. 2002. Unpredictable evolution in a 30-year study of Darwin's finches. *Science* 296:707–11.

Grant, P. R., B. R. Grant, and K. Petren. 2001. A population founded by a single pair of individuals: establishment, expansion, and evolution. *Genetica* 112–13: 359–82.

Grant, P. R., B. R. Grant, J.N.M. Smith, I. J. Abbott, and L. K. Abbott. 1976. Darwin's finches: Population variation and natural selection. *Proceedings of the National Academy of Sciences U.S.A.* 73:257–61.

Hailer, F., B. Helander, A. O. Folkestad, S. A. Ganusevich, S. Garstad, P. Hauff, C. Koren, T. Nygård, V. Volke, C. Vilá, and H. Ellegren. 2006. Bottlenecked but long-lived: High genetic diversity retained in white-tailed eagles upon recovery from population decline. *Biology Letters* 2:316–19.

Hajji, G. M., F. Charfi-Cheikrouha, R. Lorenzini, J-D. Vigne, G. B Hartl, and F. E. Zachos. 2008. Phylogeography and founder effect of the endangered Corsican red deer (*Cervus elaphus corsicanus*). *Biodiversity and Conservation* 17:659–73.

Hale, K. A., and J. V. Briskie. 2007. Decreased immunocompetence in a severely bottlenecked population of an endemic New Zealand bird. *Animal Conservation* 10:2–10.

Hansson, B., S. Bensch, D. Hasselquist, B-G. Lillandt, L. Wennerberg, and T. Von Schantz. 2000. Increase of genetic variation over time in a recently founded population of great reed warblers (*Acrocephalus arundinaceus*) revealed by microsatellites and DNA fingerprinting. *Molecular Ecology* 9:1529–38.

Hansson, B., H. Westerdahl, D. Hasselquist, M. Åkesson, and S. Bensch. 2004. Does linkage disequilibrium generate heterozygosity-fitness correlations in great reed warblers? *Evolution* 58:870–79.

Hawley, D. M., J. Briggs, A. A. Dhondt, and I. J. Lovette. 2008. Reconciling molecular signatures across markers: mitochondrial DNA confirms founder effect in invasive North American house finches (*Carpodacus mexicanus*). *Conservation Genetics* 9:637–43.

Hawley, D. M., D. Hanley, A. A. Dhondt, and I. J. Lovette. 2006. Molecular evidence of a founder effect in invasive house finch (*Carpodacus mexicanus*) populations experiencing an emergent disease epidemic. *Molecular Ecology* 15:263–75.

Hopley, D. 1982. *The Geomorphology of the Great Barrier Reef: Quaternary Development of Coral Reefs*. New York: John Wiley and Sons.

Illera, J. C., B. C. Emerson, and D. S. Richardson. 2007. Population history of Berthelot's pipit: colonization, gene flow and morphological divergence in Macaronesia. *Molecular Ecology* 16:4599–612.

Jamieson, I. G., G. P. Wallis, and J. V Briskie. 2006. Inbreeding and endangered species management: Is New Zealand out of step with the rest of the world? *Conservation Biology* 20:38–47.

Kaeuffer, R., D. W. Coltman, J-L. Chapuis, D. Pontier, and D. Réale. 2007. Unexpected heterozygosity in an island mouflon population founded by a single pair of individuals. *Proceedings of the Royal Society of London, Series B* 274: 527–33.

Keast, A. 1970. Adaptive evolution and shifts in niche occupation in island birds. *Biotropica* 2:61–75.

Keller, L. F., K. J. Jeffery, P. Arcese, M. A. Beaumont, W. M. Hochachka, J. N. M. Smith, and M W. Bruford. 2001. Immigration and the ephemerality of a natural population bottleneck: Evidence from molecular markers. *Proceedings of the Royal Society of London, Series B* 268:1387–94.

Kikkawa, J. 1976. The birds of the Great Barrier Reef. In *Biology and Geology of Coral Reefs,* ed. O. A. Jones and R. Endean, 279–341. New York: Academic Press.

———. 1980. Winter survival in relation to dominance classes among silvereyes *Zosterops lateralis chlorocephala* of Heron Island, Great Barrier Reef. *Ibis* 122:437–46.

Kinnison, M. T., and A. P. Hendry. 2001. The pace of modern life II: From rates of contemporary microevolution to pattern and process. *Genetica* 112-113: 145–64.

Kliber, A., and C. G. Eckert. 2005. Interaction between founder effect and selection during biological invasion in an aquatic plant. *Evolution* 59:1900–1913.

Lack, D. 1969. The number of bird species on islands. *Bird Study* 16:193–209.

———. 1971. *Ecological Isolation in Birds*. Oxford: Blackwell Scientific Publications.

Lambert, D. M., T. King, L. D. Shepherd, A. Livingston, S. Anderson, and J. L. Craig. 2005. Serial population bottlenecks and genetic variation: translocated populations of the New Zealand Saddleback (*Philesturnus carunculatus rufaster*). *Conservation Genetics* 6:1–14.

Lande, R. 1976. Natural selection and random genetic drift in phenotypic evolution. *Evolution* 30:314–34.

———. 1980. Genetic variation and phenotypic evolution during allopatric speciation. *American Naturalist* 116:463–79.

Latch, E. K., and O. E. Rhodes, Jr. 2005. The effects of gene flow and population isolation on the genetic structure of reintroduced wild turkey populations: Are genetic signatures of source populations retained? *Conservation Genetics* 6:981–97.

Le Corre, V., and A. Kremer. 1998. Cumulative effects of founding events during colonization on genetic diversity and differentiation in an island and stepping-stone model. *Journal of Evolutionary Biology* 11:495–512.

Le Gouar, P., F. Rigal, M. C. Boisselier-Dubayle, F. Sarrazin, C. Arthur, J. P. Chlisy, O. Hatzofe, S. Henriquet, P. Lécuyer, C. Tessier, G. Susie, and S. Samadi. 2008. Genetic variation in a network of natural and reintroduced populations of Griffon vulture (*Gyps fulvus*) in Europe. *Conservation Genetics* 9:349–59.

Le Page, S. L., R. A. Livermore, D. W. Cooper, and A. C. Taylor. 2000. Genetic analysis of a documented population bottleneck: Introduced Bennett's wallabies (*Macropus rufogriseus rufogriseus*) in New Zealand. *Molecular Ecology* 9:753–63.

Leinonen, T., R. B. O'Hara, J. M. Cano, and J. Merilä. 2007. Comparative studies of quantitative trait and neutral marker divergence: A meta-analysis. *Journal of Evolutionary Biology* 21:1–17.

Lindström, K. M., J. Foufopoulos, H. Pärn, and M. Wikelski. 2004. Immunological investments reflect parasite abundance in island populations of Darwin's finches. *Proceedings of the Royal Society of London, Series B* 271: 1513–19.

Losos, J. B. T. R. Jackman, A. Larson, K. De Queiroz, and L. Rodriguez-Schettino. 1998. Contingency and determinism in replicated adaptive radiations of island lizards. *Science* 279:2115–17.

Lynch, M. 1990. The rate of morphological evolution in mammals from the standpoint of the neutral expectation. *American Naturalist* 136:727–41.

———. 2007. The frailty of adaptive hypotheses for the origins of organismal complexity. *Proceedings of the National Academy of Sciences U.S.A.* 104: 8597–604.

MacArthur, R. H., J. M. Diamond, and J. R. Karr. 1972. Density compensation in island faunas. *Ecology* 53:330–42.

MacArthur, R. H., and E. O. Wilson. 1967. *The Theory of Island Biogeography.* Princeton, NJ: Princeton University Press.

Martin, J-L. 1992. Niche expansion in an insular bird community: An autecological perspective. *Journal of Biogeography* 19:375–81.

Matson, K. D. 2006. Are there differences in immune function between continental and insular birds? *Proceedings of the Royal Society of London, Series B* 273:2267–74.

Mayr, E. 1942. *Systematics and the Origin of Species.* New York: Colombia University Press.

———. 1954. Changes of genetic environment and evolution. In *Evolution as a Process,* ed. J. Huxley, A. C. Hardy, and E. B. Ford, 157–80. London: Allen and Unwin.

Mees, G. F. 1969. A systematic review of the Indo-Australian Zosteropidae (Part III). *Zoologische Verhandelingen* 102:1–390.

Merilä, J., M. Björklund, and A. J. Baker. 1996. The successful founder: genetics of introduced *Carduelis chloris* (greenfinch) populations in New Zealand. *Heredity* 77:410–22.

Merilä, J., L.E.B. Kruuk, and B. C. Sheldon. 2001. Natural selection on the genetical component of variance in body condition in a wild bird population. *Journal of Evolutionary Biology* 14:918–29.

Merilä, J., and B. C. Sheldon. 2001. Avian quantitative genetics. In *Current Ornithology,* ed. V. Nolan and C. F. Thompson, 179–255. New York: Kluwer.

Merilä, J., B. C. Sheldon, and H. Ellegren. 1998. Quantitative genetics of sexual size dimorphism in the collared flycatcher, *Ficedula albicollis, Evolution* 52:870–76.

Michaux, J. R., J. G. de Bellocq, M. Sarà, and S. Morand. 2002. Body size increases in insular rodent populations: A role for predators? *Global Ecology and Biogeography* 11:427–36.

Miller, H. C., and D. M. Lambert. 2004. Genetic drift outweighs balancing selection in shaping post-bottleneck major histocompatibility complex variation in New Zealand robins (Petroicidae). *Molecular Ecology* 13:3709–21.

Mock, K. E., E. K. Latch, and O. E. Rhodes, Jr. 2004. Assessing losses of genetic diversity due to translocation: long-term case histories in Merriam's turkey (*Meleagris gallopavo merriami*). *Conservation Genetics* 5:631–45.

Motro, U., and G. Thomson. 1982. On heterozygosity and the effective size of populations subject to size changes. *Evolution* 36:1059–66.

Muñoz-Fuentes, V., A. J. Green, M. D. Sorenson, J. J. Negro, and C. Vilá. 2006. The ruddy duck *Oxyura jamaicensis* in Europe: Natural colonization or human introduction? *Molecular Ecology* 15:1441–53.

Mundy, N. I., C. S. Winchell, T. Burr, and D. S. Woodruff. 1997. Microsatellite variation and microevolution in the critically endangered San Clemente Island loggerhead shrike (*Lanius ludovicianus mearnsi*). *Proceedings of the Royal Society of London, Ser. B* 264:869–75.

Nei, M., T. Maruyama, and R. Chakraborty. 1975. The bottleneck effect and genetic variability in populations. *Evolution* 29:1–10.

O'Brien, J., G. F. McCracken, L. Say, and T. J. Hayden. 2007. Rodrigues fruit bats (*Pteropus rodricensis*, Megachiroptera: Pteropodidae) retain genetic diversity despite population declines and founder events. *Conservation Genetics* 8:1073–82.

Ohnishi, N. T. Saitoh, Y. Ishibashi, and T. Oi. 2007. Low genetic diversities in isolated populations of the Asian black bear (*Ursus thibetanus*) in Japan, in comparison with large stable populations. *Conservation Genetics* 8:1331–37.

Ortego, J., J. M. Aparicio, G. Calbuig, and P. J. Cordero. 2007. Increase of heterozygosity in a growing population of lesser kestrels. *Biology Letters* 3:585–88.

Pastor, T., J. C. Garza, P. Allen, W. Amos, and A. Aguilar. 2004. Low genetic variability in the highly endangered Mediterranean monk seal. *Journal of Heredity* 95:291–300.

Phillimore, A. B. 2006. The ecological basis of speciation and divergence in birds. Ph.D. dissertation, Department of Biological Sciences, Imperial College, London.

Piper, S. D., and C. P. Catterall. 2003. A particular case and a general pattern: hyperaggressive behaviour by one species may mediate avifaunal decreases in fragmented Australian forests. *Oikos* 101:602–14.

Price, T. 1984. The evolution of sexual size dimorphism in Darwin's finches (*Geospiza fortis*). *American Naturalist* 123:500–518.

———. 1987. Diet variation in a population of Darwin's finches. *Ecology* 68:1015–28.

———. 2008. *Speciation in Birds.* Westview Village, CO: Roberts and Company.

Provine, W. B. 1989. Founder effects and genetic revolutions in microevolution and speciation: An historical perspective. In *Genetics, Speciation and the Founder Principle,* ed. L. V. Giddings, K. Y. Kaneshiro, and W. W. Anderson, 43–76. New York: Oxford University Press.

Pruett, C. L., and K. Winker. 2005. Northwestern song sparrow populations show genetic effects of sequential colonization. *Molecular Ecology* 14: 1421–34.

Rasner, C. A., P. Yeh, L. S. Eggert, K. E. Hunt, D. S. Woodruff, and T. D. Price. 2004. Genetic and morphological evolution following a founder event in the dark-eyed junco, *Junco hyemalis thurberi. Molecular Ecology* 13: 671–81.

Reiland, U., S. Hodge, and M. A. F. Noor. 2002. Strong founder effect in *Drosophila pseudoobscura* colonizing New Zealand from North America. *Journal of Heredity* 93:415–20.

Renaud, S., P. Chevret, and J. Michaux. 2007. Morphological vs. molecular evolution: ecology and phylogeny both shape the mandible of rodents. *Zoologica Scripta* 35:525–35.

Reznick, D. N., F. H. Shaw, F. H. Rodd, and R. G. Shaw. 1997. Evaluation of the rate of evolution in natural populations of guppies (*Poecilia reticulata*). *Science* 275:1934–37.

Rice, W., and E. Hostert. 1993. Laboratory experiments on speciation: What have we learned in 40 years? *Evolution* 47:1637–53.

Robinson-Wolrath, S. I., and I. P. F. Owens. 2003. Large size in an island-dwelling bird: intraspecific competition and the Dominance Hypothesis. *Journal of Evolutionary Biology* 16:1106–14.

Rodda, G. H., and Dean-Bradley K. 2002. Excess density compensation in island herpetofaunal assemblages. *Journal of Biogeography* 29:623–32.

Roques, S., and J. J. Negro. 2005. MtDNA genetic diversity and population history of a dwindling raptorial bird, the red kite (*Milvus milvus*). *Biological Conservation* 126:41–50.

Roughgarden, J. 1974. Niche width: biogeographic patterns among *Anolis* lizard populations. *American Naturalist* 108:429–42.

Rowe, G., T. J. C. Beebee, and T. Burke. 1998. Phylogeography of the natterjack toad *Bufo calamita* in Britain: Genetic differentiation of native and translocated populations. *Molecular Ecology* 7:751–60.

Rychlik, L., and R. Zwolak. 2006. Interspecific aggression and behavioural dominance among four sympatric species of shrews. *Canadian Journal of Zoology* 84:434–48.

Schlotfeldt, B. E., and S. Kleindorfer. 2006. Adaptive divergence in the Superb Fairy-wren (*Malurus cyaneus*): A mainland versus island comparison of morphology and foraging behaviour. *Emu* 106:309–19.

Schoener, T. W., and C. A. Toft. 1983. Spider populations: Extraordinarily high densities on islands without top predators. *Science* 219:1353–55.

Scott, S. N., S. M Clegg, S. P. Blomberg, J. Kikkawa, and I. P. F. Owens. 2003. Morphological shifts in island-dwelling birds: The roles of generalist foraging and niche expansion. *Evolution* 57:2147–56.

Seutin, G., J. Brawn, R. E. Ricklefs, and E. Bermingham. 1993. Genetic divergence among populations of a tropical passerine, the streaked saltator (*Saltator albicollis*). *Auk* 110:117–26.

Slatkin, M. 1996. In defense of founder-flush theories of speciation. *American Naturalist* 147:493–505.

Smith, T. B. 1987. Bill size polymorphism and intraspecific niche utilization in an African finch. *Nature* 329:717–19.

Spitze, K. 1993. Population structure in *Daphnia obtusa*: Quantitative genetic and allozymic variation. *Genetics* 135:367–74.

Stamps, J. A., and Buechner. 1985. The territorial defense hypothesis and the ecology of insular vertebrates. *Quarterly Review of Biology* 60:155–81.

Tarr, C. L., S. Conant, and R. C. Fleischer. 1998. Founder events and variation at microsatellite loci in an insular passerine bird, the Laysan finch (*Telespiza cantans*). *Molecular Ecology* 7:719–31.

Taylor, A. C., P. E. Cowan, B. L. Fricke, S. Geddes, B. D. Hansen, M. Lam, and D. W. Cooper. 2004. High microsatellite diversity and differential structuring among populations of the introduced common brushtail possum, *Trichosurus vulpecula*, in New Zealand. *Genetical Research* 83:101–11.

Taylor, S. S., and I. G. Jamieson. 2008. No evidence for loss of genetic variation following sequential translocations in extant populations of a genetically depauperate species. *Molecular Ecology* 17:545–56.

Templeton, A. R. 1996. Experimental evidence for the genetic-transilience model of speciation. *Evolution* 50:909–15.

———. 2008. The reality and importance of founder speciation in evolution. *Bioessays* 30:470–79.

Thiollay, J-M. 1993. Habitat segregation and the insular syndrome in two congeneric raptors in New Caledonia, the White-bellied Goshawk *Accipter haplochrous* and the Brown Goshawk *A. fasciatus*. *Ibis* 135:237–46.

Turelli, M., J. H. Gillespie, and R. Lande. 1988. Rate tests for selection on quantitative characters during macroevolution and microevolution. *Evolution* 42: 1085–89.

Van Valen, L. 1965. Morphological variation and width of the ecological niche. *American Naturalist* 99:377–90.

Walsh, H. E., I. L. Jones, and V. L. Friesen. 2005. A test of founder effect speciation using multiple loci in the Auklets (*Aethia* spp.). *Genetics* 171: 1885–94.

Weber, D. S., B. S. Stewart, and N. Lehman. 2004. Genetic consequences of a severe population bottleneck in the Guadalupe fur seal (*Arctocephalus townsendi*). *Journal of Heredity* 95:144–53.

Werner, T. K., and T. W. Sherry. 1987. Behavioral feeding specialization in *Pinarolaxias inornata*, the "Darwin's Finch" of the Cocos Island, Costa Rica. *Proceedings of the National Academy of Sciences U.S.A.* 84:5506–10.

Westerdahl, H., B. Hansson, S. Bensch, and D. Hasselquist. 2004. Between-year variation of MHC allele frequencies in great reed warblers: selection or drift? *Journal of Evolutionary Biology* 17:485–92.

Wright, S. 1931. Evolution in Mendelian populations. *Genetics* 16:97–159.

Wu, Z, Y. Li, and B. R. Murray. 2006. Insular shifts in body size of rice frogs in the Zhoushan Archipelago, China. *Journal of Animal Ecology* 75:1071–80.

Yeh, P. J. 2004. Rapid evolution of a sexually selected trait following population establishment in a novel habitat. *Evolution* 58:166–74.

Yonekura, R., K. Kawamura, and K. Uchii. 2007. A peculiar relationship between genetic diversity and adaptability in invasive exotic species: Bluegill sunfish as a model species. *Ecological Research* 22:911–19.

Zeisset, I., and T. J. C. Beebee. 2003. Population genetics of a successful invader: The marsh frog *Rana ridibunda* in Britain. *Molecular Ecology* 12:639–46.

Zenger, K. R., B. J. Richardson and A-M. Vachot-Griffin. 2003. A rapid population expansion retains genetic diversity within European rabbits in Australia. *Molecular Ecology* 12:789–94.

Sympatric Speciation, Immigration, and Hybridization in Island Birds

Peter R. Grant and B. Rosemary Grant

In this chapter we pay homage to Ed Wilson as Naturalist. His influence on our research on speciation has been much greater than this chapter will reveal, so we begin by making one explicit connection. In the *Theory of Island Biogeography*, MacArthur and Wilson (1967) came close to discussing speciation in chapter 7 when referring to the prevailing view, associated with Mayr (1963), that given enough time isolated populations will diverge genetically to the point at which they are incapable of exchanging genes when finally they encounter each other. They made the insightful point that if islands could be reached once they could be reached again; therefore repeated immigration (and breeding) would retard divergence and a balance would be struck between these opposing processes, rather like the immigration-extinction balance they so successfully modeled. Since then the dynamics of gene flow and selection have been thoroughly investigated (Slatkin 1975, Barton and Slatkin 1986), and they form the core of divergence-with-gene-flow ideas about how speciation occurs (e.g., Rice and Hostert 1993, Smith et al. 1997, Price 2008).

The last forty years of research on bird speciation on islands has yielded different pictures or models of the speciation process (Grant 2001, Price 2008, Grant and Grant 2008a). One elaborates the views on allopatric speciation described above. Divergence takes place in allopatry, and barriers to interbreeding arise there as a result of selection, with or without gene flow from parent to daughter population (model I; see also Clegg, this volume). Founder effects may contribute at the beginning. Speciation is both initiated and completed in allopatry. In the next two models speciation begins in allopatry and is completed in sympatry. The second (model II) lays stress on accelerated divergence at the time of secondary contact through selective reinforcement of reproductive and/or ecological trait differences that initially evolved in allopatry. A third (model III) emphasizes an exchange of genes at the sympatric stage through episodic introgressive hybridization. The exchange does not simply destroy the differences, but through selective backcrossing creates new combinations of genes that enhance responsiveness to selection. The result is speeded up divergence along existing trajectories or change to new trajectories. Fission tendencies alternate with fusion. A fourth view (model IV) holds that all changes occur sympatrically; there is no allopatric phase, and hence no secondary contact. For sympatric speciation to occur there must be assortative mating among members of two groups formed from one by disruptive selection.

The four models differ in biogeographic features, and in how selection is supposed to occur. They combine elements of ecological and nonecological speciation (Schluter 1996, 2001, Price 2008). Three are variations on the allopatric speciation theme. All involve a secondary sympatric phase through immigration, and therefore can be accommodated by the theory of island biogeography fairly simply. The fourth, sympatric speciation, is fundamentally different because it proposes in situ, within-island, origination of new species without immigration. It enhances diversity over and above the effects of immigration, and for that reason we focus on it.

One fruitful approach to the problem of understanding speciation is to study directly the processes hypothesized to be important. This complements the more often used, indirect, comparative method for inferring evolutionary history. The critical processes that need to be demonstrated to discriminate among these four models are effects of intraspecific gene flow from island to island, introgressive hybridization within islands, enhancement of differences between populations soon after secondary contact is made, mate choice, and selection, be it disruptive or directional. All these are amenable to direct study by observation, measurement and experimentation.

Species *in statu nascendi* are especially suitable for direct study of dynamical interactions in sympatry and for extrapolation to the unobserved history of species that are now completely reproductively isolated. This chapter discusses what has been learned recently about speciation through field study of two groups of such species; buntings in the Tristan da Cunha archipelago in the south Atlantic and ground finches in the Galápagos archipelago in the eastern tropical Pacific. In their isolated locations, one can be confident the species evolved where they are now found. In contrast, species in many continental regions and on less isolated islands like the Baltic islands of Gotland and Öland (Tegelström and Gelter 1990) and Britain (Newton 2003) may have evolved in one place and now, postglacially, occupy another.

Sympatric speciation (model IV) has been invoked in both of the cases we review. Serious investigation of sympatric speciation began with a theoretical analysis by Maynard Smith (1966), coincidentally at about the same time as the first synthesis of island biogeography theory (MacArthur and Wilson (1963, 1967). The theories have had largely independent lives since then. In the Discussion we explore some connections between them.

Sympatric Speciation

Solitary islands provide the strongest evidence of sympatric speciation. One species is likely to have given rise to two, sympatrically, if they occupy a single and solitary island, too small to allow for spatial segregation, and they are more related to each other than either is to a third. For example, two species of palms apparently evolved on the single, Australian, Lord Howe Island (Savolainen, Anstett et al. 2006, Savolainen, Lexer et al. 2006, Stuessy 2006, Gavrilets and Vose 2007). This example is similar to fish that have apparently undergone diversification and speciation in single bodies of water where opportunities for spatial segregation are minimal (Schliewen et al. 1994, 2006, Barluenga et al. 2006a,b, Gavrilets et al. 2007). These are essentially insular environments, solitary islands in effect. Coyne and Price (2000) surveyed the relevant bird literature and could find no such examples. Where they might have found examples they didn't. For example, the Cocos finch has been present on the well-isolated Cocos Island long enough to have given rise to other species, and its environment is varied enough to support a variety of feeding types in the population (Werner and Sherry 1987), and yet it has remained a single species under conditions suitable for sympatric, but not allopatric, speciation.

However, three recent studies have suggested that birds may indeed undergo sympatric speciation on islands. One investigated *Nesospiza* buntings on islands in the South Atlantic Tristan da Cunha archipelago (Ryan et al. 2007), and another studied a population of *Geospiza* finches in the Galápagos archipelago (Huber et al. 2007). A third one, suggesting that *Oceanodroma* petrels have speciated sympatrically as a result of breeding in the same location at different times (Friesen et al. 2007), was published after this chapter was written and is briefly mentioned in the Discussion.

Tristan Buntings

The Case for Sympatric Speciation

Two species of *Nesospiza* occur together on Inaccessible and Nightingale, two out of the three islands in the Tristan da Cunha group. One species is large (*N.w. wilkinsi*, Nightingale; *N.w. dunnei*, Inaccessible) and one is small (*N.a. questi*, Nightingale; *N.a. acunhae*, Inaccessible). They are ecologically separated on each island by bill-related food size. *N. wilkinsi* exploit *Phylica* fruits and *N. acunhae* eat grass (*Spartina*) and sedge seeds which are much smaller (Hagen 1952, Elliott 1957, Ryan et al. 2007). Reproductively they are separated by their song, plumage, and size differences (Ryan et al. 2007).

Arrival of buntings in the archipelago can be dated at ~3.3 mya on the basis of a 6.7% difference in cytochrome *b* sequences between *Nesospiza* and the presumptive sister species, *Rowettia goughensis*, on the solitary Gough Island 350 km to the south (Ryan et al. 2007). How did *Nesospiza* speciation then take place? To answer this question Ryan et al. (2007) analyzed mtDNA and microsatellite variation, and found almost complete lineage sorting by island (figure 12.1). This is consistent with *in situ* splitting of a single population into two species, on each of the two islands. Sympatric speciation is the hypothesis favored by Peter Ryan and colleagues. They support it with observations of assortative mating by size, and evidence of ecotypic variation in the smaller species on Inaccessible that is suggestive of disruptive selection and incipient speciation.

A Double-Invasion Explanation

The data are consistent with alternative hypotheses. According to one, ancestral *Nesospiza* buntings colonized the archipelago not once but twice from South America. Sequential invasions of the same lineage have been repeatedly hypothesized to explain the occurrence of two related species on some islands yet only one in the mainland source region (Grant 1968, 2001, Coyne and Price 2000). For example, two species of *Sephanoides* hummingbirds occur on the Juan Fernandez Islands off the coast of Chile, whereas there is only one on the mainland. If the island had been invaded once and the two island species had evolved sympatrically they should be sister species. Phylogenetic reconstruction by Roy et al. (1998) shows they are not. Instead, it supports the double-invasion hypothesis by showing that the mainland species is more closely related to one, presumably a relatively recent colonist, than to the other (Grant 2001).

In the case of *Nesospiza* buntings one species could have colonized the archipelago twice, or two species could have colonized once. Comparisons

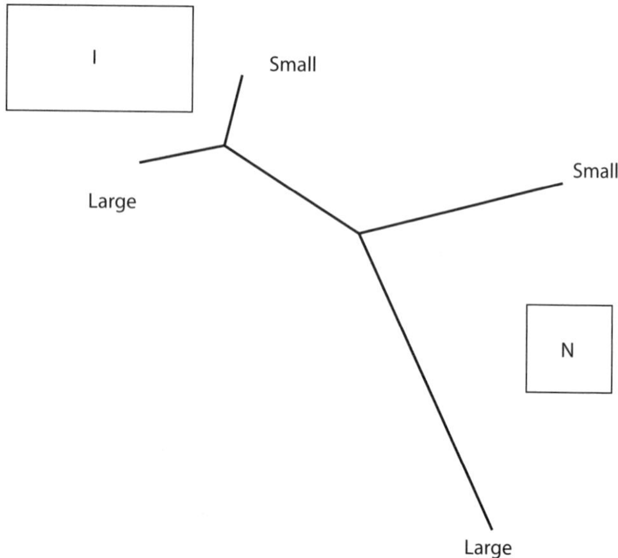

Figure 12.1. Diagram of the relationships among Tristan buntings. Inaccessible (I) and Nightingale (N) Islands are each occupied by a small species (*Nesospiza acunhae*) and a large one (*N. wilkinsi*). The unrooted dendrogram of microsatellite DNA differences placed between the islands shows each sympatric pair to be most similar to each other genetically. In contrast to this, phenotypic similarities are strongest between allopatric pairs. Adapted from Ryan et al. (2007).

with continental species and phylogenetic reconstruction performed so far suggest that Tristan da Cunha was colonized only once, and all *Nesospiza* evolution took place within the archipelago (Ryan et al. 2007).

The Allopatric Speciation Alternative

According to allopatric models of speciation, birds dispersing either from South America or from Gough Island ~3.3 mya colonized Nightingale, the oldest island in the Tristan archipelago. The population gave rise through dispersal to another on Inaccessible, and the two populations diverged, thereby beginning the process of speciation. Sympatry was subsequently established through further dispersal of members of each population to the island occupied by the other: within-archipelago double invasions after differentiation, a Darwin's finch radiation in miniature (Lowe 1923, Lack 1947). If this actually happened, why is it not reflected in the pattern of phylogenetic relationships? The answer is a well-known problem in island speciation inferred from molecular phylogenies (Clarke et al.

1998; see also Chan and Levin 2005): one sympatric lineage has "captured" another through introgressive hybridization, and the phylogenetic signal has become obliterated. Hybridization is now occurring on the younger Inaccessible (3 my), but apparently not on Nightingale (>18 my). It may have occurred on Nightingale earlier, gradually diminishing through time. If so it might be detected with coalescent methods (e.g. Peters et al. 2007).

A prediction of the allopatric hypothesis is that sympatry on Nightingale is no older than 3 my, the age of the younger island. If it is older than 3 my, the allopatric model would have to be abandoned and the sympatric alternative would be upheld. Mitochondrial data do not support such an ancient split: they yield an estimate of 0.3–0.4 my for the separation of Nightingale and Inaccessible buntings (0.7% cytochrome *b* sequence difference). Therefore the allopatric model cannot be abandoned. On the question of whether the earliest split is between species on different islands, as expected from the allopatric hypothesis, or between populations on the same island, as expected from the sympatric hypothesis, the data are equivocal. There are no mitochondrial differences between populations on the same island. This is not expected under a sympatric speciation model. One explanation among others (Ryan et al. 2007) is introgressive hybridization after initial divergence in allopatry. On the other hand, the species on Nightingale differ more in microsatellite profiles, marginally, than either does from buntings on Inaccessible (Ryan et al. 2007). This is consistent with the sympatric hypothesis.

A Long Delay in Speciation

A curious feature of Tristan buntings is that for the first 80–90% of their history on Nightingale only one species existed, to judge from the cytochrome *b* data considered at face value. Even allowing for imprecision in age estimates and the biasing effects of lineage sorting, the magnitude of the delay is remarkable. There is no comparable long delay in finch speciation in two other volcanic archipelagoes, Hawaii (Fleischer and McIntosh 2001) and Galápagos (Grant and Grant 2008a). In the first 80–90% of Darwin's finch history (2–3 my), for example, more than half of the species evolved. Galápagos differs from Tristan in that a minimum of five (volcanic) islands was always present during finch history. This may have allowed species to persist and accumulate in Galápagos even when individual populations became extinct. Note that 14 species of Darwin's finches evolved in a shorter time (2–3 my) than was available to Tristan buntings (3–4 my).

Such a long "waiting time" to speciation (Bolnick 2004) is not expected under the sympatric speciation model except under a set of restricted (genetic) conditions governing mate choice. Neither is it expected under an allopatric model, because for most of that time Inaccessible was present. It takes no more than 0.2 my for a new island to be colonized (see below). The long delay in speciation could be explained ecologically. Plants that constitute one of the niches may have arrived recently, perhaps in the last 0.5 mya. In principle this could be tested with a phylogeny of the food plants (*Phylica* trees, *Spartina* grasses, and sedges). A testable expectation under the allopatric model is that volcanic activity on Inaccessible rendered the island uninhabitable for all or part of its early history, but at the same time was less drastic on the older Nightingale. Volcanic activity occurred in the last 0.5 my on both islands, and therefore probably earlier. It may have extirpated populations on Inaccessible, and possibly also on Nightingale, thereby obscuring the history of the survivor(s).

Future Needs and Conclusions

For a better understanding of the evolutionary history of these buntings it would help to include molecular data from a population of the smaller species (*N. acunhae*) on a third island, Tristan (0.2 my old), because birds from this island may have contributed to the mixture on Inaccessible. The Tristan population is now extinct, owing to human activity; *Spartina* tussocks were destroyed (Hagen 1952) and predatory feral cats, rats, and mice were introduced (Elliott 1957). Unfortunately, it appears that only one specimen of bunting from Tristan exists in museum collections (Lowe 1923, Elliott 1957).

Second, it would help to root the tree. This might permit identification of the oldest species, thereby allowing a more precise framing of the food-niche test described above. *N. wilkinsi* on Nightingale is the best candidate, as it is genetically the most distinctive from the rest. Further, if a root is established with mainland species (e.g., *Sicalis* or *Melanodera* spp.) it might be possible to distinguish between two colonization hypotheses: separate colonizations of Tristan da Cunha and Gough Island from South America, or colonization of one followed by the other. The first hypothesis was suggested by Lowe (1923) and developed by Rand (1955). It is preferred by Ryan et al. (2007) because, among other reasons, population sizes are larger on the mainland than on the islands. Assuming *Rowettia* and *Nesospiza* are truly sister genera, we consider a single, sequential, colonization to be at least as likely as two separate ones, because the South American mainland is 3,000 km away, whereas Gough is little more than a tenth of this distance from the Tristan archipelago.

In summary, ecological, morphological, and genetic patterns among Tristan buntings display elements of all models outlined at the beginning except for one with reinforcement (II). Consistency with the model of sympatric speciation is noteworthy in view of the rarity of evidence for this mode of speciation in birds (e.g. Sorenson et al. 2003, Price 2008). We cannot draw a stronger conclusion because the issue of sympatric speciation is unresolved, and perhaps unresolvable in the light of introgressive hybridization and possible extinctions. The next example provides more evidence of sympatric speciation, of a different kind.

Darwin's Finches

Darwin's finches are a classical example of a young adaptive radiation (Grant 1986, Grant and Grant 2008a). In recent and ongoing radiations the distinction between species is often blurred because there has been insufficient time for complete discreteness to evolve and speciation is incomplete (Grant and Grant 2005). A taxonomist's nightmare is an evolutionary biologist's treasure. Incomplete speciation provides opportunities to study the process. There is no more confusing, and at the same time potentially more rewarding, situation than on Santa Cruz Island. The remainder of this chapter discusses what has been learned from field studies of finches on this and the neighboring island of Daphne Major.

Geospiza fortis *on Santa Cruz Island*

The population of medium ground finches (*Geospiza fortis*) on this island displays an unusual feature: beak sizes are bimodally distributed (figure 12.2) at some localities and at some times (Hendry et al. 2006). The bimodality is not accounted for by average size differences between males and females or between young and old birds. Phenotypic variances are unusually large, and this fact, combined with bimodality, raises the possibility of disruptive selection as a cause of the origin as well as the maintenance of the bimodality (Ford et al. 1973). The hypothesis of current disruptive selection has yet to be tested by quantifying survival and breeding success of individuals in relation to beak sizes, diets and food availability. This is difficult to do in a local area embedded within a larger region because of uncontrolled movement of birds in and out of the study area, and for that reason analysis needs to be restricted to known residents. The best evidence for disruptive selection is the nonrandom persistence of adults from one year to the next in the El Garrapatero study area (Hendry et al. 2009).

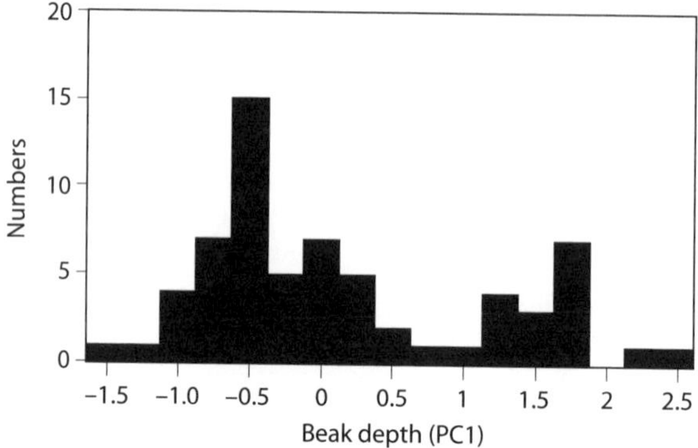

Figure 12.2. Bimodal distribution of beak depth in a sample of male *G. fortis* from the El Garrapatero locality, southern Santa Cruz Island, Galápagos, in 2004. From Hendry et al. (2006), fig. 3.

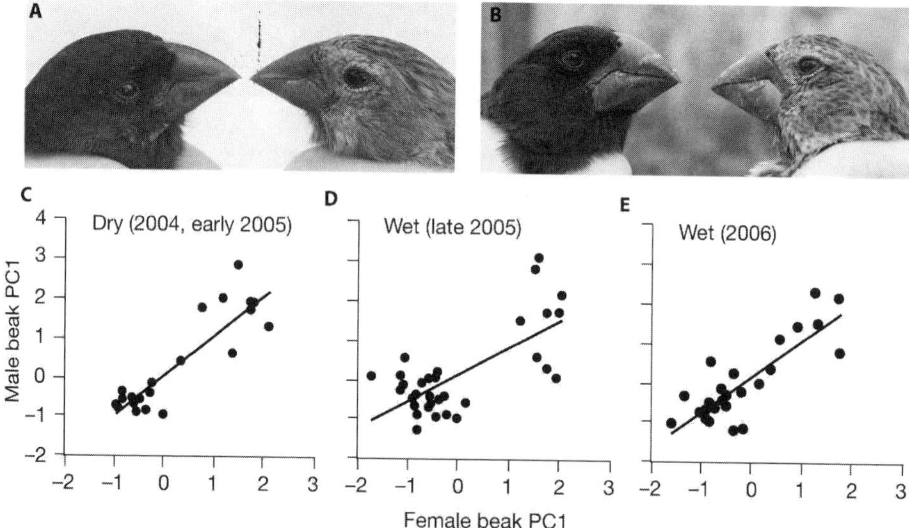

Figure 12.3. Assortative pairing of medium ground finches (*G. fortis*) at El Garrapatero on Santa Cruz Island in (A) 2004–5 (dry conditions), (B) late 2005 (very wet), and (C) 2006 (moderately wet). From Huber et al. (2007).

Another factor maintaining the bimodality is a strong tendency for birds to pair assortatively (Huber et al. 2007). The pattern of morphological variation among pairs (figure 12.3) suggests that large birds mate preferentially with large birds and small birds mate preferentially with small birds. There is no assortative mating within size groups; it is manifest only when size groups are combined. Characteristics of song vary with body and beak size (Huber and Podos 2006), so the cues used in mate choice could be provided by song, by morphology, or by both (Grant and Grant 2008a). Experiments with other populations of *Geospiza* species have demonstrated discrimination on the basis of each set of cues independent of the other (Ratcliffe and Grant 1983, 1985).

As with Tristan buntings, the origin of this interesting situation is unknown. Divergence could have originated sympatrically or allopatrically.

A bimodal beak size frequency distribution coupled with assortative pairing on the basis of beak size is consistent with the idea that the population is in the process of splitting into two, sympatrically, through disruptive selection (model IV). The split has reached the point at which large and small members of the population differ in microsatellite allele frequencies and rarely breed with each other (Huber et al. 2007).

If sympatric divergence is a correct interpretation of their origin, the process has been occurring for a century or more. Specimens of medium ground finches collected on Santa Cruz island at the beginning of the nineteenth century for museums show exactly the same positively skewed frequency distributions with bimodal tendencies as do modern samples, at both northern and southern localities, and in early (<1906) and later (>1924) samples (figures 12.4 and 12.5). Two species of ground finches that are sympatric with *G. fortis*, the small ground finch (*G. fuliginosa*) and the cactus finch (*G. scandens*), show standard normal distributions and no such skew (figure 12.6). They are a kind of "control" for the ongoing "experiment" with medium ground finches (*G. fortis*). The sample of measurements of a fourth species, the large ground finch (*G. magnirostris*), is too small for analysis.

The morphological and mating patterns are also consistent with allopatric model III, under which the population we call *G. fortis* is actually two populations. The bimodality could be the result of unusually large medium ground finches immigrating from another island where average size is large, such as San Cristóbal or Floreana to the south, and breeding with residents on Santa Cruz to some, but apparently incomplete, extent. If so, fission and fusion tendencies have yet to be resolved one way or the other. Nothing is known about current immigration to Santa Cruz. In the absence of other factors it would have to be persistent to account for the persistent bimodality and skew.

Yet another possibility is that skew and bimodality are produced by hybridization with *G. magnirostris*; either residents on Santa Cruz or immigrants from another island. The hypothesis of interbreeding on Santa Cruz is supported by one observation of a mixed pair (Huber et

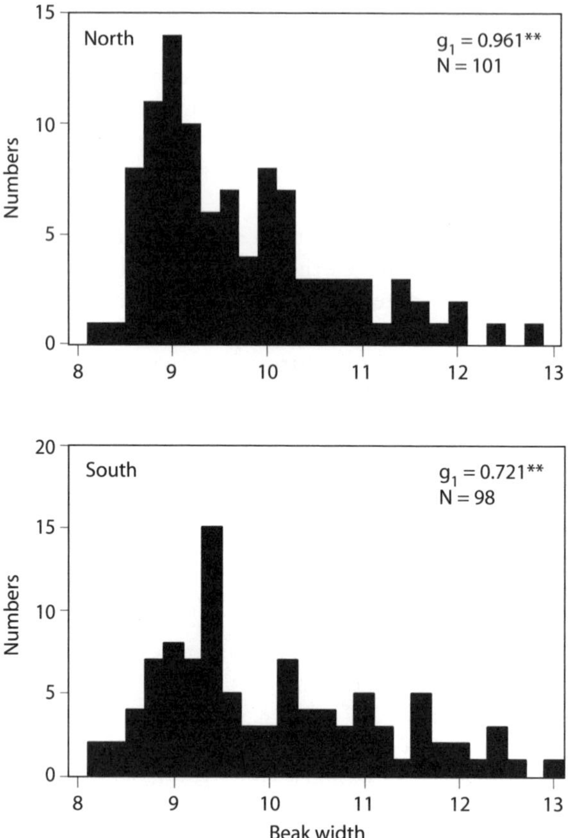

Figure 12.4. Frequency distributions of beak size of medium ground finches (*G. fortis*) collected for museums in the north (1868–1939) and south (1868–1968) of Santa Cruz Island. g_1 is a measure of skewness, *N* is sample size, and two asterisks indicate a significant departure from normality at $P<0.01$ (Snedecor and Cochran 1989). Data originally analyzed in Grant et al. (1985).

al. 2007), by the genetic (microsatellite) similarity of these species compared with allopatric pairs of the same species (Grant et al. 2005), and by the similarity in songs of *G. magnirostris* and large members of *G. fortis* (Bowman 1983, Grant and Grant 1995, 2008a, Huber and Podos 2006).

Thus there is not one but three explanations for the unusual frequency distributions of finch morphology (Grant 1986, Huber et al. 2007), and few data available to discriminate among them. An expanded array of

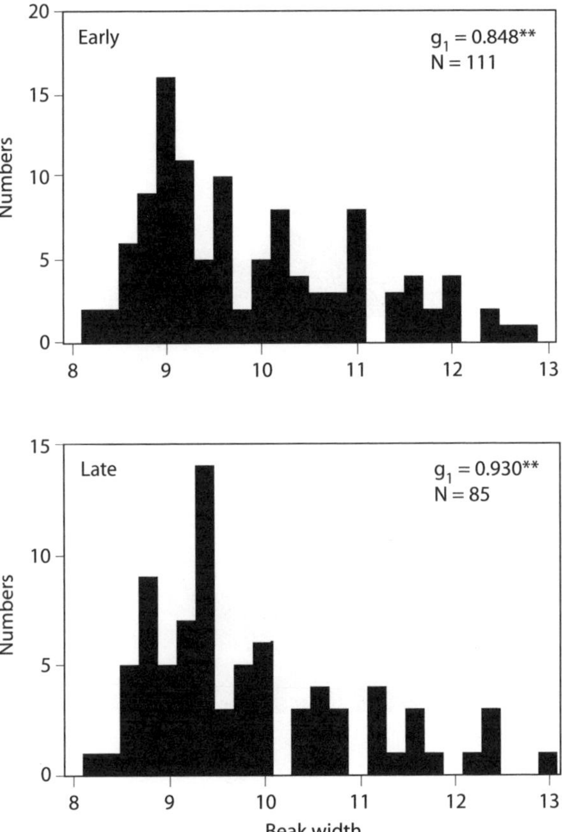

Figure 12.5. Frequency distributions of beak size of medium ground finches (*G. fortis*) collected for museums on Santa Cruz Island, early (1868–1904) and late (1924–1968). Symbols as in figure 12.3.

molecular markers is needed to detect and identify immigrants, F_1 hybrids, and backcrosses. Therefore, for a better understanding of the dynamics of immigration and hybridization, we turn to a long-term study of ground finches on the neighboring small island of Daphne Major (0.34 ha), 8 km north of Santa Cruz. We then apply the findings from Daphne to the question of *G. fortis* evolution on Santa Cruz.

For the immigration hypothesis to be supported it needs to be shown that immigrants from a morphologically differentiated population breed with residents. For the hybridization hypothesis to be supported it needs to be shown that introgressive hybridization results in a skewed distribution.

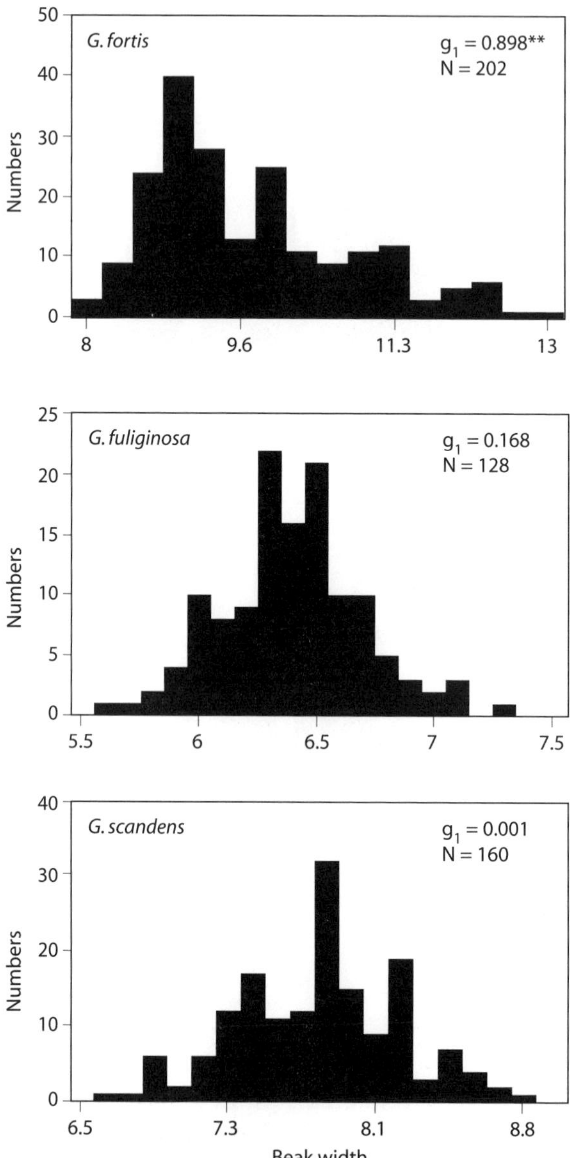

Figure 12.6. Frequency distributions of beak size of medium ground finches (*G. fortis*), small ground finches (*G. fuliginosa*) and cactus finches (*G. scandens*) collected for museums on Santa Cruz Island. Symbols as in figure 12.3.

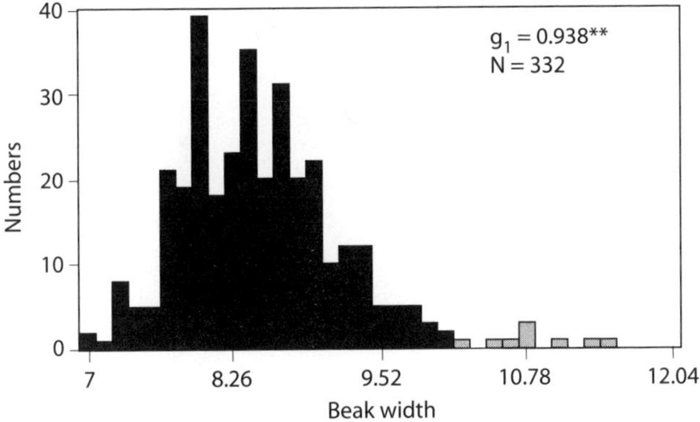

Figure 12.7. Frequency distributions of beak size of live medium ground finches (*G. fortis*) trapped and measured on Daphne Major Island. Gray bars indicate individuals, mainly immigrants and offspring, which are statistically responsible for the skew. Symbols as in figure 12.3.

Immigration of G. fortis *to Daphne Major Island*

Medium ground finches immigrate to Daphne. Their detection is made difficult by the large overlap in frequency distributions of beak and body traits between Daphne resident *G. fortis* and *G. fortis* from other islands. Moreover the 13 populations are not differentiated enough genetically (Grant et al. 2004) to enable us to identify island of origin of individuals by using assignment tests (e.g., Pritchard et al. 2000). Nevertheless, some immigrants can be detected by their phenotype. Daphne residents are smaller on average than all other conspecific populations. Therefore large birds beyond the size range of Daphne residents and within the upper size range of birds on other islands are recognizable as immigrants. They cause the frequency of beak sizes to be positively skewed. They (and their offspring) can be identified as the minimum number of individuals that must be serially deleted from the upper end of a frequency distribution to eliminate the skewness (figure 12.7).

Identified by this means, immigration of large birds is rare and intermittent. The total is 30 out of 3245 (1.0%), and they arrived at only four times. Six arrived in 1977, the year following a long breeding season in the archipelago, one arrived in 1981, 22 arrived sometime after the end of the 1983 El Niño and were captured in 1983–5, and the remaining two arrived in 2000–1. Twenty-five were never seen after their year of capture, and one was seen two years after capture. All these were in immature plumage. Therefore immigration usually ends with the disappearance

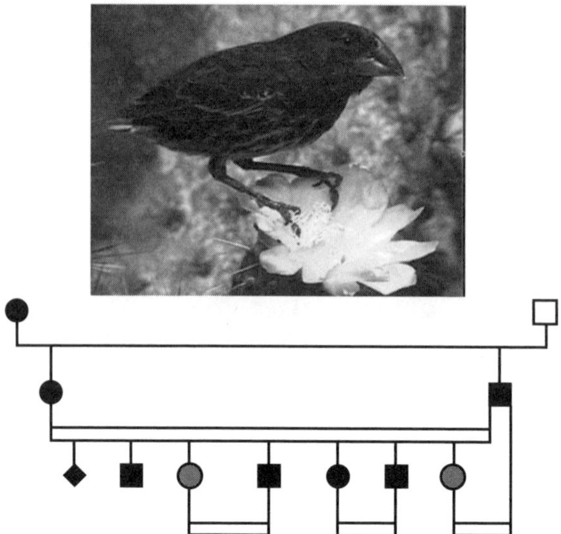

Fig. 12.8. The pedigree of large immigrants on Daphne Island below the female of the F_1 generation (from Grant and Grant 2008b). Genealogical relationships were inferred from genetic (microsatellite) data and from observations. Solid symbols are genotyped birds (circles females, squares males, diamond sex unknown). The unfilled symbol refers to an individual that was known but not genotyped. Gray symbols refer to two birds whose genetic relationships are hypothesized from their phenotypes (see text). Double lines connect the breeding of close relatives. Photo by G. B. Estes.

of the immigrants (death or emigration). We know there were none in 1991 and 1992 because all birds on the island were banded at that time.

Breeding of Immigrants on Daphne Major Island

Only five large immigrants are known to have stayed to breed; one of unknown sex arrived in the early 1970s, two males arrived at different times in the 1980s, and a male and a female arrived in 2000–1. When single birds arrived they bred successfully with residents (Grant and Grant 1996). When the male and female arrived at approximately the same time they bred with each other. Thus, as shown by this pair, some degree of reproductive isolation occurs between large immigrants and residents. This makes plausible the hypothesis of immigration as a source of bimodality, skew, and assortative mating in the Santa Cruz population of *G. fortis*.

This breeding pair is remarkable. It provides a rare example of the crucial step in the allopatric model of the establishment of sympatry.

Observations and genotypes allow us to reconstruct the pattern of events and relationships among the participants (figure 12.8). The original male was first seen in 2000 in immature plumage. It had probably hatched in 1998. It set up a territory, built a nest, and sang, but probably did not breed. The female was first seen the following year, a year of little or no breeding. They bred in 2002, and died in 2003 or early in 2004. Two offspring hatched in 2002 and bred with each other for the first time in 2005, producing at least five offspring (figure 12.8).

The original mother was captured, measured, and genotyped. Her offspring matched her genotype at all 15 microsatellite loci, and matched no other individual's complete genotype. This allowed us to exclude as the mother all *G. fortis* known or suspected to be resident in 2002–5, as well as *G. magnirostris* and *G. scandens*. The genotype of the missing father can be deduced at 12 of the loci; both alleles can be identified by default at nine of them. This enabled us to exclude all *G. magnirostris* and all *G. scandens* as possible fathers as well as all resident *G. fortis*. Altogether 263 *G. fortis*, 60 *G. magnirostris* and 100 *G. scandens* were excluded as parents. Moreover phenotypic data are also inconsistent with a hypothesis of cryptic, that is unobserved, hybridization. The large birds are not intermediate in beak proportions between those of *G. fortis* and *G. magnirostris* as they should have been if they were F_1 and F_2 hybrids (figure 12.9). Thus both the original mother and father must have immigrated. The source island is unknown. On geographical grounds Santa Cruz is the most likely candidate. Parents, offspring, and grand-offspring are above average for Santa Cruz *G. fortis*, spanning the 60th to 90th percentile range in bill characters.

Notice in figure 12.7 how few immigrants can create skewness. The degree of skewness in the frequency distribution of *G. fortis* beak sizes on Daphne in the combined samples from 2002 to 2007 (g_1=0.938, N=332, t=7.22, P<0.0001) is greatly influenced by the measured immigrant in 2000–01 and the six offspring and grand-offspring. When they are deleted from the sample the skewness is more than halved (g_1=0.421, N=325, t=3.24, P<0.0005). Statistical significance can be eliminated altogether just by deleting the next three largest birds (g_1=0.248, N=322, t=1.91, P>0.05).

Introgressive Hybridization on Daphne Major Island

The medium ground finch hybridizes with the small ground finch (*G. fuliginosa*) and the cactus finch (*G. scandens*). Hybridization is rare but persistent, carries no fitness disadvantage under favorable environmental (feeding) conditions that we have been able to discover, and, in the years

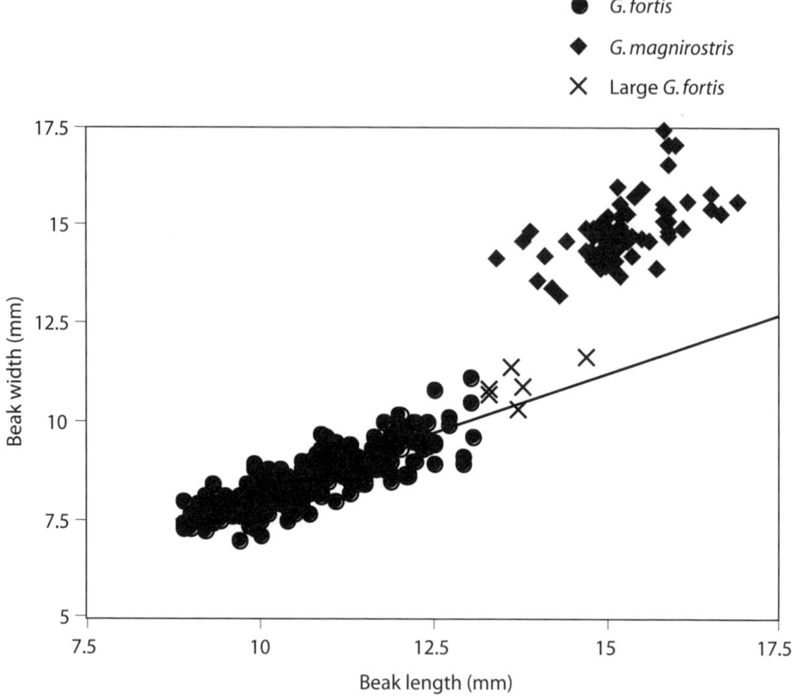

Figure 12.9. Medium ground finches (*G. fortis*) and large ground finches (*G. magnirostris*) on Daphne Island 2002–7.

following the exceptionally strong El Niño event of 1982–3 when favorable feeding conditions persisted, it resulted in a genetic and morphological convergence of the medium ground and cactus finches (figure 12.10). Introgression has the effect of increasing both variance and skewness of the recipient population (figure 12.11; Grant and Grant 2002a). Therefore skewness in the Santa Cruz frequency distributions can be plausibly explained by introgressive hybridization with large ground finches.

Santa Cruz G. fortis *Revisited*

With the known facts about immigration and hybridization on Daphne, we should expect a blurring of the morphological distinction between sympatric species on Santa Cruz. As expected, there is no clear distinction between *G. fortis* and *G. magnirostris* when large samples are analyzed (figure 12.12). Neither we, nor our colleagues, have been able to establish explicit criteria for characterizing each species and distinguishing between them. As a result, individuals between two peaks in the fre-

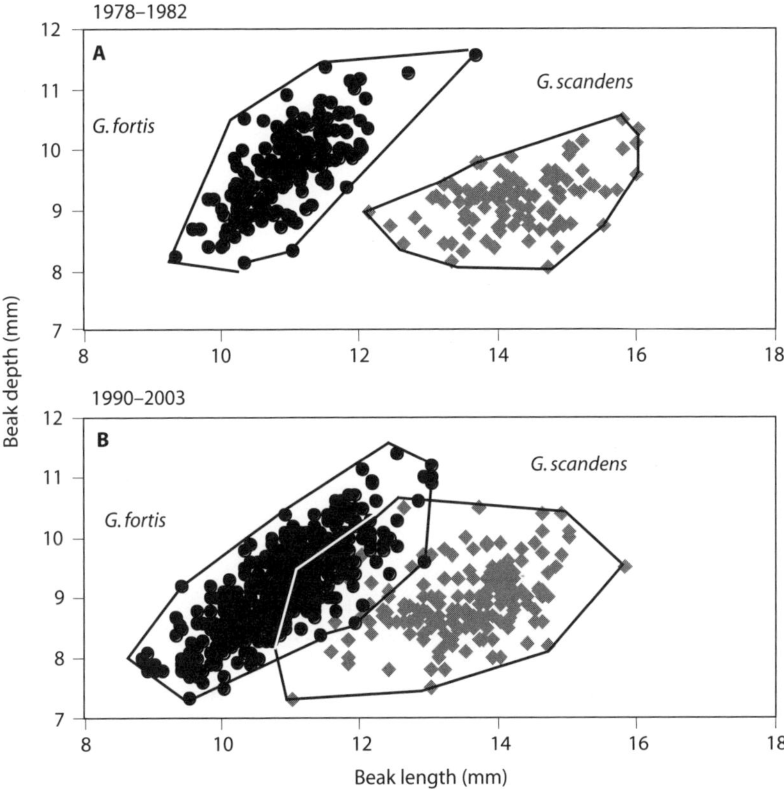

Figure 12.10. Introgressive hybridization after 1983 blurred the morphological distinction between medium ground finches (*G. fortis*) and cactus finches (*G. scandens*) on Daphne Island. Polygons enclose males that sang the species-specific songs and their mates.

quency distribution of beak sizes could be *G. fortis*, *G. magnirostris*, F_1 hybrids, or backcrosses.

However, a fortuitous circumstance enables us to identify *G. magnirostris* individuals objectively. *G. magnirostris* and *G. fortis* occur on Daphne Major without interbreeding. A breeding population of *G. magnirostris* was established on the island at the beginning of the El Niño event in 1982–83 (Gibbs and Grant 1987, Grant et al. 2001), when three female and four male immigrants stayed to breed. Numbers increased gradually as a result of breeding and local recruitment, augmented by additional immigration. Over the following 25 years, when *G. fortis* was hybridizing with *G. scandens*, large ground finches did not hybridize with *G. fortis*, probably because the morphological difference between them here

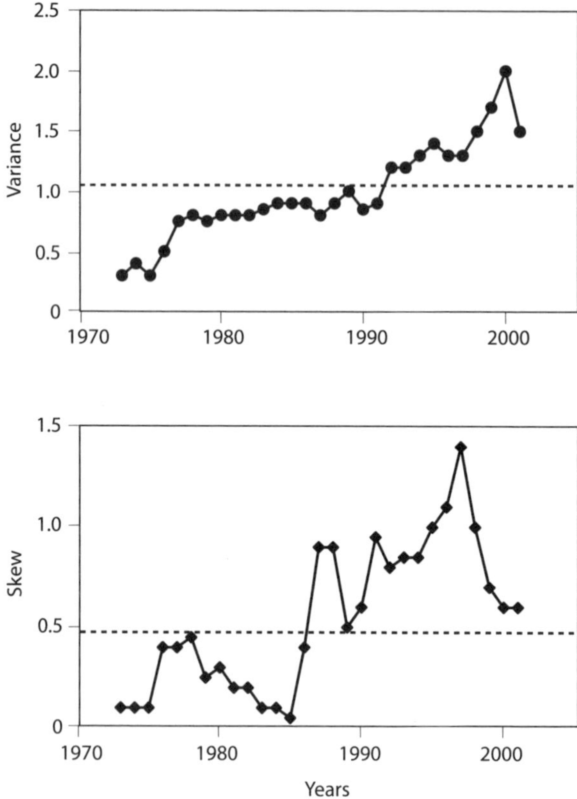

Figure 12.11. Increase in the variance (above) and skewness (below) in the frequency distribution of cactus finch (*G. scandens*) beak shape as a result of interbreeding with medium ground finches (*G. fortis*). From Grant and Grant (2002a).

is unusually large, as a result of the small average size of the *G. fortis* (figure 12.10). *G. magnirostris* on Daphne can therefore be used to identify *G. magnirostris* on neighboring Santa Cruz, on the assumption that distributions of beak sizes of *G. magnirostris* on the two islands are the same. This may not be exactly correct in view of evidence that some *G. magnirostris* immigrate to Daphne from Santiago (Grant et al. 2001). However, any bias arising from inclusion of birds from Santiago is conservative, in that large ground finches on Santiago are slightly larger on average than those on Santa Cruz (Lack 1947, Grant et al. 1985).

First, we combined measurements of live *G. fortis* and *G. magnirostris* on Santa Cruz and Daphne and performed a principal-components anal-

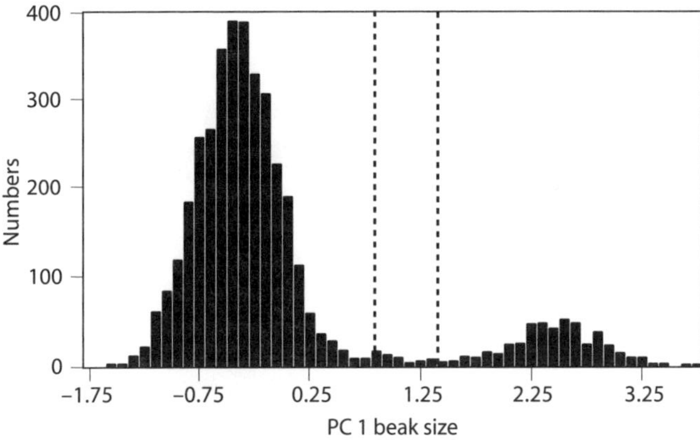

Figure 12.12. Combined frequency distributions of beak size of medium ground finches (*G. fortis*) and large ground finches (*G. magnirostris*) on Santa Cruz and Daphne Major Islands. The right hand broken line shows the lower limit of *G. magnirostris* sizes, calculated from figure 12.9. The left-hand broken line shows the upper limit of the *G. fortis* sizes on Santa Cruz, calculated by serially deleting large individuals from the Santa Cruz sample until skewness disappeared. Individuals between the lines are presumed to be F_1 hybrids and backcrosses.

ysis of three beak dimensions (length, depth, and width). We used PC 1 as an index of size because it accounts for most of the variance (97.2%), and loadings of all three beak dimensions were high (0.977–0.992). We then used the lower boundary of the Daphne *G. magnirostris* distribution as a criterion for identifying *G. magnirostris* on Santa Cruz. No adjustment for skewness was needed; there was none (g_1=0.024). In the final step we ranked the Santa Cruz birds in order of decreasing size, serially deleted birds from beyond the apparent upper end of a normal distribution, and stopped when skewness was at a minimum (g_1=0.035). Birds lying below this boundary are *G. fortis*, while birds above this boundary but below the lower *G. magnirostris* boundary belong to neither species and are therefore identified as hybrids and backcrosses (figure 12.12).

The results were as follows. Nine Santa Cruz individuals considered by us on capture to be *G. fortis* were identified as *G. magnirostris*. All were from Academy Bay in late 1973. An additional 17 were identified as hybrids. The total is 26 out of 278, or approximately 10%.

Hendry et al. (2006) plotted beak depth against beak length of *G. fortis*, measured in the same way as we did, at three localities on Santa Cruz. The results are all positively skewed. Using the classification developed for our own specimens, we estimate that 5% of the Borrero Bay

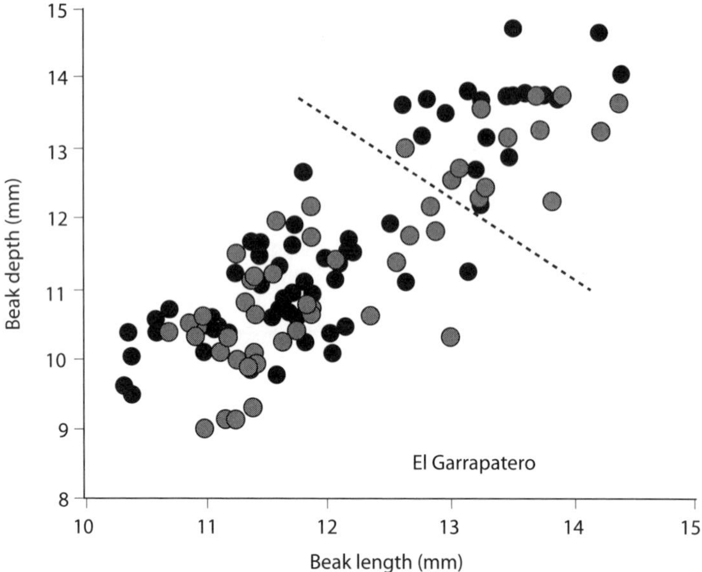

Figure 12.13. Beak sizes of medium ground finches (*G. fortis*) on Santa Cruz Island, from Hendry et al. (2006). The broken line, calculated from estimates in figure 12.12, separates large ground finches (*G. magnirostris*) and presumed hybrids (above and to the right) from *G. fortis* (below and to the left).

sample shown as *G. fortis* are in fact *G. magnirostris* and/or hybrids and backcrosses, and at the other two localities (Academy Bay and El Garrapatero), at least 25% are (figure 12.13). These numbers are approximate and could be somewhat in error; A. P. Hendry (personal communication) considers them to be too high (see Foster et al. 2008). Nevertheless there are clearly *G. magnirostris* in these samples, and probably hybrids and backcrosses too. For example, at El Garrapatero three individuals exceed 14 mm in beak depth, and four exceed 14 mm in beak length.

To summarize, large and small members of the *G. fortis* population differ in microsatellite allele frequencies, have different song characteristics on average, and rarely breed with each other (Huber and Podos 2006, Huber et al. 2007). These are characteristics of sympatric species, which suggests they could be cryptic species; a minifortis and a megafortis. The group of large *G. fortis*, the megafortis, is heterogeneous; it comprises *G. fortis*, some individuals indistinguishable from *G. magnirostris*, and probably F_1 hybrids and backcrosses. The group also appears to be reproductively isolated from larger members of the *G. magnirostris* population (Huber et al. 2007). The group of large *G. fortis* may owe its origin not to a splitting of a single population into two through disruptive selection as envisaged in models of sympatric speciation but to a pooling of genes of two species. In other words, it could be a rare example of hybrid (homoploid) speciation in birds. Ongoing studies of this population (A. P. Hendry and S. Huber, personal communication) are designed to clarify the roles of selection, competition for food, mating structure, and relationships with the small (*G. fuliginosa*) and large ground finches (*G. magnirostris*).

Discussion

As originally formulated by MacArthur and Wilson (1963, 1967), the theory of island biology was ecological and not evolutionary. Whittaker et al. (this volume) summarize efforts to extend the theory by incorporating speciation in archipelagoes (see also Gillespie and Baldwin, this volume). A biogeographically important distinction is to be made between modes of speciation. Allopatric speciation increases the number of species on an island through intra-archipelago immigration, whereas sympatric speciation increases the number on an island without immigration. Sympatric speciation is dependent on environmental heterogeneity (opportunity) within an island persisting for a long time under conditions of low rates of immigration. Logically, therefore, it is to be expected more in the middle of a radiation than early or late (Rosenzweig 1995), on large rather than small islands (Grant and Grant 1989a), and on distant rather than near islands. If sympatric speciation is common, island biogeography theory needs to be modified to allow for an increase in island diversity without immigration (Heaney 2000, Losos and Schluter 2000, Gillespie 2004, Gillespie and Baldwin, this volume). But how likely is this form of speciation for birds on islands?

Despite numerous theoretical investigations into how it might occur (Doebeli 1996, Kawecki 1997, Dieckmann and Doebeli 1999, Kondrashov and Kondrashov 1999, Doebeli and Dieckmann 2000, Dieckmann et al. 2004, Van Doorn et al. 2004, Bürger and Schneider 2006, Bürger et al. 2006, Bolnick and Fitzgerald 2007, Gavrilets et al. 2007, Gavrilets and Vose 2007) sympatric speciation is believed by many to be a rare process in nature, requiring special conditions and circumstances (Coyne and Price 2000, Coyne and Orr 2004, Gavrilets 2004, Bolnick and Fitzgerald 2007). An example of special conditions and circumstances is provided by seabirds. Like some insects (Tauber and Tauber 1989), a few have a relatively unvarying food supply, and this enables them to breed in discretely different seasons at the same place (Bourne 1957, Harris 1969a, Friesen et al. 2007). Coupled with this, some of them (petrels: procellarids) are incapable of relaying for several months if an egg is destroyed (Harris 1969b), and as a result failed breeders, after molting, are likely to return to breed out of synchrony with most of the population (see also Ashmole 1965).

Land birds lack this unusual combination of ecological opportunity and relaying constraint. Hence it is especially noteworthy that two possible cases of sympatric speciation in island land birds have been reported recently. The population of medium ground finches (*G. fortis*) on Santa Cruz Island in the Galápagos shows morphological signs of splitting into two through disruptive selection (Hendry et al. 2006), and nonrandom mating (Huber et al. 2007). The Tristan *Nesospiza* buntings display the molecular signature expected of a species that has already split into two, sympatrically, on two islands (Ryan et al. 2007).

Choosing Between Sympatric and Allopatric Alternatives

As we have discussed in this chapter, all of the observations interpreted as evidence for sympatric speciation (model IV) can be explained alternatively in terms of an allopatric phase of divergence, followed by a sympatric phase with a reversal of divergence caused by introgressive hybridization (model III). Therefore the question arises, how can one choose between them? For the Darwin's finch example, we consider the allopatric alternative to be more parsimonious because it is more strongly supported by observations of evolutionary processes. We suggest that on Santa Cruz Island there are essentially three and a half niches for granivorous finches; hybrids and backcrosses occupy the half, and this situation has persisted for at least a century and probably much more.

Where direct observation of processes is lacking, appeals to parsimony do not provide a clear answer. For example, after the Tristan da Cunha archipelago was colonized by buntings, there was either one additional island colonization and two speciations (sympatric model) or three island colonizations and one speciation (allopatric model). Since the probabilities of colonization and speciation are not known they cannot strictly be compared. Nevertheless, colonization (an event) seems to us to have a much higher likelihood of occurring than speciation (a long process), and on that basis alone the allopatric model has the stronger support.

Sympatric divergence due to selection and ecological and reproductive interactions may be identical under the two models. The crucial distinction between the models lies in the initiation of speciation. The allopatric model specifies geographical separation as the condition under which a population begins to split into two. It can be falsified with molecular data, although the scope for doing so is restricted. We attempted to falsify an allopatric speciation hypothesis for the evolution of Tristan buntings. Molecular data, showing that two islands were present when the initial split occurred, failed to reject it.

In contrast, the sympatric speciation model is difficult if not impossible to test and reject, as far as we can judge, even though some observations are not easily explained by it, e.g., the long waiting time to speciation on Nightingale. It therefore becomes a default model if there are grounds for rejecting an allopatric alternative. Where allopatric speciation cannot be rejected, we believe it is simpler than sympatric speciation because it does not have to confront the following difficulty. Disruptive selection has to be very strong to produce two morphological groups that are ecologically different enough to coexist. This can happen only if some degree of reproductive isolation allows their independent evolution. And yet mate choice of many passerine bird species is based on the learning of signals, and these must be different enough to isolate two groups reproductively. How the groups get to that point of "sufficient" difference is not clear because disruptive selection is not effective without some degree of reproductive isolation. This is the sympatric speciation dilemma.

Darwin's finches on the island of Genovesa illustrate the dilemma. In 1978 two groups of large cactus finches (*G. conirostris*), recognizable by their different songs, differed in average beak size and diets. They appeared to be undergoing a split, sympatrically, into two feeding and breeding groups (Grant and Grant 1979). However, females in this population and related ones (Grant and Grant 2002b) learn both (or all) song types sung by males. In the next generation mating was random with respect to song. Incipient ecological and morphological divergence collapsed as a result of random mating (Grant and Grant 1989b).

We know of only one example of an escape from the sympatric speciation dilemma, and it involves a discrete, rather than a graded, shift in reproductive niche. African viduine finches parasitize the nest of other finch species. As nestlings the parasites learn the characteristics of the hosts and nests. As adults they use the songs learned from their hosts to court and mate at the nests of the hosts. When they switch hosts, as they have done in the past several times, they switch mating signals as well as locations, and in so doing become reproductively isolated from the rest of the population from which they were derived, at one stroke (Payne et al. 2002, Sorenson et al 2003, Price 2008). The success of the new population may depend on overcoming deleterious effects of close inbreeding.

Allopatric Speciation in Progress

In contrast to the field studies of two putative cases of sympatric speciation on islands, recent observations of medium ground finches on Daphne Major Island (figure 12.8) have been made close to the time of origin of non-random mating. An unusually large male and a large female immigrated at approximately the same time and bred with each other, as did their offspring, and their grand-offspring. The pairing pattern of the large immigrants and their offspring reflects a degree of reproductive isolation from the rest of the population unmatched, in our 35-year experience, by any other finch family: pairing among residents on Daphne is almost always random with respect to size traits (Grant and Grant 2008b). Two members of the pedigree have not been characterized genetically, and therefore the degree of reproductive isolation is uncertain. It could be complete.

Even if the reproductive isolation is transitory, it offers two insights into the important stage in speciation when two, differentiated, populations establish sympatry. First, reproductive isolation was apparently fostered by morphological divergence in allopatry. Size, especially beak size, undergoes evolutionary change through natural selection when feeding conditions change (Grant and Grant 2002a, 2008a); hence size-based reproductive isolation is a by-product of ecological divergence under natural selection (Dobzhansky 1937, Schluter 2000, Grant and Grant 2002b). An alternative possibility, that the immigrants and offspring bred with each other and not with the residents because they differed in song, can be ruled out. The immigrant *G. fortis* sang one of the song types prevalent (but rare) among Daphne residents (Grant and Grant 2008b).

Second, small numbers of colonists imply close inbreeding in the initial stages of the sympatric phase of speciation. Colonization of Daphne by two, assortatively mating, *G. fortis* individuals parallels the establishment of a breeding population of large ground finches (*G. magnirostris*) on the same island through immigration of five individuals in the 1980s (Grant et al. 2001). In both cases the population was started with a small number of founders and underwent close inbreeding in the next two generations.

Parapatric Speciation

We conclude that sympatric speciation in island birds is likely to be rare, dwarfed in importance by the allopatric alternative. Nevertheless, there are two additional forms of within-island speciation to consider. Speciation might occur allopatrically on a single island, but only if it is very large, like Madagascar or New Zealand (Diamond 1977). Alternatively it might occur parapatrically, that is, with partial spatial isolation; this is sometimes referred to as contiguous allopatry.

Parapatric speciation can be justified theoretically (Doebeli and Dieckmann 2003, Gavrilets 2004), and models have been developed to capture the essence of well-studied field examples of speciation in palms (Gavrilets and Vose 2007) and fish (Gavrilets et al. 2007). A requirement of the models is a small number of genetic loci with large effects on mate preferences. This makes them inapplicable to the numerous bird species whose mate choice is based on sexual imprinting and not on genetic variation (but see Saether et al. 2007). To be applicable to birds, cultural, nongenetic, influences on mate preferences need to be modeled (e.g., Laland 1994, Boyd and Richerson 2002, Ihara et al. 2003).

For island birds the starting condition could be partially isolated populations along an altitudinal, ecologically varying, gradient connected by limited dispersal and gene flow (Endler 1977, Gavrilets 2004). Spatial segregation (parapatry) could allow the evolution of small, site-specific, differences in ecology and morphology, and divergence in mate preferences based on sexual imprinting. Research is needed to determine if these small morphological differences could then be subsequently magnified, perhaps as a result of divergent coevolutionary dynamics with their foods (seeds, fruits), leading to reproductive isolation of the groups when they later invade each other's ranges and become spatially intermingled. On Inaccessible Island, observations of altitudinal differentiation of *N. acunhae* bunting morphology in relation to variation in the habitat (Ryan et al. 1994, 2007) fit the parapatric speciation alternative. On Galápagos, a similar example has been found with the small ground finch (*G. fuliginosa*) on Santa Cruz Island (Kleindorfer et al. 2006). Consistent with incipient speciation, *G. fortis* on this island can discriminate between local songs and songs sung by birds only 11 km away (Podos 2007).

Nonetheless the question remains: is geographical differentiation within islands an evolutionary end point, or a stage toward completion of speciation marked by coexistence with little or no interbreeding? If within-island speciation initiated parapatrically on moderately large islands proves to be more than just feasible, but likely to occur, the fundamental relation in island biogeography will require a minor modification: at equilbrium, I (immigration) + W (within-island speciation) = E (extinction). If extinction is stochastic, species arising within an island should be just as likely to become extinct as those originating on another island, in which case the equilibrium should not be much affected by how it is reached. On the other hand, within-island speciation might be expected to affect (enhance) the rate of approach to the equilibrium. When calibrated by a measure of time, nonequilibrial communities should have more species than predicted from geography alone (see also Gillespie and Baldwin, this volume).

Acknowledgments

We thank Andrew Hendry, Sarah Huber, Jonathan Losos, Trevor Price, Bob Ricklefs, Peter Ryan, and an anonymous reviewer for their comments, advice, and suggestions.

Literature Cited

Ashmole, N. P. 1965. Adaptive variation in the breeding regime of a tropical sea bird. *Proceedings of the National Academy of Sciences U.S.A.* 53:311–18.

Barluenga, M., K. N. Stölting, W. Salzburger, M. Muschick, and A. Meyer. 2006a. Sympatric speciation in crater lake cichlid fish. *Nature* 439:719–23.

Barluenga, M., K. N. Stölting, W. Salzburger, M. Muschick and A. Meyer. 2006b. Reply: Evidence for sympatric speciation? *Nature* 444:E13.

Barton, N. H., and M. Slatkin. 1986. A quasi-equilibrium theory of the distribution of rare alleles in a subdivided population. *Heredity* 56:409–15.

Bolnick, D. 2004. Waiting for sympatric speciation. *Evolution* 58:895–99.

Bolnick, D. I., and B. M. Fitzpatrick. 2007. Sympatric speciation: models and empirical evidence. *Annual Reviews of Ecology, Evolution, and Systematics* 38: 459–87.

Bourne, W.R.P. 1957. Additional notes on the birds of the Cape Verde Islands, with particular reference to *Bulweria mollis* and *Fregata magnificens*. *Ibis* 99: 182–90.

Bowman, R. I. 1983. The evolution of song in Darwin's finches. In *Patterns of Evolution in Galápagos Organisms,* ed. R. I. Bowman, M. Berson, and A. E. Leviton, 237–537. San Francisco: American Association for the Advancement of Science, Pacific Division.

Bürger, R., and K. A. Schneider. 2006. Intraspecific competitive divergence and convergence under assortative mating. *American Naturalist* 167:190–205.

Bürger, R., K. A. Schneider, and M. Willensdorfer. 2006. The conditions for speciation through intraspecific competition. *Evolution* 60:2185–206.

Chan, K.M.A., and S. A. Levin. 2005. Prezygotic isolation and porous genomes: rapid introgression and maternally inherited DNA. *Evolution* 59:720–29.

Clarke, B., M. S. Johnson, and J. Murray. 1998. How 'molecular leakage' can mislead about island speciation. In *Evolution on Islands,* ed. P. R. Grant, 181–95. Oxford: Oxford University Press.

Coyne, J. A., and H. A. Orr. 2004. *Speciation*. Sunderland, MA: Sinauer Associates.

Coyne, J. A., and T. D. Price. 2000. Little evidence for sympatric speciation in island birds. *Evolution* 54:2166–71.

Diamond, J. M. 1977. Continental and insular speciation in Pacific island birds. *Systematic Zoology* 26:263–68.

Dieckmann, U., and M. Doebeli. 1999. On the origin of species by sympatric speciation. *Nature* 400:354–57.

Dieckmann, U., M. Doebeli, J. A. J. Metz, and D. Tautz. 2004. *Adaptive Speciation*. Cambridge: Cambridge University Press.

Dobzhansky, T. 1937. *Genetics and the Origin of Species*. New York: Columbia University Press.

Doebeli, M. 1996. A quantitative genetic competition model for sympatric speciation. *Journal of Evolutionary Biology* 9:893–909.

Doebeli, M., and U. Dieckmann. 2000. Evolutionary branching and sympatric speciation caused by different types of ecological interaction. *American Naturalist* 156 (suppl.):S77–S101.

———. 2003. Speciation along environmental gradients. *Nature* 421:259–64.

Elliott, H.F.I. 1957. A contribution to the ornithology of the Tristan da Cunha group. *Ibis* 99:545–86.

Fleischer, R. C., and C. E. McIntosh. 2001. Molecular systematics and biogeography of the Hawaiian avifauna. In *Evolution, Ecology, Conservation, and Management of Hawaiian Birds: A Vanishing Avifauna*, Studies in Avian Biology no. 22, ed. J. M. Scott, S. Conant, and C. van Riper III, 51–60. Lawrence, KS: Allen Press.

Ford, H. A., D. T. Parkin, and A. W. Ewing. 1973. Divergence and evolution in Darwin's finches. *Biological Journal of the Linnean Society* 5:289–95.

Foster, D. J., J. Podos, and A. P. Hendry. 2008. A geometric morphometric appraisal of beak shape in Darwin's finches. *Journal of Evolutionary Biology* 21:263–75.

Friesen, V. L., A. L. Smith, E. Gómez-Díaz, M. Bolton, R. W. Furness, J. González-Solis, and L. R. Monteiro. 2007. Sympatric speciation by allochrony in a seabird. *Proceedings of the National Academy of Sciences U.S.A.* 104: 18589–94.

Gavrilets, S. 2004. *Fitness Landscapes and the Origin of Species*. Princeton, NJ: Princeton University Press.

Gibbs, H. L., and P. R. Grant. 1987. Ecological consequences of an exceptionally strong El Niño event on Darwin's finches. *Ecology* 39:1735–41.

Gillespie, R. G. 2004. Community assembly through adaptive radiation in Hawaiian spiders. *Science* 303:356–59.

Grant, B. R., and P. R. Grant. 1979. Darwin's finches: population variation and sympatric speciation. *Proceedings of the National Academy of Sciences U.S.A.* 76:2359–63.

———. 1989a. *Evolutionary Dynamics of a Natural Population: The Large Cactus Finch of the Galápagos*. Chicago: University of Chicago Press.

———. 1989b. Sympatric speciation in Darwin's finches. In *Speciation and Its Consequences*, ed. D. Otte and J. A. Endler, 343–60. Sunderland, MA: Sinauer.

———. 2002b. Simulating secondary contact in allopatric speciation: an empirical test of premating isolation. *Biological Journal of the Linnean Society* 76: 545–56.

Grant, P. R. 1968. Bill size, body size and the ecological adaptations of bird species to competitive situations on islands. *Systematic Zoology* 17:319–33.

———. 1986. *Ecology and Evolution of Darwin's Finches*. Princeton, NJ: Princeton University Press.

———. 2001. Reconstructing the evolution of birds on islands: 100 years of research. *Oikos* 92:385–403.

Grant, P. R., I. Abbott, D. Schluter, R. L. Curry, and L. K. Abbott. 1985. Variation in the size and shape of Darwin's finches. *Biological Journal of the Linnean Society* 25:1–39.

Grant, P. R., and B. R. Grant 1995. The founding of a new population of Darwin's finches. *Evolution* 49:229–40.

———. 1996. Finch communities in a fluctuating environment. In *Long-Term Studies of Vertebrate Communities,* ed. M. L. Cody and J. A. Smallwood, 343–90. New York: Academic Press,.

———. 2002a. Unpredictable evolution in a 30-year study of Darwin's finches. *Science* 296:707–11.

———. 2005. Species before speciation is complete. *Annals of the Missouri Botanical Garden* 93:94–102.

———. 2008a. *How and Why Species Multiply. The Radiation of Darwin's Finches*. Princeton, NJ: Princeton University Press.

——— .2008b. Pedigrees, assortative mating and speciation in Darwin's finches. *Proceedings of the Royal Society of London, Series B* 275:661–68.

Grant, P. R., B. R. Grant, J. A. Markert, L. F. Keller, and K. Petren. 2004. Convergent evolution of Darwin's finches caused by introgressive hybridization and selection. *Evolution* 58:1588–99.

Grant, P. R., B. R. Grant, and K. Petren. 2000. The allopatric phase of speciation: the sharp-beaked ground finch (*Geospiza difficilis*) on the Galápagos islands. *Biological Journal of the Linnean Society* 69:287–317.

———. 2001. A population founded by a single pair of individuals: Establishment, expansion, and evolution. *Genetica* 112/113:359–82.

———. 2005. Hybridization in the recent past. *American Naturalist* 166:56–67.

Hagen, Y. 1952. Birds of Tristan da Cunha. *Results of the Norwegian Scientific Expedition to Tristan da Cunha 1937–38*. 20:1–248.

Harris, M. P. 1969a. The breeding seasons of sea-birds in the Galápagos Islands. *Journal of Zoology, London* 159:145–65.

———. 1969b. The biology of storm petrels in the Galápagos Islands. *Proceedings of the California Academy of Sciences* 37:95–166.

Heaney, L. R. 2000. Dynamic disequilibrium: A long-term, large-scale perspective on the equilibrium model of island biogeography. *Global Ecology and Biogeography* 9:59–74.

Hendry, A. P., P. R. Grant, B. R. Grant, H. A. Brewer, M. J. Brewer, and J. Podos. 2006. Possible human impacts on adaptive radiation: beak size bimodality in Darwin's finches. *Proceedings of the Royal Society of London, Ser. B* 273:1887–94.

Hendry, A. P., S. K. Huber, L. De León, A. Herrel and J. Podos. 2009. Disruptive selection in a bimodal population of Darwin's finches. *Proceedings of the Royal Society of London, Ser. B* 276:753–59.

Huber, S. K., L. F. De León, A. P. Hendry, E. Bermingham, and J. Podos. 2007. Reproductive isolation of sympatric morphs in a population of Darwin's finches. *Proceedings of the Royal Society of London, Series B* 274:1709–14.

Huber, S. K., and J. Podos. 2006. Beak morphology and song features covary in a population of Darwin's finches (*Geospiza fortis*). *Biological Journal of the Linnean Society* 88:489–98.

Ihara, Y., K. Aoki, and M. W. Feldman. 2003. Runaway sexual selection with paternal transmission of the male trait and gene-culture determination of the female preference. *Theoretical Population Biology* 63:53–62.

Kawecki, T. 1997. Sympatric speciation by habitat specialization driven by deleterious mutations. *Evolution* 51:1751–63.

Kleindorfer, S., T. W. Chapman, H. Winkler, and F. J. Sulloway. 2006. Adaptive divergence in contiguous populations of Darwin's small ground finch (*Geospiza fuliginosa*). *Evolutionary and Ecological Research* 8:357–72.

Kondrashov, A. S., and F. A. Kondrashov. 1999. Interactions among quantitative traits in the course of sympatric speciation. *Nature* 400:351–54.

Lack, D. 1947. *Darwin's Finches*. Cambridge: Cambridge University Press.

Laland, K. N. 1994. On the evolutionary consequences of sexual imprinting. *Evolution* 48:477–89.

Losos, J., and D. Schluter. 2000. Analysis of an evolutionary species-area relationship. *Nature* 408:84–50.

Lowe, P. R. 1923. Notes on some land birds of the Tristan da Cunha group collected by the *Quest* expedition. *Ibis* 5, series 11:519–28.

MacArthur, R. H., and E. O. Wilson. 1963. An equilibrium theory of island biogeography. *Evolution* 17:373–87.

———. 1967. *The Theory of Island Biogeography*. Princeton, NJ: Princeton University Press.

Maynard Smith, J. 1966. Sympatric speciation. *American Naturalist* 100:637–50.

Mayr, E. 1963. *Animal species and evolution*. Cambridge, MA: Belknap Press.

Mayr, E., and J. Diamond. 2001. *The Birds of Melanesia*. Cambridge, MA: Harvard University Press.

Newton, I. 2003. *The Speciation and Biogeography of Birds*. San Diego: Academic Press.

Payne, R. B., K. Hustler, R. Stjernstedt, K. M. Sefc, and M. D. Sorenson. 2002. Behavioural and genetic evidence of a recent population switch to a novel host species in brood-parasitic indigobirds *Vidua chalybeata*. *Ibis* 144:373–83.

Peters, J. L., Y. Zhuravlev, I. Fefelov, A. Logie, and K. E. Omland. 2007. Nuclear loci and coalescent methods support ancient hybridization as cause of mitochondrial paraphyly between gadwall and falcated duck (*Anas* spp.). *Evolution* 61:1992–2006.

Podos, J. 2007. Discrimination of geographical variants by Darwin's Finches. *Animal Behaviour* 73:833–44.

Price, T. 2008. *Speciation in Birds*. Greenwood Village, CO: Roberts & Co.

Pritchard, J. K., M. Stephens, and P. Donnelly. 2000. Inference of population structure using multilocus genotype data. *Genetics* 155:945–59.

Rand, A. S. 1955. The origin of the land birds of Tristan da Cunha. *Fieldiana Zoology* 37:139–66.

Ratcliffe, L. M., and P. R. Grant 1983. Species recognition in Darwin's Finches (Geospiza, Gould). I. Discrimination by morphological cues. *Animal Behaviour* 31:1139–43.

———. 1985. Species recognition in Darwin's Finches (Geospiza, Gould). III. Male responses to playback of different song types, dialects aand heterospecific songs. *Animal Behaviour* 33:290–307.

Rice, W. R., and E. E. Hostert 1993. Laboratory experiments on speciation: What have we learned in 40 years? *Evolution* 47:1637–53.

Rosenzweig, M. L. 1995. *Species Diversity in Space and Time*. Cambridge: Cambridge University Press.

Roy, M. S., J. C. Torres-Mura, and F. Herel. 1998. Evolution and history of hummingbirds (Aves: Trochilidae) from the Juan Fernandez Islands, Chile. *Ibis* 140:265–73.

Ryan, P. G., P. Bloomer, C. L. Moloney, T. J. Grant, and W. Delport. 2007. Ecological speciation in south Atlantic island finches. *Science* 315:1420–23.

Ryan, P. G., C. L. Moloney, and J. Hudon. 1994. Color variation and hybridization among *Nesospiza* buntings on Inaccessible Island, Tristan da Cunha. *Auk* 111:314–27.

Saether, S.-A., G. P. Sætre, T. Borge, C. Wiley, N. Svedin, G. Andersson, T. Veen, J. Haavie, M. R. Servedio, S. Bures, M. Kral, M. B. Hjernquist, L. Gustafsson, J. Träff, and A. Qvarnström. 2007. Sex chromosome-linked species recognition and evolution of reproductive isolation in flycatchers. *Science* 318:95–97.

Savolainen, V., M.-C. Anstett, C. Lexer, I. Hutton, J. J. Clarkson, M. V. Borup, M. P. Powell, D. Springate, N. Salamin, and W. J. Baker. 2006. Sympatric speciation in palms on an oceanic island. *Nature* 441:210–13.

Savolainen, V., C. Lexer, M.-C. Anstett, I. Hutton, J. J. Clarkson, M. V. Borup, M. P. Powell, D. Springate, N. Salamin, and W. J. Baker. 2006. Replying to T. F. Steussy. *Nature* 443:E12–E13.

Schliewen, U., T. Kocher, K. R. McKaye, O. Seehausen, and D. Tautz. 2006. Evidence for sympatric speciation? *Nature* 444:E12–E13.

Schliewen, U. K., D. Tautz, and S. Pääbo. 1994. Sympatric speciation suggested by monophyly of crater lake cichlids. *Nature* 368:629–32.

Schluter, D. 2000. *The Ecology of Adaptive Radiation*. Oxford: Oxford University Press.

———. 2001. Ecology and the origin of species. *Trends in Ecology and Evolution* 16:372–80.

Slatkin, M. 1975. Gene flow and the geographic structure of natural populations. *Science* 236:787–92.

Smith, T. B., R. K. Wayne, D. J. Girman, and M. W. Bruford. 1997. A role for ecotones in generating rainforest biodiversity. *Science* 276:1855–57.

Snedecor, G. W., and W. G. Cochran. 1989. *Statistical Methods*, 8th ed. Ames: Iowa State University Press.

Sorenson, M. D., K. M. Sefc, and R. B. Payne. 2003. Speciation by host switch in brood parasitic indigobirds. *Nature* 424:928–31.

Stuessy, T. T. 2006. Sympatric speciation in islands? *Nature* 443:E12.

Tauber, C. A., and M. J. Tauber. 1989. Sympatric speciation in insects: perception and perspective. In *Speciation and its Consequences*, ed. D. Otte and J. A. Endler, 307–44. Sunderland, MA: Sinauer Associates.

Tegelström, H., and H. P. Gelter 1990. Haldane's rule and sex-biased gene flow between two hybridizing flycatcher species (*Ficedula albicollis* and *F. hypoleuca*, Aves: Muscicapidae). *Evolution* 44:2012–21.

Van Doorn, G. S., U. Dieckmann, and F. J. Weissing. 2004. Sympatric speciation by sexual selection: a critical reevaluation. *American Naturalist* 163:709–25.

Werner, T. K., and T. W. Sherry. 1987. Behavioral feeding specialization in *Pinaroloxias inornata*, the "Darwin's Finch" of Cocos island, Costa Rica. *Proceedings of the National Academy of Sciences U.S.A.* 84:5506–10.

Island Biogeography of Remote Archipelagoes

INTERPLAY BETWEEN ECOLOGICAL AND EVOLUTIONARY PROCESSES

Rosemary G. Gillespie and Bruce G. Baldwin

THE EQUILIBRIUM THEORY OF ISLAND biogeography (ETIB) was developed around the concept of islands formed *de novo*, with species colonizing and over time reaching a balance between immigration and extinction (MacArthur and Wilson 1967, see Schoener chapter, this volume). A great challenge to the theory has been its application to remote oceanic islands—those that are formed from beneath the ocean surface and are beyond the normal limits of dispersal for a taxon, where immigration occurs relatively rarely and speciation relatively frequently. Here we examine the interaction between speciation and immigration in community assembly on remote islands. Perhaps the most significant finding is that lineages vary considerably in terms of how they colonize remote islands, and how they accumulate on those islands over time. In particular, in comparing lineages that have been in the Hawaiian archipelago for the lifespan of the current high islands (allowing island chronology to be used to assess community changes over time), some lineages seemingly accumulate species rapidly, often reaching numbers well beyond the putative equilibrium, before declining in number (see figures 13.4C and 13.4D below). Other lineages, especially those that are less diverse, appear to accumulate species more slowly, and some may not reach equilibrium within the time frame of existence of the high islands (approximately 5 my). These results have intriguing parallels to the ETIB, and lay a foundation for developing hypotheses to test the predictability of species accumulation, extinction, and invasion on remote islands.

Attributes of Remote Islands

Characteristics of communities on remote islands include (1) compositional disharmony as a result of differing abilities of lineages to disperse over long distances, leading to attenuation in the number of organismal groups represented with increasing isolation; and (2) high levels of endemism associated with rare colonization events and adaptive radiation (see Carlquist 1974). In particular, when the isolation of an island is extreme, the frequency of colonization becomes sufficiently low to allow *in situ* evolution of new species to play a role in filling the available ecological space (MacArthur and Wilson 1967), often through adaptive radiation (see Schluter 2000). The isolation necessary for the rate of speciation to exceed immigration has been termed the "radiation zone" (MacArthur and Wilson 1967): "Near the outer limit of the dispersal range of a given taxon, speciation and exchange of newly formed autochthonous species within an archipelago can outrun immigration from outside the archipelago and lead to the accumulation of species on single islands" (p. 180). The physical separation required for this effect to be manifest is clearly dependent on dispersal abilities; for example the radiation zone for mammals is much nearer the source than for many insects.

For many isolated oceanic archipelagoes, the age of each island is often known with some precision. This knowledge, coupled with molecular tools that have allowed identification of the source and frequency of colonization, has provided a chronological framework within which to examine the interplay between migration and speciation in the formation of communities over time. The Hawaiian Islands are particularly amenable to such studies, in part because they are generally considered to be the most isolated archipelago in the world. In addition, the youth of the islands (current high islands formed 0.4–5.1 mya; Clague and Dalrymple [1987]) and their linear and chronological arrangement, provide a clear-cut framework for examining how communities have been formed over recent evolutionary time. Accordingly, much of our discussion will be focused specifically on the Hawaiian archipelago. Since the 1980s, molecular studies of a wide diversity of lineages from the Hawaiian Islands have allowed for a much better characterization of dispersal patterns and timing than was previously possible and in turn allow for biogeographic insights highly relevant to the ETIB, as discussed below.

We examine four aspects of remote islands relevant to the ETIB: (1) colonization, i.e., which species reach remote islands, and how and why; (2) changes that occur subsequent to colonization on remote islands, given that the much reduced rate of colonization allows evolutionary processes to come into play; (3) mechanisms by which species are added to communities on remote islands, and comparisons between outcomes from speciation and immigration; and, finally, (4) how communities are assembled over space and time on isolated archipelagoes, in particular the interplay between ecological and evolutionary changes in dictating the composition of communities.

Colonization of Remote Islands

To understand the formation of communities on remote islands, we must first recognize the context of species arrival. What are the characteristics of successful colonization—what species arrive, in what manner, and how frequently?

Active versus Passive Dispersal

> *[I]t can be expected that stepping stones are more important to species whose propagules tend to disperse actively or on floating "rafts," such as birds, mammals, and some plants and arthropods. They are relatively less important to species whose propagules tend to be dispersed passively in the wind, such as most microorganisms and many higher plants and arthropods.*
> *—MacArthur and Wilson 1967, pp. 132–33*

A more general prediction based on this statement is that the likelihood of a species reaching a remote island, and its tendency to use intervening stepping stones, will be dictated by its propensity for active versus passive dispersal. Do recent empirical data support this prediction?

The different mechanisms and propensities for dispersal are likely to result in different biogeographic patterns, as evidenced by recent studies showing first that passively dispersive groups have colonized remote archipelagoes repeatedly and independently. For a number of these lineages, dispersal has been much reduced within each archipelago (see below), such that colonists are unlikely to use more proximate archipelagoes as stepping-stones to more remote archipelagoes. This may be simply because the chance of a highly dispersive mainland propagule reaching a remote archipelago is higher than the chance of arrival of a propagule from an intervening archipelago where evolution has resulted in reduction of dispersal ability. For example, the highly dispersive (by wind) spider genus *Tetragnatha* has colonized each of the different archipelagoes of Polynesia independently from different sources, with diversification within each archipelago from a single founder following reduction in dispersal abilities (Gillespie 2002). Drosophilid flies also seem to have colonized the different remote archipelagos of Oceania independently (P. M. O'Grady, personal communication).

Among plants that are known for passive propagule dispersal, extensive within-archipelago diversification from a common founder is unusual. For example, significantly lower levels of endemism in Hawaiian ferns compared to angiosperms—both for the entire archipelago and for individual islands—probably reflects much greater passive-dispersal ability of spores compared to seeds in general (see Fosberg 1948; Driscoll and Barrington 2007). Passive transport of fern spores in the northern subtropical jet stream is consistent with phylogenetic data from multiple Hawaiian fern lineages (Geiger et al. 2007). Molecular phylogenetic evidence supports repeated colonization of the Hawaiian archipelago by most fern genera (e.g., *Asplenium* [Ranker et al. 1994, Schneider et al. 2004]; *Dryopteris* [Geiger and Ranker 2005]; *Polystichum* [Driscoll and Barrington 2007)]). In such systems, the rate of colonization and occupancy of ecological space through dispersal may exceed or inhibit the rate of diversification (e.g., through outside gene flow) and thereby limit levels of endemicity.

In contrast, active dispersal (e.g., by birds) appears to be associated with less frequent or widespread island colonization and high levels of endemism, as in many flowering plants. Price and Wagner (2004) suggest that intermediate dispersal ability afforded by bird transport allows for plant colonization to occur across islands of an archipelago while maintaining a sufficient degree of isolation for diversification to occur. Indeed, the majority of Hawaiian angiosperm lineages have fruit or seed characteristics consistent with dispersal by birds (Carlquist 1974, Sakai et al. 1995) and those lineages are significantly more diverse than lineages with abiotic dispersal (Price and Wagner 2004); birds also appear to account for the majority of plant lineages (~90%) in the highly endemic flora of the Juan Fernandez Islands (Bernardello et al. 2006). In the genus *Cyrtandra* (Gesneriaceae), diversification on islands throughout the Pacific has been restricted to a fleshy-fruited and putatively bird-dispersed lineage within the genus, as in *Scaevola* (Howarth et al. 2003), with interisland dispersal events often associated with the origin of new species or major clades (Cronk et al. 2005). Likewise, crab spiders (Thomisidae), which are potentially bird-dispersed, are found throughout the Hawaiian, Society, and Marquesas islands, and are diverse within each archipelago, this entire lineage forming a tightly monophyletic clade (Garb and Gillespie 2006, 2009).

For taxa that undergo active dispersal, and in which the dispersal mechanism itself will not necessarily lead to loss of propagules from a remote island, it is unlikely that selection would act to dramatically reduce dispersal ability in the same manner as may occur in taxa that are passively dispersed (see below). However, selection may still reduce dispersal ability among active dispersers if habitat space is highly confined in the island environment or a shift in ecology favors changes in propagule characteristics (see Carlquist 1974). Therefore, in general, stepping-stones may play a more prominent role in the biogeography of actively-dispersed taxa than of passively dispersed taxa that are subject to strong selection against dispersal ability in an insular setting.

Niche Preemption

> *An island is closed to a particular species either when the species is excluded by competitors already in residence or else when its population size is held so low that extinction occurs much more frequently than immigration.*
>
> *—MacArthur and Wilson 1967, p. 121*

Once a niche has been filled, it appears to be more difficult for closely related and putatively ecologically similar colonizers to enter, as suggested for plants in the Canary Islands: For the full suite of endemics in each of 20 plant genera that are highly diverse in the archipelago, Silvertown (2004) noted that each lineage is monophyletic; in contrast, he provided evidence for repeated colonization of the Canary Islands by 20 genera of low insular diversity. He interpreted that pattern as possible evidence for the importance of niche pre-emption by radiating lineages and consequent failure of later arriving close relatives to become established (see also Silvertown et al. 2005). Indeed, successful independent colonizations of Macaronesia by congeneric angiosperms have occurred only when different islands are involved or when the congeneric lineages are widely divergent and putatively distinct ecologically (Carine et al. 2004).

Similar conclusions could be drawn for the flora of the Hawaiian Islands, where, as noted above, molecular phylogenetic studies have shown that all endemic angiosperm species of most individual genera or groups of related genera constitute a single endemic clade, including numerous groups that were previously thought to stem from multiple introductions, such as the extraordinarily diverse lobeliads (Givnish et al. 2008). Hawaiian angiosperm genera with indigenous taxa that stem from multiple introductions include either only one or two species in each of two endemic lineages (*Rubus*, *Santalum*) or only a single species in two of three indigenous clades (*Scaevola*) (Howarth et al. 1997, Alice and Campbell 1999, Howarth et al. 2003, Harbaugh and Baldwin 2007).

Niche preemption also may contribute to the "progression rule" of Funk and Wagner (1995): In Hawaiian plants, rarity of back migration to islands previously occupied by other members of a highly diversified island lineage and lack of diversity of such back-migrant lineages may reflect a degree of niche preemption by already present members of the same insular clade. The progression rule of successive dispersal from older to younger islands in the Hawaiian chain holds well for most well-resolved plant and animal lineages that appear to have arrived initially on older islands. In the silversword alliance, for example, no unequivocal instance of back migration has been documented (Baldwin and Robichaux 1995, B. G. Baldwin, unpublished); likewise in various spider (Hormiga et al. 2002, Gillespie 2004) and insect (Mendelson and Shaw 2005) lineages. In the lobeliad genus *Cyanea*, the only unequivocal younger to older island dispersal event (based on a cpDNA tree) involves a lineage that evidently had not previously colonized that island, instead initially dispersing east—past Oahu—from Kauai to Maui Nui and then west from Maui Nui to Oahu (Givnish et al. 1995). In the highly diverse *Schiedea*, Wagner et al.'s (2005) biogeographic hypothesis, based on molecular and morphological data, provides only one unequivocal example of a species recolonizing an ancestrally occupied island. Additional resolution of divergence times and better characterization of ecological traits of island lineages should allow for more rigorous evaluation of niche preemption within a phylogenetic context.

An example of possible niche preemption in the larger Pacific region is found in crab spiders (Thomisidae), where one lineage (apparently from the Americas) has diversified in the Hawaiian, Marquesas, and Society islands, while another (which appears to have arrived from Australasia) has diversified in the Australs (Garb and Gillespie 2006); there is no distributional overlap between the lineages. Given that these archipelagoes are separated by only ~500 km, and that molecular evidence suggests that crab spiders have been in each archipelago for ~5 million years, the lack of distributional overlap of the two lineages is likely due to priority effects (first to get there "wins") rather than lack of time for dispersal between archipelagoes.

The overall pattern of colonization of remote islands therefore has some elements of predictability, and some of stochasticity. Given propagule availability, the establishment of particular species is largely unpredictable; however, it appears that once the sweepstakes for a given niche have been "won," the chances of establishment by a distinct but closely related and ecologically similar species become considerably diminished (see also Losos et al. 1993). That interpretation is consistent with Darwin's (1859) naturalization hypothesis, of phylogenetic overdispersion in comparisons between native and invading species (see Strauss et al. 2006, Proches et al. 2008), although the causes of such overdispersion may extend beyond niche considerations to community-level factors, such as sharing of enemies (e.g., herbivores or parasites) by closely related residents and invaders (but see Parker et al. 2006a,b, Riccardi and Ward 2006).

Change Subsequent to Colonization

Following successful establishment, island taxa can be expected to undergo ecological adjustment dictated by the abiotic and biotic environment. However, on remote islands, because of the paucity of successful colonizers and the associated abundance of "open" niches, successful colonists frequently have both time and opportunity to change, adapt, and often diversify to an unusual extent, with some widespread processes and patterns evident across distantly related taxa, as discussed below.

Loss of Dispersal Ability

> *[C]onspicuous is the tendency to lose dispersal power [which can occur through the development of] flightlessness . . . increase in fruit size. . . . A second means by which dispersal power is apt to be reduced is the tendency of evolving isolates to vacate the marginal habitats that are the best staging ground for . . . arriving propagules.*
>
> *—MacArthur and Wilson 1967, pp. 157–58*

Reduction or change in dispersal abilities subsequent to colonization of oceanic islands was initially discussed by Darwin (1859), who wrote

> In some cases we might easily put down to disuse modifications of structure which are wholly, or mainly, due to natural selection. Mr. Wollaston has discovered the remarkable fact that 200 beetles, out of the 550 species inhabiting Madeira, are so far deficient in wings that they cannot fly; and that of the twenty-nine endemic genera, no less than twenty-three genera have all their species in this condition! Several facts, namely, that beetles in many parts of the world are very frequently blown to sea and perish; that the beetles in Madeira, as observed by Mr. Wollaston, lie much concealed, until the wind lulls and the sun shines; that the proportion of wingless beetles is larger on the exposed Dezertas than in Madeira itself; and especially the extraordinary fact, so strongly insisted on by Mr. Wollaston, of the almost entire absence of certain large groups of beetles, elsewhere excessively numerous, and which groups have habits of life almost necessitating frequent flight; -these several considerations have made me believe that the wingless condition of so many Madeira

> beetles is mainly due to the action of natural selection, but combined probably with disuse. For during thousands of successive generations each individual beetle which flew least, either from its wings having been ever so little less perfectly developed or from indolent habit, will have had the best chance of surviving from not being blown out to sea; and, on the other hand, those beetles which most readily took to flight will oftenest have been blown to sea and thus have been destroyed.

The above ideas have largely been substantiated, with some caveats, by more recent research, and have been discussed extensively by Carlquist (1966, 1974, 1980). At least among many taxa that are passively dispersed by wind, if dispersal ability is not reduced there is a high chance of being transported off the island and lost at sea or, within an island, being transported beyond the bounds of the narrow, stable habitats to which insular organisms often become adapted. As discussed above, these arguments do not apply as much to taxa that undergo active dispersal. Recent work continues to lend support to the arguments for loss of dispersal power among passive dispersers on remote islands (Gillespie et al. 2008). This single tendency has clearly played a major role in subsequent adaptive radiation, although selection leading to reduced dispersal ability may have diverse explanations in animals and plants (Carlquist 1966, 1974, 1980).

Ecological Release

Ecological release is a commonly documented pattern among island colonists that enter a community with a smaller fauna (MacArthur and Wilson 1967, p. 105). On very remote islands, because of the paucity of successful colonizers and the associated abundance of "open" niches, successful colonists frequently have both time and opportunity to change, adapt, and often diversify to an unusual extent (see Schluter 2000; Nosil and Reimchen 2005). Roughgarden (1972) stated that "[t]he process of faunal buildup on an island is a race between a widening of the offspring phenotype distribution of the first species there and dispersion to the island by members of some other ecologically differentiated species. If the widening of the offspring phenotype distribution, and hence the niche width, is sufficiently slow, vacant regions exist on the resource axis, which facilitate establishment of emigrants from elsewhere" (p. 117). In this scenario, ecological release, by filling the niche space, impedes faunal buildup. However, evidence also shows that broad ecological release may serve as a precursor to adaptive radiation, at least in some situations (Simpson 1953, Schluter 2000). Diamond (1970) showed that, under species-poor conditions, much of an island can become occupied by taxa

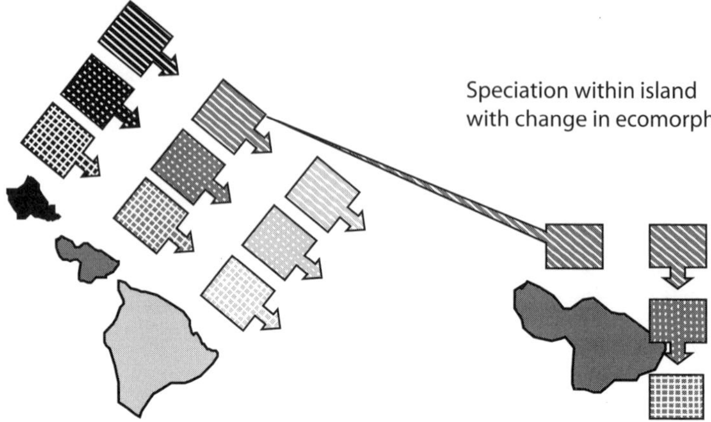

Figure 13.1. Schematic representation of the two mechanisms by which species on the Hawaiian archipelago can occupy a niche on a new island as it arises. Left. Species colonize down the chain of islands (represented by shades of gray), in the direction of open ecological space, and occupy the niche (represented by different patterns) to which they were already adapted on the older island. Right. Species colonize open ecological space on an island by speciation and an adaptive shift from one niche that allows occupation of the "new" niche.

that are relatively maladapted, at least initially, setting the stage for selection on subsequent generations of phenotypes that may lead to evolutionary specialization. During the course of adaptive radiation, at least in archipelago situations, both colonization and evolution are clearly involved in adding new species to a community (figure 13.1). Therefore, expanding on the arguments of Roughgarden (1972), in situations where colonization is extremely rare, the "race" may be between repeated colonization of different preadapted species as opposed to evolution within a single colonist lineage that expands its ecological range and subsequently specializes on a suite of different resources.

Some radiations show repeated episodes of phenotype expansion (with or without speciation), while others show a single episode of phenotype expansion at the base of the radiation, with further speciation accompanied by little ecological change. Clear examples of the latter situation have been documented in insects (plant hoppers in the genus *Nesosydne*, Delphacidae) in which extensive ecological shifts have occurred early in the radiation, with relatively minor changes subsequently (Roderick

1997). A number of flowering plant radiations show a similar pattern of early radiation into distinct ecological settings, followed by allopatric divergence in similar habitats on different islands (e.g., *Schiedea*; Sakai et al. 2006). Other flowering plant lineages undergo less frequent inter-island dispersal, with independent ecological radiation into diverse habitats on each island (e.g., the Hawaiian silversword alliance; Baldwin and Robichaux 1995) (figure 13.2). Both patterns have been commonly resolved within the same major radiation in each of several endemic Canarian angiosperm clades (e.g., woody *Sonchus* [Kim et al. 1996]; *Pericallis* [Panero et al. 1999]; *Sideritis* [Barber et al. 2000]; *Crambe* [Francisco-Ortega et al. 2002]; *Aeonium* and relatives [Mort et al. 2002]; *Lotus*, [llan et al. 2004]; *Bystropogon* [Trusty et al. 2005]). Note that the Canaries, although relatively close to a continent compared to Hawaii, are effectively remote for many taxa because the nearby (desertic) regions of Africa are climatically distinct from most of the islands. Other Canarian plant lineages display principally one pattern, of primarily allopatric divergence across similar habitats on different islands (e.g., *Argyranthemum* [Francisco-Ortega et al. 1996]) or of adaptive radiation across different habitats on the same island (e.g., *Micromeria* [Meimberg et al. 2006]), as in the Hawaiian silversword alliance (Baldwin and Robichaux 1995).

Insular adaptive radiation involves filling ecological space with limited underlying genetic diversity. A lineage is therefore constrained by the variation available for selection to act on in adaptation to a given environment. Accordingly, some niches likely remain unfilled. Also, although a given niche (e.g., "under bark," "under leaf") in different sites (e.g., different islands) may be filled by different taxa within a lineage, and not necessarily by closest relatives, those different species may use the niche in an almost identical manner (see below).

Species Addition on Remote Islands

The most pronounced difference between very isolated islands and those closer to a source is the relative contribution of evolutionary processes compared to immigration of colonists in adding species. Heaney (2000) argued that, because the interaction between speciation and colonization is complex on remote islands (the rate of the former being dictated by a fine balance with the magnitude of the latter), speciation cannot be considered simply as additive to colonization. However, the primary difference with more remote islands is that different niches within the environment will tend to remain relatively "open" for long periods, and consequently

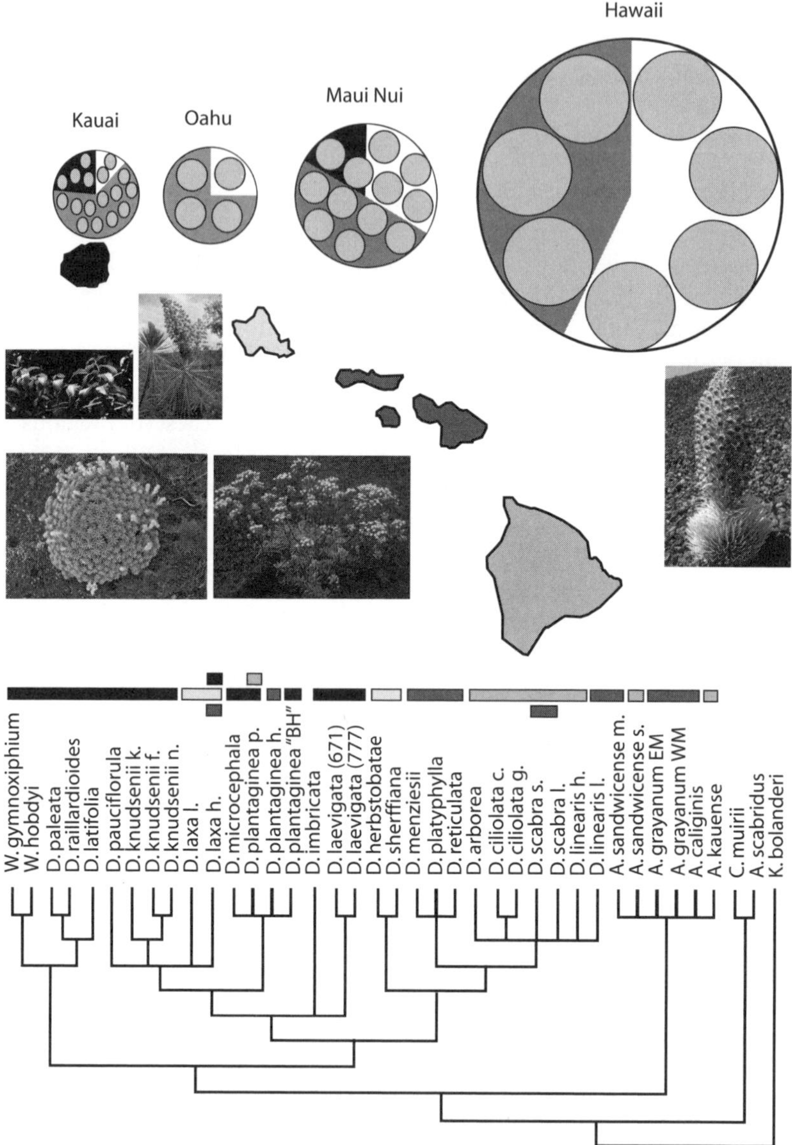

Figure 13.2. Habitat diversity in the Hawaiian silversword alliance (*Argyroxiphium*, *Dubautia*, and *Wilkesia*) across islands. As illustrated, species diversity (per unit area) is greatest on the oldest high island, Kaua'i, and least on the youngest island, Hawai'i. The proportion of species found in mesic to wet habitats (including bogs) is greatest on Kaua'i (where such habitat is relatively extensive—and highly dissected) and least on Hawai'i (where low and high elevation dry

may be filled by evolution as readily as (or more readily than) by dispersal, depending on the relative rates of these two processes.

Losos and Schluter (2000) showed that, for lizards in the Caribbean, speciation can serve as a surrogate for colonization in terms of contributing species to a system. Moreover, lizards on these islands are known to exhibit discrete ecological affinities, represented by "ecomorphs" or taxa whose appearance is determined by ecology (Williams 1972). Similar, and highly deterministic, sets of ecomorphs have evolved, almost always independently, on each island. That such evolution of similar forms can occur may be explained again by the relative roles of colonization and speciation in the context of ecological range shifts (Roughgarden 1972).

The relative rates of colonization versus adaptive differentiation will vary across and within lineages. In damselflies, for example, the rate of migration between islands, though uncommon, is likely sufficient to preclude niche shifts in the earliest colonizers (Jordan et al. 2005). In the case of host-associated insects on remote islands (Gagne 1997, Roderick 1997, Farrell and Sequeira 2004), the very few initial colonists may undergo major host shifts at the outset of the radiation, though once these shifts have occurred, subsequent differentiation within the archipelago is limited almost entirely to shifts between closely related hosts.

In assessing the parallels between colonization and speciation in forming communities, we must consider the sequence of events following colonization of remote islands. The Hawaiian Islands are particularly

---

habitats are more extensive relative to mesic and wet habitats than on the older islands). Size of pies is proportional to area of island(s). Portions of the pie indicate the habitats in which the plants occur: dry, i.e., dry forest, scrub, or barrens (no shading); bogs (heavy shading); and other wet habitats, i.e., mesic to wet forest or scrub (light shading). Circles within the pies represent species that occur in each habitat/island, with bars on the cladogram color-coded according to island: black, Kauai; white, Oahu; dark gray, Maui Nui; and light gray, Hawaii. [Note: Some species that occur predominantly in mesic to wet forest or scrub on Oahu and Hawaii also are known from bogs or bog edges there (not shown); *Anisocarpus scabridus, Carlquistia muirii*, and *Kyhosia bolanderi* are continental tarweeds.] Data are from Baldwin and Robichaux (1995) and B. G. Baldwin (unpubl.). Photos: *D. latifolia* (liana), by B. G. Baldwin (left, top); *W. gymnoxiphium* (semelparous rosette plant), by G. D. Carr (second from left, top); *D. waialealae* (cushion plant), by K. Wood (left, bottom); *D. scabra* subsp. *scabra* (mat plant), by B. G. Baldwin (second from left, bottom); *A. sandwicense* subsp. *macrocephalum* (semelparous rosette plant), by D. W. Kyhos (far right).

useful for studying this process, as the chronology of the islands allows examination of communities as they have developed on an evolutionary time scale. Among spiders in the genus *Tetragnatha*, taxa in the spiny-leg (cursorial, no webs) clade exhibit discrete ecomorphs, defined on the basis of their color and habitat use: green (associated with leaves), maroon (moss), large brown (tree trunks), and small brown (twigs) (Gillespie 2004). Among these spiders, the island chronology (comparing oldest through youngest islands) has been used to illustrate the stages of adaptive radiation: Communities on the youngest island are comprised largely of populations that descended from spiders on the older island(s). Each of the older islands contains similar numbers and ecological sets (one representative of each ecomorph) of species. However, the second youngest island (East Maui) contains a larger number of species, including multiple members of the same ecomorph, some of which appear to have arisen by colonization from an older island, others by *in situ* speciation. Thus it appears that, at least initially, subsequent to successful occupation of any given land mass, there may be no absolute limit to the number of species that can form (figure 13.3). This outcome is comparable to what has been observed in the development of arthropod communities through immigration on mangrove islets off Florida, where substantial overshoot in species numbers was found prior to equilibrium (Simberloff 1976), and was interpreted as a consequence of the small population sizes of species in the early stages of community development. The *Tetragnatha* finding also lends support to the idea that—at least in the context of community assembly—speciation can serve as a surrogate for immigration on isolated islands.

Community Assembly

Species Accumulation—Pattern and Process

A number of studies have examined the relationship between island area and/or age and species diversity for plants and animals (Peck et al. 1999, Emerson and Kolm 2005) with mixed conclusions in terms of underlying processes, although some general patterns are emerging (Whittaker et al. 2008, and this volume). Lineages that have diversified throughout the history of the modern high islands of the Hawaiian chain generally show a pattern of older-to-younger island dispersal (see Funk and Wagner 1995) and offer an excellent opportunity to examine the effects of both island age and island area on species accumulation. Here, we examined how species diversity changes with island area and time for different Hawaiian groups of animals and plants (figure 13.4). To detect the signa-

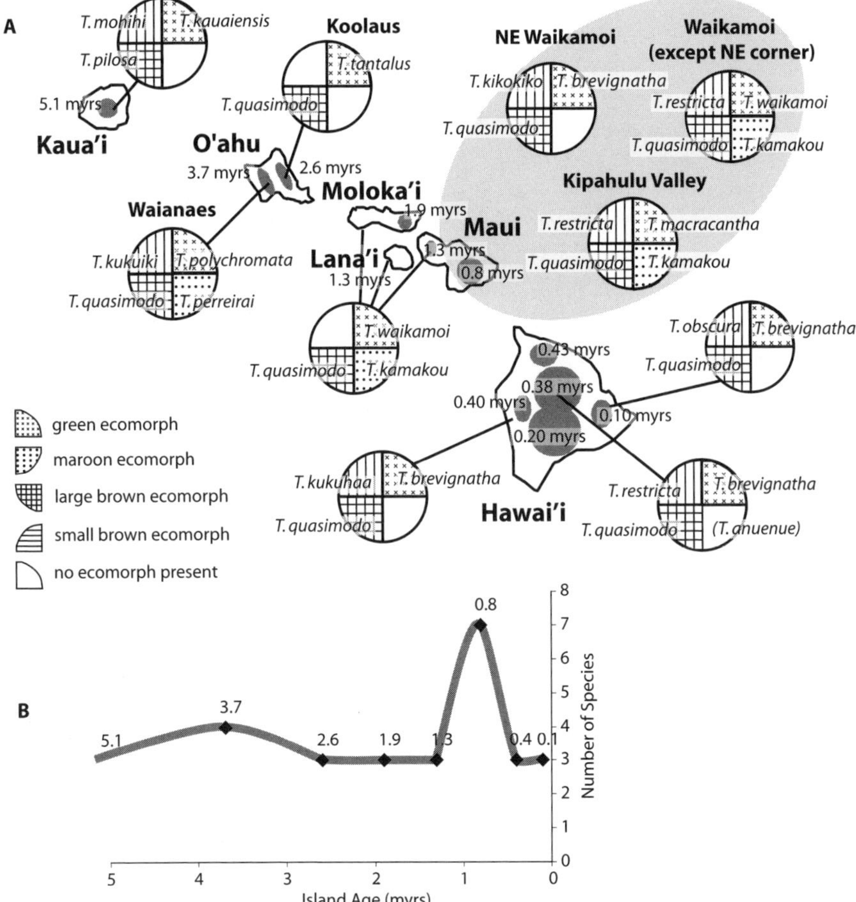

Figure 13.3. Map of the Hawaiian Islands showing how ecomorph diversity of spiny leg *Tetragnatha* changes with island age. A. Ecomorphs in different habitats (Gillespie 2004) on a given volcano (note: Kauai is made up of a single volcano; Oahu, two; Molokai and Lanai, one each; Maui, two; and Hawaii, five). Volcanoes are represented by gray circles, with age indicated in millions of years (myrs). Each section of a pie represents a different ecomorph (green, maroon, large brown, and small brown) whenever a morph is present at a site. Never are two species that share the same ecomorph found in the same locality. B. Number of species on each island against island age (simple scatterplot connected by smoothed line), showing the surprisingly large number of species on East Maui.

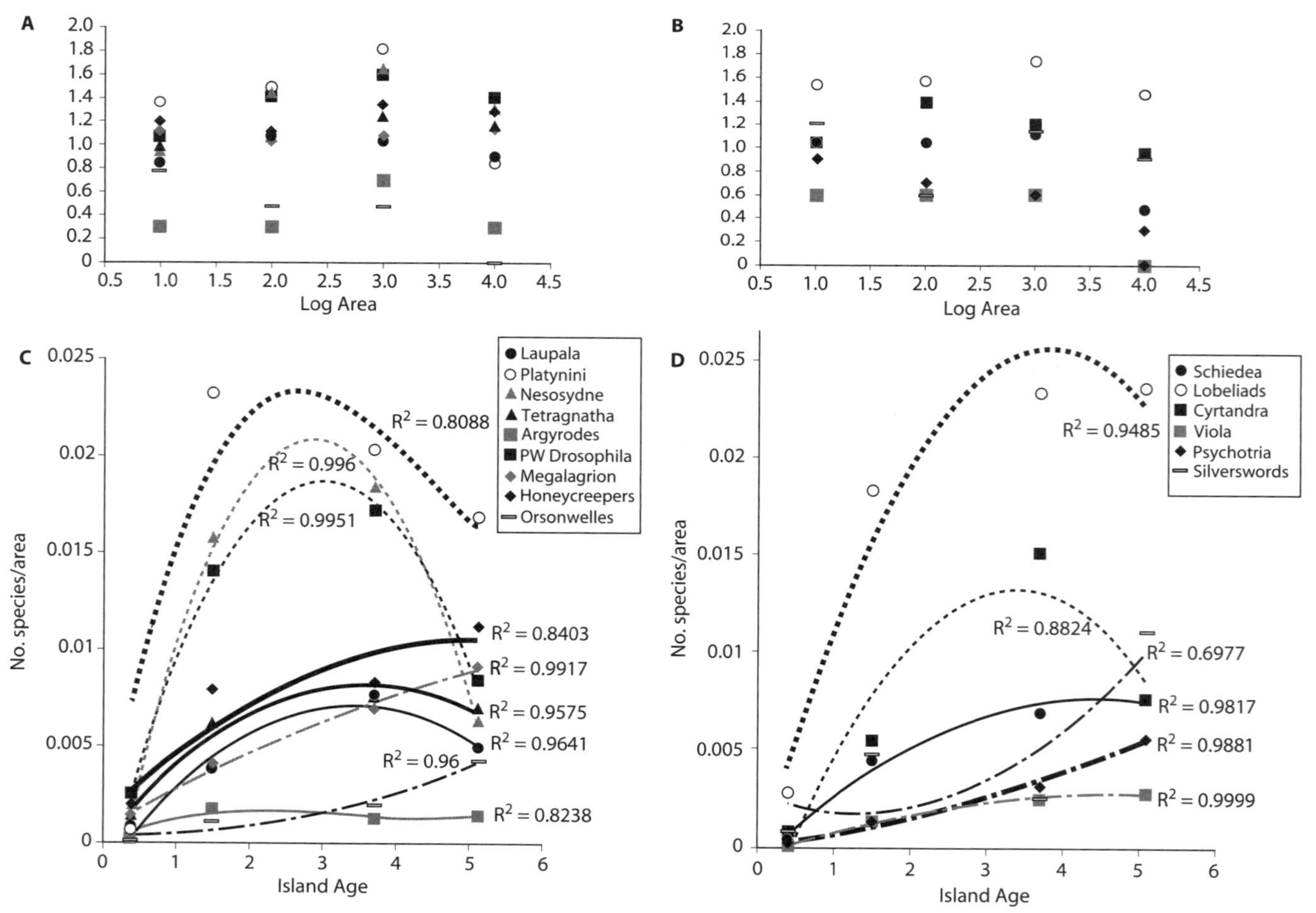
A
Log Area
B
Log Area
C
No. species/area
Island Age
Laupala
Platynini
Nesosydne
Tetragnatha
Argyrodes
PW Drosophila
Megalagrion
Honeycreepers
Orsonwelles
$R^2 = 0.8088$
$R^2 = 0.996$
$R^2 = 0.9951$
$R^2 = 0.8403$
$R^2 = 0.9917$
$R^2 = 0.9575$
$R^2 = 0.9641$
$R^2 = 0.96$
$R^2 = 0.8238$
D
No. species/area
Island Age
Schiedea
Lobeliads
Cyrtandra
Viola
Psychotria
Silverswords
$R^2 = 0.9485$
$R^2 = 0.8824$
$R^2 = 0.6977$
$R^2 = 0.9817$
$R^2 = 0.9881$
$R^2 = 0.9999$

Figure 13.4. Relationships of species numbers in the Hawaiian Islands with area of an island and time, as inferred from island age (for island ages, see Figure 13.3). The analysis is limited to groups that are estimated, based mainly on molecular data, to have been in the Hawaiian Islands at least since Kauai was the youngest island in the chain; this allows us to use the chronology of the islands to help infer process. A. Species-area relationships for different groups of arthropods. Data are for honeycreepers (Olson and James 1991, James and Olson 1991), *Megalagrion* damselflies (Jordan et al. 2003), picture-wing (PW) *Drosophila* flies (Carson 1983), *Nesosydne* planthoppers (Nishida 2002), *Laupala* crickets (Shaw 2000, Nishida 2002), and Platynine beetles (Liebherr and Zimmerman 2000), and for *Tetragnatha* (Gillespie 2004), *Orsonwelles* (Hormiga et al. 2003), and *Argyrodes* (Gillespie and Rivera 2007) spiders. There is no general picture of a relationship between species numbers and area. B. Species-area relationships for different groups of plants. Data are for lobeliads, *Viola, Cyrtandra*, the silversword alliance, *Psychotria*, and *Schiedea* (Lammers 1999, 2004, 2005, 2007; Baldwin and Robichaux 1995; Givnish et al. 1995, 2008; Wagner et al. 1999a, b, 2001, 2005; Wagner and Lorence 2000; Ballard and Sytsma 2000; Nepokroeff et al. 2003; Baldwin and Carr 2005; Cronk et al. 2005; Lorence and Perlman 2007; Havran et al. 2008; Baldwin, unpubl.). Here again, there is no general picture of a relationship between species numbers and area. C. Detailed examination of the relationship between numbers of species per unit area and time (island age) for the same groups of arthropods. Quadratic functions fitted to each data set independently, with r^2 values shown for each. All groups have the fewest number of species per unit area on Hawaii, the youngest island. The most species-rich lineages show a very high peak of diversity on islands of intermediate age (dashed lines); in most other lineages, numbers appear to level off towards the older islands (solid line), while in two (*Orsonwelles* spiders, *Megalagrion* damselflies), diversity increases constantly to the oldest island (dash-dot line). D. The same analysis for plants. Again, all groups have the fewest species on Hawaii, and the most diverse lineages have a high peak in diversity on Oahu (dashed lines). In one lineage (*Schiedea*), numbers level off towards the older islands (solid line), while diversity in the remaining lineages increases to the oldest island without evidence of any equilibrium having been achieved.

ture of time, we included only those groups that are estimated to have been in the Hawaiian Islands for at least as long as the current main islands have been in existence (otherwise the effects of island age will be confounded by recency of colonization). As can be seen (figures 13.4A and 13.4B), the relationship between species number and island area is not clear. However, when the number of species per unit area is examined over time (figures 13.4C and 13.4D), a pattern emerges. Considering only the current high islands (not the series of atolls and other islets—the remnants of previous high islands, which lie to the northwest of the main chain), in the animal groups examined (figure 13.4C), species diversification appears to occur at different rates, with the most species-rich groups examined being platynine beetles, picture-wing *Drosophila* flies, and plant hoppers. In each of these species-rich lineages, diversity per unit area is very high on islands of intermediate age, and drops off on the oldest islands. Most of the somewhat less species-rich lineages (honeycreepers, *Tetragnatha* spiders, *Argyrodes* spiders, *Laupala* crickets) show a more gradual increase to their highest diversity per unit area on islands of intermediate age, diversity appearing to level off toward the older islands. However, two lineages, *Megalagrion* damselflies and *Orsonwelles* spiders, show a steady increase in numbers per unit area from the youngest to the oldest islands, suggesting that equilibrium has not been reached in these groups.

The signature across different plant groups is remarkably similar. The two most species-rich Hawaiian plant lineages, *Cyrtandra* and the lobeliads, reach their highest diversity per unit area on one of the oldest islands (Oahu, Kauai) and may represent examples of ecological saturation (perhaps somewhat delayed relative to animal lineages) and subsequent diversity loss on the oldest island, at least in the case of *Cyrtandra* (figure 4D). In less species-rich groups (e.g., the silversword alliance, *Psychotria*, *Schiedea*, *Viola*), the highest number of species per unit area is often found on the older islands, although here diversity appears to level off in some groups (*Schiedea*, *Viola*) while the other lineages show a steady increase in diversity toward the older islands. Again, it appears that equilibrium has not been reached in these plant lineages, even on the oldest islands.

Although it would be premature to infer causal relationships as to the ecological mechanisms underlying these patterns, it is worth noting the analogy with the study of mangrove islands by Simberloff and Wilson (1969a,b), who found that on all islands but the most distant (where immigration was very low), species number rose above the predefaunation number, then fell and oscillated about that number. Correspondingly, the Hawaiian data often show an overshoot, though the effect is more pronounced in some of the large (high speciation rate) lineages. Simberloff and Wilson explained this effect as being due to the small population sizes in the early stages of colonization allowing more species to coexist. Whether analogous processes (small population sizes or small ranges allowing more species to occur within a given island) may explain the effect in the Hawaiian lineages requires further study. It is also worth pointing out that, among the Hawaiian lineages that seem not to have reached equilibrium even on the oldest island, interactions between conspecifics, and for angiosperms between plants and pollinators, may be less important generally than in at least some of the radiations that appear to have approached or reached peak diversity during the history of the modern high islands of the Hawaiian chain (Baldwin and Robichaux 1995, Hormiga et al. 2003, Jordan et al. 2003). The implications of these findings to the attainment of species equilibrium are discussed in the next section.

Equilibrium

The ETIB, in its basic form, predicts that more isolated islands will have fewer species because the immigration curve will be lower. According to MacArthur and Wilson: "The island will equilibrate at fewer species . . . [w]here the degree of isolation from the source regions that supply the species is increased, reducing immigration." (1967, pp. 22–23; reviewed in Schoener, this volume, Lomolino et al. 2005). However, no effect of isolation on the slope of the species-area relationship was found in a global analysis of birds (Kalmar and Currie 2007). Power (1972), who constructed a path diagram to model the regulation of numbers of plant and bird species on the Californian Channel Islands, also found that isolation had little effect on island plant species diversity, although Moran (1996) found that the Guadalupe Island flora has much lower species diversity by area (and much higher single-island endemism) than floras of the other, far less remote California islands.

The issue of whether remote islands support fewer species seems to hinge in part on when (and whether) equilibrium is achieved. One explanation for the reduced number of species often found on remote islands is that these islands are less likely to be at species equilibrium because of their very slow rate of acquisition of diversity through immigration+speciation compared to rates of immigration to islands near major source areas (Schoener, this volume). This nonequilibrium explanation, also noted by MacArthur and Wilson (1967, pp. 22–23), suggests that, given enough time, remote islands may be expected to attain a level of diversity comparable to otherwise similar near-source islands, although rates of island erosion and subsidence with accompanying loss of island area may be sufficiently high to prevent diversity from ever reaching such a high level. At the same time, the effect of isolation in reducing immigration might be diminished in an archipelago, where colonization between islands (coupled with speciation) would be expected to elevate species numbers, reducing (or even negating) the effect of isolation predicted in the basic formulation of the ETIB (MacArthur and Wilson 1967).

Recent findings on the number of invasive species are relevant to this discussion: Using estimated values of prehuman diversity (Burney et al. 2001, James 2004, James and Olson 2005, Paxinos and al. 2002), Sax et al. (2002) have shown that the number of bird species currently found on remote oceanic islands (native plus introduced) is no higher than the number that was there prior to human occupation. At the same time, the number of species of plants (native plus introduced) is almost double the number prior to human occupation. This finding suggests that communities may be saturated for birds, but not for plants, assuming in part that lag in time-to-extinction is no greater for plants than for birds (Sax et al. 2007). Indeed, when put in the context of the discussion above, the result is consistent with the idea that birds (and various other animal groups) may have reached equilibrium prior to arrival of Polynesians, but many plant groups had not.

Random versus Deterministic Changes

The ETIB is neutral in that it assumes that all species are equal in their probabilities of immigrating onto an island or of going extinct once there, and predicts only a diversity equilibrium, not a taxonomic equilibrium (i.e., no effect of species identity; see Hubbell, this volume). At the same time, studies have shown a degree of predictability in species composition on islands, implying a role for niche assembly in conjunction with some level of stochasticity (Roughgarden and Diamond 1986). Overall, ecological studies indicate that community formation is somewhat deterministic, even though there is a strong stochastic element to colonization. Wilson (1969) and Simberloff and Wilson (1970) termed the initial, relatively stable number of species, the "noninteractive equilibrium," and suggested that as immigrant taxa become increasingly "co-adapted," a nonrandom "interactive equilibrium" is established.

The situation on remote islands where evolutionary processes dominate in the assembly of diversity shows some intriguing parallels to those islands on which ecological processes dominate. For the Hawaiian *Tetragnatha* spiders mentioned above (Gillespie 2004), one could view the initial stages of community assembly as a "race" between *in situ* adaptive radiation producing closely related species occupying different ecomorph niches versus between-island colonization in which species pre-adapted to each of the niches arrive from elsewhere and subsequently speciate in allopatry without change in ecomorph. The phylogenetic results support a neutral model inasmuch as niche filling appears to be largely stochastic, likely dictated by the chance arrival of propagules: Whichever spider arrives first is the one that fills that niche space; but the identity of the first taxon to arrive is unpredictable, so the niche may be filled by an ancestor from another island moving into the same niche on the new island, or by adaptation and speciation from an ancestor (either on the same or a different island) occupying another niche. Moreover, it appears that a given area can initially support a large number of species while any one is rare: Before the numbers of any one species have increased beyond a critical threshold, competition is likely minimal, with little impediment to the addition of more species. However, competition likely does play a role, as species of the same ecomorph have never been found to co-occur; rather, they are separated by tight geographic (parapatric) boundaries. This particular situation may be the result of the circumstances through which a niche is filled that may be peculiar to remote islands: Because all species in the radiation are very closely related, any species of the same ecomorph is, from a niche perspective (as mentioned above), ecologically identical. An analogous situation in less remote locales is likely to allow for more distributional overlap, between less closely related and less ecologically similar species.

On the older Hawaiian Islands, almost every community is represented by a single species of each ecomorph of *Tetragnatha* spiders, suggesting that deterministic processes play a part in the equilibrium species composition. These findings implicate competitive exclusion of similar ecomorphs in the "final" species composition. In other words, the community may undergo "fine-tuning" following initial species proliferation as a result of inter- and intraspecific competition for resources (Arthur 1987). This ecological/evolutionary "jostling" may result in a predictable community structure, with single representatives filling a given niche in any one community. A similar scenario has been proposed to explain diversification in a radiation of weevils in an isolated subantarctic archipelago (Chown 1990). Indeed, such a scenario may provide a mechanistic basis for the species accumulation curves (figures 13.4C and 13.4D), and why such curves might differ between taxonomic groups.

Why do some taxa show predictable and deterministic patterns of differentiation, while no such effect has been found in other groups? As described above, some lineages tend to undergo repeated ecological release upon colonization of islands, which may be conducive to repeated evolution of specialists for a given set of resources, as found in Caribbean lizards (Losos et al. 1998) and Hawaiian spiders (Gillespie 2004). But this begs the question, why do some taxa tend to expand their ecological

Figure 13.5. Ecological plasticity in web structures of the spider *Tetragnatha stelarobusta*. A. Typical web spun by *T. stelarobusta* (D) in sites where it co-occurs with *T. eurychasma* (web shown in B, spider inset). At these sites, the web of *T. stelarobusta* (A) has tightly spun spirals while that of *T. eurychasma* (B) has large spaces between the spirals (Blackledge and Gillespie 2004). C. Typical web spun by *T. stelarobusta* at another site where *T. eurychasma* does not occur. This observation suggests that *T. stelarobusta* may change the form of its web according to the presence of congeners in a community. Photos A, B, and D by T. A. Blackledge; C by R. G. Gillespie.

amplitude while others do not? Part of the explanation for the apparent inability to utilize newly available resources may lie in the degree of initial pre-adaptation to the new resource and differences in ecological, behavioral, and/or genetic plasticity: The pattern of ancestral plasticity can determine which behavioral/ecological phenotypes are expressed in a given environment, and, therefore, which phenotypes have the potential to evolve in response to selection (West-Eberhard 2003). Indeed, such plasticity is evident in Hawaiian spiny-leg *Tetragnatha* spiders; when a given ecomorph is "missing" from a community, one species may display polymorphism (likely developmental in this case) such as to effectively fill the niche of the ecomorph that is absent (R. Carter, unpublished). Similar ecological plasticity is found in web-building representatives from the same radiation (figure 13.5). Current research is focused on quantifying the nature of the polymorphic species, and their relationship to those that are specialized for a given ecomorph.

Conclusions

We have discussed the application of ideas of the ETIB to remote archipelagoes, which have often been considered enigmatic in the context of community assembly (Webb et al. 2002). We focused in particular on the Hawaiian Islands, where the data indicate some striking parallels in the pattern of species accumulation between near-source and remote islands, with (1) often a strongly stochastic element involved in the phylogenetic identity of a taxon colonizing available niche space, and (2) some groups showing an increase in species number beyond an apparent equilibrium, prior to falling to a sometimes predictable set of "coadapted" species (Wilson 1969, Simberloff and Wilson 1970). Overall, it appears that colonization of an island may be dictated by propagule pressure and ecological opportunity: Niches are filled by the interplay between colonization of species from the same niche in another region (e.g., on another island) or by adaptive shifts from another niche on the same island. However, at least in the Hawaiian Islands, the subsequent development of the community may be dictated by an interaction between (1) the rate of speciation; and (2) the degree of ecological/behavioral plasticity.

For the Hawaiian taxa examined here (both plants and animals), some clear correlates with different patterns of diversity are evident: (1) Large overall species diversity correlates with a high peak in diversity on islands of intermediate age prior to reduced diversity on the oldest island, (2) most lineages with lower overall species diversity show a steady increase towards an apparent equilibrium on the older islands, and (3) lineages in which sister-species are largely allopatric (or microallopatric) often show a steady increase in species numbers to the oldest island with no indication of having reached a diversity equilibrium.

These observations indicate that research on remote islands must continue to recognize the evolutionary and ecological differences among lineages (e.g., Givnish et al. 2008), rather than searching for general patterns across taxa, and must incorporate hypothesized mechanisms underlying these differences in order to derive more concrete inferences on processes dictating community assembly.

Acknowledgments

We thank Robert Ricklefs and Jonathan Losos for organizing the symposium on "The Theory of Island Biogeography at 40: Impacts and Prospects." The work reported here was supported by funds from the Schlinger Foundation, the National Science Foundation, the National Tropical Botanical Garden, and UC Berkeley. We are grateful to David Ackerly, Paul Fine, Jonathan Losos, George Roderick, and two anonymous reviewers for comments.

Literature Cited

Alice, L. A., and C. S. Campbell. 1999. Phylogeny of *Rubus* (Rosaceae) based on nuclear ribosomal DNA internal transcribed spacer region sequences. *American Journal of Botany* 86:81–97.

Allan, G. J., J. Francisco-Ortega, A. Santos-Guerra, E. Boerner, and E. A. Zimmer. 2004. Molecular phylogenetic evidence for the geographic origin and classification of Canary Island *Lotus* (Fabaceae: Loteae). *Molecular Phylogenetics and Evolution* 32:123–38.

Arthur, W. 1987. *The Niche in Competition and Evolution*. New York: John Wiley and Sons.

Baldwin, B. G., and G. D. Carr. 2005. *Dubautia kalalauensis*, a new species of the Hawaiian silversword alliance (Compositae, Madiinae) from northwestern Kaua'i. *Novon* 15:259–63.

Baldwin, B. G., and R. H. Robichaux. 1995. Historical biogeography and ecology of the Hawaiian silversword alliance (Asteraceae): New molecular phylogenetic perspectives. In *Hawaiian Biogeography: Evolution on a Hot Spot Archipelago*, ed. W. L. Wagner and V. Funk, 259–87. Washington, DC: Smithsonian Institution Press.

Ballard, H. E., Jr., and K. J. Sytsma. 2000. Evolution and biogeography of the woody Hawaiian violets (*Viola*, Violaceae): Arctic origins, herbaceous ancestry and bird dispersal. *Evolution* 54:1521–32.

Barber, J. C., J. Francisco-Ortega, A. Santos-Guerra, A. A. Marrero, and R. K. Jansen. 2000. Evolution of endemic *Sideritis* (Lamiaceae) in Macaronesia: insights from a chloroplast DNA restriction site analysis. *Systematic Botany* 25:633–47.

Bernardello, G., G. J. Anderson, T. F. Stuessy, and D. J. Crawford. 2006. The angiosperm flora of the Archipelago Juan Fernandez (Chile): Origin and dispersal. *Canadian Journal of Botany* 84:1266–81.

Blackledge, T. A., and R. G. Gillespie. 2004. Convergent evolution of behavior in an adaptive radiation of Hawaiian web-building spiders. *Proceedings of the National Academy of Sciences U.S.A.* 101:16228–33.

Burney, D. A., H. F. James, L. P. Burney, S. L. Olson, W. Kikuchi, W. L. Wagner, M. Burney, M., D. McClosky, D. Kikuchi, F. V. Grady, R. Gage II, and R. Nishek. 2001. Fossil evidence for a diverse biota from Kaua'i and its transformation since human arrival. *Ecological Monographs* 74:615–41.

Carine, M. A., S. J. Russell, A. Santos-Guerra, and J. Francisco-Ortega. 2004. Relationships of the Macaronesian and Mediterranean floras: Molecular evidence for multiple colonizations into Macaronesia and back-colonization of the continent in *Convolvulus* L. (Convolvulaceae). *American Journal of Botany* 91:1070–85.

Carlquist, S. 1966. The biota of long-distance dispersal I: Principles of dispersal and evolution. *Quarterly Review of Biology* 4:247–70.

———. 1974. *Island Biology*. New York: Columbia University Press.

———. 1980. *Hawai'i, A Natural History: Geology, Climate, Native Flora and Fauna Above the Shoreline,* 2nd ed. Lawai: Pacific Tropical Botanical Garden.

Carson, H. L. 1983. Chromosomal sequences and interisland colonizations in Hawaiian *Drosophila. Genetics* 103:465–82

Chown, S. L. 1990. Speciation in the sub-antarctic weevil genus *Dusmoecetes* Jeannel (Coleoptera: Curculionidae). *Systematic Entomology* 15:283–96.

Clague, D. A., and G. B. Dalrymple. 1987. The Hawaiian-Emperor volcanic chain. In *Volcanism in Hawaii,* ed. R. W. Decker, T. L. Wright, and P. H. Stauffer, 1–54. U.S. Geological Survey Professional Paper 1350. Washington, DC: U.S. Government Printing Office.

Cronk, Q.C.B., M. Kiehn, W. L. Wagner, and J. F. Smith. 2005. Evolution of *Cyrtandra* (Gesneriaceae) in the Pacific Ocean: The origin of a supertramp clade. *American Journal of Botany* 92:1017–24.

Darwin, C. 1859. *On the Origin of Species by Means of Natural Selection.* London: John Murray.

Diamond, J. M. 1970. Ecological consequences of island colonization by Southwest Pacific birds. II. The effect of species diversity on total population density. *Proceedings of the. National Academy of Sciences U.S.A.* 67:1715–21.

Driscoll, H. E., and D. S. Barrington. 2007. Origin of Hawaiian *Polystichum* (Dryopteridaceae) in the context of a world phylogeny. *American Journal of Botany* 94:1413–24.

Emerson, B. C., and N. Kolm. 2005. Species diversity can drive speciation. *Nature* 434:1015–17.

Farrell, B. D., and A. S. Sequeira. 2004. Evolutionary rates in the adaptive radiation of beetles on plants. *Evolution* 58:1984–2001.

Fosberg, F. R. 1948. Derivation of the flora of the Hawaiian Islands. In *Insects of Hawai'i*, vol. 1, *Introduction,* ed. E. C. Zimmerman, 107–19. Honolulu: University of Hawaii Press.

Francisco-Ortega, J., J. Fuertes-Aguilar, S.-C. Kim, A. Santos-Guerra, D. J. Crawford, and R. K. Jansen. 2002. Phylogeny of the Macaronesian endemic *Crambe* section *Dendrocrambe* (Brassicaceae) based on internal transcribed spacer sequences of nuclear ribosomal DNA. *American Journal of Botany* 89: 1984–90.

Francisco-Ortega, J., R. K. Jansen, and A. Santos-Guerra. 1996. Chloroplast DNA evidence of colonization, adaptive radiation, and hybridization in the evolution of the Macaronesian flora. *Proceedings of the National Academy of Sciences U.S.A.* 93:4085–90.

Funk, V. A., and W. L. Wagner. 1995. Biogeographic patterns in the Hawaiian Islands. In *Hawaiian Biogeography: Evolution on a Hot Spot Archipelago,* ed. W. L. Wagner and V. A. Funk, 379–419. Washington, DC: Smithsonian Institution Press.

Gagne, W. C. 1997. *Insular Evolution, Speciation, and Revision of the Hawaiian Genus* Nesiomiris (*Hemiptera: Miridae*). Honolulu: Bishop Museum.

Garb, J. E., and R. G. Gillespie. 2006. Island hopping across the central Pacific: Mitochondrial DNA detects sequential colonization of the Austral Islands by crab spiders (Araneae: Thomisidae). *Journal of Biogeography* 33:201–20.

Garb, J. E., and R. G. Gillespie. 2009. Diversity despite dispersal: Colonization history and phylogeography of Hawaiian crab spiders inferred from multilocus genetic data. *Molecular Ecology* 18:1746–64.

Geiger, J.M.O., and T. A. Ranker. 2005. Molecular phylogenetics and historical biogeography of Hawaiian *Dryopteris* (Dryopteridaceae). *Molecular Phylogenetics and Evolution* 34:392–407.

Geiger, J.M.O., T. A. Ranker, J. M. Ramp Neale, and S. T. Klimas. 2007. Molecular biogeography and origins of the Hawaiian fern flora. *Brittonia* 59:142–58.

Gillespie, R. G. 2002. Colonization of remote oceanic islands of the Pacific: Archipelagos as stepping stones? *Journal of Biogeography* 29:655–62.

———. 2004. Community assembly through adaptive radiation in Hawaiian spiders. *Science* 303:356–59.

Gillespie, R. G., E. M. Claridge, and G. K. Roderick. 2008. Biodiversity dynamics in isolated island communities: Interaction between natural and human-mediated processes. *Molecular Ecology* 17:45–87.

Gillespie, R. G., and M. A. Rivera. 2007. Free-living spiders of the genus *Ariamnes* (Araneae, Theridiidae) in Hawaii. *Journal of Arachnology* 35:11–37.

Givnish, T. J., K. C. Millam, A. R. Mast, T. B. Paterson, T. J. Theim, A. L. Hipp, J. M. Henss, J. F. Smith, K. R. Wood, and K. J. Sytsma. 2009. Origin, adaptive radiation and diversification of the Hawaiian lobeliads (Asterales: Campanulaceae). *Proceedings of the Royal Society of London, Series B.* 276:407–16.

Givnish, T. J., K. J. Sytsma, J. F. Smith, and W. S. Hahn. 1995. Molecular evolution, adaptive radiation, and geographic speciation in *Cyanea* (Campanulaceae, Lobelioideae). In *Hawaiian Biogeography: Evolution on a Hot Spot Archipelago,* ed. W. L. Wagner and V. Funk, 288–337. Washington, DC: Smithsonian Institution Press.

Harbaugh, D. T., and B. G. Baldwin. 2007. Phylogeny and biogeography of the sandalwoods (*Santalum*, Santalaceae): Repeated dispersals throughout the Pacific. *American Journal of Botany* 94:1028–40.

Havran, J. C., K. J. Sytsma, and H. E. Ballard. 2008. Evolutionary relationships, interisland biogeography, and molecular evolution in the Hawaiian violets (*Viola*: Violaceae). *Botany 2008, Scientific Abstracts*:131. http://www.botany conference.org/engine/search/index.php?func=detail&aid=718.

Heaney, L. R. 2000. Dynamic disequilibrium: A long-term, large-scale perspective on the equilibrium model of island biogeography. *Global Ecology and Biogeography* 9:59–74.

Hormiga, G., M. A. Arnedo, and R. G. Gillespie. 2003. Speciation on a conveyor belt: Sequential colonization of the Hawaiian Islands by *Orsonwelles* spiders (Araneae: Linyphiidae). *Systematic Biology* 52:70–88.

Howarth, D. G., D. E. Gardner, and C. W. Morden. 1997. Phylogeny of *Rubus* subgenus *Idaeobatus* (Rosaceae) and its implication toward colonization of the Hawaiian Islands. *Systematic Botany* 22:433–41.

Howarth, D. G., M.H.G. Gustafsson, D. A. Baum, and T. J. Motley. 2003. Phylogenetics of the genus *Scaevola* (Goodeniaceae): Implication for dispersal patterns across the Pacific Basin and colonization of the Hawaiian Islands. *American Journal of Botany* 90:915–23.

James, H. F. 2004. The osteology and phylogeny of the Hawaiian finch radiation (Fringillidae: Drepanidini), including extinct taxa. *Zoological Journal of the Linnean Society* 141:207–55.

James, H. F., and S. L. Olson. 1991. Descriptions of thirty-two new species of birds from the Hawaiian Islands: Part II. Passeriformes. *Ornothological Monographs* 46:1–88.

———. 2005. The diversity and biogeography of koa-finches (Drepanidini: *Rhodacanthis*), with descriptions of two new species. *Zoological Journal of the Linnean Society* 144:527–41.

Jordan, S., C. Simon, and D. Polhemus. 2003. Molecular systematics and adaptive radiation of Hawaii's endemic damselfly genus *Megalagrion* (Odonata: Coenagrionidae). *Systematic Biology* 52:89–109.

Kalmar, A., and D. J. Currie. 2007. A unified model of avian species richness on islands and continents. *Ecology* 88:1309–21.

Kim, S.-C., J. Crawford, J. Francisco-Ortega, and A. Santos-Guerra. 1996. A common origin for woody *Sonchus* and five related genera in the Macaronesian Islands: Molecular evidence for extensive radiation. *Proceedings of the National Academy of Sciences, U.S.A.* 93:7743–48.

Lammers, T. G. 1999. A new *Lobelia* from Mexico, with additional new combinations in world Campanulaceae. *Novon* 9:381–89.

———. 2004. Five new species of the endemic Hawaiian genus *Cyanea* (Campanulaceae: Lobelioideae). *Novon* 14:84–101.

———. 2005. Revision of *Delissea* (Campanulaceae-Lobelioideae). *Systematic Botany Monographs* 73:1–75.

Lammers, T. G. 2007. Revision of *Lobelia* sect. *Galeatella* (Campanulaceae: Lobelioideae). *Journal of the Botanical Research Institute of Texas* 1:789–810.

Liebherr, J. K., and E. C. Zimmerman. 2000. Hawaiian Carabidae (Coleoptera), Part 1: Introduction and tribe Playnini. *Insects of Hawaii* 16:1–494.

Lomolino, M. V., B. R. Riddle, and J. H. Brown. 2005. *Biogeography*, 3rd ed. Sunderland, MA: Sinauer Associates.

Lorence, D. H., and S. Perlman. 2007. A new species of *Cyrtandra* (Gesneriaceae) from Hawai'i. *Novon* 17:357–61.

Losos, J. B., T. R. Jackman,, A. Larson, K. De Queiroz, and L. Rodriguez-Schettino. 1998. Contingency and determinism in replicated adaptive radiations of island lizards. *Science* 279:2115–18.

Losos, J. B., J. C. Marks, and T. W. Schoener. 1993. Habitat use and ecological interactions of an introduced and a native species of *Anolis* lizard on Grand Cayman, with a review of the outcomes of anole introductions. *Oecologia* 95:525–32.

Losos, J. B., and D. Schluter. 2000. Analysis of an evolutionary species-area relationship. *Nature* 408:847–50.

MacArthur, R. H. 1972. *Geographical Ecology*. New York: Harper and Row.

MacArthur, R. H., and E. O. Wilson. 1963. An equilibrium theory of insular zoogeography. *Evolution* 17:373–87.

———. 1967. *The Theory of Island Biogeography*. Princeton, NJ: Princeton University Press.

Meimberg, H., T. Abele, C. Bräuchler, J. K. McKay, P. L. Pérez de Paz, and G. Heubl. 2006. Molecular evidence for adaptive radiation of *Micromeria* Benth. (Lamiaceae) on the Canary Islands as inferred from chloroplast and nuclear DNA sequences and ISSR fingerprint data. *Molecular Phylogenetics and Evolution* 41:566–78.

Mendelson, T. C., and K. L. Shaw. 2005. Rapid speciation in an arthropod. *Science* 433:375–76.

Moran, R. 1996. *The Flora of Guadalupe Island, Mexico*. San Francisco: California Academy of Sciences.

Mort, M. E., D. E. Soltis, P. S. Soltis, J. Francisco-Ortega, and A. Santos-Guerra. 2002. Phylogenetics and evolution of the Macaronesian clade of Crassulaceae inferred from nuclear and chloroplast sequence data. *Systematic Botany* 27: 271–88.

Nepokroeff, M., K. S. Sytsma, W. L. Wagner, and E. A. Zimmer. 2003. Reconstructing ancestral patterns of colonization and dispersal in the Hawaiian understory tree genus *Psychotria* (Rubiaceae): A comparison of parsimony and likelihood approaches. *Systematic Biology* 52:820–38.

Nishida, G., ed. 2002. *Hawaiian Terrestrial Arthropod Checklist*. Bishop Museum Technical Report. Honolulu: Bishop Museum Press.

Nosil, P., and T. E. Reimchen. 2005. Ecological opportunity and levels of morphological variance within freshwater stickleback populations. *Biological Journal of the Linnean Society* 86:297–308.

Olson, S. L., and H. F. James. 1991. Descriptions of thirty-two new species of birds from the Hawaiian Islands: Part I. Non-passeriformes. *Ornithological Monographs* 45:1–88.

Panero J. L., J. Francisco-Ortega, R. K. Jansen, and A. Santos-Guerra. 1999. Molecular evidence for multiple origins of woodiness and a New World biogeographic connection of the Macaronesian island endemic *Pericallis* (Asteraceae: Senecioneae). *Proceedings of the National Academy of Sciences U.S.A.* 96:13886–91.

Parker, J. D., D. E. Burkepile, and M. E. Hay. 2006a. Opposing effects of native and exotic herbivores on plant invasions. *Science* 311:1459–61.

———. 2006b. Response to comment on "Opposing effects of native and exotic herbivores on plant invasions." *Science* 313:298.

Paxinos, E., E. H. James, S. L. Olson, M. D. Sorenson, J. Jackson, and R. C. Fleischer. 2002. mtDNA from fossils reveals a radiation of Hawaiian geese recently derived from the Canada goose (*Branta canadensis*). *Proceedings of the National Academy of Sciences U.S.A.* 99:1399–404.

Peck, S. B., P. Wigfull, and G. Nishida. 1999. Physical correlates of insular species diversity: The insects of the Hawaiian Islands. *Annals of the Entomological Society of America* 92:529–36.

Power, D. M. 1972. Numbers of bird species on the California Islands. *Evolution* 26:451–63.

Price, J. P., and W. L. Wagner. 2004. Speciation in Hawaiian angiosperm lineages: Cause, consequence, and mode. *Evolution* 58:2185–2200.

Proches, S., J.R.U. Wilson, D. M. Richardson, and M. Rejmanek. 2008. Searching for phylogenetic pattern in biological invasions. *Global Ecology and Biogeography* 17:5–10.

Ranker, T. A., S. K. Floyd, and P. G. Trapp. 1994. Multiple colonizations of *Asplenium adiantum–nigrum* into the Hawaiian archipelago. *Evolution* 48:1364–70.

Roderick, G. K. 1997. Herbivorous insects and the Hawaiian silversword alliance: Coevolution or cospeciation? *Pacific Science* 51:440–49.

Roughgarden, J. 1972. Evolution of niche width. *American Naturalist* 106: 683–718.

Roughgarden, J., and J. Diamond. 1986. Overview: The role of species interactions in community ecology. In *Community Ecology,* ed. J. Diamond, and T. J. Case, 333–43. New York: Harper and Row.

Sakai , A. K., W. L. Wagner, D. M. Ferguson, and D. R. Herbst. 1995. Origins of dioecy in the native Hawaiian flora. *Ecology* 76:2517–29.

Sakai, A. K., S. G. Weller, W. L. Wagner, M. Nepokroeff, and T. M. Culley. 2006. Adaptive radiation and evolution of breeding systems in *Schiedea* (Caryophyllaceae) an endemic Hawaiian genus. *Annals of the Missouri Botanical Garden* 93:49–63.

Sax, D. F., S. D. Gaines, and J. H. Brown. 2002. Species invasions exceed extinctions on islands worldwide: A comparative study of plants and birds. *American Naturalist* 160:766–83.

Sax, D. F., J. J. Stachowicz, J. H. Brown, J. F. Bruno, M. N. Dawson, S. D. Gaines, R. K. Grosberg, A. Hastings, R. D. Holt, M. M. Mayfield, M. I. O'Connor, and W. R. Rice. 2007. Ecological and evolutionary insights from species invasions. *Trends in Ecology and Evolution* 22:465–71.

Schluter, D. 2000 *The Ecology of Adaptive Radiation.* Oxford: Oxford University Press.

Schneider, H., A. R. Smith, R. Cranfill, T. E. Hildebrand, C. H. Haufler, and T. A. Ranker. 2004. Unraveling the phylogeny of polygrammoid ferns (Polypodiaceae and Grammitidaceae): Exploring aspects of the diversification of epiphytic plants. *Molecular Phylogenetics and Evolution* 31:1041–63.

Shaw, K. L. 2000. Further acoustic diversity in Hawaiian forests: Two new species of Hawaiian cricket (Orthoptera: Gryllidae: Trigonidiinae: *Laupala*). *Zoological Journal of the Linnean Society* 129:73–91.

Silvertown, J. 2004. The ghost of competition past in the phylogeny of island endemic plants. *Journal of Ecology* 92:168–73.

Silvertown, J., J. Francisco-Ortega, and M. Carine. 2005. The monophyly of island radiations: An evaluation of niche pre-emption and some alternative explanations. *Journal of Ecology* 93:653–57.

Simberloff, D. 1974. Equilibrium theory of island biogeography and ecology. *Annual Review of Ecology and Systematics* 5:161–82.

———. 1976. Species turnover and equilibrium island biogeography. *Science* 194:572–78.

Simberloff, D., and E. O. Wilson. 1969a. Experimental zoogeography of islands: Defaunation and monitoring techniques. *Ecology* 50:267–78.

———. 1969b. Experimental zoogeography of islands: The colonization of empty islands. *Ecology* 50:278–96.

———. 1970. Experimental zoogeography of islands: A two-year record of colonization. *Ecology* 51:934–37.

Simpson, G. G. 1953. *The Major Features of Evolution*. New York: Columbia University Press.

Strauss, S. Y., C. O. Webb, and N. Salamin. 2006. Exotic taxa less related to native species are more invasive. *Proceedings of the National Academy of Sciences U.S.A.* 103:5841–45.

Trusty, J. L., R. G. Olmstead, A. Santos-Guerra, S. Sá-Fontinha, and J. Francisco-Ortega. 2005. Molecular phylogenetics of the Macaronesian-endemic genus *Bystropogon* (Lamiaceae): Palaeo-islands, ecological shifts and interisland colonizations. *Molecular Ecology* 14:1177–89.

Wagner, W. L., D. R. Herbst, and S. H. Sohmer. 1999a. *Manual of the Flowering Plants of Hawai'i*. Rev. ed. Honolulu: Bishop Museum Press.

Wagner, W. L. and D. H. Lorence. 2000. A reassessment of *Cyrtandra kealiae* and *C. limahuliensis* (Gesneriaceae). *Bishop Museum Occasional Papers* 63:17–20.

Wagner, W. L., S. G. Weller, and A. K. Sakai. 2005. Monograph of *Schiedea* (Caryophyllaceae—Alsinoideae). *Systematic Botany Monographs* 72:1–169.

Wagner, W. L., K. R. Wood, and D. H. Lorence. 1999b. A new species of *Cyrtandra* (Gesneriaceae) from Kaua'i (Hawaiian Islands). *Novon* 11:146–52.

Webb, C. O., D. D. Ackerly, M. A. McPeek, and M. J. Donoghue. 2002. Phylogenies and community ecology. *Annual Review of Ecology and Systematics* 33:475–505.

West-Eberhard, M. J. 2003. *Developmental Plasticity and Evolution*. Oxford: Oxford University Press.

Whittaker, R. J., K. A. Triantis, and R. J. Ladle. 2008. A general dynamic theory of oceanic island biogeography. *Journal of Biogeography* 35:977–94.

Williams, E. E. 1972. The origin of faunas. Evolution of lizard congeners in a complex island fauna: A trial analysis. *Evolutionary Biology* 6:47–89.
Wilson, E. O. 1969. The species equilibrium. In *Diversity and Stability in Ecological Systems*. ed. G. M. Woodwell and H. H. Smith, 264. Brookhaven Symposia in Biology vol. 22. Upton, NY: Brookhaven National Laboratory.

Dynamics of Colonization and Extinction on Islands

INSIGHTS FROM LESSER ANTILLEAN BIRDS

Robert E. Ricklefs

In 1963, Robert H. MacArthur and Edward O. Wilson published a paper in *Evolution*, which they titled "An equilibrium theory of insular zoogeography" (MacArthur and Wilson 1963). In this paper, MacArthur and Wilson suggested that the number of species on islands represented a balance between the addition of new species by colonization and the loss of established species by extinction. The most radical implication of this hypothesis was that the composition of an island's biota continually changed. This view contrasted starkly with the more static concept held by David Lack (1976) and others, that islands accumulated species until they became ecologically saturated, after which the biota became stabilized. It makes sense that resident species are well adapted to local island conditions and might prevent newly arriving species from becoming established.

MacArthur and Wilson's equilibrium theory borrowed heavily from the principle of density-dependent population regulation, according to which deaths increase and births decrease with increasing population size until a balance is achieved and a population attains a steady state. The equilibrium size of a population—its carrying capacity—reflects the availability of resources in the environment. The equilibrium theory of island zoogeography simply replaces births by colonization and deaths by extinction while maintaining the essential feature of population (=species) turnover at a steady state. In this analogy, the carrying capacity of an island for species depends on both an intrinsic property: island size, and an extrinsic property: the rate of immigration to the island. Lack would have added the influence of ecological diversity on the capacity of an island to support species.

MacArthur and Wilson (1963) highlighted the consequences for species richness in their model of a decline in the rate of colonization with increasing dispersal distance and of a decrease in the rate of extinction with increasing island area. In 1967, they published a greatly expanded treatment of island biogeography (MacArthur and Wilson 1967), the fortieth anniversary of which was celebrated at the symposium at Harvard University from which this book developed (see chapters in this volume by Wilson, Lomolino and Brown, and Schoener). The dynamic view articulated by MacArthur and Wilson clearly marked the beginning of modern biogeography. Their equilibrium theory was supported early on by experimental manipulations of island diversity (Simberloff and Wilson 1969), inferences of faunal relaxation on land-bridge islands (Diamond 1972, Wilcox 1978), and observations of historical turnover on islands (Diamond 1969, 1971) and the repopulation of islands destroyed by volcanic eruption, such as Krakatau (Whittaker et al. 1989, and this volume).

However, in spite of widespread acceptance of the equilibrium theory, most of the confirming work involved small islands close to sources of colonization, which are characterized by rapid dynamics (Schoener, this volume). Until recently, colonization and extinction on longer time scales have been inaccessible. Islands supporting endemic species and adaptive radiations, representing colonization-extinction-speciation dynamics on evolutionary time scales, defied interpretation until the advent of molecular phylogenetic analyses (Losos and Parent, this volume, Whittaker et al., this volume). Equipped with these new tools, and reflecting on David Lack's views, we might reasonably ask whether island biotas have attained steady states, if species turnover is indeed an appropriate perspective, and whether the relationship of diversity to island size reflects variation in extinction rate as opposed to the capacity of islands to support species through ecological diversity. Reconstructed evolutionary relationships of island and continental species indicate sources and timing of colonization events and the buildup of island biotas. Relationships among species within island archipelagoes can provide information on diversification and extinction.

In this chapter, I shall illustrate some applications of phylogenetic data to understanding island biotal dynamics, focusing primarily on work conducted during the past twenty years with my collaborator Eldredge Bermingham and many of our colleagues. Our research program focuses on small land birds of the Lesser Antilles, many of which are old, endemic residents, for which we have reconstructed the phylogeographic history of the majority of the species. These data allow us to address the history of colonization, both from external sources and from within the archipelago, and provide first estimates of island-specific and archipelago-wide rates of extinction. Unlike the Hawaiian and Galápagos avifaunas (Amadon 1950, Grant 1986, Fleischer and McIntosh 2001, Grant and Grant 2002, and this volume, Gillespie and Baldwin, this volume), few lineages (perhaps only one: the endemic thrashers, Mimidae) of Lesser

Antillean birds have undergone a limited adaptive radiation within the island chain (Hunt et al. 2001, Ricklefs and Bermingham 2007a). For this reason, we focus on colonization-extinction dynamics and leave alone the issue of species proliferation within the archipelago (see Losos and Parent, this volume, Whittaker et al., this volume).

MacArthur and Wilson (1967), the first of the "Monographs in Population Biology" from Princeton University Press, organized their treatise around seven chapters, which provide a framework for reflecting on the legacy of their insights for modern biogeography. I shall use this framework for my remarks in this chapter. Appropriately, MacArthur and Wilson began with the importance of islands.

The Importance of Islands

In their chapter 1, MacArthur and Wilson (pp. 3–4) made three points about islands. First, "Insularity is moreover a universal feature of biogeography." Although the discrete nature of islands greatly facilitates the study of colonization and extinction, MacArthur and Wilson recognized that processes inherent to islands also operate on continents. As I point out below, archipelagoes provide a unique window on the dynamics of species distributions, which has provided insights into the variation in range size and ecological distributions of continental taxa (e.g., Brown 1995, Gaston 2003). Second, MacArthur and Wilson noted that "The same principles apply, and will apply to an accelerating extent in the future, to formerly continuous natural habitats now being broken up by the encroachment of civilization." Indeed, island biogeography theory has provided a foundation for much of conservation biology (e.g., Terborgh 1974, Laurance, this volume). Third, "the fundamental processes, namely dispersal, invasion, competition, adaptation, and extinction, are among the most difficult in biology to study and understand." How true this has proven to be!

Area and the Number of Species

Although the species-area relationship is one of the most fundamental patterns in ecology and biogeography (Lomolino 2000), we lack a theoretical foundation for the particular form and slope of the relationship (Scheiner et al. 2000, Scheiner 2003). MacArthur and Wilson paid particular attention to the species-area relationship predicted by Preston's (1962) lognormal species-abundance curve, but Preston's theory does not yield a consistent relationship between number of species and area (i.e., sample size). MacArthur and Wilson (p. 14) showed that the lognormal abundance relationship predicted the log-log slope (z) between species and area to be about 0.26, which matches empirical values quite well. However, this value was calculated only for the increase in species in the lower tail of Preston's lognormal distribution, in which the total number of individuals (J) exceeds 10^6 multiples of the number of individuals (m) in the rarest species. This range almost certainly is inappropriate for birds of the Lesser Antilles, and perhaps for other groups as well. It is difficult enough to determine how much of the species-island area relationship derives from increasing ability to sample rare species, which is the core of Preston's model, but one must also incorporate the influence of island size on ecological heterogeneity and the influence of population size on resistance to extinction.

The slope of the log(species)–log(island area) regression for the Lesser Antillean avifauna is about 0.22 (entire West Indies, Ricklefs and Cox 1972) or 0.21 (Lesser Antilles only, Ricklefs and Lovette 1999). The avifauna of the Lesser Antilles (figure 14.1) is well known and sampling is not a consideration. Both human-caused extinctions and species introductions are a potential problem in the Lesser Antilles, but Holocene fossil deposits and historical records of extinctions suggest that extinction has been a minor factor for small land-bird populations in the archipelago (Ricklefs and Bermingham 2004b).

The relative contributions of area per se and habitat diversity to species richness in the Lesser Antilles were considered by Ricklefs and Lovette (1999), who quantified habitat heterogeneity based on the proportions of an island's area occupied by major vegetation types. Although area and habitat diversity covary, the Lesser Antilles present sufficient orthogonal variation in these attributes between islands, that one can separate their effects statistically, to some extent. For birds, both area and habitat diversity contributed to the species-area relationship. Parallel analyses yielded a similar result for butterflies, but indicated that bats are insensitive to habitat, perhaps because they largely forage above vegetation, and reptiles and amphibians, which tend to be habitat specialists, are insensitive to island area. Thus, Lack's idea that species diversity reflects ecological diversity holds, to some degree, particularly for reptiles and amphibians. Of course, these correlations do not address the stability of the species roster.

Further Explanations of the Area-Diversity Pattern

MacArthur and Wilson had more to say about the species-area relationship, and here is where they set out the core of their thesis: "there might

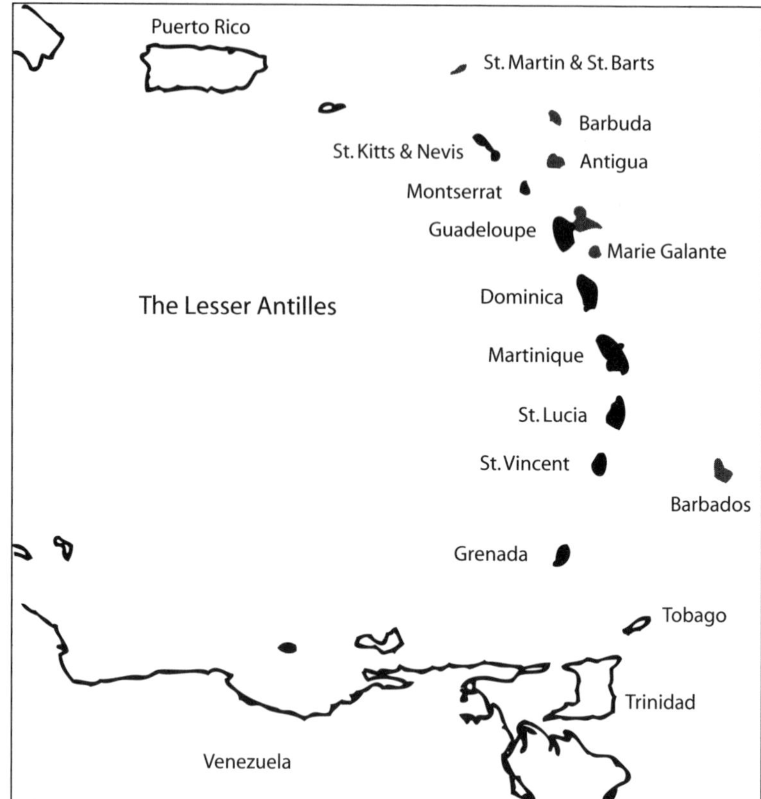

Figure 14.1. Map of the major islands of the Lesser Antilles showing volcanic islands with high elevation in black and lower islands of uplifted marine sediments and coral reefs in gray.

be a balance of immigration by extinction so that the diversity of at least some biotas could be understood as an equilibrium" (p. 20). Testing this idea depends on being able to determine that extinction is recurrent and balanced by immigration of new species (see Simberloff and Collins, this volume). With respect to the species-area relationship, it is critical to show that the rate of extinction is higher on smaller islands. In this regard, Lesser Antillean birds provide strong evidence for the critical role of extinction in producing the species-area relationship because old colonists tend to be absent from small islands (Ricklefs and Cox 1972, Ricklefs and Bermingham 2004b).

Based on subspecific differentiation between islands and gaps in geographic distribution, Ricklefs and Cox (1972) assigned West Indian birds

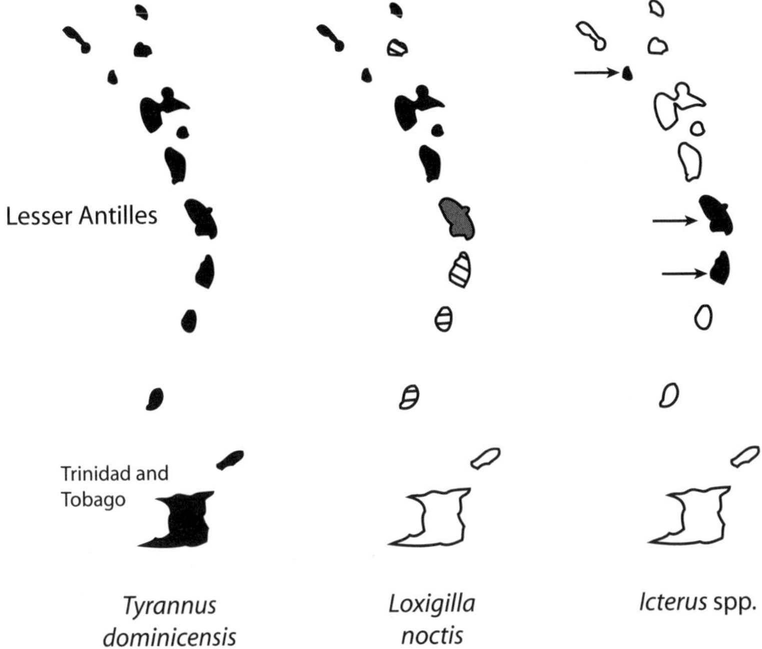

Figure 14.2. Distributions of birds in the Lesser Antilles representing three taxon-cycle stages. The gray kingbird (*Tyrannus dominicensis*, stage I) is a recent colonist from the mainland and is widely distributed throughout the Antilles in open, low-elevation habitats; the species is taxonomically undifferentiated across the archipelago. The Lesser Antillean Bullfinch (*Loxigilla noctis*, stage II) is widespread within the Antilles but exhibits subspecific differentiation between islands. *L. noctis* is endemic to the Lesser Antilles; the close phylogenetic relationship of the island populations compared to their sister taxon *L. portoricensis* indicates that the species has undergone a secondary expansion within the archipelago. The endemic Lesser Antillean orioles (*Icterus laudabilis/oberi/bonana*, stage III) are a highly differentiated monophyletic clade and present large-island gaps in their distribution (Lovette et al. 1999).

to one of four categories: I=widespread and undifferentiated, II=widespread and differentiated, III=fragmented distribution and differentiated, and IV=single island endemic (figure 14.2). They noted that the slope of the species-area relationship was low (0.075) for group-I species and increased to 0.15, 0.32, and 0.42 for groups II through IV.

Because no populations with fragmented distributions were undifferentiated, Ricklefs and Cox reasoned that the categories I through IV represented a time sequence (colonization → differentiation → extinction and gap formation→single island endemic), and referred to these groups as "stages" of the taxon cycle, following upon E. O. Wilson's work on Melanesian ants (Wilson 1959, 1961). Wilson suggested that ant taxa progress through a regular sequence of geographic expansion from continental areas to islands of progressively lower diversity, colonizing from marginal island habitats and gradually invading forest environments and losing their ability to expand further. Wilson also suggested that endemic island forms might undergo secondary expansions; hence the designation "cycle."

Greenslade (1968, 1969) applied the taxon cycle concept to birds and insects of the Solomon Islands, recognizing the same patterns. When Ricklefs and Cox (1972) suggested a similar scenario for the birds of the West Indies, they received a variety of responses, not all of them positive (e.g., Pielou 1979, Pregill and Olson 1981). Since then, molecular phylogenetic analyses have confirmed that Ricklefs and Cox's (1972) taxon cycle stages for birds in the West Indies represent a time sequence and that taxa can embark on secondary cycles of expansion within the West Indies (Ricklefs and Bermingham 1999, 2001, 2002) (see figure 14.2). Specifically, the relative ages of colonization or secondary expansion events for species increase monotonically with stage of the taxon cycle from I to IV.

Having confirmed the temporal sequence of the taxon-cycle stages, Ricklefs and Bermingham (2004b) divided species of birds in the Lesser Antilles into young taxa (taxon cycle stage I), old taxa that had undergone secondary spread (stages I or II, depending on interisland differentiation), and old endemic taxa (generally stages III and IV). They found, similarly to Ricklefs and Cox (1972), that the slope of the species-area relationship for the old endemics (0.72 ± 0.11) greatly exceeded that of young species (0.066 ± 0.016) or old expanded species (slope not significantly different from 0). These results imply that young species or recently spread older species can inhabit islands regardless of their size, but that old endemics whose populations are not restocked by frequent immigration (Brown and Kodric-Brown 1977) disappear from smaller islands. Thus, the species-area relationship for birds in the Lesser Antilles is established by the extinction of old populations that fail to undergo secondary expansions within the archipelago.

Knowing the relative ages of colonization, one can determine whether a biota has reached equilibrium species richness ($\hat{S}$) and also estimate rates of colonization and extinction, assuming time homogeneity. The classic depiction of the MacArthur-Wilson model shows the relationship of colonization rate (C) and extinction rate (E) to the number of species (S) present on an island and the size of the potential pool (P) of colonists (see Schoener, this volume). When these relationships are linear, the rate of accumulation of species on an island $dS/dt = C(P-S) - ES$ decreases linearly with increasing S, and the number of species on an island increases toward its equilibrium,

$$S = \frac{CP}{(C+E)}(1 - \exp(-Et)) \qquad (14.1)$$

where t is the time since origin of the island. Eventually (large t), the number of species reaches an asymptotic value of $CP/(C+E)$ (MacArthur and Wilson 1967, equation 3-1). If the potential species pool were very large compared to the number of species on the island, the colonization rate would be independent of the number of species on the island, the increase in species would become

$$S = (C/E)(1 - \exp(-Et)), \qquad (14.2)$$

and the steady state diversity would be $\hat{S} = C/E$.

The relationship between S and time (t) is prospective in that it describes change in number of species over time since the origin of an island ($t=0$, $S=0$). We cannot observe this development directly in the case of large islands that have accumulated species over long periods. However, when probabilities of colonization and extinction are constant over time, the relationships in equations (14.1) and (14.2) also are retrospective and describe the accumulated number of species on an island having ages up to t (Ricklefs and Bermingham 2001). Thus, by fitting the constants of these equations to the accumulation of species as a function of their time since colonization, one can estimate colonization and extinction rates and determine whether an island biota has reached equilibrium. This approach is illustrated for a particularly complete set of colonization times for the ferns and lycophytes of New Zealand, which appear to represent a time-homogeneous process (figure 14.3).

The accumulation of pteridophyte lineages over time shown in figure 14.3 is fit reasonably well by an exponential approach to an asymptote, but pteridophyte diversity appears not to have reached a steady state, at least according to the underlying model. Alternatively, if the probability of extinction increased with the age of the lineage on the island or in the island archipelago (see above), or the oldest lineages were constrained by the age of the island, the exponential approach to an asymptotic steady state might be truncated. In addition, a change in the rate of colonization resulting from gradual exhaustion of strong potential colonists (MacArthur and Wilson 1967, p. 21) would alter the shape of the early part of the accumulation curve.

Applied to lineages of endemic reptiles and amphibians in the West Indies as a whole, equation (14.2) yields fitted $C = 1.31\ \text{my}^{-1}$ and $E = 0.029\ \text{my}^{-1}$

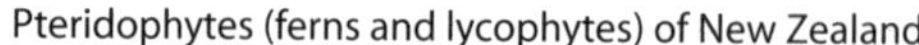

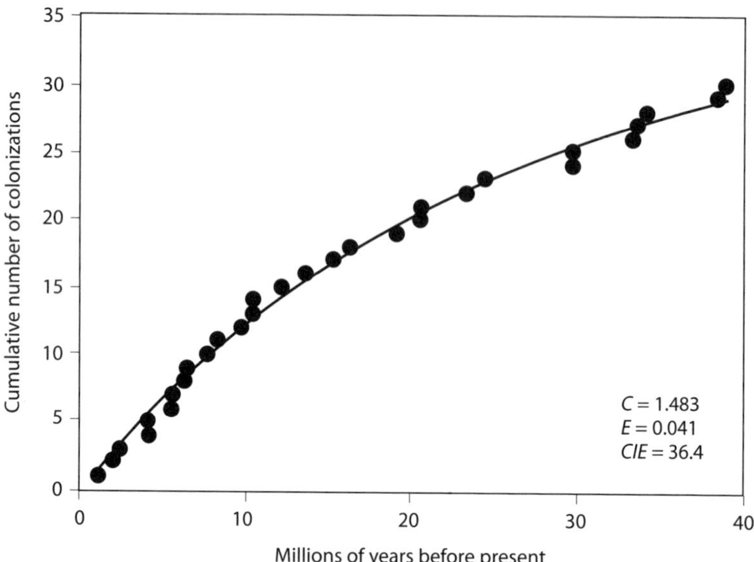

Figure 14.3. Accumulation curve for times since colonization of lineages of pteridophytes of New Zealand (Perrie and Brownsey 2007). The data consisted of Bayesian estimates of ages<50 my and are fitted by equation (14.2) with values of $C=1.483$ my^{-1}, $E=0.041$ my^{-1}, and $C/E=36.4$. Waiting times between colonization and extinction events are $1/C=0.67$ my and $1/E=24.4$ my, respectively. The potential pool of colonizing species might be minimally estimated by the pteridophyte flora of the south coast of New South Wales, Australia <http://www.anbg.gov.au/fern/prh_taxa.html>, which has 96 species. Entering this value into equation (14.1), the data are fitted with values of $C=0.025$, $E=0.041$, and $CP/(C+E)$ of 36.4. The initial rate of colonization ($S=0$) would be $CP=2.39$ my^{-1}.

(waiting times of 0.76 and 34.5 my, respectively) indicating an equilibrium of about 45 lineages (based on data in Hedges 1996, Ricklefs and Bermingham 2007b). By coincidence, these values are similar to those for the ferns and lycophytes of New Zealand. In both cases, colonization is infrequent and expected times to extinction are long. Ricklefs and Bermingham (2004a) applied a similar analysis to birds of the Hawaiian Islands and estimated $C=7.8$ my^{-1} and $E=0.3$ my^{-1}, implying a colonization waiting time of 0.13 my and an average residence time of avian lineages in the archipelago of about 3 my. We should not be surprised that birds exhibit a relatively higher rate of colonization compared to reptiles and amphibians, in spite of the distance of the Hawaiian archipelago from source areas. That extinction rates of birds are an order of

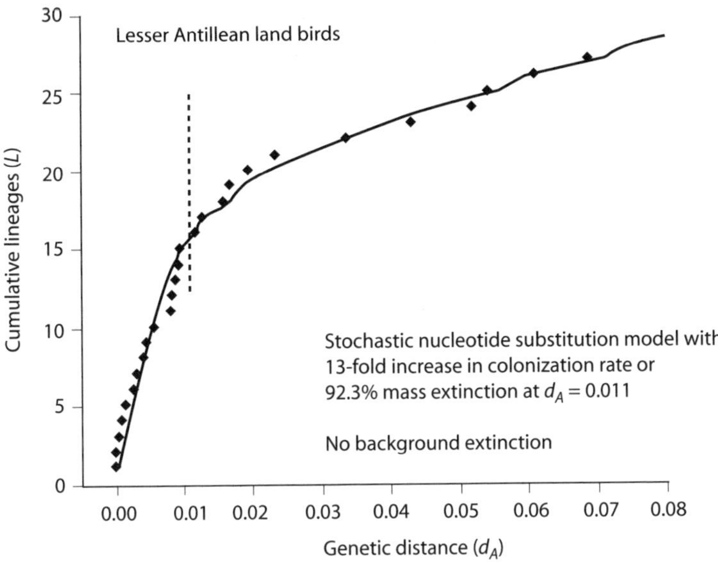

Figure 14.4. Accumulation of lineages of Lesser Antillean nonraptorial land birds with increasing genetic distance between island and external source area populations in South America and the Greater Antilles. The data are fitted by equation (14.2), with a stochastic nucleotide substitution model for genetic distance, and a 13-fold increase in colonization rate or 92.3% mass extinction at a time equivalent to a genetic distance of 0.011 (Ricklefs and Bermingham 2001).

magnitude higher also makes sense, considering the relatively smaller population sizes (Ricklefs and Lovette 1999), and perhaps poorer resistance to stressful environmental conditions by birds, compared to reptiles, amphibians, and ferns and their allies. Although the details of the lineage accumulation curves are undoubtedly influenced by many sources of nonhomogeneity in colonization and extinction rates, and the estimates depend on the accuracy of the time calibrations, this approach nonetheless tells us much about the dynamics of colonization and extinction.

Unlike the other examples, the lineage accumulation curve for small land birds of the Lesser Antilles exhibits a dramatic change in slope at about 0.5–1 my, which can be modeled by a nearly tenfold increase in colonization rates, or a mass extinction event impacting about 90% of the avifauna of the Lesser Antilles at this time (Ricklefs and Bermingham 2001) (figure 14.4). Cherry et al. (2002) suggested that this pattern might instead represent continuing gene flow between the continental source and island populations, making older populations appear "young," that is, undifferentiated, until a speciation threshold is reached, after which genetic divergence—our only measure of "time"—can begin (Johnson et al. 2000). In response, Ricklefs and Bermingham (2004a) pointed out that under the speciation-threshold model, populations close to the continental source (i.e., Grenada and St. Vincent at the southern end of the archipelago) should be more similar genetically to mainland populations than populations on more distant islands, such as Dominica and Guadeloupe. However, no species conforms to this expectation. Rather, the data indicate a possible mass extinction event or a change in conditions that greatly facilitated colonization. One could raise the specter of localized bolide impacts (Maurrasse and Sen 1991, Crother and Guyer 1996) or massive tidal waves caused by underwater landslides (Krastel et al. 2001). However, even with proper restraint in this matter, it is evident that the colonization-extinction dynamics of birds of the Lesser Antilles have not been uniform over time.

Another attribute of the lineage-with-age accumulation curve for Lesser Antillean birds is that it remains linear to the age of the oldest colonists in the islands, at a genetic distance of about 0.14, or about 7–10 my. Recurring extinction should produce a curved line approaching an equilibrium number of species, as in figure 14.3. Linearity implies an absence of extinction in lineages of birds within the Lesser Antilles. Individual island populations do suffer extinction (see below), but the missing populations apparently are replaced by colonization from adjacent islands, that is, by secondary expansion phases within the archipelago. Several species (e.g., the Lesser Antillean bullfinch *Loxigilla noctis*, figure 14.2) are endemic to the archipelago but inhabit virtually every island. The closest relatives of *Loxigilla* are West Indian endemics, and so the distribution of *Loxigilla* in the Lesser Antilles represents a relatively recent phase of expansion. If some island populations of this species had suffered extinction at an earlier time, recolonization would have obliterated the evidence (Bellemain et al. 2008). Recolonization between islands can explain the apparent absence of extinction *within the archipelago as a whole.*

The Strategy of Colonization

In MacArthur and Wilson's words, "In this chapter we shall attempt to relate the properties of the life history of a colonizing species to its chances for success and, if it fails, to the length of time it persists before going extinct." Here MacArthur and Wilson primarily address "the probability that a propagule of a given species will establish a successful colony" (p. 68). Stochasticity and propagule size play a primary role in these considerations, and this chapter had a seminal influence on estimating the probability of extinction of small populations (population viability analysis; Gilpin and Soulé 1986, Boyce 1992) and the probability of establishment of introduced populations (Duncan 1997, Duncan and Young 1999, Cassey et al. 2005). Although propagule size might be a consideration for some types of organisms, particularly for the colonization of islands across vast distances, it appears to be unimportant with respect to established species in the Lesser Antilles. I say this because, where it has been observed, colonization appears to occur in a stepping-stone fashion, from one island to the next, in a nearly deterministic fashion (Bond 1956, Raffaele et al. 1998). If propagule size were marginal for establishment, one might expect more haphazard patterns of island colonization.

Considering how important colonization is to the geographical distribution of species, few generalizations can be made. Colonization is unlikely to be an evolved strategy for the express purpose of reaching remote points because most dispersers fail to colonize, and the traits of the successful colonist are propagated only in the receiving population and not in the source population. Thus, the qualities that incidentally make a good colonizer evolve because of their fitness consequences within the source population.

Organisms can spread rapidly within a large region, particularly in the case of highly mobile organisms, such as birds. Two species, the bared-eyed thrush *Turdus nudigenis* and the shiny cowbird *Molothrus bonairiensis,* have colonized most of the Lesser Antilles during the twentieth century (Bond 1956). Certainly colonization was aided by the creation of abundant human-altered habitat—both species are at home in gardens and agricultural lands—but these birds crossed the water gaps between islands apparently without assistance. Spread involved repeated cycles of arrival on an island (starting from the South American continent), buildup of populations over a few decades, and subsequent colonization of the next island up the chain. These historical records, combined with DNA evidence on a longer time scale, suggest that once a population enters a phase of colonization, it spreads rapidly relative to the mitochondrial DNA differentiation of island populations. Thus, colonization in the Lesser Antilles is hardly haphazard; the only probabilistic aspects seem to be which species will undergo an expansion phase, and when.

Water gaps between islands are selective barriers: among birds, few inner forest species have colonized the Lesser Antilles (Ricklefs and Cox 1972, Terborgh 1973; Terborgh et al. 1978). Continental source populations that have colonized the Lesser Antilles tend to be widespread and abundant. Colonization success also depends on the source area. Species that have invaded the Lesser Antilles from the Greater Antilles in the north generally penetrate the island chain further than species coming from the South American continent to the south. Ricklefs and Bermingham (2007b) suggested that Greater Antillean birds already are successful island colonists, selected for their dispersal and colonization abilities. In addition, populations that invade the Lesser Antilles from Puerto Rico must cross a large water gap and colonize several small islands before reaching the core of large islands in the Lesser Antilles, beginning with Guadeloupe. This places another stringent filter on species with respect to colonizing ability before they reach the Lesser Antilles.

Why some species enter phases of expansion and colonization remains unanswered. Ricklefs and Cox (1972) pointed out that species in expansion phases and species in contraction phases cannot be distinguished by particular ecological characteristics, such as food resources. For example, different insectivorous species might be expanding or contracting at the same time. In addition, because of the lack of synchrony among species, expansion and contraction phases do not appear to be driven by cyclical climate change during the Pleistocene. Warm and cold phases are mirrored by wet and dry phases in the Neotropics (Curtis et al. 2001), which would have influenced the relative proportions of forest versus open habitat. Most species invade the islands through open habitat at low elevation, and so this might have influenced rates of colonization. If true, however, the connection is not strong.

MacArthur and Wilson's concern about the size of propagules and the probability of establishing a colonizing population is generally not amenable to direct observation, except for introduced species (Duncan 1997). However, this can now be addressed in part by quantifying the genetic diversity among individuals in recently established populations. Founder effects are expressed in the proportion of genetic variation in a source population that is sampled by the colonists to an island. Although such analyses have not been undertaken with respect to birds in the Lesser Antilles, one would expect from the relative determinism of stepping-stone colonization that propagule size is relatively large in this system and genetic diversity does not diminish appreciably as colonists move through the island chain (see Clegg, this volume).

Invasibility and the Variable Niche

During the 1960s, ecologists widely believed that communities could become saturated with species, largely following theoretical work by Robert MacArthur and Richard Levins (1967) on limiting similarity and on the stability of the community matrix of species interactions (Vandermeer 1972, May 1975). MacArthur and Wilson (1967, p. 121) asserted

"There is a limit to the number of species persisting on a given island. An island is closed to a particular species either when the species is excluded by competitors already in residence or else when its population size is held so low that extinction occurs much more frequently than immigration."

The first part of this explanation for exclusion parallels Lack's ideas about species filling the ecological space on islands after which further colonization is precluded. However, Lesser Antillean birds provide little evidence of ecological saturation (cf. Terborgh and Faaborg 1980). Censuses of birds across matched habitats on several islands in the West Indies, Trinidad, and central Panama, show that additional species are accommodated within the Caribbean Basin both by habitat compression and by within-habitat niche compression, resulting in lower abundances of individual species (Cox and Ricklefs 1977, Wunderle 1985, Ricklefs 2000).

As populations become more narrowly distributed and less abundant, the probability of their extinction could increase, in which case the rising rate of extinction with increasing numbers of species would contribute to the regulation of species numbers in a steady state. This is the second part of MacArthur and Wilson's explanation for a limit to island diversity. Based on an analysis of avian species richness, island area, and colonization distance in the Solomon Islands, Gilpin and Diamond (1976) estimated that the rate of extinction increased at least as the square of the number of species on an island, indicating that increasing diversity accelerates the extinction of individual island populations (but see Simberloff and Collins, this volume). This result has not, to my knowledge, been confirmed in other systems, but it is supported by our estimates of extinction rates on islands of different size in the Lesser Antilles (see below).

The compression of within- and between-habitat components of the niche appears to have no upper limit, at least within the range of diversity occurring within the Lesser Antilles. Whether or how much ecological compression contributes to extinction probability in Lesser Antillean birds has not been determined. The average time in generations to the stochastic extinction of a population in the absence of density dependence is on the order of the population size. The numbers of Lesser Antillean birds are not well known, but probably few populations exceed 10^5 individuals—about 1 individual per hectare on the largest islands. As indicated below, the estimated time to extinction of individual island populations averages about 2 my, much too long to have been caused by purely demographic stochasticity. Other causes of extinction, such as storms, volcanic eruptions, and introduced predators and pathogens would act faster, and might eradicate small populations more readily than large ones. Accordingly, higher diversity would lead to increased rate of extinction per population.

Recent colonists tend to be widespread and locally abundant. Thus, these species are seemingly not constrained by the local established populations on an island, and the effects of population compression through competition are therefore felt unevenly within the biota. Old populations that have not recently expanded through the archipelago exhibit reduced ecological ranges, often being restricted to forest environments at higher elevation, and having lower local abundance (Cox and Ricklefs 1977, Ricklefs and Bermingham 2004b). Thus, adaptation to the local island environment appears not to give resident species advantages over recent colonists.

One of the important lessons for community ecologists from consideration of island species is that local diversity, both on an island and locally within particular island habitats, is sensitive to the external pressure of colonization and the consequences of extinction for whole island diversity (Terborgh and Faaborg 1980, Ricklefs 1987, Srivastava 1999, Ricklefs 2000). As MacArthur and Wilson (1967, p. 105) pointed out, niches are compressible and communities can be invaded, at least within the range of diversity on most islands. One of the ironies of the development of ecology during the 1960s was that niche saturation was discussed as an intrinsic property of ecological communities without considering the implications of ecological release and compression on islands in response to colonization pressure from outside the system (Kingsland 1985, Ricklefs 1987). Although these two views have largely been reconciled, they have yet to be fully assimilated into ecological thinking.

Stepping-Stones and Biotic Interchange

MacArthur and Wilson were primarily concerned here about the role of stepping-stone islands in promoting dispersal between regions (Clegg, this volume). The concept of stepping-stone colonization also was related to developing ideas about metapopulations and the persistence of populations in subdivided habitats (Hanski 1997, this volume). MacArthur and Wilson showed theoretically that the rate of exchange between two areas is increased when a stepping-stone is placed between them. Because the probability of dispersal decreases faster than the distance between source and recipient areas, placing a stepping-stone between two areas increases the chance of colonization. For example, if the probability of colonization drops off exponentially with the square of the distance, i.e., $P(d)=\exp(-d^2)$, the probability of successfully colonizing across two segments each having half the distance is higher: $P(d/2)\times P(d/2)=\exp(-d^2/2)$.

The idea of stepping-stones as highways of dispersal is relevant to island chains such as the Lesser Antilles, particularly because the distance between islands is not far and probably within the field of vision of most birds. As pointed out above, most colonists to the Lesser Antilles extend their distributions over much of the archipelago. A few species have even dispersed from the Greater Antilles through the island chain to the mainland of South America: *Icterus* orioles (Omland et al. 1999), *Myiarchus* flycatchers (Joseph et al. 2004), and the bananaquit *Coereba flaveola* (Seutin et al. 1994, Bellemain and Ricklefs 2008).

The idea that colonization might proceed in a stepping-stone fashion also suggests a useful tool for the study of extinction. As we have seen, birds in the Lesser Antilles colonize islands in stepping-stone fashion (Ricklefs and Bermingham 2008). That is, recent colonists occupy virtually every island within the Lesser Antilles, regardless of size or ecological diversity, without gaps in their distributions (Bond 1956). Therefore, the absence of a species from an island can be inferred to represent the extinction of an island population. Ricklefs and Bermingham (1999) used this inference to estimate extinction rates in two ways. First, in an adaptation of survival analysis, we assigned each species an "age" reflecting the most recent phase of expansion, whether this was the original colonization event from the continental source area or secondary (or later) expansion within the archipelago. We then fitted an exponential survival curve, $S=\exp(-Et)$, to the proportion of islands occupied as a function of age. We assumed that all islands are occupied ($S=1$) at age 0, and so the exponential rate of decrease in proportion of islands occupied is the extinction rate (E). Applied to small land birds of the Lesser Antilles, we calculated a rate of approximately 0.25 (±0.02 SE) per % mtDNA (ATPase6,8) sequence divergence. Using a common calibration of 2% mtDNA sequence divergence per My (Weir and Schluter 2008), this is equivalent to an extinction rate of about 50% per million years, or an expected time to extinction of about 2 my. Considering the uncertainty concerning this calibration (Lovette 2004), a range of extinction rates between 0.25 and 1 (1 to 4 my between events) probably brackets the average value.

Our second approach was to estimate a common extinction rate for populations on a single island from both the presence of species and the absence of species occurring elsewhere in the Lesser Antilles, under the assumption that missing species were formerly present. Having estimated the relative ages of the most recent expansion phases of most species within the Lesser Antilles, we set the probability that a species would be present on a particular island (i) as $P(\text{present})=\exp(-E_i t)$ and the

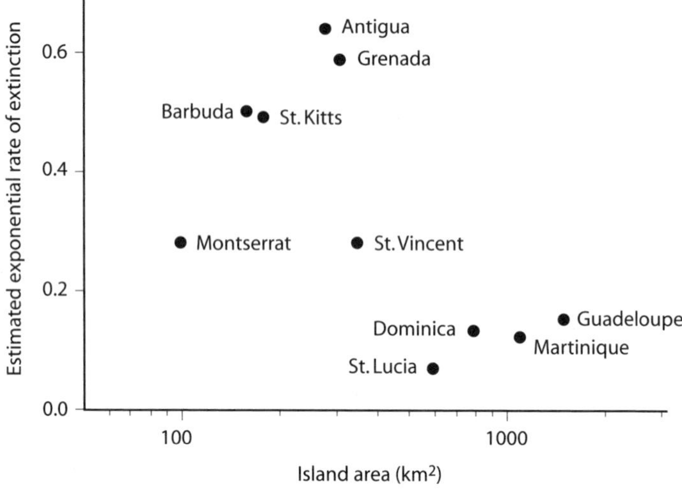

Figure 14.5. Estimated rates of extinction as a function of island size for the major islands in the Lesser Antilles (from Ricklefs and Bermingham 2007b).

probability that the species would have gone extinct as $P(\text{absent}) = 1 - P(\text{present}) = 1 - \exp(-E_i t)$. We then solved for the value of E_i that gave the maximum likelihood ($\Sigma \ln P$). For the island of St. Vincent, for example, we estimated E_{SV} to be 0.28 per % mtDNA sequence divergence, or about the same as the value estimated from survivorship over the archipelago as a whole.

For other islands, E_i varied from low values (0.09–0.23) for the core islands in the center of the chain (St. Lucia, Martinique, Dominica, Guadeloupe) to higher values for low lying islands in the north of the archipelago (Antigua = 1.36, Barbuda = 0.74) islands. As predicted by equilibrium biogeography, extinction rate was inversely related to island size (figure 14.5). However, two anomalous values are of interest. Montserrat (0.30) has a low apparent extinction rate for its small size, yet it lies close to Guadeloupe, which might result in frequent recolonization of the island. Grenada (0.57) has a high apparent extinction rate for its size (as does the low-lying Antigua), lacking several relatively young lineages that are distributed elsewhere in the archipelago and harboring few old endemics. This might be related to the proximity of Grenada to South America and the presence of a high proportion of recent South American colonists on the island, which might have either caused extinctions or prevented colonization from the north.

Estimates of extinction rates depend on the assumption of homogeneity of rates over time. However, it is clear that over the Lesser Antilles as

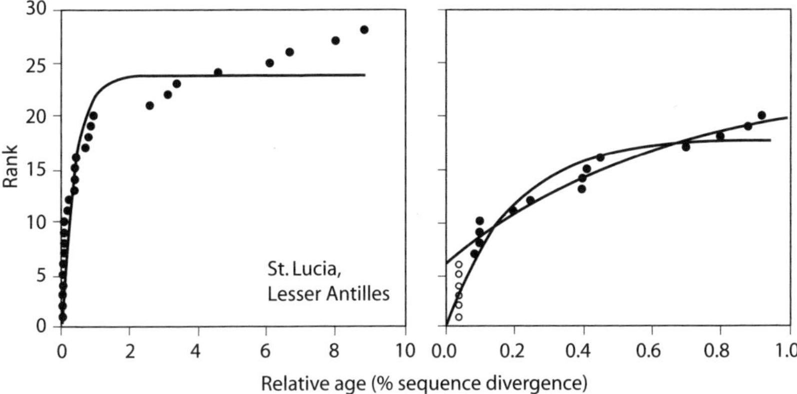

Figure 14.6. Species accumulation as a function of genetic differentiation of St. Lucia populations of birds. Left: All species for which data are available (C=60.9, E=2.56, C/E=23.7). Right: Species with genetic distances <1% including undifferentiated populations (C=94.68, E=5.31, C/E=17.83), or excluding them (data with open symbols, divergence=0%; C=28.46, E=1.69, C/E=11.57).

a whole, rates of colonization and/or extinction have changed dramatically in the past (Ricklefs and Bermingham 2001). The same heterogeneity appears when considering the ages of lineages on individual islands, such as St. Lucia (figure 14.6, left). The lineage accumulation curve exhibits a marked break between 1 and 2% sequence divergence, and a model incorporating an exponential approach to equilibrium fails to describe the data adequately by a homogeneous process. The pattern in figure 14.6 could reflect several kinds of heterogeneity, which evidently have affected islands throughout the Lesser Antilles.

At one extreme, a mass extinction event might have decimated populations older than 2% sequence divergence (Ricklefs and Bermingham 2001), and the subsequent rate of colonization might have increased in response to the opening of ecological space on the islands. The onset of glacial cycles and periods of drier climate (Curtis et al. 2001) also might have opened up lowland habitats and facilitated colonization. The accumulation of lineages on St. Lucia with <1% sequence divergence from other island or external source populations (figure 14.2, right) can be fitted by exponential models reasonably well, but suggest low equilibrium numbers and high turnover rates, even when the very recent colonists, which are undifferentiated from mainland sources and might have appeared after humans, are excluded from the analysis. Species on St. Lucia that are undifferentiated relative to other islands are common in, or even restricted to, human-altered habitats, including gardens and agricultural areas.

Evolutionary Changes Following Colonization

It makes sense that a population of recent colonists should evolve over time to become better adapted to the local conditions on an island. Incumbency should be a powerful position, and one might expect that an island population's probability of extinction decreases with continued evolution. David Lack thought that species, once established on an island, would exclude ecologically similar potential colonists. However, in the case of birds in the West Indies, it is the old populations that are at risk of extinction.

Wilson (1959, 1961) noted that ants colonize islands in Melanesia primarily through coastal habitats and subsequently invade interior forests over time. On islands of any size, interior forests are vastly larger environments than coastal fringes and the adaptive advantage of being able to exploit these environments is evident. Based on the relative ages of island populations in the West Indies, birds follow this pattern, as well. Most species colonize islands through lowland, open environments (possible exceptions are the primarily montane quail-dove *Geotrygon montana* and solitaire *Myadestes genibarbis*) and then with time expand into the interior, high-elevation forests.

In the absence of recurring expansion phases, habitat distribution becomes more restricted and population density decreases (Ricklefs and Cox 1978). It is also possible that extinction rates increase for these populations with their tenure on an island. Although the absence of old populations on some islands suggests extinction, this could represent simply the stochastic loss of populations at a constant rate over time. Nonetheless, historical extinction of individual populations, or their decline to a threatened or endangered status owing to human-caused changes in the environment, involve primarily old populations, a pattern that also appears to apply to birds in the Galapagos and Hawaiian archipelagoes (Ricklefs and Cox 1972, Ricklefs and Bermingham 1999).

The increase in vulnerability with time is puzzling in view of adaptation to local conditions until one considers that the biota of the island also is evolving in response to new colonists. Colonizing species already possess properties that lead to broad distribution and high population density in the source area. Otherwise, they would not be colonists. When these species become established on islands, their populations grow rapidly. Selection inevitably refines the relationships of these populations to the new environments encountered on islands. However, the colonists also exert selective pressure on other island populations with which they interact—predators, pathogens, food resources, mutualists. George Cox and I (Ricklefs 1970; Ricklefs and Cox 1972) referred to this as "counteradaptation" to emphasize the special role of antagonists, and we suggested that counteradaptation (i.e., "counterevolution") drives the taxon cycle.

The scenario is this. New colonists, whether from outside or within the archipelago, have large, widely distributed populations, which exert strong selection on other populations. In particular, local predators and parasites encounter a new, unused potential resource population and benefit from adaptations to exploit this resource. This in turn reduces the productivity of the colonizing population, leading to reduced ecological distribution and population density. New colonists not exploited by endemic predators and pathogens have a competitive advantage and apply further pressure on the older colonists, potentially hastening their extinction. However, as old colonists become less abundant, they are no longer important players in the ecology and evolution of the island biota, potentially leading to disassociation by the local biota, and they become better adapted themselves to resist local antagonists. Over time, this might lead to increasing productivity of such populations and new phases of expansion. The cyclic nature of this interaction derives from the evolutionary time lags built into the responses of antagonistic populations to each other, either because of the time required for changes in allele frequencies or, especially in the case of relationships with pathogens, the waiting times to useful new genetic variation.

The concept of the taxon cycle developed by Ricklefs and Cox extends Wilson's evolutionary scenario by postulating feedbacks between colonists and the resident biotas of islands. Two implications of this scenario are (1) that populations are unlikely themselves to attain steady states, and (2) that changes in populations on islands will occur over evolutionary time scales independently of change in the island environment. In other words, taxon cycles are intrinsic to biological systems. The stability that Lack envisioned is not permitted because of these evolutionary interactions. The steady state that MacArthur and Wilson envisioned incorporates an evolutionary dynamic, in addition to stochastic colonization and extinction, at least on large islands that hold populations long enough to exhibit ecologically meaningful evolutionary change.

As mentioned earlier, the taxon-cycle idea espoused by Ricklefs and Cox (1972) was strongly criticized at the time by some authors. Pielou (1979, p. 198) said that ". . . the whole taxon cycle may simply be the effect of sporadically occurring climatic 'bad years' on species-populations too isolated for losses to be quickly made good from nearby populations." This strongly ecological viewpoint assigned a primary role to the environment for population change. Our molecular phylogenetic framework for birds of the Lesser Antilles shows that expansion and contraction cycles occur over evolutionary time periods and are largely unsynchronized among populations, which would not be expected for strong climatic drivers.

Pregill and Olson (1981, p. 91) disputed our interpretation of data concerning distribution and taxonomic differentiation: "The concept of 'counteradaptation' is an artificial construct needed to explain a nonexistent phenomenon—the taxon cycle." But they also promoted a view that evolution inevitably increases fitness in a local environment when they said "Ecological doctrine and good sense revolt at the idea that a species with a long history of adaptation to a particular environment would be at a competitive disadvantage with newly arriving colonists."

In fact, among species that persist for long periods on large, isolated islands, evolutionary processes likely drive the most interesting colonization-extinction dynamics. The phylogeography of the bananaquit *Coereba flaveola*, which has been worked out in detail by Eva Bellemain from mitochondrial and nuclear markers (Bellemain et al. 2008), shows particularly well the complex history of a taxon, including repeated phases of expansion within a region. The bananaquit currently is distributed throughout the West Indies, except for Cuba, and is widespread in Central and South America. The geographic distribution of genetic variation suggests a long history in the West Indies, with the earliest nodes in the phylogeny being rooted in the Greater Antilles. Nuclear and mitochondrial alleles provide evidence of many phases of expansion, at least three of which passed through islands in the Lesser Antilles (figure 14.7), and at least one case of introgression of a mitochondrial genome into an established island population. Many of the details are lost in history, no fossil record exists, and the causes of expansion phases are unknown. However, it is clear that bananaquits have had a dynamic history in the West Indies.

The genius of MacArthur and Wilson was to recognize the dynamic nature of island biotas and to emphasize the processes of colonization and extinction that shaped patterns of biodiversity. They did not have the advantages of modern molecular methods for reconstructing past history, but they nonetheless used inferences from taxonomy and biogeography to infer the importance of history to understanding present distributions and patterns of diversity. Their vision has been substantially confirmed and enlarged over the past four decades.

Phylogeographic analyses of birds in the Lesser Antilles have contributed to our understanding of island biogeographic processes: dispersal, invasion, competition, adaptation, and extinction, which MacArthur and Wilson thought to be "among the most difficult in biology to study and understand." Colonization times of a complete fauna or flora enable

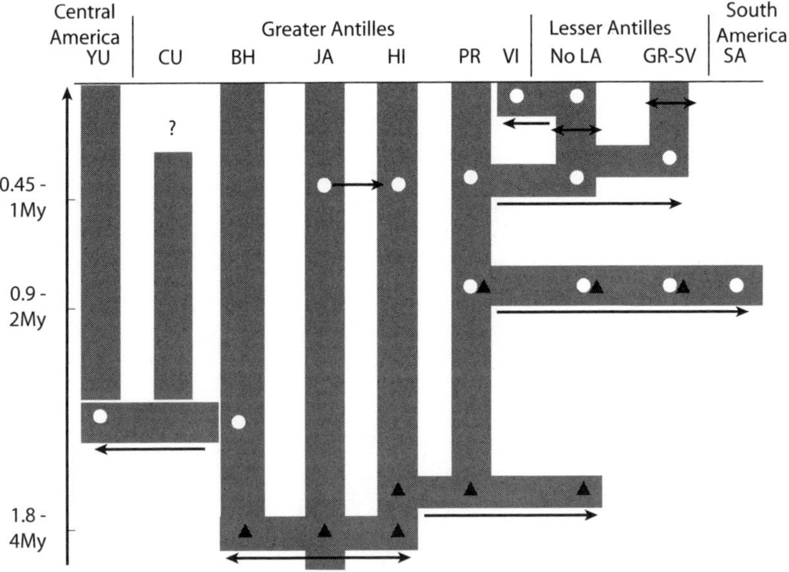

Figure 14.7. Inferred phylogeographic history of the bananaquit within the West Indies and surrounding continental areas of the Caribbean Basin. Based on Bellemain et al. (2008). Vertical bars represent presence on a particular island or in a region for long periods. Horizontal arrows represent phases of geographic expansion inferred from shared genetic variation for mitochondrial (open circles) or nuclear (black triangles) markers. The presence of an extinct population of the bananaquit on Cuba is inferred. Age estimates are based on a range of calibrations for mitochondrial DNA. Locations are YU=Quintana Roo on the Yucatan Peninsula, Mexico; CU=Cuba, BH=Bahama Islands; JA=Jamaica; HI=Hispaniola; PR=Puerto Rico; VI=Virgin Islands; No LA=northern Lesser Antilles, from St. Lucia northward; GR-SV=Grenada and St. Vincent.

one to test the time homogeneity of processes, assess the turnover prediction of the MacArthur-Wilson model, determine whether a system is in equilibrium with respect to species number, and produce preliminary estimates of whole-archipelago colonization and extinction rates. The phylogeographic demonstration of stepping-stone colonization in the Lesser Antilles allows one to estimate single-island extinction rates from gaps in species distributions. Furthermore, detailed phylogeographic analyses of individual species uncover the dynamic taxon-cycle nature of species distributions, with alternating phases of geographic expansion and contraction. The apparent independence of these phases across species suggests that they are driven by species-specific coevolutionary relationships between bird populations and their enemies, whether predators or pathogens. As MacArthur and Wilson clearly perceived, these insights derived from island studies also inform our understanding of diversity and distribution of species on continental land masses (Ricklefs 2004).

We know that Ed Wilson has been very pleased by these developments (Wilson, this volume) and we suspect that Robert MacArthur also would have been gratified by the way things have turned out. The equilibrium theory of island biogeography has inspired generations of ecologists, evolutionary biologists, and biogeographers, but also provided the basis for analytical approaches to untangling historical processes. These approaches have only recently become a part of the biogeographer's toolbox, but they promise continued vitality of the field and further integration of biogeography with ecology and evolutionary biology.

Acknowledgments

My research on West Indian birds over the past twenty years has been carried out in collaboration with Dr. Eldredge Bermingham at the Smithsonian Tropical Research Institute. We have had the benefit of working with many students and postdoctoral fellows over this period, particularly Dr. Gilles Seutin and Dr. Irby Lovette. Earlier collaboration with Dr. George W. Cox laid the foundation for this work. Funding has been generously provided by the National Geographic Society, the National Science Foundation, the Smithsonian Institution, and the University of Missouri. None of this would have been possible without the help of numerous field assistants, colleagues, government officials, and government agencies in countries throughout the Caribbean Basin. Jonathan Losos, Irby Lovette, and an anonymous reviewer commented on the manuscript.

Literature Cited

Bellemain, E., E. Bermingham, and R. E. Ricklefs. 2008. The dynamic evolutionary history of the bananaquit (*Coereba flaveola*) in the Caribbean revealed by a multigene analysis. *BMC Evolutionary Biology* 8:240.

Bellemain, E., and R. E. Ricklefs. 2008. Are islands the end of the colonization road? *Trends in Ecology and Evolution* 23:461–68.

Bond, J. 1956. *Checklist of Birds of the West Indies*. Philadelphia: Academy of Natural Sciences.

Boyce, M. S. 1992. Population viability analysis. *Annual Review of Ecology and Systematics* 23:481–506.

Brown, J. H. 1995. *Macroecology*. Chicago: University of Chicago Press.

Brown, J. H., and A. Kodric-Brown. 1977. Turnover rates in insular biogeography: effect of immigration on extinction. *Ecology* 58:445–49.

Cassey, P., T. M. Blackburn, R. P. Duncan, and J. L. Lockwood. 2005. Lessons from the establishment of exotic species: a meta-analytical case study using birds. *Journal of Animal Ecology* 74:250–58.

Cherry, J. L., F. R. Adler, and K. P. Johnson. 2002. Islands, equilibria, and speciation. *Science* 296:975a.

Cox, G. W., and R. E. Ricklefs. 1977. Species diversity, ecological release, and community structuring in Caribbean land bird faunas. *Oikos* 29:60–66.

Crother, B. I., and C. Guyer. 1996. Caribbean historical biogeography: Was the dispersal-vicariance debate eliminated by an extraterrestrial bolide. *Herpetologica* 52:440–65.

Curtis, J. H., M. Brenner, and D. A. Hodell. 2001. Climate change in the circum-Caribbean (Late Pleistocene to Present) and implications for regional biogeography. In *Biogeography of the West Indies:Patterns and Perspectives*, ed. C. A. Woods and F. E. Sergile, 35–54. Boca Raton, FL: CRC Press.

Diamond, J. M. 1969. Avifauna equilibria and species turnover rates on the Channel Islands of California. *Proceedings of the National Academy of Sciences U.S.A.* 67:1715–21.

———. 1971. Comparison of faunal equilibrium turnover rates on a tropical island and a temperate island. *Proceedings of the National Academy of Sciences U.S,A.* 68:2742–45.

———. 1972. Biogeographic kinetics: Estimation of relaxation times for avifaunas of Southwest Pacific islands. *Proceedings of the National Academy of Sciences U.S.A.* 69:3199–203.

Duncan, R. P. 1997. The role of competition and introduction effort in the success of passeriform birds introduced to New Zealand. *American Naturalist* 149:903–15.

Duncan, R. P., and J. R. Young. 1999. The fate of passeriform introductions on oceanic islands. *Conservation Biology* 13:934–36.

Gaston, K. J. 2003. *The Structure and Dynamics of Geographic Ranges*. Oxford: Oxford University Press.

Gilpin, M. E., and J. M. Diamond. 1976. Calculations of immigration and extinction curves from the species-area distance relation. *Proceedings of the National Academy of Sciences U.S.A.*73:4130–34.

Gilpin, M. E., and M. E. Soulé. 1986. Minimum viable populations: processes of species extinction. In *Conservation Biology: The Science of Scarcity and Diversity*, ed. M. E. Soulé, 19–34.. Sunderland, MA: Sinauer Associates.

Greenslade, P.J.M. 1968. Island patterns in the Solomon Islands bird fauna. *Evolution* 22:751–61.

———. 1969. Insect distribution patterns in the Solomon Islands. *Philosophical Transactions of the Royal Society of London, Series B* 255:271–84.

Hanski, I. 1997. Metapopulation dynamics: From concepts and observations to predictive models. In *Metapopulation Biology: Ecology, Genetics, and Evolution*, ed. I. A. Hanski and M. E. Gilpin, 69–91. San Diego: Academic Press.

Hedges, S. B. 1996. The origin of West Indian amphibians and reptiles. In *Contributions to West Indian Herpetology: A Tribute to Albert Schwartz*, ed. R. Powell and R. W. Henderson, 95–128. Ithaca, NY: Society for the Study of Reptiles and Amphibians.

Hunt, J. S., E. Bermingham, and R. E. Ricklefs. 2001. Molecular systematics and biogeography of Antillean thrashers, tremblers, and mockingbirds (Aves: Mimidae). *Auk* 118:35–55.

Johnson, K. P., F. R. Adler, and J. L. Cherry. 2000. Genetic and phylogenetic consequences of island biogeography. *Evolution* 54:387–96.

Joseph, L., T. Wilke, E. Bermingham, D. Alpers, and R. Ricklefs. 2004. Towards a phylogenetic framework for the evolution of shakes, rattles, and rolls in *Myiarchus* tyrant-flycatchers (Aves: Passeriformes: Tyrannidae). *Molecular Phylogenetics and Evolution* 31:139–52.

Kingsland, S. E. 1985. *Modeling Nature. Episodes in the History of Population Ecology*. Chicago: University of Chicago Press.

Krastel, S., H. U. Schmincke, C. L. Jacobs, R. Rihm, T. P. Le Bas, and B. Alibes. 2001. Submarine landslides around the Canary Islands. *Journal of Geophysical Research—Solid Earth* 106:3977–97.

Lack, D. 1976. *Island Biology Illustrated by the Land Birds of Jamaica*. Berkeley: University of California Press.

Lomolino, M. V. 2000. Ecology's most general, yet protean pattern: The species-area relationship. *Journal of Biogeography* 27:17–26.

Lovette, I. J. 2004. Mitochondrial dating and mixed support for the "2% rule" in birds. *Auk* 121:1–6.

Lovette, I. J., E. Bermingham, and R. E. Ricklefs. 1999. Mitochondrial DNA phylogeography and the conservation of endangered Lesser Antillean *Icterus* orioles. *Conservation Biology* 15:1088–96.

MacArthur, R. H., and R. Levins. 1967. The limiting similarity, convergence, and divergence of coexisting species. *American Naturalist* 101:377–85.

MacArthur, R. H., and E. O. Wilson. 1963. An equilibrium theory of insular zoogeography. *Evolution* 17:373–87.

———. 1967. *The Theory of Island Biogeography*. Princeton, NJ: Princeton University Press.

Maurrasse, F.J.-M.R., and G. Sen. 1991. Impacts, tsunamis, and the Haitian Cretaceous-Tertiary boundary layer. *Science* 252:1690–93.

May, R. M. 1975. Patterns of species abundance and diversity. In *Ecology and Evolution of Communities*, ed. M. L. Cody and J. M. Diamond, 81–120. Cambridge, MA: Harvard University Press, Belknap Press.

Omland, K. E., S. M. Lanyon, and S. J. Fritz. 1999. A molecular phylogeny of the new world orioles (*Icterus*): The importance of dense taxon sampling. *Molecular Phylogenetics and Evolution* 12:224–39.

Perrie, L., and P. Brownsey. 2007. Molecular evidence for long-distance dispersal in the New Zealand pteridophyte flora. *Journal of Biogeography* 34: 2028–38.

Pielou, E. C. 1979. *Biogeography*. New York: Wiley.

Pregill, G. K., and S. L. Olson. 1981. Zoogeography of West Indian vertebrates in relation to Pleistocene climatic cycles. *Annual Review of Ecology and Systematics* 12:75–98.

Preston, F. W. 1962. The canonical distribution of commonness and rarity. *Ecology* 43:185–215, 410–32.

Raffaele, H., J. Wiley, O. Garrido, A. Keith, and J. Raffaele. 1998. *A Guide to the Birds of the West Indies*. Princeton, NJ: Princeton University Press.
Ricklefs, R. E. 1970. Clutch-size in birds: Outcome of opposing predator and prey adaptations. *Science* 168:599–600.
———. 1987. Community diversity: relative roles of local and regional processes. Science 235:167–71.
———. 2000. The relationship between local and regional species richness in birds of the Caribbean Basin. *Journal of Animal Ecology* 69:1111–16.
———. 2004. A comprehensive framework for global patterns in biodiversity. *Ecology Letters* 7:1–15.
Ricklefs, R. E., and E. Bermingham. 1999. Taxon cycles in the Lesser Antillean avifauna. *Ostrich* 70:49–59.
———. 2001. Nonequilibrium diversity dynamics of the Lesser Antillean avifauna. *Science* 294:1522–24.
———. 2002. The concept of the taxon cycle in biogeography. *Global Ecology and Biogeography* 11:353–61.
———. 2004a. Application of Johnson et al.'s speciation threshold model to apparent colonization times of island biotas. *Evolution* 58:1664–73.
———. 2004b. History and the species-area relationship in Lesser Antillean birds. *American Naturalist* 163:227–39.
———. 2007a. The causes of evolutionary radiations in archipelagoes: passerine birds in the Lesser Antilles. *American Naturalist* 169:285–97.
———. 2007b. The West Indies as a laboratory of ecology and evolution. *Philosophical Transactions of the Royal Society of London, Series B* 363:2393–413. DOI:10.1098/rstb.2007.2068.
———. 2008. Likely pre-Columbian introduction of the red-legged thrush to Dominica, West Indies. *Auk* 125:299–303.
Ricklefs, R. E., and G. W. Cox. 1972. Taxon cycles in the West Indian avifauna. *American Naturalist* 106:195–219.
Ricklefs, R. E., and I. J. Lovette. 1999. The roles of island area *per se* and habitat diversity in the species-area relationships of four Lesser Antillean faunal groups. *Journal of Animal Ecology* 68:1142–60.
Scheiner, S. M. 2003. Six types of species-area curves. *Global Ecology and Biogeography* 12:441–47.
Scheiner, S. M., S. B. Cox, M. Willig, G. G. Mittelbach, C. Osenberg, and M. Kaspari. 2000. Species richness, species-area curves and Simpson's paradox. *Evolutionary Ecology Research* 2:791–802.
Seutin, G., N. K. Klein, R. E. Ricklefs, and E. Bermingham. 1994. Historical biogeography of the bananaquit (*Coereba flaveola*) in the Caribbean region: a mitochondrial DNA assessment. *Evolution* 48:1041–61.
Simberloff, D., and E. O. Wilson. 1969. Experimental zoogeography of islands: the colonization of empty islands. *Ecology* 50:278–96.
Terborgh, J. 1973. Chance, habitat and dispersal in the distribution of birds in the West Indies. *Evolution* 27:338–49.
———. 1974. Preservation of natural diversity: The problem of species extinction. *BioScience* 24:715–22.

Terborgh, J. W., and J. Faaborg. 1980. Saturation of bird communities in the West Indies. *American Naturalist* 116:178–95.

Terborgh, J., J. Faaborg, and H. J. Brockman. 1978. Island colonization by Lesser Antillean birds. *Auk* 95:59–72.

Vandermeer, J. H. 1972. Niche theory. *Annual Review of Ecology and Systematics* 3:107–32.

Weir, J. T., and D. Schluter. 2008. Calibrating the avian molecular clock. *Molecular Ecology* 17:2321–28.

Whittaker, R. J., M. B. Bush, and K. Richards. 1989. Plant recolonization and vegetative succession on the Krakatau Islands, Indonesia. *Ecological Monographs* 59:59–123.

Wilcox, B. A. 1978. Supersaturated island faunas: A species-age relationship for lizards on post-Pleistocene land-bridge islands. *Science* 199:996–98.

Wilson, E. O. 1959. Adaptive shift and dispersal in a tropical ant fauna. *Evolution* 13:122–44.

———. 1961. The nature of the taxon cycle in the Melanesian ant fauna. *American Naturalist* 95:169–93.

Wunderle, J. M. 1985. An ecological comparison of the avifaunas of Grenada and Tobago, West Indies. *Wilson Bulletin* 97:356–65.

The Speciation-Area Relationship

Jonathan B. Losos and Christine E. Parent

> *The species-area relationship is often referred to as the closest thing to a rule in ecology (Schoener 1976). . . . The pattern appears to be so common that it would be much more expedient to report the few exceptions . . . than the many hundreds, and possibly thousands of studies reporting this pattern.*
>
> —Lomolino 2000

Inspired in part by the species-area relationship, MacArthur and Wilson (1967) proposed the equilibrium theory of island biogeography, which relied on the ecological processes of colonization and extinction to determine the species diversity of islands. Although widely influential, theirs was not the only ecologically oriented explanation of insular species richness. Lack (1976), for example, believed that island diversity was determined by the habitat diversity on islands; more distant islands had lower diversity because they tend to be impoverished in terms of habitat heterogeneity. These ideas, particularly the MacArthur and Wilson theory, dominated thinking about island species diversity throughout the latter part of the twentieth century.

Islands are also widely recognized as natural laboratories of evolution, ideal localities in which to study evolutionary processes and their long-term consequences (e.g., Carlquist 1974, Grant 1998, Losos and Ricklefs, 2009). One area that has been particularly influenced by research on islands is the study of adaptive radiation, the idea that a single ancestral species diversifies, producing descendant species that occupy a wide variety of ecological niches. Many of the most famous cases of adaptive radiations—Darwin's finches, Hawaiian silverswords, African Rift Lake cichlids—occur on islands or islandlike settings. One consequence of adaptive radiation, if it occurs *in situ*, is that the diversity of an island is a result not just of colonization and extinction, but also of the evolutionary input of species resulting from within-island or within-archipelago speciation (cladogenesis); for example, the tiny island of Rapa in the South Pacific (size=40 km^2) harbors 67 species of *Miocalles* weevils, all the presumed descendants of a single ancestral colonist (Paulay 1985).

MacArthur and Wilson (1963, 1967) were not unaware of the potential significance of evolutionary processes occurring on islands. Indeed, the final chapter of their monograph was entitled, "Evolutionary Changes Following Colonization." Nonetheless, it's fair to say that for more than three decades after the book's publication, little attention was paid to evolutionary issues as research focused on the ecological factors affecting species richness.

However, times have changed and in recent years researchers have begun to pay attention to the role of evolutionary factors in generating and maintaining insular species richness. This work—like much of the renaissance in macroevolutionary thinking—has been sparked by the increased availability of phylogenies and of comparative methods based on phylogenetic information.

The goal of this chapter is simple: to investigate the extent to which evolutionary diversification may be responsible for generating species-area relationships. Few explicitly phylogenetic studies have addressed such questions, and we will focus here on two case studies, Caribbean lizards in the genus *Anolis* and Galápagos snails in the genus *Bulimulus*. These two groups occur on island groups that differ greatly in age, size, and isolation. Moreover, the two groups have been studied with different approaches (though using the same conceptual methodology discussed below). Despite these differences, the similarity in general pattern of evolutionary diversification is striking. In addition to these case studies, we will discuss evolutionary processes that may serve to obscure species-area relationships.

Methods

Our approach is straightforward: by examination of the geography of species in a phylogenetic context, we can estimate the extent to which the species on an island arrived there by colonization versus originating *in situ* by a speciation event in which one ancestral species divided into two descendant species. For islands with more than one species, colonization is indicated by the existence of distantly related species on the same island (figure 15.1A). The most parsimonious explanation of such a pattern is that the species are the descendants of independent colonization events. Conversely, the existence of a clade of species on an island suggests that the clade originated by the colonization of a single species that subsequently diversified *in situ*, producing many descendant species on that island (figure 15.1B). A clade of n species on an island would suggest the occurrence of at least $n-1$ speciation events—"at least" because, of course,

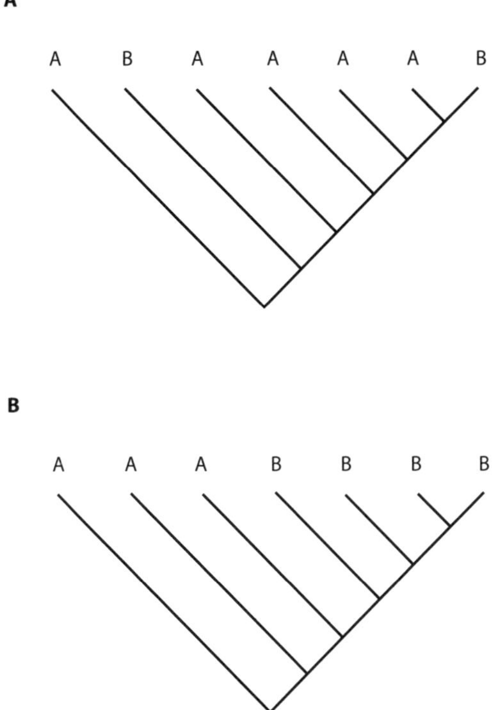

Figure 15.1. Reconstruction of colonization and within-island speciation from a phylogeny. In (A), two distantly related species occur on island B. The most parsimonious explanation is that they independently colonized the island. By contrast, in (B), a clade of species all occur on island B. The most parsimonious explanation in this case is that B was colonized once, followed by a number of speciation events on that island.

evidence of some speciation events may have been lost due to subsequent extinction.

Of course, this method is not infallible. For example, multiple colonization events followed by extinction of related species in the external source area may leave the descendants as sister taxa among extant taxa, thus erroneously implying the occurrence of within-island speciation (figure 15.2A). Although this alternative explanation is a possibility, it seems unlikely to account for the existence of large clades, such as in figure 15.1B, because it would require the extinction of so many related species on the ancestral island. On the other hand, within-island diversification could be mistaken for colonization if members of the radiating clade

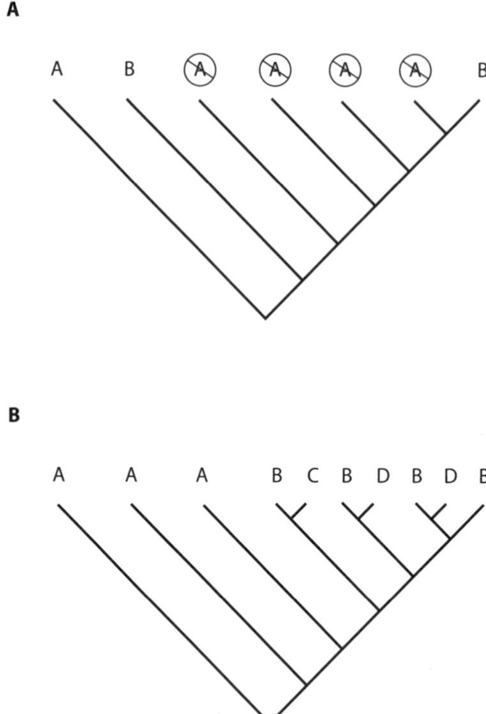

Figure 15.2. Examples of how phylogenetic interpretations can be misleading. In (A), two species independently colonized island B, as in figure 15.1. However, subsequently, other species on island A went extinct, thus leaving the two species on island B as sister taxa among extant taxa, which would falsely suggest that they were the result of within-island speciation subsequent to a single colonization event. In (B), within-island speciation on island B produced a number of species as in Figure 1. Subsequently, however, several of these species sent off colonists to other islands, so that the species on B are no longer each other's closest relatives. As a result, evidence for within-island speciation on island B becomes less clear-cut.

send off colonists to other islands or back to the source area, thus breaking the monophyly of the species on the ancestral island (figure 15.2B).

The hypothesis we wish to test is simple: the extent of within-island diversification is a function of island area, with larger areas experiencing higher rates of diversification. A related question, should such a pattern be detected, concerns the cause of the relationship. Ecological investigations of the species-area relationship focus primarily on two explanations: increased habitat heterogeneity on larger islands may create a corresponding increase in the number of different ecological types that can be supported, and island size per se may directly affect the number of species on an island, for example by the decreased extinction rate of the larger populations that may occur on larger islands (Ricklefs and Lovette 1999).

In a corresponding way, the same two factors may lead to a speciation-area relationship: On one hand, large islands may have more ecological heterogeneity, either directly sparking increased rates of speciation, as suggested by current ideas commonly referred to as "ecological speciation" (e.g., Rundle and Nosil 2005, Funk et al. 2006), or as a result of greater persistence (lower extinction rates) of newly speciated taxa which can avoid competitive exclusion by adapting to different habitats. Alternatively, larger islands may provide greater opportunity for populations to become isolated by geographical means (rivers, mountains, inhospitable habitat, etc.); the resulting higher speciation rates may lead to higher species richness, irrespective of any ecological differences among islands.

Anolis Lizards of the West Indies

Anolis is the second most species-rich genus of vertebrate, only slightly surpassed by frogs in the genus *Pristimantis* (Hedges et al. 2008). Currently, approximately 361 species are recognized, of which 155 occur in the West Indies and the rest in mainland Central and South America, plus one species native to the southeastern United States. In the West Indies, anoles are found on almost every emergent landmass more than a few square meters in area; species diversity ranges from one on many islands to more than 60 on Cuba (for a review of anole ecology and evolution, see Losos [2009]).

Examination of the species-area relationship for West Indian *Anolis* reveals a significant relationship, but one not well fit by linear regression (Rand 1969; Losos 1996). By contrast, a breakpoint regression indicates the existence of two lines, one which covers the majority of the range of island sizes and which fits the data poorly, and the second which includes the four large islands of the Greater Antilles and which fits the data extremely well (figure 15.3).

Species-Area Relationship on Smaller Islands

The poor fit of the regression for the smaller islands is readily explainable. This is a heterogeneous group of islands that have different underlying mechanisms determining their species richness. For example, the islands of the Great Bahama Bank were connected into one enormous landmass, almost the size of Cuba, during the last Ice Age. On the now

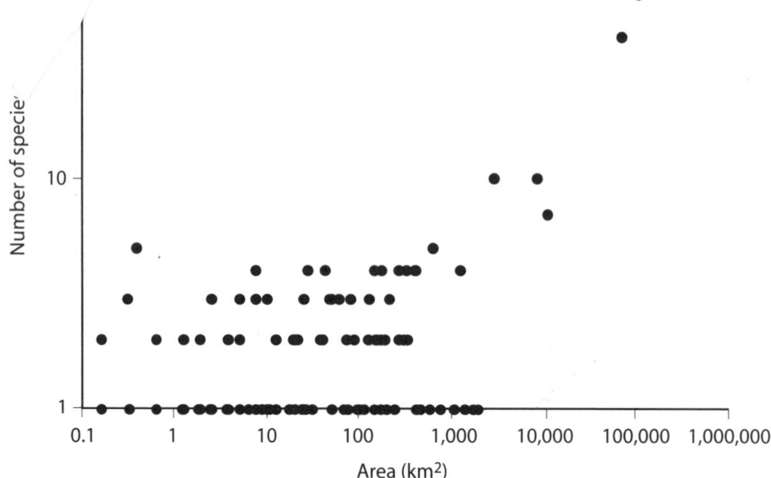

Figure 15.3. Relationship between island area and number of *Anolis* species in the West Indies. A linear regression on ln-transformed data is significant, but only explains 27% of the variation. However, a breakpoint regression finds two lines, one with a shallow slope over the majority of the area range which explains only 11% of the variation, and a second line with a much greater slope that fits the four largest islands and explains 93% of the variation in species numbers in those islands.

fragmented islands of this Bank (termed "land-bridge" islands because they used to be connected to a larger landmass), a very regular species-area relationship exists (figure 15.4A). The occurrence of species on these islands is strongly nested with respect to island area and is a classic example of faunal relaxation: when islands are fragmented, the smaller the island, the greater the number of species that become extinct (Wilcox 1978, Richman et al. 1988). To a large extent, this pattern of extinction is driven by habitat; as islands get smaller, they becoming increasingly less vegetated, and as a result, the more arboreal species vanish. A similar explanation accounts for the species-area relationship among islands that are located near, and formerly were connected to, Puerto Rico, Cuba, and Hispaniola (figure 15.4B). However, because the Greater Antilles have higher species diversity than the large islands of the Bahamas, small islands near the Greater Antilles have greater species richness than islands of similar size on the Bahamas Bank; this difference partially explains the poor predictive ability of area when these islands are considered together.

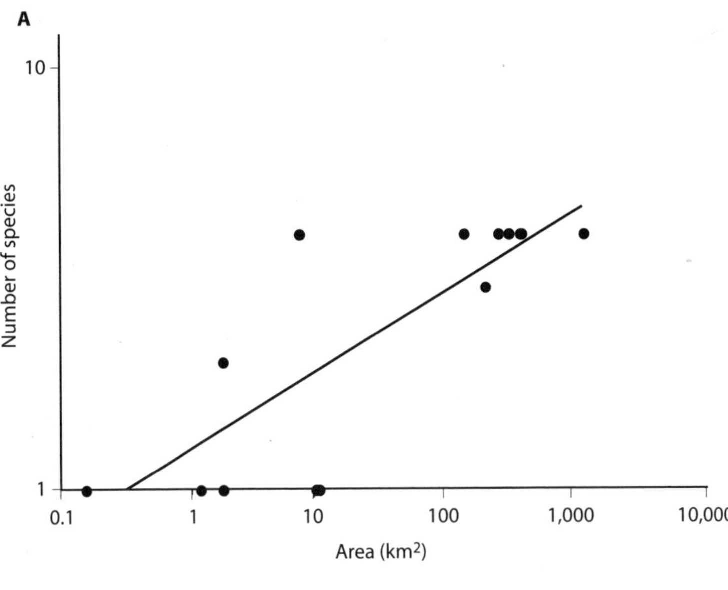

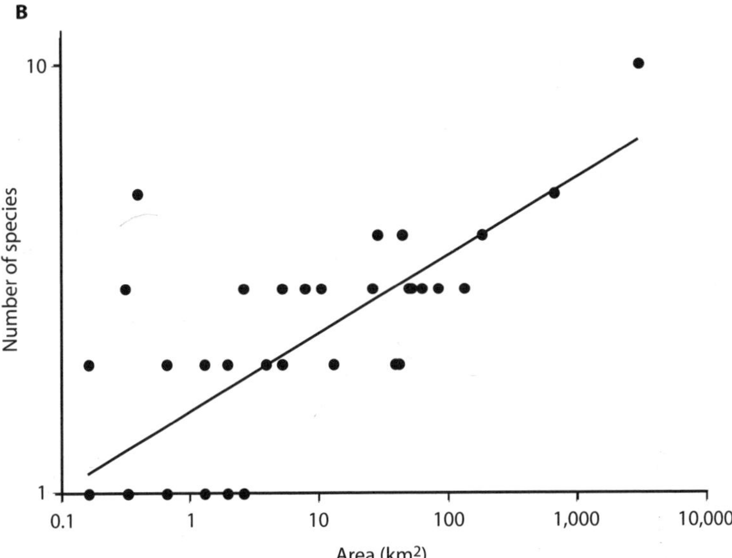

Figure 15.4. Relationship between island area and number of *Anolis* species on subsets of West Indian islands. A. Islands on the Great Bahama bank. B. Islands located near Cuba, Puerto Rica, and Hispaniola which were connected to these larger landmasses at times of lower sea levels (termed "land-bridge" islands). C. Oceanic islands in the West Indies.

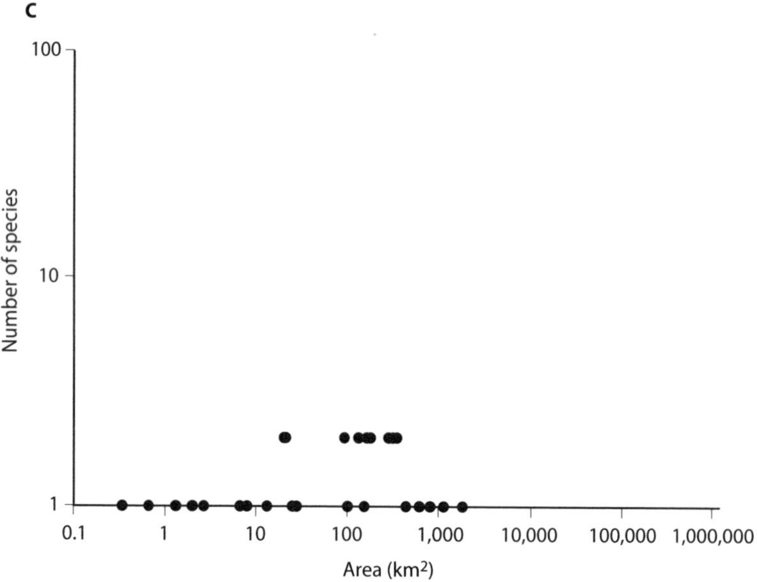

Figure 15.4. (*continued*)

More important in disrupting the species-area relationship, however, are the oceanic islands of the Caribbean, islands that were never connected to larger landmasses and thus must have received their anoles as a result of overwater colonization. These islands include not only the Lesser Antilles chain, but also islands scattered throughout the northern Caribbean, such as the Cayman Islands, St. Croix, Mona, Navassa, and others. In contrast to the extinction-structured diversity of land-bridge islands, oceanic islands never contain more than two species. Although the distance to the Lesser Antilles might account for this low number, even oceanic islands near the Greater Antilles only contain at most two species, even though some of these islands are much larger than some land-bridge islands that contain 3–4 species. Moreover, no area effect is evident among oceanic islands; in fact, the largest islands in the Lesser Antilles only harbor a single anole species (figure 15.4C).

Speciation-Area Relationship in the Greater Antilles

In contrast to the weak relationship among smaller islands, the four largest Caribbean islands, the Greater Antilles, exhibit a tight species-area relationship with a much elevated slope. What might explain this difference? One obvious possibility is that the species richness of larger islands is being augmented by *in situ* evolutionary diversification.

Examination of the phylogeny of *Anolis* (Nicholson et al., 2005; for review of anole phylogenetics, see Losos [2009]) reveals several patterns. First, ample evidence exists for within-island speciation in the Greater Antilles. For example, six of the seven species on Jamaica are closely related members of the *grahami* series, indicating the occurrence of at least five speciation events on Jamaica.[1] Puerto Rico contains three anole lineages, two of which only contain one species, but the third has eight species. Similarly, Hispaniola and Cuba harbor many lineages, containing 1–33 species.

By contrast, almost no evidence suggests the occurrence of within-island speciation on any island smaller in size than Puerto Rico (Losos and Schluter 2000). This pattern suggests an area threshold below which speciation occurs rarely, if at all. Such a high threshold—approximately 9,000 km^2—is unexpected. Although some smaller islands have only been recently isolated, others are geologically very old and, according to molecular estimates, have harbored their anoles for many millions of years (e.g., Malhotra and Thorpe 1994, Schneider 1996). Moreover, many of these islands exhibit great habitat heterogeneity and substantial elevational relief, and even have offshore islets upon which populations might be isolated. Why speciation has not occurred on any of these islands is a mystery, although not a unique one: in a consideration of isolated oceanic islands in the Pacific, Diamond (1977) noted that no island smaller than New Guinea had experienced a within-island speciation event among birds, a finding later corroborated by Coyne and Price (2000).

Examination of the Greater Antilles indicates that the vast majority of the species diversity is the result of within-island diversification; at most 11 between island speciation events are required to explain the distribution of species across islands (unfortunately, uncertainty about Caribbean geology precludes distinguishing between colonization and vicariance as the cause of between island speciation [see Losos 2009]); consequently, more than 90% of the 121 species in the Greater Antilles have resulted from *in situ* evolutionary diversification.

Second, quantitative analysis indicates that the rate of speciation on Greater Antillean islands is strongly correlated with island area (figure 15.5; Losos and Schluter 2000). This result is based on a regression of

[1]The *grahami* series has one other member, *A. conspersus* on Grand Cayman. This species is clearly derived from within *A. grahami* itself (Jackman et al. 2002) and thus is clearly a case of a colonization event from Jamaica to Grand Cayman. Thus, the *grahami* series is not, technically, monophyletic on Jamaica. Nonetheless, the phylogeny clearly indicates the existence of multiple speciation events on Jamaica. Similar examples occur in several other of the island clades discussed here (e.g., the *carolinensis* species group on Cuba [Glor et al. 2005]).

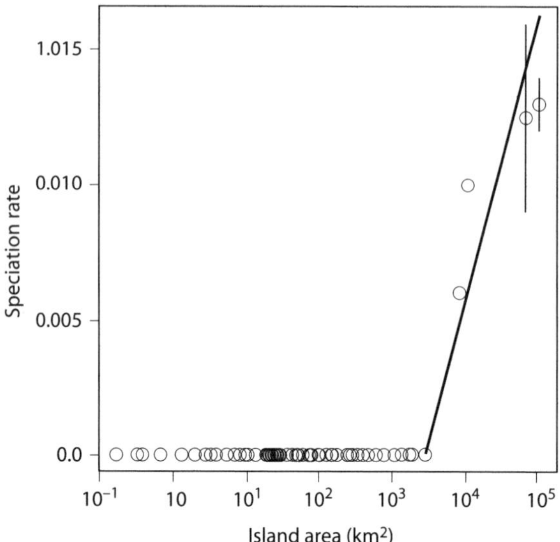

Figure 15.5. Speciation-area relationship in the Greater Antilles. Rates of speciation were estimated from the phylogenetic relationships of Greater Antillean anoles, assuming that the presence of sister taxa on the same island indicates the existence of a within-island speciation events. Error bars on the two larger islands reveal uncertainty in assignment of ancestral locations deep in the phylogeny—essentially, either Hispaniola or Cuba could be the ancestral home from which the other island was colonized multiple times (see discussion in Losos 2009); this uncertainty changes the identification of within-island speciation events deep in the phylogeny, leading to slight changes in estimated speciation rates (Losos and Schluter 2000).

the minimum number of speciation events inferred to have occurred on each island versus island area and is scaled to branch lengths on the phylogeny. These analyses also suggest, using maximum likelihood methods, that the diversification-area relationship results primarily, if not exclusively, from a relationship between speciation and island area, rather than from area effects on extinction rates (Losos and Schluter 2000).

Determinants of the Speciation-Area Relationship

Given that a speciation-area relationship exists for West Indian *Anolis*, the next question is: what drives that relationship? One possibility is that greater ecological specialization occurs on larger islands. To assess this possibility, we consider three components of anole diversity:

1. *Ecomorphs.*The Greater Antilles are famous for the evolution of the same set of habitat specialists on each island, called "ecomorphs" and named for the part of the structural habitat they utilize (e.g., trunk-ground, trunk-crown, twig). Phylogenetic analysis indicates that the ecomorphs on each island have evolved for the most part independently, producing much the same evolutionary end point (Williams 1983, Losos et al. 1998). Although six ecomorph types are recognized (reviewed in Losos 2009), only four occur on all four islands in the Greater Antilles; the grass-bush ecomorph is absent from Jamaica and the trunk ecomorph from Jamaica and Puerto Rico. Consequently, the ecomorph phenomenon contributes in a small way to the species-area relationship, as the larger two islands have six ecomorphs, whereas the smaller two islands have only four and five.

2. *Unique habitat specialists.* In contrast to the ecomorphs, a number of habitat specialists have evolved on only a single island, with no parallel on other islands. These unique types include a species that always occurs near streams and both escapes and forages in the water (*A. vermiculatus,* Cuba); giant, chameleon-like species specialized for eating mollusks and other hard-bodied prey (Chamaeleolis clade, formerly considered its own genus, but now recognized to have evolved within *Anolis*, Cuba); and a leaf-litter specialist from Hispaniola (*Chamaelinorops* clade, also formerly considered its own genus, Hispaniola). Six such types occur on Hispaniola, five on Cuba, one on Jamaica, and none on Puerto Rico. Thus, the evolution of unique types has also contributed somewhat to the species-area relationship.

3. *Within-habitat specialist clade diversification.* By far the largest component of anole species diversity in the Greater Antilles has resulted from species proliferation within clades of microhabitat specialists. Thus, the *sagrei* clade of trunk-ground anoles is comprised of 14 trunk-ground species on Cuba, the *alutaceus* clade of grass-bush anoles contains 14 grass-bush anoles on Cuba, and the *cybotes* clade of trunk-ground anoles sports nine species on Hispaniola. Several unique habitat specialists also have speciated extensively, most notably the five members of the Chamaeleolis clade in Cuba. Overall, 52 of the species on Cuba represent multiple species within clades of habitat specialists (i.e., 63 species on Cuba, minus the 11 independent evolutionary instances of habitat specialization), compared to 29 such species on Hispaniola, five on Puerto Rico, and one on Jamaica.

Some of this within-habitat specialist diversification has involved adaptation to different parts of the environment to permit resource partitioning and coexistence. For example, on all four islands, large and small species of trunk-crown anole have evolved, presumably to minimize competition for food. In addition, within trunk-ground and grass-bush clades on several islands, species have evolved different thermal physiological tolerances, permitting partitioning of thermal microclimates within localities (reviewed in Losos 2009).

Most within-specialist clade speciation, however, appears more related to geography than adaptive divergence. Many—perhaps most—of these species are allo- or parapatrically distributed and have small geographic ranges, often centered on different mountain ranges. In the *alutaceus* clade, for example, two species are island-wide in Cuba, but the other 12 have very small distributions, mostly in mountainous eastern Cuba. The natural history of many of these species is poorly known, but it seems likely that much of this diversity is the result of the allopatric speciation across the rugged landscape of these islands. The greater speciation rate of these clades on larger islands, then, may primarily be a consequence of the greater opportunity afforded for geographical isolation and speciation on the larger—and very mountainous—landmasses of Cuba and Hispaniola.

To summarize the anole story: islands smaller than the size of Puerto Rico have not experienced within-island speciation; their diversity is solely the result of the ecological processes of colonization and extinction. By contrast, more than 90% of the species on the four large islands of the Greater Antilles have arisen *in situ*. The strong species-area relationship for these islands is thus a consequence of a speciation-area relationship. Although some of this relationship results because more habitat specialists occur on larger islands, the majority of diversity stems from a greater rate of within-habitat specialist speciation on larger islands; this increased rate may result primarily from the greater opportunity for speciation to occur on larger islands.

Bulimulid Land Snails of the Galápagos

With over 70 described species (Chambers 1991), the bulimulid land snails of the Galápagos represent the most species-rich radiation of these islands. The entire group *Bulimulus*, subgenus *Naesiotus* (sometimes considered its own genus) to which Galápagos bulimulids belong includes 162 known species distributed in South America, from Venezuela to Argentina (mostly in the Andean region) and in the southern half of Brazil (Breure 1979). All Galápagos bulimulid species are endemic, and current phylogenetic evidence based on multiple independent molecular markers suggests that all species studied in detail are single island endemics (Parent and Crespi 2006).

Bulimulids have colonized all of the major Galápagos islands, and they are found at all elevations except on the shoreline, which is composed mainly of lava boulders and sandy beaches. Vegetation on the Galápagos can be separated into 6–7 altitudinal zones (Wiggins and Porter 1971, van der Werff 1979), and the plant-species composition of each zone is a reflection of the humidity level of the zone, with moisture level increasing with elevation (McMullen 1999). Galápagos bulimulid species vary remarkably in shell size, shape, color, and color pattern, and this morphological variation in shell morphology is related at least partly to aspects of ecological variation, including vegetation zones, related moisture levels, and microhabitat (Coppois and Glowacki 1983). Furthermore, a significant positive correlation between shell shape (degree of shell roundness) and elevation suggests that snail species have adapted morphologically to the varying moisture levels (Parent, unpublished data). Since plants provide food and shelter, and probably most importantly, habitat structure, land snails have potentially adapted to different plant species for feeding, hiding, or resting.

The geological history of the Galápagos archipelago is relatively well understood, with individual islands formed as the Nazca plate moved over a single active hot spot, presumably currently located under the volcanically active Fernandina Island (White et al. 1993). Española, Santa Fe, and San Cristobal are the oldest islands of the archipelago (2.35–3.90 million years [my] old); Santa Cruz, Floreana, Pinzon and Santiago islands form a middle-aged group (0.77–1.52 my old), and finally the most recent group of islands includes Isabela and Fernandina islands (less than 0.7 my old). Isabela Island is formed by six major volcanoes that are separated by extensive barren lava flows. Bulimulid land snails cannot survive without a minimum of vegetation for food and shelter, and thus they are not found at low elevations between the volcanoes forming Isabela. Therefore, each volcano forming Isabela Island can be considered as a separate island with regard to bulimulid land snail distribution.

The colonization sequence of the Galápagos bulimulid lineage (inferred from a molecular-based phylogeny) was found to roughly parallel the geological order of the islands (Parent and Crespi 2006), supporting the progression rule hypothesis with species found on older emerged islands connecting at deeper nodes (see chapters in this volume by Whittaker, and by Gillespie and Baldwin).

We determined the importance of island area, habitat diversity (measured as the number of native plant species), island insularity (measured as distance from the nearest older major island), and island age on island species diversity. We used data from Parent and Crespi (2006) to reanalyze the role that island area has in combination with island habitat diversity, insularity and age on (1) total island species diversity, (2) diversity due to within-island speciation, and (3) diversity due to between-island colonization.

We used a phylogeny based on multiple independent DNA markers (Parent and Crespi 2006) to distinguish species that arose *in situ* on an island from those that arrived by colonization from another island. In these analyses, we also included species that are the sole inhabitant of islands, because even if they are not represented on the phylogeny, we can safely infer that they arose by between-island colonization. Following this method we inferred 25 colonization and 15 speciation events for a total of 40 species distributed over 14 islands.

An examination of the species diversity for Galápagos bulimulids highlights several points:

1. Variation in the total bulimulid species richness among Galápagos islands is significantly explained by island area when considering the total number of species on islands or including only species resulting from *in situ* speciation. However, the species-area relationship is not significant when only species resulting from between-island colonization are considered (figure 15.6).

2. The speciation-area relationship in bulimulid land snails suggests that there is an island area threshold below which *in situ* speciation rarely occurs (figure 15.6, solid circles). Neither of the islands smaller than Pinzón (18.1 km^2) has experienced an *in situ* speciation event, and very little *in situ* speciation occurred on islands smaller than Floreana (172.5 km^2). Interpretation of these trends is complicated due to the confounding factors of island age and vegetational diversity. For example, the four islands larger than Pinzón that do not have species resulting from *in situ* speciation have particularly low plant species diversity for their area due to either their low elevation (Marchena and Española) or geologically young age (Fernandina, and volcanoes Darwin and Wolf on Isabela Island).

3. The young islands of Fernandina and Isabela together form over 60% of Galápagos total land area, but only 12 of the 71 described bulimulid land snail species (about 17%) are found on these islands. Although the total species-island age relationship is marginally nonsignificant (adjusted $R^2=0.15$, $p=0.097$), this pattern suggests that at least some of the youngest islands have not reached their equilibrial species diversity (cf. chapter by Gillespie and Baldwin).[2]

4. When we focus on the total species-area relationship, linear regression does not fit the total species diversity data very well (figure 15.6). However, the species-area relationship fits the data much better when cor-

[2]Unfortunately, the geological history of the Caribbean is too complicated and poorly understood to allow comparable analyses for *Anolis*. However, phylogenetic information indicates that three of the four Greater Antillean islands have been occupied for long and roughly similar amounts of time, which suggests that age effects may not be of primary importance in determining species richness, a point further reaffirmed by the failure to speciate of some old lineages in the Lesser Antilles (for review, see Losos 2009).

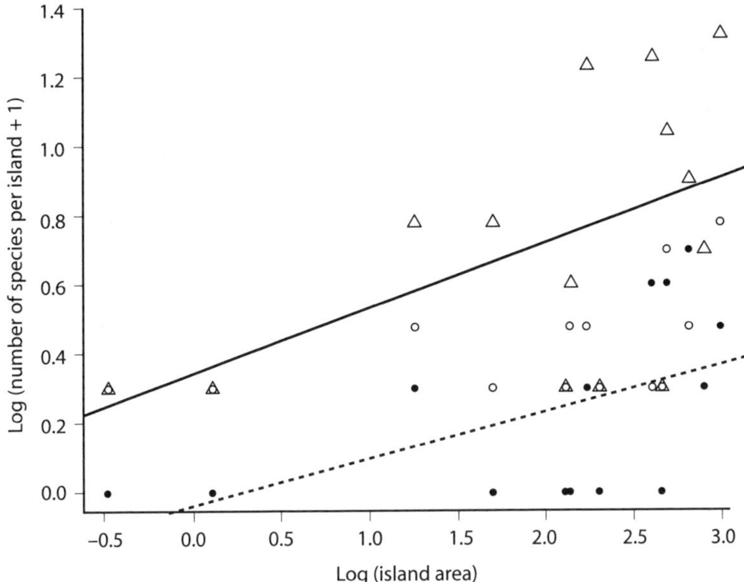

Figure 15.6. Regression of island area against the total number of bulimulid land snail species (open triangle; adjusted $R^2=0.20$, $p=0.062$), the number of species inferred to have resulted from within-island speciation (filled circles; adjusted $R^2=0.22$, $p=0.053$), and the number of species inferred to have resulted from between-island colonization on the Galápagos Islands (open circles; adjusted $R^2=0.071$, $p=0.183$).

recting total species diversity for island age (using the standardized residuals of island species diversity against age; figure 15.7; table 15.1).Likewise, the speciation-area relationship fits the data better once the number of speciation events is corrected for island age (figure 15.7; table 15.1).

5. Although island insularity does not have a significant effect on total island species richness on its own or combined with other biogeographical factors, we found that it does contribute to the species richness resulting from interisland colonization (figure 15.7; table 15.1), as predicted by MacArthur and Wilson. Indeed, we find that the colonization-area relationship fits the data better once the number of colonization events is corrected for island insularity (table 15.1).

6. The species-habitat diversity and speciation-habitat diversity relationships (both corrected for island age as above) provide an even better fit to the species diversity data (figure 15.8, table 15.1) than the species-area relationship corrected for island age (figure 15.7). Island area is often related to habitat diversity (Ricklefs and Lovette 1999, Whittaker and Fernández-Palacios 2007), but the number of plant species can provide a

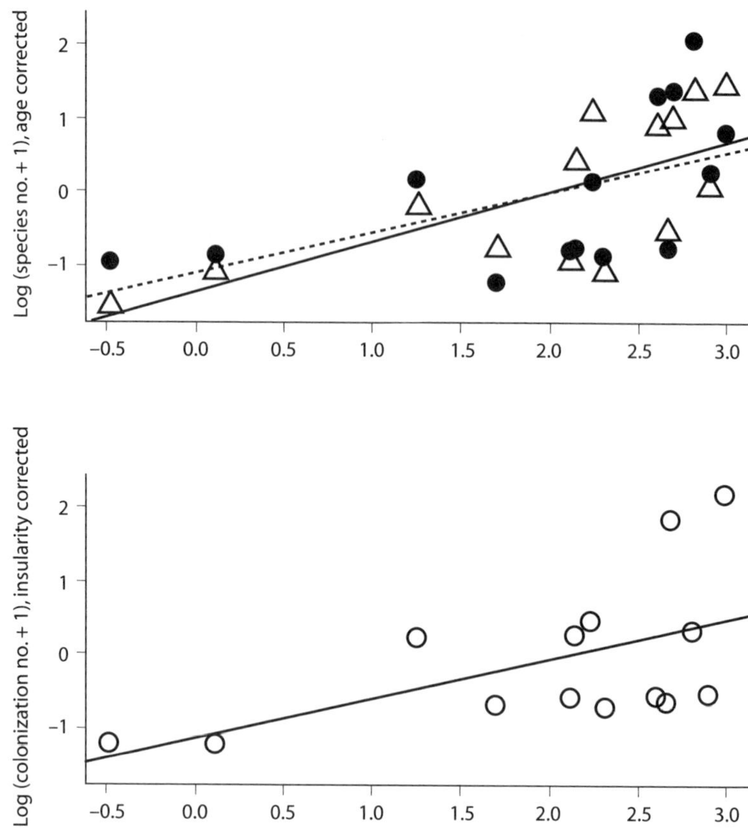

Figure 15.7. Regression of island area against the total number of Galápagos bulimulid land snail species (top panel; open triangle; adjusted $R^2=0.43$, $p=0.006$), the number of species inferred to have resulted from within-island speciation (top panel; filled circles; adjusted $R^2=0.24$, $p=0.044$), both corrected for island age, and the number of species inferred to have resulted from between-island colonization, corrected for island insularity (bottom panel; adjusted $R^2=0.24$, $p=0.043$).

more direct measure of habitat diversity for land snails or other animals whose ecology is directly related to plant diversity.

The significant "speciation-area" relationship compared to the much weaker "colonization-area" suggests that the overall species-area relationship is primarily the result of the contribution of within-island speciation to total island species diversity. Nevertheless, habitat diversity explains a

TABLE 15.1
Results of Forward Stepwise Multiple Regression Analyses for Different Models

Model	*Independent variables*	n	β	*SE of* β	t
Overall species diversity	Island area	26	0.23***	0.058	3.93
Adjusted R^2=0.37**	Island age		0.23*	0.093	2.45
Within-island speciation	Island area	14	0.036**	0.011	3.14
Adjusted R^2=0.39*	Island age		0.59	0.44	1.34
Overall species diversity	Island habitat diversity	26	0.54***	0.12	4.41
Adjusted R^2=0.43***	Island age		0.11	0.084	1.34
Within-island speciation	Island habitat diversity	14	0.0044**	0.0012	3.46
Adjusted R^2=0.44*	Island age		0.10	0.41	0.24
Between-island speciation	Island area	14	0.026*	0.0091	2.86
Adjusted R^2=0.35*	Island insularity		–0.00001*	0.0000042	–2.42

Notes: The number of bulimulid land snail species is used as dependent variable for all models considered. All variables were transformed to meet the assumptions of parametric statistics. The sample size (n) is provided for each model, as well as the standardized regression coefficient (β) and the test statistic (t) for each independent variable entered in each model. P values for adjusted R^2 and β values are indicated as follows: $^*P<0.05$; $^{**}P<0.01$; $^{***}P<0.001$.

greater proportion of the variation in number of age-corrected speciation events than island area (figure 15.8). In fact, multiple regression analysis reveals no area effect once habitat diversity is considered (Parent and Crespi 2006). This result, combined with the lack of detection of a habitat diversity effect for between-island colonization diversity, implies that an island with more plant diversity will accumulate more species mainly because it will be more likely to provide more opportunity for species differentiation and speciation, rather than offering more suitable habitat for colonizing

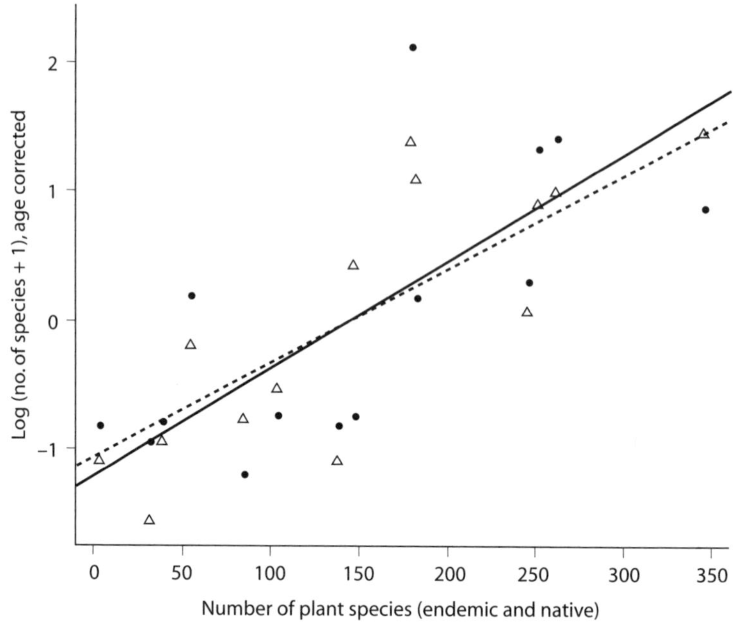

Figure 15.8. Regression of the total number of Galápagos bulimulid land snail species (open triangle; adjusted R^2=0.65, p=0.0003) and the number of species inferred to have resulted from within-island speciation (filled circles; adjusted R^2=0.45, p=0.005), both corrected for island age, against island habitat diversity measured as the number of endemic and native plant species.

species. Different faunal groups can differ in their responses to area and habitat diversity, so that area, habitat diversity or a combination of both have strong effects on species richness depending on the biological traits of the different taxonomic groups (Ricklefs and Lovette 1999). Galápagos bulimulid land snail species have adapted to the different vegetation zones, and most species are found on specific plants or defined microhabitats (Coppois and Glowacki 1983, Parent, unpublished data). Thus adaptation to specific vegetation types apparently provides the opportunity for bulimulid snails to differentiate within-island and partition the niche space to allow species to co-occur and accumulate on a given island.

To summarize the bulimulid story: in contrast to *Anolis*, colonization accounts for more species diversity than speciation in bulimulid species diversity. Nonetheless, like *Anolis*, the species area relationship primarily results from greater amounts of within-island diversification on larger and more habitat diverse islands. Unlike *Anolis*, habitat diversity, rather than area effects on speciation per se, seem to account for the speciation-area relationship.

Other Evolutionary Factors Besides Area Affecting Species Diversity on Islands

In addition to island area, a number of other factors could affect the rate of species diversification on islands. For the most part, these factors are independent of area; thus, to the extent that they are important, these factors may tend to diminish the relationship between island area and rates of within-island diversification.

Isolation

Classically, island isolation acts to decrease the rate of colonization and thus, in the MacArthur and Wilson theory, to lower the equilibrium number of species expected on an island, as we found with bulimulid snails. Moreover, to the extent that distant islands are impoverished faunistically and, particularly, floristically, they may be less able to support other species (Lack 1976).

By contrast, for several reasons, isolation may serve to increase the rate of within-island speciation (e.g., MacArthur and Wilson 1963, 1967, Heaney 2000, 2007):

1. More distant islands are likely to experience less ongoing gene flow, thus increasing the likelihood that an island population could diverge and speciate from its ancestral population and subsequently diversify *in situ*.
2. More distant islands, being impoverished biotically, may harbor fewer competitor species, thus facilitating adaptive radiation of a successful colonist.
3. Moreover, the disharmonic nature of distant island faunas—with some mainland taxa represented and others not—is likely to substantially alter patterns of natural selection stemming from interspecific interactions, thus driving ecological adaptation, and enhancing the likelihood of speciation.

We are unaware of any studies that have demonstrated any of these phenomena, but they certainly are plausible in principle.

Island Configuration

MacArthur and Wilson (1963) noted: "adaptive radiation takes place as species are generated within archipelagoes, disperse between islands, and, most importantly, accumulate on individual islands to form diversified associations of sympatric species." This view encompasses the classic view of adaptive radiation, exemplified by Darwin's finches of the Galápagos: allopatric speciation occurring on separate islands in an archipelago followed by secondary reinvasion can result in the build-up of species richness and adaptive radiation. The importance of island configuration is clearly seen in Darwin's finches (e.g., Grant and Grant 2008, this volume). In contrast to the great diversity of this clade in the Galápagos, the only other Darwin's finch, on isolated and vegetationally diverse Cocos Island in the Pacific Ocean, has failed to diversify, despite having occurred there for a long period of time. Presumably, the lack of opportunity for allopatric speciation has prevented speciation on Cocos and facilitated it in the Galápagos.

We are unaware of any quantitative treatment of this subject, but again it seems likely that diversification would be greater in an archipelagic setting, particularly for species, such as birds, which are unlikely to speciate on single, isolated islands.

Such speciation would not constitute within-island speciation; rather, in the MacArthur and Wilson framework, it would result in an evolutionarily increasing pool of potential colonists.[3] That is, usually the source pool of species is considered to be the number of species on the nearby mainland. With intra-archipelagic speciation, the source pool for each island in the archipelago would be enhanced by the species that evolved on the other islands, thus leading to a higher equilibrium number than would occur on isolated islands.

Age

Time might be related to evolutionary diversification in several different ways. On one hand, one might expect that the longer a clade has occurred on an island, the greater the opportunities for speciation and the greater the number of resulting species would be (Heaney 2000), as appears to be the case for bulimulid snails. A similar positive association between *in situ* speciation and island age has been suggested in a preliminary study of diversification in Galápagos beetle genus *Galapaganus* (Sequeira et al. 2008). This expectation seems reasonable, particularly for isolated islands which might not reach their carrying capacity of species.

On the other hand, long-jawed spiders (*Tetragnatha*) in the Hawaiian Islands reach their greatest diversity on middle-aged (and middle-sized) islands. Gillespie (2004) suggested the following scenario: young islands have not had enough time to generate their equilibrium number of species.

[3]As figure 15.2b demonstrates, understanding the geography of speciation in an archipelago may be challenging. If an ancestral species diverges into two species on one island, but then the two species each colonize other islands in the archipelago, then the initial within-island speciation event may not be evident on a phylogeny because each species has as its closest relative a species on another island.

Middle-aged islands have been colonized by a number of lineages, each of which has diversified. However, the descendants of such species have not had time to disperse widely, and as a result, the descendants of different colonists have not yet come into contact. Given enough time, as has occurred on older islands, these species do increase their range and come into contact with each other. At that point, interspecific interactions—probably primarily competition—lead to the elimination of some species (see the chapter by Gillespie and Baldwin for a broader discussion of age effects in diverse Hawaiian lineages).

Although direct evidence of interaction-driven extinction on older islands is not available, such an ecological overshoot is also seen in recolonization experiments over ecological time (Simberloff and Wilson 1969); other workers have suggested that a similar evolutionary phenomenon occurs in other systems (Gillespie 2004, Gavrilets and Vose 2005, Seehausen 2006).

Finally, islands themselves evolve through time. Whittaker (this volume) suggests that volcanic islands may go through a life cycle in which species number is maximized early in the history of the island and then decreases through time as the island subsides and erodes. Such a phenomenon is seen in *Tarphius* beetles in the Canary Islands (Emerson and Oromi 2005). Interestingly, the hump-shaped pattern of species richness through time in these beetles is entirely determined by the number of species generated by within-island speciation, as the number of colonization-derived species per island is roughly constant. Situations such as this and the Hawaiian *Tetragnatha* also are among that small group of case studies alluded to by Lomolino in the opening quote that do not exhibit a species-area relationship; they represent situations in which age-dependent effects on species richness outweigh area effects.

Conclusion

This is an exciting time for those interested in the determinants of species richness on islands. Thanks in large part to the great current interest in applying phylogenetic approaches to the understanding of macroevolutionary questions, as well as to a desire by many to integrate ecological and evolutionary thinking (e.g., Whittaker and Fernández-Palacios 2007; Emerson and Gillespie, 2008), the opportunity to understand the evolutionary aspect of species richness has never been greater.

At this point few studies exist in which we can quantitatively assess the relative significance of ecological and evolutionary processes, but this situation is likely to change soon with the flood of phylogenetic information that is rapidly becoming available. With this information, we will be able to answer questions such as:

- What circumstances determine the relative importance of ecological and evolutionary processes?
- How common are island threshold sizes below which *in situ* diversification does not occur?
- Under what circumstances do isolation, configuration, and age effects predominate?

By the time the MacArthur and Wilson theory reaches its fiftieth anniversary, we predict that a rich and varied data base will exist to provide answers to these questions and many more, and thus to fully integrate evolutionary considerations into island biogeography theory, a goal clearly articulated by MacArthur and Wilson (1963, 1967), but only now being realized.

Acknowledgments

We thank Rosemary Gillespie and Bob Ricklefs for helpful comments.

Literature Cited

Breure, A.S.H. 1979. Systematics, phylogeny and zoogeography of Bulimulinae (Mollusca), Leiden: E.J. Brill.

Carlquist, S. 1974. *Island Biology*. New York: Columbia University Press.

Chambers, S. M. 1991. Biogeography of Galápagos land snails. In *Galápagos Marine Invertebrates*, ed. M. J. James, 307–25. New York: Plenum.

Coppois, G., and C. Glowacki. 1983. Bulimulid land snails from the Galapagos: 1. Factor analysis of Santa Cruz Island species. *Malacologia* 23:2009–219.

Coyne, J. A., and T. D. Price. 2000. Little evidence for sympatric speciation in island birds. *Evolution* 54:2166–71.

Diamond, J. M. 1977. Continental and insular speciation in Pacific land birds. *Systematic Zoology* 26:263–68.

Emerson, B. C., and R. G. Gillespie. 2008. Phylogenetic analysis of community assembly and structure over space and time. *Trends in Ecology and Evolution* 23:619–30.

Emerson, B. C., and P. Oromi. 2005. Diversification of the forest beetle genus *Tarphius* on the Canary Islands, and the evolutionary origins of island endemics. *Evolution* 59:586–98.

Funk, D. J., P. Nosil, and W. J. Etges. 2006. Ecological divergence exhibits consistently positive associations with reproductive isolation across disparate taxa. *Proceedings of the National Academy of Sciences U.S.A.* 103:3209–13.

Gavrilets, S., and A. Vose. 2005. Dynamic patterns of adaptive radiation. *Proceedings of the National Academy of Sciences U.S.A.* 102:18040–45.

Gillespie, R. 2004. Community assembly through adaptive radiation in Hawaiian spiders. *Science* 303:356–59.

Grant, P. R., ed. 1998. *Evolution on Islands.* New York: Oxford University Press.

Grant, P. R., and B. R. Grant. 2008. *How and Why Species Multiply: The Radiation of Darwin's Finches.* Princeton, NJ: Princeton University Press.

Heaney, L. R. 2000. Dynamic disequilibrium: A long-term, large-scale perspective on the equilibrium model of island biogeography. *Global Ecology and Biogeography* 9:59–74.

———. 2007. Is a new paradigm emerging for oceanic island biogeography? *Journal of Biogeography* 34:753–57.

Hedges, S. B., W. E. Duellman, and M. P. Heinicke. 2008. New World direct-developing frogs (Anura: Terrarana): Molecular phylogeny, classification, biogeography, and conservation. *Zootaxa* 1737:1–182.

Lack, D. 1976. *Island Biology, Illustrated by the Land Birds of Jamaica.* Berkeley: University of California Press.

Lomolino, M. V. 2000. Ecology's most general, yet protean pattern: The species-area relationship. *Journal of Biogeography* 27:17–26.

Losos, J. B. 1996. Ecological and evolutionary determinants of the species-area relation in Caribbean anoline lizards. *Philosophical Transactions of the Royal Society of London, Series B* 351:847–54.

———. 2009. *Lizards in an Evolutionary Tree: Ecology and Adaptive Radiation of* Anolis. Berkeley: University of California Press.

Losos, J. B., T. R. Jackman, A. Larson, K. de Queiroz, and L. Rodríguez-Schettino. 1998. Contingency and determinism in replicated adaptive radiations of island lizards. *Science* 279:2115–18.

Losos, J. B., and R. E. Ricklefs. 2009. Adaptation and diversification on islands. *Nature* 457:830–36.

Losos, J. B., and D. Schluter. 2000. Analysis of an evolutionary species-area relationship. *Nature* 408:847–50.

MacArthur, R. H., and E. O. Wilson. 1963. An equilibrium theory of insular zoogeography. *Evolution* 17:373–87.

———. 1967. *The Theory of Island Biogeography.* Princeton, NJ: Princeton University Press.

Malhotra, A., and R. S. Thorpe. 1994. Parallels between island lizards suggests selection on mitochondrial DNA and morphology. *Proceedings of the Royal Society of London, Series B* 257:37–42.

McMullen, C. K. 1999. *Flowering Plants of the Galápagos.* Ithaca, NY: Cornell University Press.

Nicholson, K. E., R. E. Glor, J. J. Kolbe, A. Larson, S. B. Hedges, and J. B. Losos. 2005. Mainland colonization by island lizards. *Journal of Biogeography* 32: 929–38.

Parent, C. E., and B. J. Crespi. 2006. Sequential colonization and diversification of Galápagos endemic land snail genus *Bulimulus* (Gastropoda, Stylommatophora). *Evolution* 60:2311–28.

Paulay, G. 1985. Adaptive radiation on an isolated oceanic island —the Cryptorhynchinae (Curculioinidae) of Rapa revisited. *Biological Journal of the Linnean Society* 26:95–187.

Price, T. D. 2007. *Speciation in Birds*. Greenwood Village, CO: Roberts and Company.

Richman, A. D., T. J. Case, and T. D. Schwaner. 1988. Natural and unnatural extinction rates of reptiles on islands. *American Naturalist* 131:611–30.

Ricklefs, R. E., and I. J. Lovette. 1999. The roles of island area per se and habitat diversity in the species-area relationships of four Lesser Antillean faunal groups. *Journal of Animal Ecology* 68:1142–60.

Rundle, H. D., and P. Nosil. 2005. Ecological speciation. *Ecology Letters* 8:336–52.

Schneider, C. J. 1996. Distinguishing between primary and secondary intergradation among morphologically differentiated populations of *Anolis marmoratus*. *Molecular Ecology* 5:239–49.

Schoener, T. W. 1976. The species-area relationship within archipelagoes: models and evidence from island birds. *Proceedings of XVI International Ornithological Congress* 6:629–42.

Seehausen, O. 2006. African cichlid fish: A model system in adaptive radiation research. *Proceedings of the Royal Society of London, Series B* 273:1987–98.

Sequeira, A. S., A. A. Lanteri, L. Roque-Albelo, S. Bhattacharya, and M. Sijapati. 2008. Colonization history, ecological shifts and diversification in the evolution of endemic Galápagos weevils. *Molecular Ecology* 17:1089–107.

Simberloff, D., and E. O. Wilson. 1969. Experimental zoogeography of islands: The colonization of empty islands. *Ecology* 50:278–96.

van der Werff, H. 1979. Conservation and vegetation of the Galápagos Islands. In *Plants and Islands*, ed. D. Bramwell, 391–404. London: Academic Press.

White, W. M., A. R. McBirney, and R. A. Duncan. 1993. Petrology and geochemistry of the Galapagos-Islands—Portrait of a pathological mantle plume. *Journal of Geophysical Research—Solid Earth* 98:19533–63.

Whittaker, R. J., and J. M. Fernández-Palacios. 2007. *Island Biogeography: Ecology, Evolution, and Conservation*, 2nd ed. Oxford: Oxford University Press.

Wiggins, I. L., and D. M. Porter. 1971. *Flora of the Galápagos Islands*. Stanford, CA: Stanford University Press.

Wilcox, B. A. 1978. Supersaturated island faunas: A species-age relationship for lizards on post-Pleistocene land-bridge islands. *Science* 199:996–98.

Williams, E. E. 1983. Ecomorphs, faunas, island size, and diverse end points in island radiations of *Anolis*. In *Lizard Ecology: Studies of a Model Organism*, ed. R. B. Huey, E. R. Pianka, and T. W. Schoener, 326–70. Cambridge, MA: Harvard University Press.

Ecological and Genetic Models of Diversity

LESSONS ACROSS DISCIPLINES

Mark Vellend and John L. Orrock

Ecology and evolutionary biology have been linked to varying degrees throughout their histories as scientific disciplines (Collins 1986, Holt 2005). As recognized by Darwin and countless biologists since, evolutionary change can hardly be understood without knowledge of ecological context, and many of our most cherished ecological patterns, such as relationships between species diversity and area or latitude, ultimately require evolutionary explanations, at least in part (Dobzhansky 1964, Schluter 2000, Ricklefs 2004). The degree of integration between ecological and evolutionary studies has waxed and waned over the years, but in response to the rise of molecular biology during the 1960s, a group of leading researchers in ecology and evolution, including Robert MacArthur, Richard Levins, Richard Lewontin, and Edward Wilson, made a concerted effort to draw the two disciplines together under the unifying banner of population biology (Wilson 1994, Odenbaugh 2006). One of the defining contributions of this era was *The Theory of Island Biogeography* (MacArthur and Wilson 1967), which emerged from the integration and synthesis of seemingly disparate branches of organismal biology, in large part thanks to MacArthur and Wilson's "faith in the ultimate unity of population biology" (p. xi).

The arguments laid out by MacArthur and Wilson (1967) are broadly representative of the way in which ecology and evolution were being integrated at the time, and indeed to some degree the way in which ecology and evolution have been brought together over the past forty years. As in the 1960s (Birch 1960, Levins 1968, MacArthur and Wilson 1967), a major focus today remains on how ecological and evolutionary processes combine to produce the patterns of species distributions, traits, and diversity over space and time (Collins 1986, Ricklefs and Schluter 1993, Johnson and Stinchcombe 2007, Fussmann et al. 2007). The last decade in particular has produced a steady stream of studies demonstrating the necessity of considering both ecological and evolutionary processes to understand phenomena ranging from the outcome of species interactions in small-scale experiments (Yoshida et al. 2003, Lankau and Strauss 2007) to broad-scale patterns of species diversity (Ricklefs 2004). This has spawned a number of recent viewpoints on how to reinvigorate the effort to more fully integrate ecology and evolution (Holt 2005, Vellend and Geber 2005, Johnson and Stinchcombe 2007, Fussmann et al. 2007). While we are enthusiastic proponents of these efforts, we also believe that there is a different and equally important way in which ecology and evolution might be integrated. Not only do ecological and evolutionary processes act in concert, but even if we consider ecological and evolutionary dynamics in isolation, some of the processes involved show remarkable parallels across disciplines (Vellend and Geber 2005). These parallels are strongest in the subdisciplines of ecology and evolution that are focused specifically on species diversity and genetic diversity—namely, community ecology and population genetics, respectively.

The processes that drive changes in the frequencies of alleles or genotypes in populations—mutation, drift, migration, and selection—are much the same as the processes that drive changes in the relative abundances (and therefore composition and diversity) of species in communities (Antonovics 1976, 2003, McPeek and Gomulkiewicz 2005). Compared to treatments of population genetics, ecological texts typically offer a much longer list of processes that drive changes in communities, including the usual suspects of competition, predation, dispersal, succession, and so on. However, as we will argue in more detail later, these processes can be readily grouped, as in population genetics, into four parallel categories: speciation, drift, migration, and selection (see also Vellend and Geber 2005). Examining parallel models in population genetics and community ecology more carefully may be quite useful in that portions of theory in these two disciplines could potentially be merged.

The similarities between processes underlying patterns of species diversity and genetic diversity have been repeatedly noted (Antonovics 1976, 2003, Hubbell 2001, Chase and Leibold 2003, McPeek and Gomulkiewicz 2005, Vellend and Geber 2005), but this recognition has not permeated the two disciplines, as evidenced by repeated, independent developments within ecology and genetics of separate models with essentially the same underlying processes. This is clearly illustrated in the central model of *The Theory of Island Biogeography*, in which the diversity of species on an island is modeled as a balance between a rate of input (colonization) and a rate of output (extinction). The underpinning of the model is illustrated with a cartoon showing islands of variable size and distance from a mainland, the two key island characteristics assumed to determine rates of extinction and colonization (figure 16.1A). The MacArthur-Wilson model is widely admired as a landmark, original

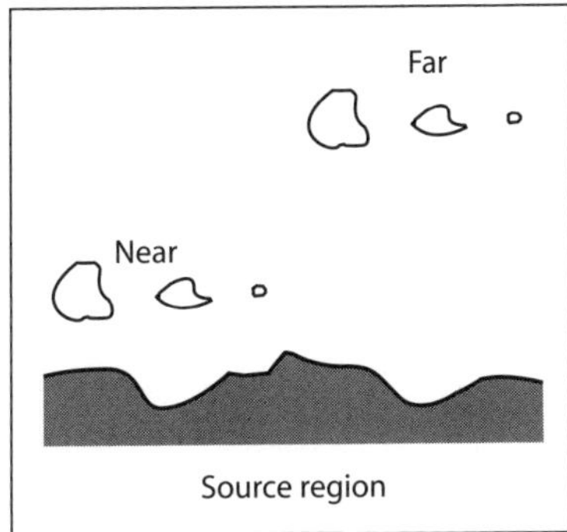

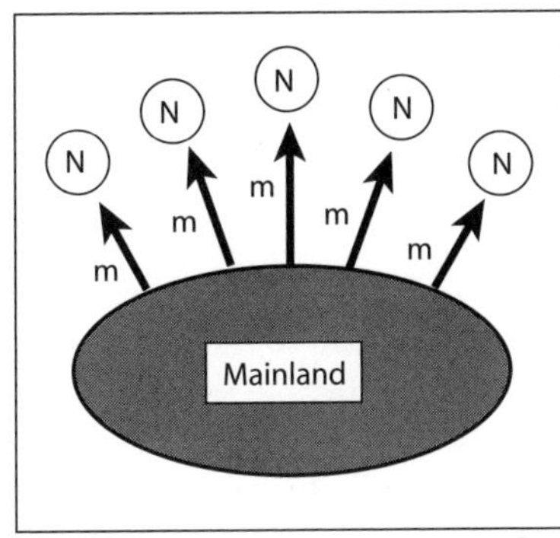

Figure 16.1. Schematic diagrams illustrating the central model in *The Theory of Island Biogeography* (A) and Wright's (1940) mainland-island model of population genetics (B). The two diagrams are adapted from MacArthur and Wilson (1967) and Hedrick (2000), respectively.

contribution to biology, but more than a quarter of a century earlier, Sewall Wright (1940) developed a strikingly similar model, the mainland-island model of population genetics, which predicts patterns of genetic diversity in a set of "island" populations based on a rate of input (immigration from the mainland) and a rate of output (genetic drift). The figure often used to illustrate Wright's model (figure 16.1B) bears a close resemblance to the MacArthur and Wilson model (figure 16.1A), with the parameter N representing the size of each island population and therefore the importance of drift, and the parameter m representing the rate of immigration. It is easy to point out differences in the specifics of the two models, but qualitatively they are much the same if we simply exchange the two words "species" and "allele." So why, given the explicit integration of ecological and evolutionary thinking in the work of MacArthur, Wilson, and others at the time, was the earlier and extremely similar island model from genetics not drawn upon, or even acknowledged in the citations of MacArthur and Wilson (1967; there is no citation of any of Wright's papers, though he is mentioned once in a different context)? Are there other models in population genetics that could be drawn on to provide novel insights into pressing questions in community ecology? We return to these questions in subsequent sections.

We argue that a broader recognition of the conceptual parallels between community ecology and population genetics can contribute to significant advances in these disciplines in at least three ways. First, consideration of processes acting in parallel on species diversity and genetic diversity can generate novel, testable predictions concerning patterns of biodiversity in nature. This topic has been treated in depth elsewhere (Vellend 2003, 2005, Vellend and Geber 2005), and is not discussed further here. Second, theoretical progress can be accelerated by drawing on existing models in one discipline whose counterparts have yet to be developed in the other. And third, the way we structure and organize the different facets of the two disciplines can benefit from consideration of the success or lack thereof of different organizational frameworks within community ecology and population genetics. Community ecology in particular has struggled to define its identity (Lawton 1999, Simberloff 2004), and might glean some useful lessons from the structure of population genetics as a discipline. Here we focus largely on ideas and concepts in population genetics that might be put to use in community ecology. We begin by briefly outlining the key parallels between processes in population genetics and community ecology. We proceed with a treatment of genetic analogues of two key models presented in *The Theory of Island Biogeography*, with an emphasis on novel lessons concerning (1) effects of size and migration on diversity, and (2) the conditions that influence successful establishment of new variants. We then draw on the organizational structure of population genetics to suggest how a similar structure might help provide a greater degree of coherence and order to what has been referred to as a bit of a "mess" in community ecology (Lawton 1999).

Parallels between Population Genetics and Community Ecology

Many aspects of the evolutionary process, such as epistasis, pleiotropy, inbreeding, and recombination, have either no parallels, or only very loose parallels at best, in community ecology. However, if we narrow our focus to the bare bones of population genetics—single-locus haploid models—the parallels with community ecology are striking.

Both population genetics and community ecology are essentially concerned with variation over space and time in the relative abundance and diversity of discrete biological variants: alleles or species, respectively. Four logically distinct processes can change the abundances and diversity of biological variants (Vellend and Geber 2005). First, due to the finite number of individuals in a population or community, the relative frequencies of alleles or species will to some degree change stochastically. This is genetic or ecological drift. When an individual organism moves between localities (i.e., migration), it may introduce novel alleles to a population or it may represent a new species in the recipient community. Selection occurs when particular alleles or species are deterministically favored over others. These three processes—drift, migration, and selection—act in closely analogous ways in genetic and ecological models of diversity. Finally, mutation and speciation are the analogous processes that create globally novel alleles or species, respectively, but admittedly the parallel here is not as strong. Nonetheless, as detailed in subsequent sections, mutation is often treated in models simply as the appearance of a new allele, which is much the same way species invasion or immigration is often treated in ecological models.

Area and Isolation in Genetic and Ecological Models of Diversity

Chapter 3 in *The Theory of Island Biogeography* presents the now famous crossing colonization and extinction curves as functions of the number of species on an island. The model was inspired by empirical patterns demonstrating a positive effect of island area, and a negative effect of distance from a mainland, on species richness. While the model does indeed predict these patterns, these were not new predictions. One of the novel predictions offered by the MacArthur-Wilson model was that the slope of the species-area curve in log-log space should be steeper on archipelagoes far from the mainland than on those close to the mainland (or on sections of the mainland itself) (figure 16.2A). A data set on the species richness of ants in insular faunas of different parts of the world vs. nested sections of the large island of New Guinea was offered as support for this prediction, with a steeper species-area curve in the former than in the latter (figure 16.2B). Some data sets agree with this prediction, but meta-analyses have not found general evidence that the slopes of species-area relationships are steeper on more isolated islands (Schoener 1976, Connor and McCoy 1979, Williamson 1988). One possible explanation is that islands in distant archipelagos receive considerable interisland dispersal, and are thus not as effectively isolated as distance-to-mainland calculations imply (Schoener, this volume).

The prediction of steeper species-area relationships on far versus near islands is intuitively appealing in that it essentially states that isolation and small size should act multiplicatively rather than additively to reduce species diversity. However, from another perspective the crossing-curves model is difficult to intuit because the unit of analysis in the mathematics is the species, with no underlying population dynamics of these species, either explicitly or implicitly. Thus, despite the widespread appeal of the *The Theory of Island Biogeography*, Hubbell (2001) has argued that in fact "there is no theoretical foundation for species-area curves that derives from fundamental processes of population dynamics" (Hubbell 2001, but see Hanski, this volume). Hubbell (2001) also makes the important point

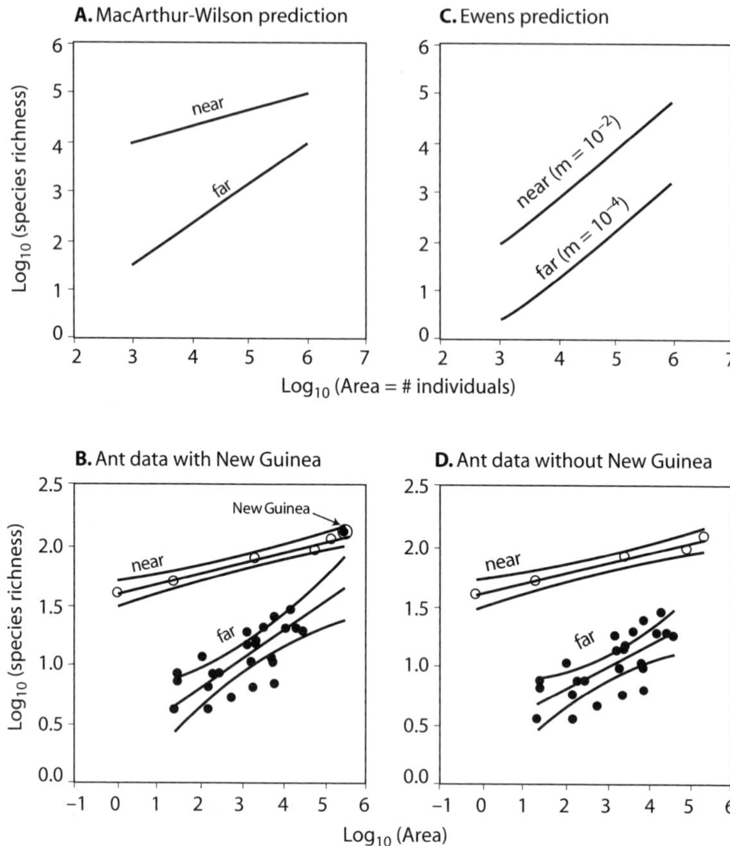

Figure 16.2. Predictions for the slopes of species area relationships with high and low rates of immigration (near and far islands, respectively). The theory of island biogeography predicts a steeper slope at a lower level of immigration (A) and the haploid version of Ewen's sampling formula predicts equal slopes in the two cases (C). Using data on ants in insular faunas (filled symbols) and on mainland portions of New Guinea (open symbols), the two data sets show significantly different slopes ($p<0.01$ for the interaction term with area in a general linear model) if the full island of New Guinea is included in both data sets (B), but the slopes are not significantly different ($p>0.06$) if New Guinea is excluded (D).

that despite the many verbal arguments in *The Theory of Island Biogeography* for why differences among species matter, in fact the crossing-curves model itself is essentially neutral, requiring no functional differences among species to produce its predictions (see also Hubbell, this volume).

The development of neutral theory in population genetics began in the first half of the twentieth century, and underwent a major period of refinement and elaboration during the 1960s and 1970s when emerging molecular data sparked the debate on whether widespread polymorphisms were the product of neutral or non-neutral processes (Lewontin 1974, Kimura 1983). As described already, the mainland-island model of Wright (1940) is essentially the genetic analogue of the crossing-curves model of MacArthur and Wilson, with one important difference being the inclusion of implicit population dynamics in the genetic theory. Another difference is that Wright's model makes explicit predictions for levels of heterozygosity and population differentiation, but not the number of alleles (the analogue of species) expected in populations of different sizes. At least for one specific scenario, this problem was solved by Ewens (1972).

Assuming an isolated population of constant size and a given rate of mutation (under the infinite alleles model), the expected number of alleles in a sample of individuals of a given size can be calculated using the famed Ewens sampling formula (Ewens 1972). By taking the haploid version of the model, inserting metacommunity size, J_M, in place of population size, and inserting a speciation rate, ν, in place of the mutation rate, Hubbell (2001) applied this model to predict the number of species, $E(S)$, expected in a sample of J individuals from the metacommunity:

$$E(S) = \theta/\theta + \theta/(\theta+1) + \cdots + \theta/(\theta+J-1).$$

Apart from the number of individuals in the sample, this equation has only one composite parameter, $\theta = 2J_M\nu$. By considering portions of the metacommunity of different sizes and degrees of migration among them, Hubbell (2001) calculated expected species-area relationships at different levels of migration. On the surface, this seems comparable to a prediction for islands with different levels of immigration, but in fact it is very different. The MacArthur-Wilson model concerns discrete islands of different sizes and different rates of input via immigration from outside, while Hubbell calculated species accumulation curves for larger and larger areas of the same landmass. These two kinds of relationships are quite different (Whittaker and Fernández-Palacios 2007), and this point is made obvious by the fact that Hubbell finds lower, rather than higher, species richness as the rate of migration is increased.

Surprisingly, despite some highly sophisticated elaborations of neutral theory in ecology (e.g., Etienne and Alonso 2005, Etienne 2007), we still lack quantitative predictions based on underlying population dynamics for what *island* species-area relationships might look like with variable levels of immigration. As a first approximation, one way to do this is to assume that each island is its own isolated (meta)community; to think of immigration from the mainland as akin to speciation (the introduction of new variants); and then to apply the Ewens sampling formula to calculate the expected number of species on islands of different size, J_M, and rates of immigration, m (which we insert in place of the mutation rate, ν). In contrast to the MacArthur-Wilson prediction of different species-area slopes at different rates of immigration (figure 16.2A), the modified Ewens formula predicts precisely equivalent slopes above a threshold community size of $J \approx 10^3$ (figure 16.2C).

In light of this altered prediction, it is worth taking a second look at the ant data initially presented in support of the MacArthur-Wilson model. The different slopes shown in figure 16.3B are in fact highly dependent on one data point for the island of New Guinea. Since New Guinea anchors the right end of the "mainland" curve, and is also the largest island in the insular curve, then as long as species richness is generally lower in small insular faunas than on mainland portions of New Guinea (a separate issue), this data point will force the slopes to be different (figure 16.2B). Excluding New Guinea, the difference in slopes is quite modest, and indeed not statistically significant (figure 16.2D). Of course this is only one empirical case study of hundreds, but it helps highlight the point that the evaluation of island species-area curves at different degrees of isolation need not be viewed simply as a test of the MacArthur-Wilson model, but as a test among competing alternatives models (see also Schoener, this volume).

The more general point of this section is that a simple and elegant theory of diversity with underlying population dynamics was sitting on the genetic shelf for decades before Hubbell (2001) imported it into ecology, the result being arguably the most influential contribution to ecology in the last decade (Hubbell's book has been cited more than 1100 times in only eight years). Neutral theory still has many unexplored angles that may provide new insights into ecological patterns (island species-area curves being one example), and the genetic shelf has got lots of other models that ecologists might make great use of. A number of ecological studies have done this already (e.g., Norberg et al.2001, McPeek and Gomulkiewicz 2005, Orrock and Fletcher 2005, Fox 2006), but we feel there is still plenty of untapped potential. At the same time, many ecological models—including recent elaborations of neutral theory that go beyond traditional population-genetic models—might be imported into population genetics with the same potential benefits.

Genetic and Ecological Models of the Establishment of New Variants

Chapter 4 of *The Theory of Island Biogeography* is concerned with estimation of the "probability that a propagule of a given species will establish a successful colony" (MacArthur and Wilson 1967, p. 92). Assuming density-dependent population growth and overlapping generations with birth and death rates of λ and μ, respectively, MacArthur and Wilson estimated the probability of establishment starting from single propagule (e.g., a single seed or pregnant female) to be approximately r/λ, where r is the intrinsic rate of population increase, and $r=\lambda-\mu$. Throughout this paper we refer to r as the expected rate of population growth, because even in the absence of competition the expectation for r will depend on the environment (and is therefore not really "intrinsic"). With the birth rate in the denominator, according to this model the ideal colonist achieves success largely via a particularly low death rate. This somewhat odd result appears to us to be a consequence of overlapping generations in the model, with the most likely fate of a new propagule (i.e., extinction) avoided only if the first individual does not die before reproducing. The more general result would seem to be that the probability of establishment is proportional to the rate of population growth when rare, r.

As with their colonization-extinction model of diversity, MacArthur and Wilson's species establishment model has some close analogues in population genetics. From a mathematical viewpoint, establishment of a new species in a community is similar to the establishment of a new allele in a population. When a new mutation arises in a population, ultimately it must disappear, rise in frequency to fixation, or be maintained in a stable polymorphism. Mutation fixation models, as the name implies, focus on estimating the probability of fixation, but as originally formulated using the mathematics of branching processes, they can at least loosely be interpreted as addressing "the course of events in a population where the new factor is present in such numbers as to be in no danger of extinction by mere bad luck" (Haldane 1927, p. 838). The specific "new factor" of interest to Haldane was a mutation, but his language makes clear that it could be any kind of new variant, such as an immigrant. Haldane assumed a population of infinite size and a positive selection coefficient s, and estimated the probability of fixation to be approximately $2s$ for small values of s. Recognizing that in a population or community of fixed size, the selection coefficient, s, is equivalent to the expected rate of population growth, r, we can see a clear parallel between the two models. The probability of establishment of an allele in a population or a species in a community is approximately proportional to the degree of deterministic advantage when rare, despite different assumptions concerning the underlying population dynamics in the different models. Haldane's model preceded the MacArthur-Wilson model by forty years, again begging the question of why it was not drawn upon or acknowledged in *The Theory of Island Biogeography* (Haldane is not cited but is acknowledged in a different context).

Over the past forty years the clear parallel between the establishment of species in communities and alleles in populations has been recognized (e.g., Haccou and Iwasa 1996, McPeek and Gomulkiewicz 2005, Orrock and Fletcher 2005), but not fully explored. Indeed many of the qualitative lessons that might be taken from fixation models would not represent particularly deep insights in ecology, such as the positive effect of r or initial population size on the probability of establishment (Kimura 1962). However, most community models do not include a parameter for the total community size, J, and when this enters into fixation models things can get ecologically quite interesting. A general model for the probability of fixation of both beneficial and deleterious mutations was presented by Kimura (1962), the haploid version of which we can apply to the probability of invasion, *Pr(inv)*, of new species in communities:

$$Pr(inv) = (1 - \exp(-2J_e rp))/(1 - \exp(-2J_e r)).$$

In this equation, r is the expected rate of population growth as before, p is the initial frequency of the invader (i.e., the initial population size divided by the census community size, N_{init}/J), and J_e is the effective community size. J_e represents the community-level equivalent of the effective population size, and can differ from J according to factors such as fluctuating community size, which reduces J_e relative to the arithmetic mean of J over time. This model is appealing in that it captures a key characteristic of the invading species (p), a key characteristic of the recipient community (J_e), and a parameter summarizing the interaction between the species' traits and the local abiotic and biotic conditions (r). Admittedly the model treats as a black box the details of many of the ecological interactions (e.g., competition and predation) that determine r, and also makes an assumption of constant (or at least extrinsically determined) community size, which may apply only under fairly restrictive conditions (Houlahan et al. 2007). Nonetheless, we feel it provides an appropriate point of departure for an initial consideration of the potential ecological consequences of finite community size.

Analysis of this model reveals some interesting and, in our opinion, non-intuitive lessons for community ecology and invasion biology. We focus here on selected results for species with positive r (i.e., those deterministically favored to invade). First, if initial population size is relatively small, failure to invade is the most likely outcome even for species with a large value of r (figure 16.3), as all invaders are susceptible to stochastic loss when rare (see also McPeek and Gomulkiewicz 2005). This model prediction is consistent with the many cases in which highly successful invasions were preceded by repeated failures (Sax and Brown 2000, Sakai et al. 2001). Second, for a given expected rate of increase, r, and

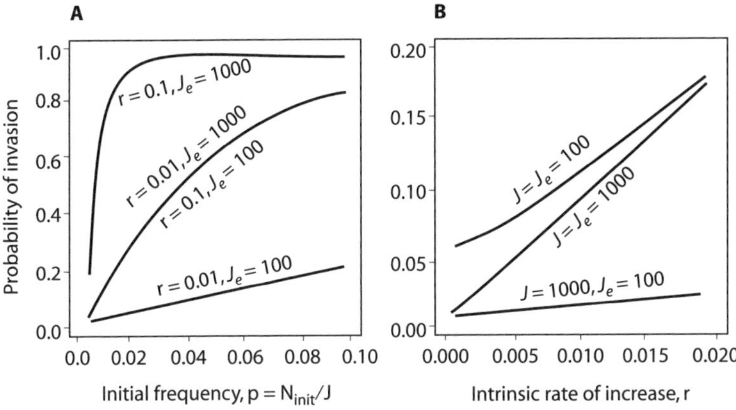

Figure 16.3. Invasion probabilities, *Pr(inv)*, from the haploid version of the Kimura (1962) model as a function of: (A) the initial frequency ($p=N_{init}/J$) of the introduced species with different expected rates of increase, r, in communities of different effective size, J_e; and (B) the expected rate of increase, r, for communities of different census size, J, and effective size, J_e, and an invader with initial population size, $N_{init}=5$.

initial frequency, p, an introduced species is more likely to invade a large community than a small one (figure 16.3A). The reason for this is that for a given p, the initial population size is higher in large than small communities, and therefore much less likely to go extinct. We can imagine this comparison applying to lakes of different sizes in which boat traffic is the main vector of exotic species propagules: if boat traffic is approximately proportional to lake size, then exotic species invade large lakes with larger initial population sizes (N_{init}) but approximately equivalent relative frequencies ($p=N_{init}/J$) as in small lakes. In contrast, for a given initial population size, just the opposite is true—an invasive species is more like to invade a small community than a large one, at least for small values of r and J. Five exotic birds released on a small island with 100 birds in the same guild starts at an initial frequency of $p=0.05$, whereas on a larger island with 1000 birds it starts at an initial frequency of $p=0.005$, and is therefore more likely to go quickly extinct.

Finally, any factor that decreases J_e for a given J, will decrease the probability of invasion for species with positive r because the relative importance of drift increases relative to selection (figure 16.3; see also Orrock and Fletcher 2005). Fluctuating or declining community size due to periodic disturbance is one such factor that reduces J_e, although disturbance is also likely to have a strong influence on r that may outweigh its effects on J_e. On the surface this prediction would seem to contradict intuition and empirical observations that habitat loss (declining community size) and repeated disturbance (fluctuating community size) seem to increase susceptibility to invasion (Elton 1958, Williamson 1996). However, this seeming contradiction may well help clarify the aspects of human-mediated disturbance that increase susceptibility to invasion. In all likelihood disturbance changes *r* for different species, specifically by increasing it for many introduced species. The effects of disturbance per se (i.e., the destruction of biomass [Grime 2001]) is unlikely to be an important factor in and of itself, and in fact may have the opposite effect if its influence were not swamped out by changes in *r*.

This analysis of the fairly simple Kimura model has revealed some lessons that are of clear relevance to community ecology, although admittedly the conditions under which the effect of J_e is important relative to the effects of initial population size or *r* may be quite limited. But really we have just scratched the surface. Mutation fixation models have been elaborated in a number of important ecological contexts, including spatially structured populations (Whitlock 2003), spatially heterogeneous environments (Whitlock and Gomulkiewicz 2005), and changing population sizes (Otto and Whitlock 1997). Each of these cases has clear analogues in community ecology, presenting some potentially fruitful avenues for theoretical elaborations.

In sum, we hope to have demonstrated in the last two sections that a close examination of analogous models in community ecology and population genetics, and subsequent importation of models where appropriate, can open up a range of new avenues for theoretical and empirical progress within these disciplines. Such opportunities abound.

A New Structure for Community Ecology

In this final section, we want to step back from thinking about specific genetic models that might be imported into ecology, and consider the broader structure of community ecology as a discipline and how it compares to population genetics. Community ecology is a particularly vibrant subfield of ecology, with ongoing, lively debates concerning issues such as the important determinants of species diversity and composition (Hubbell 2001, Chase and Leibold 2003), the relative importance of positive versus negative interspecific interactions (e.g., Bruno et al. 2003), and the causes and consequences of the architecture of complex interaction networks (e.g., Bascompte and Jordano 2007). But what exactly is community ecology, and how do these various topics fit together under a single framework?

We can start by defining community ecology most generally as the study of the diversity, abundance, and composition within groups of species that co-occur in arbitrarily delineated units of space and time. Thus, the central question of community ecology is: why do we find different numbers, kinds, and abundances of species in different places and at different times? Ecologists—community ecologists in particular—have been highly critical of their own discipline (Simberloff 1980, Peters 1991, Weiner 1995, Lawton 1999), to a degree that does not seem to be matched in population genetics and evolutionary biology. The frustration many ecologists have with the seeming lack of general (and nontrivial) ecological laws, and with the difficulty in making accurate predictions about future states of ecological systems given their great complexity, is often cited as a chronic symptom of "physics envy" (Cohen 1971). But since the subject matter of physics seems so distinct from that of ecology, the comparison between ecology and physics does not seem particularly illuminating (see also Simberloff 2004). Population genetics, and evolutionary biology more generally, seem like much more appropriate foils for community ecology, given that ecology, genetics, and evolution are in many ways sister disciplines (Chase and Leibold 2003). Should ecologists have "evolution envy"? Can the perceived "mess" in community ecology (Lawton 1999) be tidied up if we learn some lessons from the basic structure of population genetics as a whole?

Some general perceptions about the theoretical underpinnings of these disciplines seem clear. Many evolutionary biologists work within a common conceptual framework that traces back to key insights from Darwin and Mendel on mechanisms for adaptive evolution and inheritance, respectively, and the subsequent integration of these ideas into a broader framework by the architects of the modern synthesis (Kutschera and Niklas 2004). This is of course an oversimplification, as the modern synthesis is open to criticism for leaving aside important parts of evolutionary biology (Pigliucci 2007), but certainly if we narrow the scope to population genetics there is a strong and general perception that the discipline lies on a foundation of general theoretical principles. To be sure, community ecology has towering historical figures of its own and a rich tradition of theoretical developments (Kingsland 1995), but community ecologists do not in any obvious way work within a similarly coherent, unified framework based on general theoretical principles (Chase and Leibold 2003). Colyvan and Ginzburg (2003) stated the situation succinctly: "ecology lacks a grand, widely accepted, explanatory theory such as Darwinian evolution." This difference between population genetics and community ecology presents a conundrum, if, as we have argued, the processes underlying changes in allele frequencies and species abundances are much the same.

TABLE 16.1
Simplified Representation of the Points of Emphasis in General Treatments of the Disciplines of Community Ecology and Population Genetics

Community ecology	*Population genetics*
Patterns of diversity and abundance	Measurement of genetic diversity
Space and time	Mutation
Competition	Drift
Predation	Migration
Food webs	Selection
Niches	
General approach: Pattern first	General approach: Process first

Note: The rows of the table do not represent a one-to-one correspondence of analogous processes across the two disciplines (the lists are independent).

Some progress can be made in solving this conundrum by examining how practitioners organize the various parts of their respective disciplines, as reflected in the tables of contents of textbooks on community ecology (e.g., Putman 1994, Morin 1999) or population genetics (e.g., Hartl and Clark 1997, Hedrick 2000). Our interpretation of the organization of such texts, while admittedly and necessarily oversimplified, reveals distinct points of emphasis in the two fields. Treatments of community ecology typically emphasize community patterns, issues of space and time, competition, predation, food webs, and the concept of niches in one form or another (table 16.1). Treatments of population genetics emphasize how genetic variation is measured, and then the four main processes that cause genetic change: mutation, drift, migration and selection (table 16.1). The key difference, first pointed out to us by Joan Roughgarden (personal communication), is that in community ecology there is a tendency to approach things "pattern first," whereas in population genetics the approach is "process first.". Both pattern and process certainly feature prominently in both disciplines, but this difference in emphasis seems quite clear. Community ecologists ask: what broad and general patterns of species diversity, distribution, and abundance do we find in nature? They then seek theoretical explanations for the patterns. MacArthur (1972) was explicit in promoting the pattern-first approach: "To do science is to search for repeated patterns, not simply to accumulate facts." In contrast, population geneticists ask: what is the basic set of processes capable of producing changes in gene frequencies, and how do they interact to drive evolution? With a firm understanding of potentially important processes, specific models can then be tailored to any particular situation in nature.

This simple difference between community ecology and population genetics—the pattern-first vs. process-first approach —likely traces to the origins of the two disciplines. Many ecological patterns, such as variation in species richness with latitude or area, were well known before ecology was even a named discipline (Brown and Lomolino 1998), so it would have been nearly impossible to approach theoretical analyses in community ecology without patterns in mind. Population genetics was initially developed as a theoretical discipline, with process-based models derived before empirical data on the frequency and diversity of alleles in real populations were even available to any great extent (Provine 1971). The pattern-process difference no doubt belies a more complex set of differences as well, but nonetheless, to us it highlights two important points. First, although population genetics appears to rest on a firmer theoretical foundation than community ecology, we are not actually any better at predicting broad scale patterns of genetic diversity than we are at predicting broad scale patterns in communities. If anything, the opposite is true. The difference is that in population genetics this is not considered a shortcoming given the coherent set of basic models that can be successfully tailored to meet the inherently contingent specifics of any particular case, whereas in ecology we are set up for disappointment when we hope for grand all-encompassing theories to make the contingencies disappear. Second, perhaps community ecology already has all of the building blocks to achieve the same level of theoretical generality as in population genetics, but simply lacks an organizational framework that emphasizes the generality of process over the contingency of pattern (see also Scheiner and Willig 2008).

Building on suggestions from Joan Roughgarden, here we propose an organizational structure for community ecology that is modeled after population genetics. We are not proposing that all of the important processes and phenomena in community ecology can be captured by population-genetic models, though the previous two sections demonstrate that there are particular situations for which this may be the case. Rather, regardless of how well the specifics translate across disciplines, we feel that the organizational framework of population genetics can be adapted for community ecology, with some important benefits. Specifically, our hope is to provide a framework that is simple in concept and terminology, but that also embraces all of what community ecologists actually do, thereby allowing clearer relationships to be defined between the many facets of the discipline. In our caricature of how community ecology is organized at present (table 16.1), patterns come first, and the rest of the items on

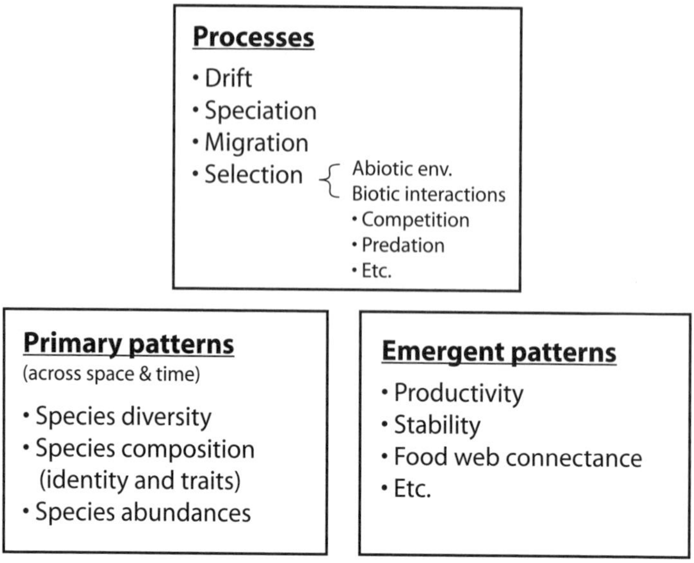

Figure 16.4. An organizational framework for community ecology.

the list are a mixture of selected processes on which a lot of work has been done (competition, predation) and items that are not really patterns or processes but rather concepts or phenomena (food webs, niches). It is difficult if not impossible to logically organize this list. Our main goal here, and really the only substantive difference between our framework and others, is to organize the processes that influence community change into a logical hierarchy, whereby particular processes are grouped according to their similarity to one another, regardless of their relative importance in nature or the degree to which they've been studied in the past.

A preliminary sketch of our proposed framework is illustrated in figure 16.4, where we recognize three main components of community ecology: primary patterns, underlying processes, and emergent patterns. The one thing we think most community ecologists would agree on is that the primary patterns we are interested in understanding are of species diversity, abundance, and composition over space and time. The real challenge is to organize the giant morass of specific processes that can influence community patterns.

Applying the framework of population genetics, four classes of process can influence community patterns: speciation, drift, migration, and selection. We feel this simple framework can provide some much-needed order to community ecology because unlike existing organizational schemes, these four categories represent logically distinct classes, and they provide a way to classify more specific processes at subsequent levels of the hierarchy. Speciation is the only process that can add new species to the global pool. As a necessary consequence of communities containing finite numbers of individuals, some changes in species abundances will occur due to random drift. The movement of individuals from place to place (i.e., migration) can impact communities in a variety of ways. Finally, any process involving deterministic differences among species in their rates of survival or reproduction can be grouped under the heading of selection.

Selection, of course, can arise in a myriad of ways, including differential effects of abiotic conditions or resources, competition, predation, mutualism, facilitation, disturbance, and so on. While the term selection has not traditionally been used this way in community ecology (but see Bell et al. 2006), it is now widely used in studies of the ecosystem consequences of species diversity (e.g., Loreau and Hector 2001), and it is the appropriate term for this general class of processes as it implies only differential success of individual organisms (Nowak 2006). Most models of interspecific interactions are essentially models of frequency- or density-dependent selection among types of individuals. In community ecology individuals are defined by their species identity rather than by allelic states as in genetics. Selection is also arguably where the vast majority of research in community ecology is focused, perhaps reflecting its overwhelming importance in determining community patterns. As such, it is important to emphasize that this framework is agnostic on the topic of which processes are more or less important. It only aims to provide organization and structure to the discipline. For example, competition is further down the hierarchy than drift not because it is any less important (it may or may not be), but because it shares a logical similarity with other kinds of biotic interactions in the same category, and is logically distinct from drift, migration, or speciation.

Finally, a wide variety of community properties can be measured that are neither primary patterns of diversity, abundance, and composition, nor processes. We refer to these as emergent properties of communities, and include common measurements of interest such as productivity summed over species, indices of the number, strength, and direction of species interactions, stability of any other primary or emergent pattern over time, and so on. Again, this framework in no way implies that these measurements are less important than the primary patterns, simply that considering them in a separate category can help provide some order to an otherwise disorderly discipline.

Unfortunately we haven't the space to flesh out this framework in any detail, but it is easy to see how general treatments of community ecology (e.g., in undergraduate courses) might present the subject matter in a more coherent and organized way. Incorporating the importance of spatial scale, we can envision how these processes interact across local, regional,

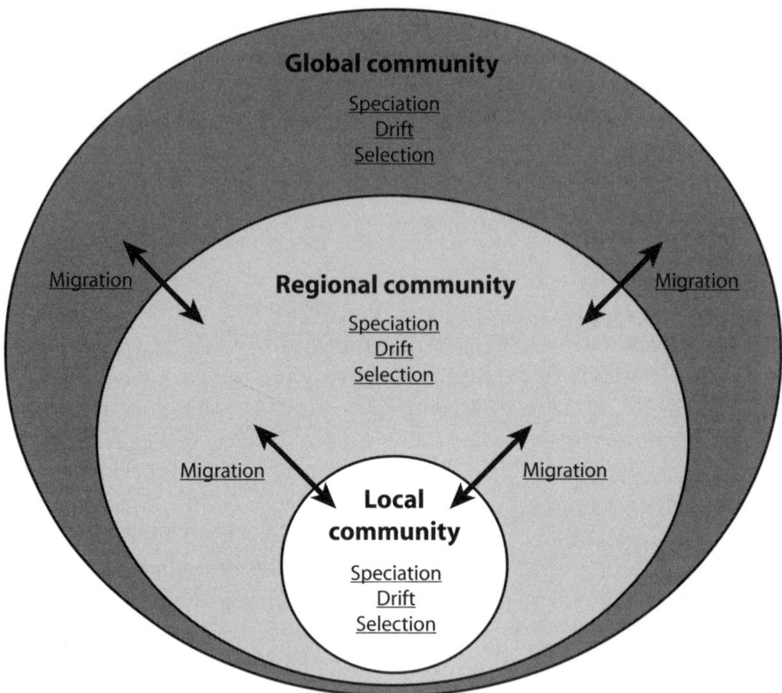

Figure 16.5. The interaction of processes acting across spatial scales in community ecology. We include all processes (underlined) at each scale to illustrate the full range of theoretical possibilities, even if some processes are unlikely to be important at particular scales (e.g., speciation at the local scale).

and global scales (figure 16.5), and it is fairly straightforward to place existing community models within this framework. For example, Hubbell's neutral model explores the balance between speciation and drift, while island biogeography explores the balance between immigration and drift. As mentioned earlier, models of local species interactions essentially address different kinds of frequency- or density-dependent selection among species, while the effects of abiotic gradients on community patterns treat the case of constant local selection. Stochastic versions of such models represent simultaneous selection and drift. Environmental heterogeneity among local habitat patches, which are connected by dispersal, represents selection-migration balance at the local scale, and spatially variable selection with migration at the regional scale. The list could go on (see also McPeek and Gomulkiewicz 2005). This framework also makes clear how community ecology links with other branches of ecology and biology more generally. Communities link to ecosystems via their emergent properties and effects of energy and nutrient fluxes on selection. Evolution creates new species, modifies ecologically relevant traits of existing ones, and itself depends on community processes and patterns. Population ecology is in many ways a subset of community ecology in which all important community processes affecting an individual species are abstracted into the parameters of a population model such as r and K.

In sum, we hope this organizational framework can contribute to an ongoing discussion about how best to advance the science of community ecology. Advances in basic research are of course the cornerstone of such efforts, but reevaluation of the way we frame the discipline to begin with might go a long way as well.

Conclusions

We would like to end this contribution by returning to the question of why the contributions of Wright, Haldane, Kimura, and others were not drawn upon in the construction of analogous models in *The Theory of Island Biogeography*. At the conclusion of the presentation of the contents of this chapter in October 2007, E. O. Wilson himself provided the answer when he said, "it never occurred to us." Given that at the time of writing their book MacArthur and Wilson were as much evolutionary biologists as ecologists, this seems to be in need of explanation. We speculate that the different starting points of the two bodies of theory may be part of the explanation. As keen naturalists, MacArthur and Wilson approached the problem with a deep appreciation of the complexity of nature, and the importance of species differences and their myriad manifestations. As such, population-genetic models may have seemed far too simplistic to encapsulate the key elements of ecological communities, even if ultimately the two sets of models ended up in more or less exactly the same place. In the end, perhaps we are richer for having both, but hopefully the process of cross-fertilization between models and concepts in community ecology and population genetics can itself provide the inspiration for future advances in these fields.

Acknowledgments

We especially thank Joan Roughgarden for sharing ideas on evolution and ecology, Mike Whitlock for input on mutation fixation models, and the editors for inviting this contribution. Richard Gomulkiewicz, Mark Urban, Bob Holt, Jon Shurin, Ben Gilbert, Joe Bennett, Will Cornwell,

Ilkka Hanski, Jonathan Losos, and two anonymous reviewers provided valuable input on different parts and stages of this project. Portions of this work were conducted while JLO was a postdoctoral associate at the National Center for Ecological Analysis and Synthesis, a Center funded by NSF (Grant No. DEB-0072909), the University of California, and the Santa Barbara Campus. MV was funded by the Natural Sciences and Engineering Research Council, Canada.

Literature Cited

Antonovics, J. 1976. The input from population genetics: "The new ecological genetics." *Systematic Botany* 1:233–45.

Antonovics, J. 2003. Toward community genomics? *Ecology* 84:598–601.

Bascompte, J., and P. Jordano. 2007. Plant-animal mutualistic networks: the architecture of biodiversity. *Annual Review of Ecology, Evolution, and Systematics* 38:567–93.

Bell, G., M. J. Lechowicz, and M. J. Waterway. 2006. The comparative evidence relating to functional and neutral interpretations of biological communities. *Ecology* 87:1378–86.

Birch, L. C. 1960. The genetic factor in population ecology. *American Naturalist* 94:5–24.

Brown, J. H., and M. V. Lomolino. 1998. *Biogeography*, 2nd ed. Sunderland, MA: Sinauer Associates.

Bruno, J. F., J. J. Stachowicz, and M. D. Bertness. 2003. Inclusion of facilitation into ecological theory. *Trends in Ecology and Evolution* 18:119–25.

Chase, J. M., and M. A. Leibold. 2003. *Ecological Niches: Linking Classical and Contemporary Approaches*. Chicago: University of Chicago Press.

Cohen, J. E. 1971. Mathematics as metaphor. *Science* 172:674–75.

Collins, J. P. 1986. Evolutionary ecology and the use of natural selection in ecological theory. *Journal of the History of Biology* 19:257–88.

Colyvan, M., and L. R. Ginzburg. 2003. Laws of nature and laws of ecology. *Oikos* 101:649–53.

Connor, E. F., and E. D. McCoy. 1979. The statistics and biology of the species-area relationship. *American Naturalist* 113:791–833.

Dobzhansky, T. 1964. Biology, molecular and organismic. *American Zoologist* 4:443–52.

Elton, C. S. 1958. *The Ecology of Invasion by Plants and Animals*. Chicago: University of Chicago Press.

Etienne, R. S. 2007. A neutral sampling formula for multiple samples and an "exact" test of neutrality. *Ecology Letters* 10:608–18.

Etienne, R. S., and D. Alonso. 2005. A dispersal-limited sampling theory for species and alleles. *Ecology Letters* 8:1147–56.

Ewens, W. J. 1972. The sampling theory of selectively neutral alleles. *Theoretical Population Biology* 3:87–112.

Fox, J. W. 2006. Using the Price Equation to partition the effects of biodiversity loss on ecosystem function. *Ecology* 87:2687–96.

Fussmann, G. F., M. Loreau, and P. A. Abrams. 2007. Eco-evolutionary dynamics of communities and ecosystems. *Functional Ecology* 21:465–77.

Grime, J. P. 2001. *Plant Strategies, Vegetation Processes, and Ecosystem Properties*, 2nd ed. Chichester, U.K.: John Wiley & Sons.

Haccou, P., and Y. Iwasa. 1996 . Invasion probability in fluctuating environments: A branching process model. *Theoretical Population Biology* 50:254–80.

Haldane, J.B.S. 1927. A mathematical theory of natural and artificial selection. V. Selection and mutation. *Proceedings of the Cambridge Philosophical Society* 23:838–44.

Hartl, D. L., and A. G. Clark. 1997. *Principles of Population Genetics*, 3rd ed. Sunderland, MA: Sinauer Associates.

Hedrick, P. W. 2000. *Genetics of Populations*. Sudbury, MA: Jones and Barlett Publishers.

Holt, R. D. 2005. On the integration of community ecology and evolutionary biology: historical perspectives and current prospects. In *Ecological Paradigms Lost: Routes of Theory Change,* ed. K. Cuddington and B. Beisner, 235–71. Amsterdam: Elsevier.

Houlahan, J. E., D. J. Currie, K. Cottenie, G. S. Cumming, S.K.M. Ernest, C. S. Findlay, S. D. Fuhlendorf, U. Gaedke, P. Legendre, J. J. Magnuson, B. H. McArdle, E. H. Muldavin, D. Noble, R. Russell, R. D. Stevens, T. J. Willis, I. P. Woiwod, and S. M. Wondzell. 2007. Compensatory dynamics are rare in natural ecological communities. *Proceedings of the National Academy of Sciences U.S.A.* 104:3273–77.

Hubbell, S. P. 2001. *The Unified Neutral Theory of Biogeography and Biodiversity*. Princeton, NJ: Princeton University Press.

Kimura, M. 1962. On the probability of fixation of mutant genes in a population. *Genetics* 47:713–19.

———. 1983. *The Neutral Theory of Molecular Evolution*. Cambridge: Cambridge University Press.

Kingsland, S. 1995. *Modeling Nature: Episodes in the History of Population Ecology*, 2nd ed. Chicago: University of Chicago Press.

Kutschera, U., and K. J. Niklas. 2004. The modern theory of biological evolution: an expanded synthesis. *Naturwissenschaften* 91:255–76.

Johnson, M.T.J., and J. R. Stinchcombe. 2007. An emerging synthesis between community ecology and evolutionary biology. *Trends in Ecology and Evolution* 22:250–57.

Lankau, R. A., and S. Y. Strauss. 2007. Mutual feedbacks maintain both genetic and species diversity in a plant community. *Science* 317:1561–63.

Lawton, J. H. 1999. Are there general laws in ecology? *Oikos* 84:177–92.

Levins, R. 1968. *Evolution in Changing Environments: Some Theoretical Explorations*. Princeton, NJ: Princeton University Press.

Lewontin, R. C. 1974. *The Genetic Basis of Evolutionary Change*. New York: Columbia University Press.

Loreau, M., and A. Hector. 2001. Partitioning selection and complementarity in biodiversity experiments. *Nature* 412:72–76.
MacArthur, R. H. 1972. *Geographical Ecology*. Princeton, NJ: Princeton University Press.
MacArthur, R. H., and E. O. Wilson. 1967. *The Theory of Island Biogeography*. Princeton, NJ: Princeton University Press.
McPeek, M. A., and R. Gomulkiewicz. 2005. Assembling and depleting species richness in metacommunities: Insights from ecology, population genetics and macroevolution. In *Metacommunities: Spatial Dynamics and Ecological Communities,* ed. M. A. Leibold, M. Holyoak, and R. D. Holt, 355–73. Chicago: University of Chicago Press.
Morin, P. J. 1999. *Community Ecology*. Oxford: Blackwell Science.
Norberg J., D. P. Swaney, J. Dushoff, J. Lin, R. Casagrandi, and S. A. Levin. 2001. Phenotypic diversity and ecosystem functioning in changing environments: A theoretical framework. *Proceedings of the National Academy of Sciences U.S.A.* 98:11376–81.
Nowak, M. 2006. *Evolutionary Dynamics: Exploring the Equations of Life*. Cambridge, MA: Harvard University Press.
Odenbaugh, J. 2006. The strategy of "The strategy of model building in population biology." *Biology and Philosophy* 21:607–21.
Orrock, J. L., and R. J. Fletcher, Jr. 2005. Changes in community size affect the outcome of competition. *American Naturalist* 166:107–11.
Otto, S. P., and M. C. Whitlock. 1997. The probability of fixation in populations of changing size. *Genetics* 146:723–33.
Peters, R. H. 1991. *A Critique for Ecology*. Cambridge: Cambridge University Press..
Pigliucci, M. 2007. Do we need an extended evolutionary synthesis? *Evolution* 61:2743–49.
Provine, W. B. 1971. *The Origins of Theoretical Population Genetics*. Chicago: University of Chicago Press.
Putman, R. J. 1994. *Community Ecology*. New York: Chapman and Hall.
Ricklefs, R. E. 2004. A comprehensive framework for global patterns in biodiversity. *Ecology Letters* 7:1–15.
Ricklefs, R. E., and D. Schluter. 1993. *Species Diversity in Ecological Communities: Historical and Geographical Perspectives*. Chicago: University of Chicago Press.
Sakai, A. K., F. W. Allendorf, J. S. Holt, D. M. Lodge, J. Molofsky, K. A. With, S. Baughman, R. J. Cabin, J. E. Cohen, N. C. Ellstrand, D. E. McCauley, P. O'Neil, I. M. Parker, J. N. Thompson, and S. G. Weller. 2001. The population biology of invasive species. *Annual Review of Ecology and Systematics* 32: 305–32.
Sax, D. F., and J. H. Brown. 2000. The paradox of invasion. *Global Ecology and Biogeography* 9:363–71.
Scheiner, S. M., and M. R. Willig. 2008. A general theory of ecology. *Theoretical Ecology* 1:21–28.
Schluter, D. 2000. *The Ecology of Adaptive Radiation*. Oxford: Oxford University Press.

Schoener, T. W. 1976. The species-area relation within archipelagos: Models and evidence from island land birds. In *Proceedings of the 16th International Ornithological Conference,* ed. H. J. Firth and J. H. Calaby, 629–42. Canberra: Australian Academy of Science.

Schrader-Frechette, K. S., and E. D. McCoy. 1993. *Method in Ecology.* Cambridge: Cambridge University Press.

Simberloff, D. 1980. A succession of paradigms in ecology: Essentialism to materialism to probabilism. *Synthese* 43:3–39.

———. 2004. Community ecology: Is it time to move on? *American Naturalist* 163:787–99.

Vellend, M. 2003. Island biogeography of genes and species. *American Naturalist* 162:358–65.

———. 2005. Species diversity and genetic diversity: Parallel processes and correlated patterns. *American Naturalist* 166:199–215.

Vellend, M., and M. A. Geber. 2005. Connections between species diversity and genetic diversity. *Ecology Letters* 8:767–81.

Weiner, J. 1995. On the practice of ecology. *Journal of Ecology* 83:153–58.

Whitlock, M. C. 2003. Fixation probability and time in subdivided populations. *Genetics* 164:767–79.

Whitlock, M. C., and R. G. Gomulkiewicz. 2005. Probability of fixation in a heterogenous environment. *Genetics* 171:1407–17.

Whittaker, R., and J. M. Fernández-Palacios. 2007. *Island Biogeography: Ecology, Evolution, and Conservation.* Oxford: Oxford University Press.

Williamson, M. 1988. Relationship of species number to area, distance and other variables. In *Analytical Biogeography,* ed. A. A. Myers and P. S. Giller, 91–115. New York: Chapman and Hall.

———. 1996. *Biological Invasions.* London: Chapman and Hall.

Wilson, E. O. 1994. *Naturalist.* New York: Island Press.

Wright, S. 1940. Breeding structure of populations in relation to speciation. *American Naturalist* 74:232–48.

Yoshida T., L. E. Jones, S. P. Ellner, G. F. Fussman, and N. G. Hairston Jr. 2003. Rapid evolution drives ecological dynamics in a predator-prey system. *Nature* 424:303–30

Index